NASA SPACE TECHNOLOGY ROADMAPS AND PRIORITIES

Restoring NASA's Technological Edge and Paving the Way for a New Era in Space

Steering Committee for NASA Technology Roadmaps

Aeronautics and Space Engineering Board

Division on Engineering and Physical Sciences

NATIONAL RESEARCH COUNCIL
OF THE NATIONAL ACADEMIES

THE NATIONAL ACADEMIES PRESS
Washington, D.C.
www.nap.edu

THE NATIONAL ACADEMIES PRESS **500 Fifth Street, NW** **Washington, DC 20001**

NOTICE: The project that is the subject of this report was approved by the Governing Board of the National Research Council, whose members are drawn from the councils of the National Academy of Sciences, the National Academy of Engineering, and the Institute of Medicine. The members of the committee responsible for the report were chosen for their special competences and with regard for appropriate balance.

This study is based on work supported by Contract NNH10CD04B between the National Academy of Sciences and the National Aeronautics and Space Administration. Any opinions, findings, conclusions, or recommendations expressed in this publication are those of the authors and do not necessarily reflect the views of the agency that provided support for the project.

Cover: Design by Tim Warchocki.

International Standard Book Number-13: 978-0-309-25362-8
International Standard Book Number-10: 0-309-25362-4

Copies of this report are available free of charge from:

Aeronautics and Space Engineering Board
National Research Council
500 Fifth Street, NW
Washington, DC 20001

Additional copies of this report are available from the National Academies Press, 500 Fifth Street, NW, Keck 360, Washington, DC 20001; (800) 624-6242 or (202) 334-3313; http://www.nap.edu.

Printed in the United States of America

OTHER REPORTS OF THE AERONAUTICS AND SPACE ENGINEERING BOARD

An Interim Report on NASA's Draft Space Technology Roadmaps (Aeronautics and Space Engineering Board [ASEB], 2011)
Limiting Future Collision Risk to Spacecraft: An Assessment of NASA's Meteoroid and Orbital Debris Programs (ASEB, 2011)
Preparing for the High Frontier—the Role and Training of NASA Astronauts in the Post-Space Shuttle Era (ASEB, 2011)
Recapturing a Future for Space Exploration: Life and Physical Sciences Research for a New Era (Space Studies Board [SSB] with ASEB, 2011)
Summary of the Workshop to Identify Gaps and Possible Directions for NASA's Meteoroid and Orbital Debris Programs (ASEB, 2011)

Advancing Aeronautical Safety: A Review of NASA's Aviation Safety-Related Research Programs (ASEB, 2010)
Capabilities for the Future: An Assessment of NASA Laboratories for Basic Research (Laboratory Assessments Board with ASEB, 2010)
Defending Planet Earth: Near-Earth-Object Surveys and Hazard Mitigation Strategies: Final Report (SSB with ASEB, 2010)
Final Report of the Committee to Review Proposals to the 2010 Ohio Third Frontier (OTF) Wright Projects Program (WPP) (ASEB, 2010)

America's Future in Space: Aligning the Civil Space Program with National Needs (SSB with ASEB, 2009)
Approaches to Future Space Cooperation and Competition in a Globalizing World: Summary of a Workshop (SSB with ASEB, 2009)
An Assessment of NASA's National Aviation Operations Monitoring Service (ASEB, 2009)
Final Report of the Committee for the Review of Proposals to the 2009 Engineering and Physical Science Research and Commercialization Program of the Ohio Third Frontier Program (ASEB, 2009)
Fostering Visions for the Future: A Review of the NASA Institute for Advanced Concepts (ASEB, 2009)
Launching Science: Science Opportunities Enabled by NASA's Constellation System (SSB with ASEB, 2009)
Near-Earth Object Surveys and Hazard Mitigation Strategies: Interim Report (SSB with ASEB, 2009)
Radioisotope Power Systems: An Imperative for Maintaining U.S. Leadership in Space Exploration (SSB with ASEB, 2009)

Assessing the Research and Development Plan for the Next Generation Air Transportation System: Summary of a Workshop (ASEB, 2008)
A Constrained Space Exploration Technology Program: A Review of NASA's Exploration Technology Development Program (ASEB, 2008)
Launching Science: Science Opportunities Provided by NASA's Constellation System (SSB with ASEB, 2008)
Managing Space Radiation Risk in the New Era of Space Exploration (ASEB, 2008)
NASA Aeronautics Research: An Assessment (ASEB, 2008)
Review of NASA's Exploration Technology Development Program: An Interim Report (ASEB, 2008)
Science Opportunities Enabled by NASA's Constellation System: Interim Report (SSB with ASEB, 2008)
United States Civil Space Policy: Summary of a Workshop (SSB with ASEB, 2008)
Wake Turbulence: An Obstacle to Increased Air Traffic Capacity (ASEB, 2008)

Limited copies of ASEB reports are available free of charge from:

Aeronautics and Space Engineering Board
National Research Council
The Keck Center of the National Academies
500 Fifth Street, N.W., Washington, DC 20001
(202) 334-2858/aseb@nas.edu
www.nationalacademies.org/aseb

STEERING COMMITTEE FOR NASA TECHNOLOGY ROADMAPS

RAYMOND S. COLLADAY, RC Space Enterprises, Inc., *Chair*
JOHN D. ANDERSON, JR., Smithsonian Institution
JAMES B. ARMOR, JR., ATK Spacecraft System & Services
EDWARD F. CRAWLEY, Massachusetts Institute of Technology
RAVI B. DEO, EMBR
WALT FAULCONER, Strategic Space Solutions, LLC
PHILIP D. HATTIS, The Charles Stark Draper Laboratory, Inc.
TAMARA E. JERNIGAN, Lawrence Livermore National Laboratory
JOHN C. KARAS, Lockheed Martin Space Systems Company
JOHN M. KLINEBERG, Loral Space and Communications, Ltd. (retired)
IVETT A. LEYVA, Air Force Research Laboratory
LESTER L. LYLES, The Lyles Group
H. JAY MELOSH, Purdue University
DANIEL R. MULVILLE, Independent Consultant
DAVA J. NEWMAN, Massachusetts Institute of Technology
RICHARD R. PAUL, Independent Consultant
LISELOTTE J. SCHIOLER, National Institute of Aerospace
GERALD SCHUBERT, University of California, Los Angeles

PROPULSION AND POWER PANEL

JOHN R. ROGACKI, Florida Institute for Human and Machine Cognition, *Chair*
DOUGLAS M. ALLEN, Independent Consultant
HENRY W. BRANDHORST, JR., Carbon-Free Energy, LLC
DAVID E. CROW, University of Connecticut
ALEC D. GALLIMORE, University of Michigan
MARK W. HENLEY, Boeing Research and Technology
ANTHONY K. HYDER, University of Notre Dame
IVETT A. LEYVA, Air Force Research Laboratory
PAULO C. LOZANO, Massachusetts Institute of Technology
JOYCE A. McDEVITT, Independent Consultant
ROGER M. MYERS, Aerojet General Corporation
LAWRENCE J. ROSS, Aerospace Engineering Associates, LLC
RAYMOND J. SEDWICK, University of Maryland
GEORGE F. SOWERS, United Launch Alliance

ROBOTICS, COMMUNICATIONS, AND NAVIGATION PANEL

STEPHEN P. GOREVAN, Honeybee Robotics, Ltd., *Chair*
JULIE A. ADAMS, Vanderbilt University
EDWARD J. GROTH III, Princeton University
PHILIP D. HATTIS, The Charles Stark Draper Laboratory, Inc.
JONATHAN P. HOW, Massachusetts Institute of Technology
JAMES W. LOWRIE, Lockheed Martin Missile and Fire Control-Orlando
DAVID P. MILLER, University of Oklahoma
JONATHAN R. SALTON, Sandia National Laboratories
DONNA L. SHIRLEY, Managing Creativity
GEORGE W. SWENSON, JR., University of Illinois, Urbana-Champaign

INSTRUMENTS AND COMPUTING PANEL

JAMES L. BURCH, Southwest Research Institute, *Chair*
PHILIP E. ARDANUY, Raytheon Intelligence and Information Systems
WEBSTER CASH, University of Colorado, Boulder
JOHN A. HACKWELL, The Aerospace Corporation
ROBERT J. HANISCH, Space Telescope Science Institute
DAVID Y. KUSNIERKIEWICZ, Johns Hopkins University Applied Physics Laboratory
JOEL R. PRIMACK, University of California, Santa Cruz
GERALD SCHUBERT, University of California, Los Angeles
DANIEL A. SCHWARTZ, Harvard-Smithsonian Center for Astrophysics
ALAN M. TITLE, Lockheed Martin Advanced Technology Center
DANIEL WINTERHALTER, Jet Propulsion Laboratory
CARL WUNSCH, Massachusetts Institute of Technology

HUMAN HEALTH AND SURFACE EXPLORATION PANEL

BONNIE J. DUNBAR, Independent Consultant, *Chair*
DAVID L. AKIN, University of Maryland, College Park
DALLAS G. BIENHOFF, The Boeing Company
ROBERT L. CURBEAM, JR., ARES Corporation
GREGORY J. HARBAUGH, Sigma Chi Foundation
TAMARA E. JERNIGAN, Lawrence Livermore National Laboratory
DANIEL R. MASYS, University of Washington, Seattle
ERIC E. RICE, Orbital Technologies Corporation
RONALD E. TURNER, ANSER

MATERIALS PANEL

MOOL C. GUPTA, University of Virginia, *Chair*
GREGORY R. BOGART, Sandia National Laboratories
DONALD M. CURRY, The Boeing Company
JOHN R. HOWELL, University of Texas, Austin
GEORGE A. LESIEUTRE, Pennsylvania State University
LISELOTTE J. SCHIOLER, National Institute of Aerospace
ROBERT E. SKELTON, University of California, San Diego
GEORGE W. SUTTON, SPARTA (retired)

ENTRY, DESCENT, AND LANDING PANEL

TODD J. MOSHER, Sierra Nevada Corporation, *Chair*
JOHN D. ANDERSON, JR., Smithsonian Institution
TYE M. BRADY, The Charles Stark Draper Laboratory, Inc.
BASIL HASSAN, Sandia National Laboratories
STEPHEN RUFFIN, Georgia Institute of Technology
ROBERT J. SINCLAIR, Airborne Systems of North America
BYRON D. TAPLEY, University of Texas, Austin
BETH E. WAHL, Independent Consultant
GERALD D. WAHLBERG, North Carolina State University (retired)

Staff

ALAN C. ANGLEMAN, Senior Program Officer, *Study Director*
JOSEPH K. ALEXANDER, Senior Program Officer
IAN W. PRYKE, Senior Program Officer
ROBERT L. RIEMER, Senior Program Officer
JOHN WENDT, Senior Program Officer
MAUREEN MELLODY, Program Officer
CATHERINE A. GRUBER, Editor
AMANDA R. THIBAULT, Research Associate
DIONNA WILLIAMS, Program Associate
TERRI BAKER, Senior Project Assistant
RODNEY HOWARD, Senior Project Assistant
LINDA WALKER, Senior Project Assistant
ANNA B. WILLIAMS, Christine Mirzayan Science and Technology Policy Fellow

MICHAEL H. MOLONEY, Director, Aeronautics and Space Engineering Board

AERONAUTICS AND SPACE ENGINEERING BOARD

RAYMOND S. COLLADAY, RC Space Enterprises, Inc., *Chair*
LESTER LYLES, The Lyles Group, *Vice Chair*
ELLA M. ATKINS, University of Michigan
AMY L. BUHRIG, Boeing Commercial Airplanes Group
INDERJIT CHOPRA, University of Maryland, College Park
JOHN-PAUL B. CLARKE, Georgia Institute of Technology
RAVI B. DEO, EMBR
VIJAY DHIR, University of California, Los Angeles
EARL H. DOWELL, Duke University
MICA R. ENDSLEY, SA Technologies
DAVID GOLDSTON, Harvard University
R. JOHN HANSMAN, Massachusetts Institute of Technology
JOHN B. HAYHURST, The Boeing Company (retired)
WILLIAM L. JOHNSON, California Institute of Technology
RICHARD KOHRS, Independent Consultant
IVETT LEYVA, Air Force Research Laboratory, Edwards Air Force Base
ELAINE S. ORAN, Naval Research Laboratory
ALAN G. POINDEXTER, Naval Postgraduate School
HELEN R. REED, Texas A&M University
ELI RESHOTKO, Case Western Reserve University
EDMOND SOLIDAY, United Airlines (retired)

Staff

MICHAEL H. MOLONEY, Director
CARMELA J. CHAMBERLAIN, Administrative Coordinator
TANJA PILZAK, Manager, Program Operations
CELESTE A. NAYLOR, Information Management Associate
CHRISTINA O. SHIPMAN, Financial Officer
SANDRA WILSON, Financial Assistant

Preface

In late 2010, NASA developed a set of 14 draft roadmaps to guide the development of space technologies under the leadership of the NASA Office of the Chief Technologist (OCT).[1,2] Each of these draft roadmaps focuses on a particular technology area (TA). The roadmaps are intended to foster the development of advanced technologies and concepts that address NASA's needs and contribute to other aerospace and national needs. In June of 2010, Robert Braun, NASA's Chief Technologist at the time, requested that the National Research Council (NRC) conduct a study to review the roadmaps. The role of the study was to gather and assess relevant community input, make recommendations and suggest priorities to inform NASA's decisions as it finalizes its roadmaps, and undertake a time-sequenced and prioritized advanced space technology development program that lays the technical foundation for future NASA missions. The full statement of task appears in Appendix A of this report. Specific elements of the statement of task include the following:

- Establish a set of criteria to enable prioritization of technologies within each and among all of the technology areas that the NASA technology roadmaps should satisfy;
- Consider technologies that address the needs of NASA's exploration systems, Earth and space science, and space operations mission areas, as well as those that contribute to critical national and commercial needs in space technology;
- Integrate the outputs to identify key common threads and issues and to summarize findings and recommendations; and
- Prioritize the highest-priority technologies from all 14 roadmaps.

In response to this request, the NRC appointed the 18-member Steering Committee for NASA Technology Roadmaps and six study panels with a total of 56 additional experts. The study panels were organized by technical area, based on the organization of the 14 roadmaps, as follows:

[1] The draft roadmaps are available at http://www.nasa.gov/offices/oct/home/roadmaps/index.html.

[2] This study (and the 14 draft roadmaps) does not cover aeronautics technologies except to the extent that they are needed to achieve NASA and national needs in space. Guidance on the development of core aeronautics technologies is already available in the National Aeronautics Research and Development Plan, which was published in 2010 by the White House National Science and Technology Council and Office of Science and Technology Policy. It is available at http://www.whitehouse.gov/sites/default/files/microsites/ostp/aero-rdplan-2010.pdf.

- Panel 1: Propulsion and Power
  - TA01 Launch Propulsion Systems
  - TA02 In-Space Propulsion Technologies
  - TA03 Space Power and Energy Storage
  - TA13 Ground and Launch Systems Processing
- Panel 2: Robotics, Communications, and Navigation
  - TA04 Robotics, TeleRobotics, and Autonomous Systems
  - TA05 Communication and Navigation
- Panel 3: Instruments and Computing
  - TA08 Science Instruments, Observatories, and Sensor Systems
  - TA11 Modeling, Simulation, and Information Technology and Processing
- Panel 4: Human Health and Surface Exploration
  - TA06 Human Health, Life Support, and Habitation Systems
  - TA07 Human Exploration Destination Systems
- Panel 5: Materials Panel
  - TA10 Nanotechnology
  - TA12 Materials, Structures, Mechanical Systems, and Manufacturing
  - TA14 Thermal Management Systems
- Panel 6: Entry, Descent, and Landing Panel
  - TA09 Entry, Descent, and Landing Systems

After initial discussions by the study chair and a few members of the steering committee and staff to plan committee meetings and draft a uniform set of evaluation criteria, an initial meeting of the steering committee and all six panels was held in Washington, D.C. The January 2011 meeting reviewed and approved the evaluation criteria and study process and also served as a forum to discuss the content of the roadmaps with NASA staff. The steering committee subsequently held three additional meetings between January and September 2011 for information gathering, deliberations, and report writing. During that same time period, each of the six panels also held two additional meetings and hosted a 1-day public workshop for each roadmap under its purview. At each public workshop, the study panels engaged with invited speakers, guests, and members of the public in a dialogue on the technology areas and their value to NASA based on the common evaluation criteria established by the steering committee. More detailed information on each workshop, including a complete agenda and copies of many presentations, can be viewed at http://sites.nationalacademies.org/DEPS/ASEB/DEPS_060733.

Broad community input was also solicited from a public website where 144 individuals provided 244 sets of comments on the draft roadmaps in terms of criteria (such as benefit, risk and reasonableness, and alignment with NASA and national goals) that the steering committee established. The individuals providing these inputs included 91 personnel from NASA (including the Jet Propulsion Laboratory), 6 from other government organizations, 26 from industry, 16 from academia, and 5 from other organizations or no organization at all. (The data provided in the public input forms can be found at http://www8.nationalacademies.org/asebsurvey/tabs/publicview.aspx.) In addition, 87 sets of general comments were received via e-mail from 7 individuals who completed the public input forms noted above and from 68 individuals who did not. These individuals included 47 personnel from NASA (including the Jet Propulsion Laboratory), 1 from another government agency, 7 from industry, 4 from academia, 5 from other organizations, and 11 whose organization is unknown.

Based on the important input from the community and its own deliberations, the steering committee prepared a brief interim report that makes high-level observations on the roadmaps and addresses the advisability of modifying the technologies included within each of the existing draft roadmaps, as well as technology gaps that cut across multiple roadmaps. This interim report is available at http://www.nap.edu/catalog.php?record_id=13228.

From these various forms of public input, as well as their own internal deliberations, the study panels prioritized technologies for each of their assigned roadmaps into high, medium, and low categories; described the value of the high-priority technologies; identified gaps in the draft roadmaps; identified development or schedule changes of the technologies covered; and summarized the public workshop that focused on the draft roadmap. The results of the panels' work are summarized in this report in 14 appendixes (D through Q; one for each roadmap). This input from the panels was then integrated by the steering committee and documented in the main body of this report.

The steering committee and panels would like to acknowledge the significant contributions of the following staff members of the Aerospace Corporation who assisted the steering committee, the panels, and the NRC staff in this effort: Torrey Radcliffe, Dean Bucher, Robert Kinsey, Kristina Kipp, Marcus Lobbia, and Gregory Richardson. Finally, I wish to personally recognize the hard work and dedicated efforts of the steering committee, and of the panel members and their chairs, and the outstanding support from the NRC staff, without which we would not have been able to meet our delivery milestones. In particular, the tireless and professional attention to all aspects of the study by Alan Angleman and Michael Moloney supported by Maureen Mellody was exceptional.

Raymond S. Colladay, *Chair*
Steering Committee for NASA Technology Roadmaps

Acknowledgment of Reviewers

This report has been reviewed in draft form by individuals chosen for their diverse perspectives and technical expertise, in accordance with procedures approved by the National Research Council's (NRC's) Report Review Committee. The purpose of this independent review is to provide candid and critical comments that will assist the institution in making its published report as sound as possible and to ensure that the report meets institutional standards for objectivity, evidence, and responsiveness to the study charge. The review comments and draft manuscript remain confidential to protect the integrity of the deliberative process. We wish to thank the following individuals for their review of this report:

Ellen Bass, University of Virginia,
Robert H. Bishop, Marquette University,
David C. Byers, Consultant,
Elizabeth Cantwell, Lawrence Livermore National Laboratory,
Robert L. Crippen, U.S. Navy (retired),
Joseph H. Koo, University of Texas, Austin,
Kurt Kreiner, Boeing Space and Intelligence Systems,
Jonathan Lunine, Cornell University,
Alfred U. MacRae, MacRae Technologies,
Bruce D. Marcus, TRW (retired),
Edward D. McCullough, The Boeing Company (retired),
Joseph Nainiger, Alphaport Incorporated,
Michael Norman, University of California, San Diego,
Robert Pinkerton, Orbital Sciences Corporation,
George H. Rieke, University of Arizona,
Stephen M. Rock, Stanford University, and
Al Sacco, Jr., Texas Tech University.

Although the reviewers listed above have provided many constructive comments and suggestions, they were not asked to endorse any conclusions or recommendations, nor did they see the final draft of the report before its release. The review of this report was overseen by Martha P. Haynes, Cornell University, and Ronald M. Sega, Colorado State University. Appointed by the NRC, they were responsible for making certain that an independent examination of this report was carried out in accordance with institutional procedures and that all review comments were carefully considered. Responsibility for the final content of this report rests entirely with the authoring committee and the institution.

Contents

APPENDIXES[1]

[1] Note that Appendixes D through R are not included in the print version of this report. They appear in the electronic version of the report, which is available on the CD-ROM that accompanies the printed report and is posted as a free PDF on the National Academies Press website at http://www.nap.edu/catalog.php?record_id=13354.

Summary

Success in executing future NASA space missions will depend on advanced technology developments that should already be underway. It has been years since NASA has had a vigorous, broad-based program in advanced space technology development, and NASA's technology base is largely depleted. As noted in a recent National Research Council report on the U.S. civil space program:

> Future U.S. leadership in space requires a foundation of sustained technology advances that can enable the development of more capable, reliable, and lower-cost spacecraft and launch vehicles to achieve space program goals. A strong advanced technology development foundation is needed also to enhance technology readiness of new missions, mitigate their technological risks, improve the quality of cost estimates, and thereby contribute to better overall mission cost management. . . . Yet financial support for this technology base has eroded over the years. The United States is now living on the innovation funded in the past and has an obligation to replenish this foundational element. (NRC, 2009, pp. 56-57)

NASA has developed a draft set of technology roadmaps to guide the development of space technologies under the leadership of the NASA Office of the Chief Technologist.[1] The NRC appointed the Steering Committee for NASA Technology Roadmaps and six panels to evaluate the draft roadmaps, recommend improvements, and prioritize the technologies within each and among all of the technology areas as NASA finalizes the roadmaps. The steering committee is encouraged by the initiative NASA has taken through the Office of the Chief Technologist (OCT) to develop technology roadmaps and to seek input from the aerospace technical community with this study.

TECHNOLOGY DEVELOPMENT PROGRAM RATIONALE AND SCOPE

In February 2011, NASA issued an updated strategic plan outlining agency goals and plans for achieving those goals in the 2011-2021 decade and beyond (NASA, 2011). The strategic plan highlights six strategic goals. Five of them relate directly to the scope of this study. The other one deals directly with the agency's aeronautics mission, which, as mentioned in the preface, is outside the statement of task for this study. The 14 draft space technology roadmaps identify a number of critical enabling technologies that the steering committee and panels evaluated and prioritized. Together they represent a foundation upon which to build and achieve the strategic goals outlined in the *2011 NASA Strategic Plan:*

[1] The draft space technology roadmaps are available at http://www.nasa.gov/offices/oct/strategic_integration/technology_roadmap.html.

1. Extend and sustain human activities across the solar system.
2. Expand scientific understanding of Earth and the universe in which we live.
3. Create the innovative new space technologies for our exploration, science, and economic future.
4. Advance aeronautics research for societal benefit.
5. Enable program and institutional capabilities to conduct NASA's aeronautics and space activities.
6. Share NASA with the public, educators, and students to provide opportunities to participate in our Mission, foster innovation, and contribute to a strong national economy.

As part of the effort to develop a detailed plan for implementing the Space Technology Program, OCT developed a set of 14 draft technology roadmaps. These roadmaps establish time sequencing and interdependencies of advanced space technology research and development over the next 5 to 30 years for the following 14 technology areas (TAs):

- TA01. Launch Propulsion Systems
- TA02. In-Space Propulsion Technologies
- TA03. Space Power and Energy Storage
- TA04. Robotics, TeleRobotics, and Autonomous Systems
- TA05. Communication and Navigation
- TA06. Human Health, Life Support, and Habitation Systems
- TA07. Human Exploration Destination Systems
- TA08. Science Instruments, Observatories, and Sensor Systems
- TA09. Entry, Descent, and Landing Systems
- TA10. Nanotechnology
- TA11. Modeling, Simulation, and Information Technology and Processing
- TA12. Materials, Structures, Mechanical Systems, and Manufacturing
- TA13. Ground and Launch Systems Processing
- TA14. Thermal Management Systems

These draft roadmaps represented the starting point and point of departure for the steering committee to evaluate and prioritize technologies and recommend areas for improvement. The roadmaps are organized through a technology area breakdown structure, which in turn served as the structure for evaluating the technologies for this study. Level 1 represents the technology area (TA), which is the title of the roadmap. Each roadmap describes level 2 subareas and level 3 technologies.

TECHNOLOGY EVALUATION PROCESS AND CRITERIA

A set of criteria was established by the steering committee to enable the prioritization of technologies within each and, ultimately, among all of the technology areas of the NASA technology roadmaps. These criteria were chosen to capture the potential benefits, breadth, and risk of the various technologies and were used as a guide by both the panels and the steering committee to determine the final prioritization of the technologies.

Broad community input was solicited from a public website where more than 240 public comments were received on the draft roadmaps using the established steering committee evaluation criteria and other descriptive factors. The public and panels were given the same rubrics to evaluate the technologies so that the various inputs could be more fairly compared against each other.

A series of public workshops were held to solicit input from the members of the community who were interested in contributing to the discussion of the technology roadmaps. The workshops were organized by the various panels, and all included speakers specifically invited by the panel members. The workshops were open to the public and included times for open discussion by all members of the audience. The views expressed during the workshops were considered by the panel members as they assessed the level 3 technologies.

The panels identified a number of challenges for each technology area that should be addressed for NASA to improve its capability to achieve its strategic goals. These top technical challenges were generated to assist in the prioritization of the level 3 technologies. The challenges were developed to identify the general needs NASA has within each technology area, whereas the technologies themselves address how those needs will be met.

The individual panels were tasked with categorizing the individual level 3 technologies into high-, medium-, and low-priority groups. The panels generated a weighted decision matrix based on quality function deployment (QFD) techniques for each technology area. In this method, each criterion and sub-criterion was given a numerical weight by the steering committee. The steering committee based the criteria weighting on the importance of the criteria to meeting NASA's goals of technology advancement.

HIGH-PRIORITY TECHNOLOGIES BY ROADMAP

The study panels produced an assessment of each roadmap that defined top technical challenges for that technical area; prioritized the level 3 technologies for the assigned roadmap into high, medium, and low categories; described the value of the high-priority technologies; identified gaps in the draft roadmaps; identified development or schedule changes of the technologies covered; and summarized the public workshop that focused on the draft roadmap. The results of the panels' work are summarized in this report in 14 appendixes (D through Q; one for each roadmap). This input from the panels was then integrated by the steering committee and documented in the main body of this report.

The high-priority technologies identified by the panels are shown in Table S.1. The panels identified a total of 83 high-priority technologies from a total of 295 possible technologies. In subsequent prioritizations, the steering committee used only these 83 technologies from which to make its technology assessments.

TECHNOLOGY OBJECTIVES

The technology priorities recommended in this report were generated with an awareness of NASA's current mission plans, but those priorities are not closely linked to any particular set of future NASA missions because the goals and schedules of individual missions frequently change. As described above, NASA's 2011 strategic plan formed the foundation for the panel's process of setting technology priorities, and defining top technical challenges was an important intermediate step for setting the panels' technology priorities.

In selecting the highest-priority technologies among all 14 roadmaps, the steering committee took the additional step of established an organizing framework that addressed balance across NASA mission areas, relevance in meeting the highest-priority technical challenges, and expectations that significant progress could be made in the next 5 years of the 30-year window of the roadmaps. Furthermore, the steering committee constrained the number of highest-priority technologies to be included in the final list in the belief that in the face of probable scarce resources, focusing initially on a small number of the highest-priority technologies offers the best chance to make the greatest impact, especially given that agency mission areas, particularly in exploration, are being refined and can be shaped by technology options. Within this organizing framework, technology objectives were defined by the steering committee to address the breadth of NASA missions and group related technologies.

- **Technology Objective A:** *Extend and sustain human activities beyond low Earth orbit.* Technologies to enable humans to survive long voyages throughout the solar system, get to their chosen destination, work effectively, and return safely.
- **Technology Objective B:** *Explore the evolution of the solar system and the potential for life elsewhere.* Technologies that enable humans and robots to perform in situ measurements on Earth (astrobiology) and on other planetary bodies.
- **Technology Objective C:** *Expand our understanding of Earth and the universe in which we live.* Technologies for remote measurements from platforms that orbit or fly by Earth and other planetary bodies, and from other in-space and ground-based observatories.

TABLE S.1 The 83 High-Priority Level 3 Technologies Selected by the Panels

TA01 Launch Propulsion Systems
- 1.3.1 Turbine Based Combined Cycle (TBCC)
- 1.3.2 Rocket Based Combined Cycle (RBCC)

TA02 In-Space Propulsion Technologies
- 2.2.1 Electric Propulsion
- 2.4.2 Propellant Storage and Transfer
- 2.2.3 (Nuclear) Thermal Propulsion
- 2.1.7 Micro-Propulsion

TA03 Space Power and Energy Storage
- 3.1.3 Solar Power Generation (Photovoltaic and Thermal)
- 3.1.5 Fission Power Generation
- 3.3.3 Power Distribution and Transmission
- 3.3.5 Power Conversion and Regulation
- 3.2.1 Batteries
- 3.1.4 Radioisotope Power Generation

TA04 Robotics, TeleRobotics, and Autonomous Systems
- 4.6.2 Relative Guidance Algorithms
- 4.6.3 Docking and Capture Mechanisms/Interfaces
- 4.5.1 Vehicle System Management and FDIR
- 4.3.2 Dexterous Manipulation
- 4.4.2 Supervisory Control
- 4.2.1 Extreme Terrain Mobility
- 4.3.6 Robotic Drilling and Sample Processing
- 4.2.4 Small Body/Microgravity Mobility

TA05 Communication and Navigation
- 5.4.3 Onboard Autonomous Navigation and Maneuvering
- 5.4.1 Timekeeping and Time Distribution
- 5.3.2 Adaptive Network Topology
- 5.5.1 Radio Systems

TA06 Human Health, Life Support, and Habitation Systems
- 6.5.5 Radiation Monitoring Technology
- 6.5.3 Radiation Protection Systems
- 6.5.1 Radiation Risk Assessment Modeling
- 6.1.4 Habitation
- 6.1.3 Environmental Control and Life Support System (ECLSS) Waste Management
- 6.3.2 Long-Duration Crew Health
- 6.1.2 ECLSS Water Recovery and Management
- 6.2.1 Extravehicular Activity (EVA) Pressure Garment
- 6.5.4 Radiation Prediction
- 6.5.2 Radiation Mitigation
- 6.4.2 Fire Detection and Suppression
- 6.1.1 Air Revitalization
- 6.2.2 EVA Portable Life Support System
- 6.4.4 Fire Remediation

TA07 Human Exploration Destination Systems
- 7.1.3 In Situ Resource Utilization (ISRU) Products/Production
- 7.2.1 Autonomous Logistics Management
- 7.6.2 Construction and Assembly
- 7.6.3 Dust Prevention and Mitigation
- 7.1.4 ISRU Manufacturing/Infrastructure etc.
- 7.1.2 ISRU Resource Acquisition
- 7.3.2 Surface Mobility
- 7.2.4 Food Production, Processing, and Preservation
- 7.4.2 Habitation Evolution
- 7.4.3 Smart Habitats
- 7.2.2 Maintenance Systems

TA08 Science Instruments, Observatories, and Sensor Systems
- 8.2.4 High-Contrast Imaging and Spectroscopy Technologies
- 8.1.3 Optical Systems (Instruments and Sensors)
- 8.1.1 Detectors and Focal Planes
- 8.3.3 In Situ Instruments and Sensors
- 8.2.5 Wireless Spacecraft Technology
- 8.1.5 Lasers for Instruments and Sensors
- 8.1.2 Electronics for Instruments and Sensors

TA09 Entry, Descent, and Landing (EDL) Systems
- 9.4.7 GN&C Sensors and Systems (EDL)
- 9.1.1 Rigid Thermal Protection Systems
- 9.1.2 Flexible Thermal Protection Systems
- 9.1.4 Deployment Hypersonic Decelerators
- 9.4.5 EDL Modeling and Simulation
- 9.4.6 EDL Instrumentation and Health Monitoring
- 9.4.4 Atmospheric and Surface Characterization
- 9.4.3 EDL System Integration and Analysis

TA10 Nanotechnology
- 10.1.1 (Nano) Lightweight Materials and Structures
- 10.2.1 (Nano) Energy Generation
- 10.3.1 Nanopropellants
- 10.4.1 (Nano) Sensors and Actuators

TA11 Modeling, Simulation, and Information Technology and Processing
- 11.1.1 Flight Computing
- 11.1.2 Ground Computing
- 11.2.4a Science Modeling and Simulation
- 11.3.1 Distributed Simulation

TA12 Materials, Structures, Mechanical Systems, and Manufacturing
- 12.2.5 Structures: Innovative, Multifunctional Concepts
- 12.2.1 Structures: Lightweight Concepts
- 12.1.1 Materials: Lightweight Structure
- 12.2.2 Structures: Design and Certification Methods
- 12.5.1 Nondestructive Evaluation and Sensors
- 12.3.4 Mechanisms: Design and Analysis Tools and Methods
- 12.3.1 Deployables, Docking, and Interfaces
- 12.3.5 Mechanisms: Reliability/Life Assessment/ Health Monitoring
- 12.4.2 Intelligent Integrated Manufacturing and Cyber Physical Systems

TA14 Thermal Management Systems
- 14.3.1 Ascent/Entry Thermal Protection Systems
- 14.1.2 Active Thermal Control of Cryogenic Systems

NOTE: Technologies are listed by roadmap/technology area (TA01 through TA14; there are no high-priority technologies in TA13). Within each technology area, technologies are listed by the QFD score assigned by the panels, in descending order. This sequencing may be considered a rough approximation of the relative priority of the technologies within a given technology area.

The technology objectives are not independent, and more than one objective may be addressed by a single mission, such as a human mission to explore planetary bodies, and some technologies support more than one of these objectives. Furthermore, these three technology objectives helped categorize similar technologies with similar drivers (i.e., technologies driven by keeping humans alive, productive, and transported; in situ measurements; and remote measurements) and enabled prioritization among diverse technologies on a meaningful basis.

Balance

One of the steering committee's basic assumptions was that NASA would continue to pursue a balanced space program across its mission areas of human exploration, space science, space operations, space technology, and aeronautics. Therefore, since OCT's technology program should broadly support the breadth of the agency's missions and serve to open up options for future missions, the steering committee established priorities in each of the three technology objective areas, A, B, and C, independently. No one technology objective area was given priority over another.

TOP TECHNICAL CHALLENGES

With the three technology objectives defined, the steering committee evaluated the top technical challenges from the panels' prioritized list of challenges for each roadmap TA01 through TA14. The top 10 technical challenges for each of the three technology objectives are described in Table S.2.

HIGHEST-PRIORITY LEVEL 3 TECHNOLOGIES ACROSS ALL ROADMAPS

Using the panel results, which established a high degree of correlation between high-priority level 3 technologies and the respective technical challenges for each roadmap (see the correlation matrices in the third figure in each of the Appendixes D through Q), the steering committee was able to relate high-priority technologies that aligned with each of the three technology objectives.

The steering committee determined that, in several instances, technologies on the original list of 83 high-priority technologies were highly coupled. During the prioritization process, these highly coupled technologies were grouped together and considered as one unit. There are a total of five unified technologies (designated X.1 through X.5). Each one consists of three to five original technologies as follows:

- X.1 Radiation Mitigation for Human Spaceflight
    - 6.5.1 Radiation Risk Assessment Modeling
    - 6.5.2 Radiation Mitigation
    - 6.5.3 Radiation Protection Systems
    - 6.5.4 Radiation Prediction
    - 6.5.5 Radiation Monitoring Technology
- X.2 Lightweight and Multifunctional Materials and Structures
    - 10.1.1 (Nano) Lightweight Materials and Structures
    - 12.1.1 Materials: Lightweight Structures
    - 12.2.1 Structures: Lightweight Concepts
    - 12.2.2 Structures: Design and Certification Methods
    - 12.2.5 Structures: Innovative, Multifunctional Concepts
- X.3 ECLSS
    - 6.1.1 Air Revitalization
    - 6.1.2 ECLSS Water Recovery and Management
    - 6.1.3 ECLSS Waste Management
    - 6.1.4 Habitation

TABLE S.2 Top Technical Challenges by Technology Objective

Top Technical Challenges for Technology Objective A: Extend and sustain human activities beyond low Earth orbit.	Top Technical Challenges for Technology Objective B: Explore the evolution of the solar system and the potential for life elsewhere (in situ measurements).	Top Technical Challenges for Technology Objective C: Expand our understanding of Earth and the universe in which we live (remote measurements).
A1) **Improved Access to Space:** Dramatically reduce the total cost and increase reliability and safety of access to space.	B1) **Improved Access to Space:** Dramatically reduce the total cost and increase reliability and safety of access to space.	C1) **Improved Access to Space:** Dramatically reduce the total cost and increase reliability and safety of access to space.
A2) **Space Radiation Health Effects:** Improve understanding of space radiation effects on humans and develop radiation protection technologies to enable long-duration space missions.	B2) **Precision Landing:** Increase the ability to land more safely and precisely at a variety of planetary locales and at a variety of times.	C2) **New Astronomical Telescopes:** Develop a new generation of astronomical telescopes that enable discovery of habitable planets, facilitate advances in solar physics, and enable the study of faint structures around bright objects by developing high-contrast imaging and spectroscopic technologies to provide unprecedented sensitivity, field of view, and spectroscopy of faint objects.
A3) **Long-Duration Health Effects:** Minimize the crew health effects of long-duration space missions (other than space radiation).	B3) **Robotic Maneuvering:** Enable mobile robotic systems to autonomously and verifiably navigate and avoid hazards and increase the robustness of landing systems to surface hazards.	C3) **Lightweight Space Structures:** Develop innovative lightweight materials and structures to reduce the mass and improve the performance of space systems such as (1) launch vehicle and payload systems; (2) space and surface habitats that protect the crew, including multifunctional structures that enable lightweight radiation shielding, implement self-monitoring capability, and require minimum crew maintenance time; and (3) lightweight, deployable synthetic aperture radar antennas, including reliable mechanisms and structures for large-aperture space systems that can be stowed compactly for launch and yet achieve high-precision final shapes.
A4) **Long-Duration ECLSS:** Achieve reliable, closed-loop Environmental Control and Life Support Systems (ECLSS) to enable long-duration human missions beyond low Earth orbit.	B4) **Life Detection:** Improve sensors for in-situ analysis to determine if synthesis of organic matter may exist today, whether there is evidence that life ever emerged, and whether there are habitats with the necessary conditions to sustain life on other planetary bodies.	C4) **Increase Available Power:** Eliminate the constraint of power availability for space missions by improving energy generation and storage with reliable power systems that can survive the wide range of environments unique to NASA missions.
A5) **Rapid Crew Transit:** Establish propulsion capability for rapid crew transit to and from Mars or other distant targets.	B5) **High-Power Electric Propulsion:** Develop high-power electric propulsion systems along with the enabling power system technology.	C5) **Higher Data Rates:** Minimize constraints imposed by communication data rate and range.

TABLE S.2 Continued

Top Technical Challenges for Technology Objective A: Extend and sustain human activities beyond low Earth orbit.	Top Technical Challenges for Technology Objective B: Explore the evolution of the solar system and the potential for life elsewhere (in situ measurements).	Top Technical Challenges for Technology Objective C: Expand our understanding of Earth and the universe in which we live (remote measurements).
A6) **Lightweight Space Structures:** Develop innovative lightweight materials and structures to reduce the mass and improve the performance of space systems such as (1) launch vehicle and payload systems; (2) space and surface habitats that protect the crew, including multifunctional structures that enable lightweight radiation shielding, implement self-monitoring capability, and require minimum crew maintenance time; and (3) lightweight, deployable synthetic aperture radar antennas, including reliable mechanisms and structures for large-aperture space systems that can be stowed compactly for launch and yet achieve high-precision final shapes.	B6) **Autonomous Rendezvous and Dock:** Achieve highly reliable, autonomous rendezvous, proximity operations, and capture of free-flying space objects.	C6) **High-Power Electric Propulsion:** Develop high-power electric propulsion systems along with the enabling power system technology.
A7) **Increase Available Power:** Eliminate the constraint of power availability for space missions by improving energy generation and storage with reliable power systems that can survive the wide range of environments unique to NASA missions.	B7) **Increase Available Power:** Eliminate the constraint of power availability for space missions by improving energy generation and storage with reliable power systems that can survive the wide range of environments unique to NASA missions.	C7) **Design Software:** Advance new validated computational design, analysis, and simulation methods for design, certification, and reliability of materials, structures, and thermal, EDL, and other systems.
A8) **Mass to Surface:** Deliver more payload to destinations in the solar system.	B8) **Mass to Surface:** Deliver more payload to destinations in the solar system.	C8) **Structural Monitoring:** Develop means for monitoring structural health and sustainability for long-duration missions, including integration of unobtrusive sensors and responsive on-board systems.
A9) **Precision Landing:** Increase the ability to land more safely and precisely at a variety of planetary locales and at a variety of times.	B9) **Lightweight Space Structures:** Develop innovative lightweight materials and structures to reduce the mass and improve the performance of space systems such as (1) launch vehicle and payload systems; (2) space and surface habitats that protect the crew, including multifunctional structures that enable lightweight radiation shielding, implement self-monitoring capability, and require minimum crew maintenance time; and (3) lightweight, deployable synthetic aperture radar antennas, including reliable mechanisms and structures for large-aperture space systems that can be stowed compactly for launch and yet achieve high-precision final shapes.	C9) **Improved Flight Computers:** Develop advanced flight-capable devices and system software for real-time flight computing with low-power, radiation-hard, and fault-tolerant hardware that can be applied to autonomous landing, rendezvous, and surface hazard avoidance.
A10) **Autonomous Rendezvous and Dock:** Achieve highly reliable, autonomous rendezvous, proximity operations, and capture of free-flying space objects.	B10) **Higher Data Rates:** Minimize constraints imposed by communication data rate and range.	C10) **Cryogenic Storage and Transfer:** Develop long-term storage and transfer of cryogens in space using systems that approach near-zero boil-off.

- X.4 GN&C
  - 4.6.2 Relative Guidance Algorithms
  - 5.4.3 Onboard Autonomous Navigation and Maneuvering
  - 9.4.7 GN&C Sensors and Systems (EDL)
- X.5 EDL TPS
  - 9.1.1 Rigid Thermal Protection Systems
  - 9.1.2 Flexible Thermal Protection Systems
  - 14.3.1 Ascent/Entry TPS

To develop as short a list as is reasonable in the face of anticipated constrained budgets, several rounds of prioritization were conducted to determine the highest-priority technologies to emphasize over the next 5 years. The resulting short list of the highest-priority technologies to emphasize over the next 5 years is shown in ranked order in Table S.3 (three columns with 16 different technologies). Again, the steering committee assumes NASA will pursue enabling technology related to all three objectives in a balanced approach, and the steering committee does not recommend or advocate support for one objective over another.

Finally, the steering committee reasoned that this intentionally limited set of recommended high-priority technologies comprised a scope that could reasonably be accommodated within the most likely expected funding level available for technology development by OCT (in the range of $500 million to $1 billion annually). Also considered within the scope of a balanced technology development program is the importance of low technology readiness level (TRL; 1 and 2) exploratory concept development and high-TRL flight demonstrations. The steering committee consensus is that low-TRL, NASA Institute for Advanced Concepts-like funding should be on the order of 10 percent of the total, and that the research should quickly weed out the least competitive concepts, focusing on those that show the greatest promise in addressing the top technical challenges. At the high-TRL end of the spectrum, flight demonstrations, while expensive, are sometimes essential to reach a readiness level required for transition of a technology to an operational system. Such technology flight demonstrations are considered on a case-by-case basis when there is ample "pull" from the user organization, including a reasonable level of cost sharing. Also, there were two technologies, Advanced Stirling Radioisotope Generators and On-Orbit Cryogenic Storage and Transfer, that the steering committee considered to be at a "tipping point," meaning that a relatively small increase in the research effort could produce a large advance in its technology readiness.

Recommendation. ***Technology Development Priorities.*** During the next 5 years, NASA technology development efforts should focus on (1) the 16 identified high-priority technologies and associated top technical challenges, (2) a modest but significant investment in low-TRL technology (on the order of 10 percent of NASA's technology development budget), and (3) flight demonstrations for technologies that are at a high TRL when there is sufficient interest and shared cost by the intended user.

Recommendation. ***Advanced Stirling Radioisotope Generators.*** The NASA Office of the Chief Technologist should work with the Science Mission Directorate and the Department of Energy to help bring Advanced Stirling Radioisotope Generator-technology hardware to flight demonstration on a suitable space mission beyond low Earth orbit.

Finding. ***Plutonium-238.*** Consistent with findings of previous National Research Council reports on the subject of plutonium-238 (NRC 2010, 2011), restarting the fuel supply is urgently needed. Even with the successful development of Advanced Stirling Radioisotope Generators, if the funds to restart the fuel supply are not authorized and appropriated, it will be impossible for the United States to conduct certain planned, critical deep-space missions after this decade.

Recommendation. ***Cryogenic Storage and Handling.*** Reduced-gravity cryogenic storage and handling technology is close to a "tipping point," and NASA should perform on-orbit flight testing and flight demonstrations to establish technology readiness.

TABLE S.3 Final Prioritization of the Top Technologies, Categorized by Objective

Highest-Priority Technologies for Technology Objective A	Highest-Priority Technologies for Technology Objective B	Highest-Priority Technologies for Technology Objective C
Radiation Mitigation for Human Spaceflight (X.1)	GN&C (X.4)	Optical Systems (Instruments and Sensors) (8.1.3)
Long-Duration Crew Health (6.3.2)	Solar Power Generation (Photovoltaic and Thermal) (3.1.3)	High-Contrast Imaging and Spectroscopy Technologies (8.2.4)
ECLSS (X.3)	Electric Propulsion (2.2.1)	Detectors and Focal Planes (8.1.1)
GN&C (X.4)	Fission Power Generation (3.1.5)	Lightweight and Multifunctional Materials and Structures (X.2)
(Nuclear) Thermal Propulsion (2.2.3)	EDL TPS (X.5)	Active Thermal Control of Cryogenic Systems (14.1.2)
Lightweight and Multifunctional Materials and Structures (X.2)	In Situ Instruments and Sensors (8.3.3)	Electric Propulsion (2.2.1)
Fission Power Generation (3.1.5)	Lightweight and Multifunctional Materials and Structures (X.2)	Solar Power Generation (Photovoltaic and Thermal) (3.1.3)
EDL TPS (X.5)	Extreme Terrain Mobility (4.2.1)	

CROSSCUTTING FINDINGS AND RECOMMENDATIONS

In reviewing and evaluating the draft roadmaps and considering the purpose and strategic goals for the advanced technology development program managed by OCT, the steering committee formed some general observations concerning the program as a whole and reached some conclusions on how the effectiveness of the program can be maintained or enhanced. The topics dealt with tend to address multiple roadmaps.

Recommendation. ***Systems Analysis.*** NASA's Office of the Chief Technologist (OCT) should use disciplined system analysis for the ongoing management and decision support of the space technology portfolio, particularly with regard to understanding technology alternatives, relationships, priorities, timing, availability, down-selection, maturation, investment needs, system engineering considerations, and cost-to-benefit ratios; to examine "what-if" scenarios; and to facilitate multidisciplinary assessment, coordination, and integration of the roadmaps as a whole. OCT should give early attention to improving systems analysis and modeling tools, if necessary to accomplish this recommendation.

Recommendation. ***Managing the Progression of Technologies to Higher Technology Readiness Levels (TRLs).*** OCT should establish a rigorous process to down select among competing technologies at appropriate milestones and TRLs to assure that only the most promising technologies proceed to the next TRL.

Recommendation. ***Foundational Technology Base***. OCT should reestablish a discipline-oriented technology base program that pursues both evolutionary and revolutionary advances in technological capabilities and that draws upon the expertise of NASA centers and laboratories, other federal laboratories, industry, and academia.

Recommendation. ***Cooperative Development of New Technologies.*** OCT should pursue cooperative development of high-priority technologies with other organizations to leverage resources available for technology development.

Recommendation. ***Flight Demonstrations and Technology Transition.*** OCT should collaborate with other NASA mission offices and outside partners in defining, advocating, and where necessary co-funding flight demonstrations of technologies. OCT should document this collaborative arrangement using a technology transition plan or similar agreement that specifies success criteria for flight demonstrations as well as budget commitments by all involved parties.

Finding. ***Facilities.*** Adequate research and testing facilities are essential to the timely development of many space technologies. In some cases, critical facilities do not exist or no longer exist, but defining facility requirements and then meeting those requirements fall outside the scope of NASA's OCT (and this study).

Finding. ***Program Stability.*** Repeated, unexpected changes in the direction, content, and/or level of effort of technology development programs have diminished the programs' productivity and effectiveness. In the absence of a sustained commitment to address this issue, the pursuit of OCT's mission to advance key technologies at a steady pace will be threatened.

Recommendation. ***Industry Access to NASA Data.*** OCT should make the engineering, scientific, and technical data that NASA has acquired from past and present space missions and technology development more readily available to U.S. industry, including companies that do not have an ongoing working relationship with NASA and that are pursuing their own commercial goals apart from NASA's science and exploration missions. To facilitate this process in the future, OCT should propose changes to NASA procedures so that programs are required to archive data in a readily accessible format.

Recommendation. ***NASA Investments in Commercial Space Technology.*** While OCT should focus primarily on developing advanced technologies of high value to NASA's own mission needs, OCT should also collaborate with the U.S. commercial space industry in the development of precompetitive technologies of interest to and sought by the commercial space industry.

Finding. ***Crosscutting Technologies.*** Many technologies, such as those related to avionics and space weather beyond radiation effects, cut across many of the existing draft roadmaps, but the level 3 technologies in the draft roadmaps provide an uneven and incomplete list of the technologies needed to address these topics comprehensively.

Recommendation. ***Crosscutting Technologies.*** OCT should review and, as necessary, expand the sections of each roadmap that address crosscutting level 3 technologies, especially with regard to avionics and space weather beyond radiation effects. OCT should assure effective ownership responsibility for crosscutting technologies in each of the roadmaps where they appear and establish a comprehensive, systematic approach for synergistic, coordinated development of high-priority crosscutting technologies.

In summary, the draft set of 14 roadmaps produced by NASA contained 320 level 3 technologies. The panels assessed the technology area breakdown structure of the 14 roadmaps and developed a revised structure containing 295 level 3 technologies. Of those 295 technologies, 83 were considered high priority by the panels. The steering committee then evaluated those 83 technologies. Through an organizing framework relating objectives, challenges, and individual technologies, the prioritization process across all roadmaps identified 7 or 8 technologies for each of three independent technology objectives, for a total of 16 unique technologies that this report recommends be emphasized over the next 5 years of the 5- to 30-year window of the technology roadmaps.

Technological breakthroughs have been the foundation of virtually every NASA success. The Apollo landings on the Moon are now an icon for the successful application of technology to a task that was once regarded as a distant dream. NASA science missions that continue to unlock the secrets of our solar system and universe, and human and robotic exploration of the solar system are inherently high-risk endeavors and require new technologies, new ideas, and bold applications of technology, engineering, and science to create the required vehicles, support systems, and space operations infrastructure. NASA has led in the development and application of many critically important space technologies. In addition, technological advances have yielded benefits far beyond space itself in down-to-Earth applications.

The technologies needed for the Apollo program were generally self-evident and driven by a clear and well-defined goal. In the modern era, the goals of the country's broad space mission include multiple objectives, extensive involvement from both the public and private sectors, choices among multiple paths to different destinations, and very limited resources. As the breadth of the country's space mission has expanded, the necessary technological developments have become less clear, and more effort is required to evaluate the best path for a forward-looking technology development program. NASA has now entered a transitional stage, moving from the past era in which desirable technological goals were evident to all to one in which careful choices among many conflicting alternatives must be made. This report provides specific guidance and recommendations on how the effectiveness of the technology development program managed by NASA's Office of the Chief Technologist can be enhanced in the face of scarce resources by focusing on the highest-priority technologies.

REFERENCES

NASA (National Aeronautics and Space Administration). 2011. *2011 NASA Strategic Plan.* NASA Headquarters, Washington, D.C. Available at http://www.nasa.gov/pdf/516579main_NASA2011StrategicPlan.pdf.

NRC (National Research Council). 2009. *America's Future in Space: Aligning the Civil Space Program with National Needs.* The National Academies Press, Washington, D.C.

NRC. 2010. *Radioisotope Power Systems: An Imperative for Maintaining U.S. Leadership in Space Exploration.* The National Academies Press, Washington, D.C.

NRC. 2011. *Vision and Voyages for Planetary Science in the Decade 2013-2022.* The National Academies Press, Washington, D.C.

1

Introduction

Success in executing future NASA space missions will depend on advanced technology developments that should already be underway. However, it has been years since NASA has had a vigorous, broad-based program in advanced space technology. NASA's technology base is largely depleted, and few new, demonstrated technologies (that is, at high technology readiness levels) are available to help NASA execute its priorities in exploration and space science. As noted in a recent National Research Council report on the U.S. civil space program:

> Future U.S. leadership in space requires a foundation of sustained technology advances that can enable the development of more capable, reliable, and lower-cost spacecraft and launch vehicles to achieve space program goals. A strong advanced technology development foundation is needed also to enhance technology readiness of new missions, mitigate their technological risks, improve the quality of cost estimates, and thereby contribute to better overall mission cost management. . . . Yet financial support for this technology base has eroded over the years. The United States is now living on the innovation funded in the past and has an obligation to replenish this foundational element. (NRC, 2009, pp. 56-57)

Currently available technology is insufficient to accomplish many intended space missions. Consider the following examples:

- To send humans to the Moon, Mars, or other destinations beyond low Earth orbit (LEO), new technologies are needed to (1) mitigate the effects of space radiation from both the cosmic ray background and from solar flares; (2) advance the state of the art in environmental control and life support systems (ECLSS) so that they are highly reliable, can be easily repaired in space, and feature closed-loop water, air, and food cycles; and (3) provide advanced fail-safe mobile pressure suits, lightweight rovers, improved human-machine interfaces, in situ resource utilization (ISRU) systems, and other mechanical systems that can operate in dusty, reduced-gravity environments.
- NASA's future capabilities would also benefit greatly from new technologies to build robotic vehicles that can maneuver over a wider range of gravitational, environmental, surface, and subsurface conditions with a sufficient degree of autonomy to enhance operation at large distances from Earth.
- Commercial space activities in LEO and deep-space exploration would benefit from advanced launch and space transportation systems, some of which may need to store and transfer cryogenic propellants in space. In addition, deep-space exploration options could be opened up with high-thrust electric or nuclear upper-stage propulsion systems.

- To enhance the ability of spacecraft to land on a wide variety of surfaces in our solar system, new technologies are needed to provide guidance, navigation, and control (GN&C) systems with greater precision, and real-time recognition with trajectory adaptation for surface hazard avoidance.
- Future space science missions capable of addressing the highest-priority goals in astrophysics will need a new generation of lower-cost astronomical telescopes that can utilize advanced coolers and camera systems, improved focal-plane arrays, and low-cost, ultra-stable, large-aperture mirrors. Likewise, high-contrast exoplanet imaging technologies with unprecedented sensitivity, field of view, and spectroscopy of faint objects are needed to enable discovery and characterization of exoplanets orbiting in the habitable zones of their host stars.

A robust space technology base is urgently needed. The Steering Committee for NASA Technology Roadmaps is encouraged by the initiative NASA has taken through the Office of the Chief Technologist (OCT) to develop technology roadmaps and seek input from the aerospace technical community via this study.[1]

TECHNOLOGY DEVELOPMENT PROGRAM RATIONALE AND SCOPE

The 2010 NASA Authorization Act, signed into law on October 11, 2010, directed NASA to create a program to maintain its research and development base in space technology:

> It is critical that NASA maintain an agency space technology base that helps align mission directorate investments and supports long term needs to complement mission-directorate funded research and support, where appropriate, multiple users, building upon its Innovative Partnerships Program and other partnering approaches. (Public Law 111-267, Sec. 904)

On February 14, 2011, NASA issued its *2011 NASA Strategic Plan* outlining agency goals and plans for achieving those goals in the 2011-2021 decade and beyond (NASA, 2011). The strategic plan highlights five strategic goals that relate directly to the scope of this study. The sixth strategic goal deals directly with the agency's aeronautics mission, which as mentioned in the preface is outside the statement of task for this study. The 14 draft space technology roadmaps identify a number of critical enabling technologies that the steering committee and panels evaluated and prioritized. Together they represent a foundation upon which to build and achieve the strategic goals outlined in the 2011 strategic plan:

1. Extend and sustain human activities across the solar system.
2. Expand scientific understanding of the Earth and the universe in which we live.
3. Create the innovative new space technologies for our exploration, science, and economic future.
4. Advance aeronautics research for societal benefit.
5. Enable program and institutional capabilities to conduct NASA's aeronautics and space activities.
6. Share NASA with the public, educators, and students to provide opportunities to participate in our Mission, foster innovation, and contribute to a strong national economy.

DRAFT TECHNOLOGY ROADMAPS

As part of the effort to develop a detailed plan for implementing the Space Technology Program, OCT developed a set of 14 draft technology roadmaps. These roadmaps establish time sequencing and interdependencies of advanced space technology research and development over the next 5 to 30 years for the following 14 technology areas (TAs):

- TA01. Launch Propulsion Systems
- TA02. In-Space Propulsion Technologies
- TA03. Space Power and Energy Storage
- TA04. Robotics, TeleRobotics, and Autonomous Systems

[1] The draft space technology roadmaps are available at http://www.nasa.gov/offices/oct/strategic_integration/technology_roadmap.html.

- TA05. Communication and Navigation
- TA06. Human Health, Life Support, and Habitation Systems
- TA07. Human Exploration Destination Systems
- TA08. Science Instruments, Observatories, and Sensor Systems
- TA09. Entry, Descent, and Landing Systems
- TA10. Nanotechnology
- TA11. Modeling, Simulation, and Information Technology and Processing
- TA12. Materials, Structures, Mechanical Systems, and Manufacturing
- TA13. Ground and Launch Systems Processing
- TA14. Thermal Management Systems

For each TA, OCT established a cross-agency team to draft each of the 14 technology roadmaps. They were released to the public in November 2010 (see http://www.nasa.gov/offices/oct/home/roadmaps/index.html). The draft technology roadmaps identified a wide variety of opportunities to revitalize NASA's advanced space technology development program. The draft roadmaps represented the starting point and point of departure for the steering committee to evaluate and prioritize technologies and recommend areas for improvement. Also, there were a number of common themes across the roadmaps where recommendations are made that if dealt with collectively would lead to improvements as a whole.

The roadmaps are organized through a technology area breakdown structure (see Appendix B), which in turn served as the structure for evaluating the technologies for this study. Level 1 represents the technology area (TA), which is the title of the roadmap. Each roadmap describes level 2 subareas and level 3 technologies.[2] The draft set of 14 roadmaps produced by NASA contained 320 level 3 technologies. The panels assessed the technology area breakdown structure of the 14 roadmaps and developed a revised structure containing 295 level 3 technologies.[3] (The full revised technology area breakdown structure is shown in Appendix B.) Of those 295 technologies, 83 were considered high priority by the panels and are summarized in Chapter 2. The steering committee then evaluated only those 83 technologies in its prioritization. In its first round of prioritization, the steering committee developed an interim list of 11-15 technologies per objective, for a total of 28 unique technologies. The final round of prioritization resulted in 7-8 technologies per objective, for a total of 16 unique technologies. These steps in the prioritization process are described in Chapter 3.

The purpose of the roadmaps is to establish a sustained collection of technology development goals for the next 5 to 30 years. In the process of defining level 3 technologies of interest, NASA mission directorates helped identify "pull" technologies that could contribute to specific future missions. The roadmaps also include emerging "push" technologies that may enable mission capabilities that lie outside the baseline requirements of planned missions and which may enable missions not yet envisioned.

This report is the second of two reports produced by this study. An interim report, released in August 2011, defines a modified set of level 3 technologies for many of the roadmaps. It also makes high-level observations associated with the roadmaps and identifies technology gaps that cut across multiple roadmaps. The interim report is available online at http://www.nap.edu/catalog.php?record_id=13228.

STAKEHOLDERS: RESEARCH AND DEVELOPMENT PARTNERS AND END USERS

Most of the technologies included in the roadmaps have multiple stakeholders where cooperative research and technology development is beneficial to all parties involved and combines resources where appropriate to achieve greater progress. Other agencies and departments in the government, such as the Department of Defense, as well as parallel efforts in industry and universities, have ongoing technology development efforts. NASA program managers

[2] Many of the roadmaps also list and/or describe level 4 technology topics in text and/or figures. Per the statement of task, this report is focused at the level 3 technologies, of which there were more than 300. For the most part, space does not allow this report to address the even more numerous level 4 technology topics.

[3] The evaluation process established by the steering committee was designed to focus on assessing the individual technologies and ranking them in priority order rather than on how the technologies were grouped into the 14 roadmaps.

and researchers need to work cooperatively with their peers outside the agency in a collaborative research and development partnership, where appropriate. Similarly, in the interest of expediting technology transition when OCT believes it is ready to hand off the technology for application, coordination with the end user needs to occur early and often. End users internal to NASA are the mission directorates in the agency in science, exploration, and operations. They are also partners. With a proactive culture of collaboration, OCT will encourage technology transition to end users in industry or other government agencies and departments, or in universities that might pursue new avenues outside NASA space objectives. The scope of this study included space technology needs of industry for commercial space and space technologies that address national needs like energy, medicine, etc. on a broader scale.

ORGANIZATION OF THIS REPORT

This report represents the compiled technical input, assessment, and prioritization of NASA's draft roadmaps by six study panels and the steering committee. The panels, which were composed of subject-matter experts, were each responsible for evaluating one to four draft roadmaps. The steering committee was responsible for providing guidance to the panels, coordinating their work, and compiling both the interim report and this final report.

Chapter 2 describes the process used by each panel and summarizes their key results in the form of a prioritized list of top technical challenges and a description of high-priority technologies for each of the 14 draft roadmaps. A more detailed description of the results of each panel's deliberations for each roadmap appears in Appendices D (for TA01) through Q (for TA14). Specifically, those appendixes contain the following:

- A description of the draft roadmap's technology area, including changes made by the panels to the list of level 3 technologies associated with each technology area,
- The top technical challenges determined by the panel,
- A detailed numerical assessment of the level 3 technologies,
- A description and assessment of each of the highest-priority technologies,
- A brief explanation of medium- and low-priority technology ratings,[4]
- A discussion of development and schedule changes for technologies in the roadmap,
- Other general comments, and
- A summary of the public workshop held on the draft roadmap.

Chapter 3 describes the process used by the steering committee to take the inputs from the panels on each roadmap and develops recommendations on the highest-priority technologies for emphasis in the next 5 years of the 30-year window considered. Chapter 3 prioritizes the most important top technical challenges using an organizing framework defined by three technology objectives:

- Extend and sustain human activities beyond low Earth orbit.
- Explore the evolution of the solar system and the potential for life elsewhere (in situ measurements).
- Expand our understanding of Earth and the universe in which we live (remote measurements).

Chapter 4 addresses observations and develops additional recommendations for topics that transcend a single roadmap, including many of the topics addressed in the interim report (which did not include recommendations).

REFERENCES

NASA (National Aeronautics and Space Administration). 2011. *2011 NASA Strategic Plan.* NASA Headquarters, Washington, D.C.

NRC (National Research Council). 2009. *America's Future in Space: Aligning the Civil Space Program with National Needs.* The National Academies Press, Washington, D.C.

[4] In Chapter 2 and the appendixes, the report focuses on providing detailed information and explanations only for technologies ranked as high priority.

2

Top Technical Challenges and High-Priority Technologies by Roadmap

TECHNOLOGY EVALUATION PROCESS AND CRITERIA

A set of criteria was established by the steering committee to enable the prioritization of technologies within each and, ultimately, among all of the technology areas of the NASA technology roadmaps.[1] These criteria were chosen to capture the potential benefits, breadth, and risk of the various technologies and were used as a guide by both the panels and the steering committee to determine the final prioritization of the technologies. In addition to the primary criteria used to prioritize the technologies, an additional set of secondary descriptive factors were also assessed for each technology. These descriptive factors were added to provide a complete picture of the panels' assessments of the technologies and assisted in the evaluations.

Broad community input was solicited through a public website, where more than 240 public comments were received on the draft roadmaps using the established steering committee criteria and other descriptive factors. The public and panels were given the same rubrics to evaluate the technologies so that the various inputs could be more fairly compared against each other. These views, along with those expressed during the public workshops, were taken into account by the panel members as they assessed the technologies. The panels then came to a consensus view for each criterion for each technology.

In evaluating and prioritizing the technologies identified, the steering committee made a distinction between technology development and engineering development. Technology development, which is the intended focus of the draft roadmaps, addresses the process of understanding and evaluating capabilities needed to improve or enable performance advantages over current state-of-the-art space systems. Technologies of interest include both hardware and software, as well as testing and evaluation of hardware (from the component level to the systems level) and software (including design tools) at various levels of technology readiness for application in future space systems. In contrast, engineering development, which generally attempts to implement and apply existing or available technology, is understood for the purposes of this study to be hardware, software, design, test, verification, and validation of systems in all phases of NASA's acquisition process. The high-priority technologies do not include items for which engineering development is the next step in advancing capabilities.

---

[1] The draft space technology roadmaps are available at http://www.nasa.gov/offices/oct/strategic_integration/technology_roadmap.html.

Top Technical Challenges

The panels identified a number of challenges for each technology area that should be addressed for NASA to improve its capability to achieve its objectives. These top technical challenges were generated to provide some focus for technology development and to assist in the prioritization of the level 3 technologies. The challenges were developed to identify the general needs NASA has within each technology area, whereas the technologies themselves address how those needs will be met. Once the top technical challenges were identified, the panels then determined the relative importance of the challenges within each technology area to put them in priority order.

Descriptive Factors

The steering committee identified three descriptive factors that helped characterize each technology. While these factors were not primary in the determination of technology prioritization, they did assist in generating a better understanding of the current status or state of the art of the technology.

- **Technology Readiness Level (TRL):** This factor describes the current state of advancement of the technology using NASA's TRL scale. The TRL scale is defined in Table 2.1. It was determined that TRL should not be a basis for prioritizing technologies, as NASA should be investing across all levels of technology readiness. In assessing TRL levels, the panels were directed to evaluate the most promising developments that should receive attention. For example, electric propulsion systems are commonly used today, so as a whole, they would be assessed as TRL 9; however, the promising area of advancement of high power electric propulsion is less advanced, and thus 2.2.1 Electric Propulsion was assessed as TRL 3.

- **Tipping Point:** The tipping point factor was used to determine if the technology was at a state such that a relatively small additional effort (compared to that which advanced the technology to its current state) could produce a significant advance in technology readiness that would justify increasing the priority associated with this technology.

- **NASA Capabilities:** This factor captured how NASA research in this technology aligns with the expertise, capabilities, and facilities of NASA and/or other organizations cooperating with NASA in this area. It also indicated how much value NASA research in this technology would add to ongoing research by other organizations. This was not a primary consideration in assessing which technologies should be prioritized. Instead it was a factor in considering whether the technology should be developed by NASA, or if NASA should support other current efforts. The factor also addressed whether or not NASA should invest in improving its own capability for pursuing the high-priority technologies.

Evaluation Criteria

The steering committee identified three main criteria on which the technologies were to be judged for evaluation. The three criteria were benefit, alignment with NASA's goals and objectives, and technical risk and challenge. Each of these is described in further detail below. For the latter two criteria, three further sub-criteria were created to assist in evaluating the technologies.

For each evaluated criterion or sub-criterion, a set of four (or in one case five) grades or bins were established, and the public and panel members were asked to determine what grade each technology should receive for that criterion. For consistency, a set of definitions were generated for each grade. The grading definitions were provided as guidelines to help the panel and steering committee members assign an appropriate range of grades necessary to prioritize the technologies in question. They were generated such that most technologies would be placed into one of the middle bins, while placement at the upper/lower bounds would need significant justification. The grades were assigned numeric scores on a non-linear scale (e.g., 0-1-3-9) to accentuate the spread of the summed

TABLE 2.1 NASA Technology Readiness Levels

TRL	Definition	Hardware Description	Software Description	Exit Criteria
1. Basic principles observed and reported.	Lowest level of technology readiness. Scientific research begins to be translated into applied research and development. Examples might include paper studies of a technology's basic properties.	Scientific knowledge generated underpinning hardware technology concepts/applications.	Scientific knowledge generated underpinning basic properties of software architecture and mathematical formulation.	Peer reviewed publication of research underlying the proposed concept/application.
2. Technology concept and/ or application formulated.	Invention begins. Once basic principles are observed, practical applications can be invented. The application is speculative, and there is no proof or detailed analysis to support the assumption. Examples are still limited to paper studies.	Invention begins, practical application is identified but is speculative, no experimental proof or detailed analysis is available to support the conjecture.	Practical application is identified but is speculative, no experimental proof or detailed analysis is available to support the conjecture. Basic properties of algorithms, representations and concepts defined. Basic principles coded. Experiments performed with synthetic data.	Documented description of the application/ concept that addresses feasibility and benefit.
3. Analytical and experimental critical function and/or characteristic proof of concept.	At this step in the maturation process, active research and development (R&D) is initiated. This must include both analytical studies to set the technology into an appropriate context and laboratory-based studies to physically validate that the analytical predictions are correct. These studies and experiments should constitute "proof-of-concept" validation of the applications/concepts formulated at TRL 2.	Analytical studies place the technology in an appropriate context and laboratory demonstrations, modeling and simulation validate analytical prediction.	Development of limited functionality to validate critical properties and predictions using non-integrated software components.	Documented analytical/ experimental results validating predictions of key parameters.
4. Component and/ or breadboard validation in laboratory environment.	Following successful "proof-of-concept" work, basic technological elements must be integrated to establish that the pieces will work together to achieve concept-enabling levels of performance for a component and/or breadboard. This validation must be devised to support the concept that was formulated earlier and should also be consistent with the requirements of potential system applications. The validation is relatively "low-fidelity" compared to the eventual system: it could be composed of ad hoc discrete components in a laboratory.	A low fidelity system/ component breadboard is built and operated to demonstrate basic functionality and critical test environments, and associated performance predictions are defined relative to the final operating environment.	Key, functionally critical software components are integrated, and functionally validated, to establish interoperability and begin architecture development. Relevant environments defined and performance in this environment predicted.	Documented test performance demonstrating agreement with analytical predictions. Documented definition of relevant environment.

TABLE 2.1 Continued

TRL	Definition	Hardware Description	Software Description	Exit Criteria
5. Component and/or breadboard validation in relevant environment.	At this level, the fidelity of the component and/or breadboard being tested has to increase significantly. The basic technological elements must be integrated with reasonably realistic supporting elements so that the total applications (component-level, subsystem-level, or system-level) can be tested in a "simulated" or somewhat realistic environment.	A medium fidelity system/component brassboard is built and operated to demonstrate overall performance in a simulated operational environment with realistic support elements that demonstrates overall performance in critical areas. Performance predictions are made for subsequent development phases.	End-to-end software elements implemented and interfaced with existing systems/simulations conforming to target environment. End-to-end software system, tested in relevant environment, meeting predicted performance. Operational environment performance predicted. Prototype implementations developed.	Documented test performance demonstrating agreement with analytical predictions. Documented definition of scaling requirements.
6. System/subsystem model or prototype demonstration in a relevant environment.	A major step in the level of fidelity of the technology demonstration follows the completion of TRL 5. At TRL 6, a representative model or prototype system or system, which would go well beyond ad hoc, "patch-cord," or discrete component level breadboarding, would be tested in a relevant environment. At this level, if the only relevant environment is the environment of space, then the model or prototype must be demonstrated in space.	A high fidelity system/component prototype that adequately addresses all critical scaling issues is built and operated in a relevant environment to demonstrate operations under critical environmental conditions.	Prototype implementations of the software demonstrated on full-scale realistic problems. Partially integrate with existing hardware/software systems. Limited documentation available. Engineering feasibility fully demonstrated.	Documented test performance demonstrating agreement with analytical predictions.
7. System prototype demonstration in an operational environment.	Prototype near or at planned operational system. TRL 7 is a significant step beyond TRL 6, requiring an actual system prototype demonstration in a space environment. The prototype should be near or at the scale of the planned operational system, and the demonstration must take place in space. Examples include testing the prototype in a test bed.	A high fidelity engineering unit that adequately addresses all critical scaling issues is built and operated in a relevant environment to demonstrate performance in the actual operational environment and platform (ground, airborne, or space).	Prototype software exists having all key functionality available for demonstration and test. Well integrated with operational hardware/software systems demonstrating operational feasibility. Most software bugs removed. Limited documentation available.	Documented test performance demonstrating agreement with analytical predictions.

continued

TABLE 2.1 Continued

TRL	Definition	Hardware Description	Software Description	Exit Criteria
8. Actual system competed and "flight qualified" through test and demonstration.	Technology has been proven to work in its final form and under expected conditions. In almost all cases, this level is the end of true system development for most technology elements. This might include integration of new technology into an existing system.	The final product in its final configuration is successfully demonstrated through test and analysis for its intended operational environment and platform (ground, airborne, or space).	All software has been thoroughly debugged and fully integrated with all operational hardware and software systems. All user documentation, training documentation, and maintenance documentation completed. All functionality successfully demonstrated in simulated operational scenarios. Verification and Validation (V&V) completed.	Documented test performance verifying analytical predictions.
9. Actual system flight proven through successful mission operations	Actual application of the technology in its final form and under mission conditions, such as those encountered in operational test and evaluation. In almost all cases, this is the end of the last "bug fixing" aspects of true system development. This TRL does not include planned product improvement of ongoing or reusable systems.	The final product is successfully operated in an actual mission.	All software has been thoroughly debugged and fully integrated with all operational hardware/ software systems. All documentation has been completed. Sustaining software engineering support is in place. System has been successfully operated in the operational environment.	Documented mission operational results.

SOURCE: NASA Procedural Requirements 7120.8, Appendix J (http://nodis3.gsfc.nasa.gov/displayDir.cfm?Internal_ID= N_PR_7120_0008_&page_name=AppendixJ) and NASA Procedural Requirements 7123.1A, Table G.19 (http://esto.nasa.gov/ files/TRL.dochttp:/nodis3.gsfc.nasa.gov/displayDir.cfm?Internal_ID=N_PR_7123_001A_&page_name=AppendixG).

final scores. Higher numeric scores implied greater ability to meet NASA's goals. Negative numbers indicated characteristics that were not desirable.

Benefit: Would the technology provide game-changing, transformational capabilities in the timeframe of the study? What other enhancements to existing capabilities could result from development of this technology?

1. *The technology is unlikely to result in a significant improvement in performance or reduction in life cycle cost of missions during the next 20 years. Score: 0*
2. *The technology is likely to result in (a) a minor improvement in mission performance (e.g., less than a 10 percent reduction in system launch mass); (b) a minor improvement in mission life cycle cost; or (c) less than an order of magnitude increase in data or reliability of missions during the next 20 years. Score: 1*
3. *The technology is likely to result in (a) a major improvement in mission performance (e.g., a 10 percent to 30 percent reduction in mass); or (b) a minor improvement in mission life cycle cost or an order of magnitude increase in data or reliability of missions during the next 20 years. Score: 3*

4. *The technology is likely to provide game-changing, transformational capabilities that would enable important new projects or missions that are not currently feasible during the next 20 years. Score: 9*

Alignment: Three sub-criteria were created to evaluate the alignment with NASA's goals and objectives criterion.

***Alignment with NASA Needs*:** How does NASA research in this technology improve NASA's ability to meet its long-term needs? For example, which mission areas and which missions listed in the relevant roadmap would directly benefit from development of this technology, and what would be the nature of that impact? What other planned or potential missions would benefit?

1. *Technology is not directly applicable to NASA. Score: 0*
2. *Technology will impact one mission in one of NASA's mission areas. Score: 1*
3. *Technology will impact multiple missions in one of NASA's mission areas. Score: 3*
4. *Technology will impact multiple missions in multiple NASA mission areas. Score: 9*

Alignment with Non-NASA Aerospace Technology Needs: How does NASA research in this technology improve NASA's ability to address non-NASA aerospace technology needs?

1. *Little or no impact on aerospace activities outside of NASA's specific needs. Score: 0*
2. *Impact will be limited to niche roles. Score 1*
3. *Will impact a large subset of aerospace activities outside of NASA's specific needs (e.g., commercial spacecraft). Score: 3*
4. *Will have a broad impact across the entire aerospace community. Score: 9*

Alignment with Non-Aerospace National Goals: How well does NASA research in this technology improve NASA's ability to address national goals from broader national perspective (e.g., energy, transportation, health, environmental stewardship, or infrastructure)?

1. *Little or no impact outside the aerospace industry. Score: 0*
2. *Impact will be limited to niche roles. Score: 1*
3. *Will be useful to a specific community outside aerospace (e.g., medicine). Score: 3*
4. *Will be widely used outside the aerospace community (e.g., energy generation or storage). Score: 9*

Technical Risk and Challenge*:* Three sub-criteria were created to evaluate the technical risk and challenge criterion. In this criterion, the grades created were not as straightforward as those for benefit and alignment. They were developed to capture the steering committee's view on the appropriate risk posture for NASA technology developments.

Technical Risk and Reasonableness: What is the overall nature of the technical risk and/or the reasonableness that this technology development can succeed in the timeframe envisioned? Is the level of risk sufficiently low that industry could be expected to complete development of this technology without a dedicated NASA research effort, or is it already available for commercial or military applications? Regarding the expected level of effort and timeframe for technology development: (a) are they believable given the complexity of the technology and the technical challenges to be overcome; and (b) are they reasonable given the envisioned benefit vis-à-vis possible alternate technologies?

1. *The technical risk associated with development of this technology is very low, such that it is feasible for industry or a specific NASA mission office to complete development (without additional NASA technology funding if a mission need arises). Score: 1*

2. *The technical risk associated with development of this technology is low, and the likely cost to NASA and the timeframe to complete technology development are not expected to substantially exceed those of past efforts to develop comparable technologies. Score: 3*
3. *The technical risk associated with development of this technology is moderate to high, which is a good fit to NASA's level of risk tolerance for technology development, but the likely cost to NASA and the timeframe to complete technology development are expected to substantially exceed those of past efforts to develop comparable technologies. Score: 3*
4. *The technical risk associated with development of this technology is moderate to high, which is a good fit to NASA's level of risk tolerance for technology development, and the likely cost to NASA and the timeframe to complete technology development are not expected to substantially exceed those of past efforts to develop comparable technologies. Score: 9*
5. *The technical risk associated with development of this technology is extremely high, such that it is unreasonable to expect any operational benefits over the next 20 years without unforeseen revolutionary breakthroughs and/or an extraordinary level of effort. Score: 1*

Sequencing and Timing: Is the proposed timing of the development of this technology appropriate relative to when it will be needed? What other new technologies are needed to enable the development of this technology, have they been completed, and how complex are the interactions between this technology and other new technologies under development? What other new technologies does this technology enable? Is there a good plan for proceeding with technology development? Is the technology development effort well connected with prospective users?

1. *This is an extremely complex technology and/or is highly dependent on multiple other projects with interfaces that are not well thought out or understood. Score: –9*
2. *The development of this technology is just roughly sketched out and there are no clearly identified users (i.e., missions). Score: –3*
3. *There is a clear plan for advancing this technology. While there is an obvious need, there are no specifically identified users. Score: –1*
4. *There is a clear plan for advancing this technology, there is an obvious need, and joint funding by a user seems likely. Score: +1*

Time and Effort to Achieve Goals: How much time and what overall effort are required to achieve the goals for this technology?

1. *National endeavor: Likely to require more than 5 years and substantial new facilities, organizations, and workforce capabilities to achieve; similar to or larger in scope than the Shuttle, Manhattan Project, or Apollo Program. Score: –9*
2. *Major project: Likely to require more than 5 years and substantial new facilities to achieve; similar in scope to development of the Apollo heat shield or the Orion environmental systems. Score: –3*
3. *Moderate effort: Can be achieved in less than 5 years with a moderately sized (less than 50 people) team (e.g., Mars Pathfinder's airbag system). Score: –1*
4. *Minimal effort: Can be achieved in a few years by a very small (less than 10 people) team (e.g., graduate student/faculty university project). Score: 0*

Evaluation Methodology

The individual panels were tasked with binning the individual technologies into high, medium, and low priority for level 3 technologies. This was done primarily by grading the technologies using the criteria described above. The panels generated a weighted decision matrix based on quality function deployment (QFD) techniques for each technology area. In this method, each criterion was given a numerical weight by the steering committee, described

TABLE 2.2 Numerical Weighting Factors Given to Evaluation Criteria in Panel Assessments

Criterion	Numerical Weight
Benefit (27)	27
Alignment (9)	
Alignment with NASA needs	5
Alignment with non-NASA aerospace needs	2
Alignment with non-aerospace national goals	2
Technical risk and challenge (18)	
Technical risk and reasonableness	10
Sequencing and timing	4
Time and effort	4

below. By multiplying the panel grades by the criteria weighting factor and summing the results, a single score was calculated for each technology.

The steering committee based the criteria weighting on the importance of the criteria to meeting NASA's goals of technology advancement. It determined that the potential benefit of the technology was the most important factor in prioritizing, with the risk and challenges being second, and alignment being third in importance of the three main criteria. To allow for weighting at the sub-criteria level, the steering committee assigned a total weighting of 9 to alignment, 18 to risk and challenges, and 27 to benefits. It then divided those values among the sub-criteria to generate the values shown in Table 2.2.

This method provided an initial assessment of how technologies met NASA's goals via the criteria evaluation. After each panel came to a consensus on the grades for all criteria for each technology, a total QFD score was computed for each technology. Consider the example shown in Figure 2.1. The QFD score for technology 1.1.1 Propellants is computed using the score for each criterion and the corresponding multiplier as follows:

$$1 \times 27 + 3 \times 5 + 3 \times 2 + 0 \times 2 + 3 \times 10 - 1 \times 4 - 1 \times 4 = 70$$

The technologies were then sorted by their total QFD scores. In Figure 2.1, technology 1.3.1 TBCC has the highest score, and thus it is the highest priority of the three technologies shown.

Once the panels had ordered the technologies by their total scores, they then divided the list into high, medium, and low priority technology groups.[2] This division was subjectively performed by each panel for each technology area for which it was responsible, seeking where possible natural break points. For instance, in the case of the assessment of TA01, the panel decided that the split between high and medium priority technologies should occur at a score of 150, and that the split between medium and low priority technologies should occur at a score of 90.

To add flexibility to the assessment process, the panels were also given the option of identifying key technologies that they believed should be high priority but that did not have a numerical score that achieved a high priority rank. These override technologies were deemed by the panels to be high priority irrespective of the numerical scores. As such, by allowing the panels to use this override provision, the numerical scoring process could be used effectively without the evaluation becoming a slave to it. Based on the raw QFD scoring of the 295 level 3 technologies, 64 were initially classified as high priority, 128 as medium priority, and 103 as low priority. The panels subsequently decided to override the QFD scores to elevate 18 medium priority technologies and 1 low priority technology (6.4.4 Remediation) to the high priority group. The final result was to have 83 high priority technologies, 110 medium

[2] The panels were tasked with designating each technology as high, medium, or low priority only. The high-priority technologies are listed in this section of each appendix by QFD score, in descending order; this sequencing may be considered a rough approximation of the relative priority of the technologies within each technology area. Also, this ordering places the override technologies (which were designated as high priority despite their relatively low QFD scores) as least among the high-priority technologies, although that is not necessarily the case.

	Benefit	Alignment with NASA Needs	Alignment with Non-NASA Aerospace Technology Needs	Alignment with Non-Aerospace National Goals	Technical Risk and Reasonableness	Sequencing and Timing	Time and Effort	QFD Score (Weighted)	Panel Priority
Multiplier	**27**	**5**	**2**	**2**	**10**	**4**	**4**		
	0/1/3/9	**0/1/3/9**	**0/1/3/9**	**0/1/3/9**	**1/3/9**	**-9/-3/-1/1**	**-9/-3/-1/0**		
Technology Name	**Benefit**	**Alignment**			**Risk/Difficulty**				
1.1.1. (Solid Rocket) Propellants	1	3	3	0	3	-1	-1	70	L
1.2.1. LH2/LOX Based	1	9	9	0	3	1	-3	112	M
1.3.1.Turbine Based Combined Cycle (TBCC)	3	9	9	0	3	-3	-3	150	H

FIGURE 2.1 Sample QFD matrix, showing three technologies from TA01 and their resulting QFD scores.

priority technologies, and 102 low priority technologies. The steering committee believes that the results of the panel scoring validate the design of the QFD scoring process and the decision to allow the panels to override those scores as appropriate.

The panels also assessed which of the technologies have the greatest chance of meeting the identified top technical challenges. While many of the technologies within a technology area could potentially address one or more of the challenges, the panels only labeled those where investment would have a major or moderate impact. This assessment was used to verify the proper identification of the high-priority technologies and occasionally as validation for using the override option.

Public Workshops

A series of workshops were held to solicit input for the members of the community who were interested in contributing to the discussion of the technology roadmaps. The workshops were organized by the various panels and all included speakers specifically invited by the panel members. The workshops were open to the public and included times for open discussion by all members of the audience. The views expressed during the public workshops were considered by the panel members as they assessed the level 3 technologies. Detailed summaries of each workshop can be found at the end of each roadmap report (the roadmap reports can be found in Appendixes D-Q). Table 2.3 lists information on each public workshop.

SUMMARY OF TOP TECHNICAL CHALLENGES AND HIGH-PRIORITY TECHNOLOGIES BY ROADMAP

The methods described above were applied to all 14 draft roadmaps by the six technical panels. Using the various forms of public input as well as their own internal deliberations, the study panels produced reports for the steering committee that prioritized the level 3 technologies into high, medium, and low categories; described the value of the high-priority technologies; identified gaps in the draft roadmaps; identified development or schedule changes of the technologies covered; and summarized the public workshop that focused on the draft roadmap. Each panel report, one per draft roadmap, is included as an appendix to this report (see Appendixes D-Q). The top technical challenges and high-priority technologies for each roadmap are summarized below, along with any other high-level summary information for each roadmap.

It should be noted that the top technical challenges for each roadmap have been prioritized by the panels and are listed here in priority order. The panels were not instructed to prioritize level 3 technologies, other than to categorize them into high, medium, and low priority "bins." All high-priority technologies are described below; the order is determined by the QFD score the technology received.

TABLE 2.3 Summary Information on Public Workshops Held on Each Roadmap

Roadmap	Workshop Date	Workshop Location	Responsible Panel
TA01 Launch Propulsion Systems	March 23, 2011	California Institute of Technology, Pasadena, CA	Panel 1: Propulsion and Power
TA02 In-Space Propulsion Technologies	March 21, 2011	California Institute of Technology, Pasadena, CA	Panel 1: Propulsion and Power
TA03 Space Power and Energy Storage	March 24, 2011	California Institute of Technology, Pasadena, CA	Panel 1: Propulsion and Power
TA04 Robotics, TeleRobotics, and Autonomous Systems	March 30, 2011	Keck Center, Washington, DC	Panel 2: Robotics, Communication, and Navigation
TA05 Communication and Navigation	March 29, 2011	Keck Center, Washington, DC	Panel 2: Robotics, Communication, and Navigation
TA06 Human Health, Life Support, and Habitation Systems	April 26, 2011	The Lunar and Planetary Institute, Houston, TX	Panel 4: Human Health and Surface Exploration
TA07 Human Exploration Destination Systems	April 27, 2011	The Lunar and Planetary Institute, Houston, TX	Panel 4: Human Health and Surface Exploration
TA08 Science Instruments, Observatories, and Sensor Systems	March 29, 2011	Beckman Center, Irvine, CA	Panel 3: Instruments and Computing
TA09 Entry, Descent, and Landing Systems	March 23-24, 2011	Beckman Center, Irvine, CA	Panel 6: EDL
TA10 Nanotechnology	March 9, 2011	Keck Center, Washington, DC	Panel 5: Materials
TA11 Modeling, Simulation, and Information Technology and Processing	May 10, 2011	Keck Center, Washington, DC	Panel 3: Instruments and Computing
TA12 Materials, Structures, Mechanical Systems, and Manufacturing	March 10, 2011	Keck Center, Washington, DC	Panel 5: Materials
TA13 Ground and Launch Systems Processing	March 24, 2011	California Institute of Technology, Pasadena, CA	Panel 1: Propulsion and Power
TA14 Thermal Management Systems	March 11, 2011	Keck Center, Washington, DC	Panel 5: Materials

TA01 Launch Propulsion Systems

TA01 includes all propulsion technologies required to deliver space missions from the surface of Earth to Earth orbit or Earth escape, including solid rocket propulsion systems, liquid rocket propulsion systems, air breathing propulsion systems, ancillary propulsion systems, and unconventional/other propulsion systems. The Earth-to-orbit launch industry is currently reliant on very mature technologies, to which only small incremental improvements are possible. Breakthrough technologies are not on the near horizon; therefore research and development efforts will require both significant time and financial investments.

TA01 Top Technical Challenges

1. Reduced Cost: Develop propulsion technologies that have the potential to dramatically reduce the total cost and to increase the reliability and safety of access to space.

High launch costs currently serve as a major barrier to any space mission, limiting both the number and the scope of NASA's space missions. Even in light of major monetary investments in launch over the last several decades, the cost of launch has not decreased and in fact continues to increase. Reliability and safety are essential concerns for NASA's space missions. Finding ways to improve reliability and safety without significantly effecting cost is a major technical challenge.

2. Upper Stage Engines: Develop technologies to enable lower cost, high specific impulse upper stage engines suitable for NASA, DOD, and commercial needs, applicable to both Earth-to-orbit and in-space applications.

The RL-10 engine is the current upper stage engine in use but is based on 50-year-old technology and is both expensive and difficult to produce. Alternative engine cycles and designs with the promise of reducing cost and improving reliability are a major challenge. Additionally, because high-rate production substantially lowers costs, technologies which are amenable to a wide range of applications are desirable.

TA01 High-Priority Technologies

Two high-priority technologies were identified from TA-01. In both cases high-priority status was identified because of the wide range of applications that they offered for NASA's missions. However, a significant number of challenges were also identified for each, and the steering committee believes that it will take decades of research and development and a large and sustained financial investment to makes these technologies feasible.

Technology 1.3.1, Turbine Based Combined Cycle (TBCC)

Turbine Based Combined Cycle (TBCC) propulsion systems have the potential to combine the advantages of gas turbines and rockets in order to enable lower launch costs and more responsive operations. NASA has been investigating rocket-air breathing cycles for many years, and its commitment to and expertise in hypersonic air breathing cycles is exemplified by NASA's experimental X-43 program.

Technology 1.3.2, Rocket Based Combined Cycle (RBCC)

Rocket Based Combined Cycle (RBCC) propulsion systems combine the high specific impulse of the air breathing ramjet and scramjet engines with the high thrust/weight ratio of a chemical rocket. They promise to deliver launch systems with much lower costs than present launch systems. As noted above for TBCC, NASA has been investigating rocket-air breathing cycles for many years, and its commitment to and expertise in hypersonic air breathing cycles is exemplified by NASA's experimental X-43 program.

Additional Comments

The development timeline for launch propulsion technologies will be critically dependent on the overall strategy and architecture chosen for exploration and the funding available. Of particular relevance is launch economics, particularly with regard to the launch rate and the mass of missions being launched. Additionally, there are technologies included in other roadmaps, especially TA02 (In-Space Propulsion) and TA04 (Robotics, Tele-Robotics, and Autonomous Systems) that open the trade space to other architecture options, such as fuel depots requiring on-orbit propellant transfer technologies. For example, one may be able to disaggregate some large space missions to be launched by larger numbers of smaller, lower cost launch vehicles. These technologies may allow more dramatic reductions in launch costs than would specific launch technologies themselves.

TA02 In-Space Propulsion Technology

TA02 includes all propulsion-related technologies required by space missions after the spacecraft leaves the launch vehicle from Earth, consisting of four level 2 technology subareas: chemical propulsion, non-chemical propulsion, advanced propulsion technologies, and supporting technologies. This technology area includes propulsion for such diverse applications as fine pointing an astrophysics satellite in low Earth orbit (LEO), robotic science and Earth observation missions, high-thrust Earth orbit departure for crewed vehicles, low-thrust cargo transfer for human exploration, and planetary descent, landing, and ascent propulsion, and results in diverse set of technologies including traditional space-storable chemical, cryogenic chemical, various forms of electric propulsion, various forms of nuclear propulsion, chemical and electric micropropulsion, solar sails, and space tethers.

Before prioritizing the technologies in TA02, several technologies were renamed, deleted, or moved. The steering committee deleted 2.4.1 Engine Health Monitoring and Safety, 2.4.3 Materials and Manufacturing Technologies, 2.4.4 Heat Rejection, and 2.4.5 Power, because these technologies did not fall under the scope of TA02. In each case, the reader is referred to other sections of the roadmap (2.4.1 to TA04, 2.4.3 to TA12, 2.4.4 to TA14, and 2.4.5 to TA03) to learn the details of what should be done in these areas.

TA02 Top Technical Challenges

1. High-Power Electric Propulsion Systems: Develop high-power electric propulsion systems technologies to enable high ΔV missions with heavy payloads.

Electric propulsion systems have a higher propellant efficiency than other in-space propulsion technologies that will be available in the foreseeable future, with applications to all NASA, DOD, and commercial space mission areas. Development of high power electric propulsion systems will enable larger scale missions with heavy payloads, and demonstration of large scale electric propulsion vehicles is required to ensure adequate control during autonomous rendezvous and docking operations necessary for either cargo or small body proximity operations.

2. Cryogenic Storage and Transfer: Enable long-term storage and transfer of cryogens in space and reliable cryogenic engine operation after long dormant periods in space.

Deep-space exploration missions will require high performance propulsion for all mission phases, including Earth departure, destination arrival, destination departure, and Earth return, occurring over the entire mission duration. Both high-thrust propulsion options, LOX/H_2 chemical and LH2 nuclear thermal rockets, will require storage of cryogens for well over a year to support all mission phases, and the engines must also operate reliably after being dormant for the same period. This technical challenge must be overcome if humans are ever to explore destinations beyond the Moon.

3. Microsatellites: Develop high performance propulsion technologies for high-mobility microsatellites (<100 kg).

The broader impact of small satellites is hindered by the lack of propulsion systems with performance levels similar to those utilized in larger satellites. Most existing propulsion systems are not amenable for miniaturization and work is needed to develop concepts that scale and perform favorably. Miniature propulsion would also provide functionality in different applications, such as controlling large flexible structures. Many of these high performance propulsion technologies are near a tipping point, and moderate investment would be required to validate their applicability to small satellites.

4. Rapid Crew Transit: Establish propulsion capability for rapid crew transit to/from Mars.

Developing high performance, high thrust propulsion systems to reduce transit times for crewed missions will mitigate concerns about impacts to crew health from radiation, exposure to reduced gravity, and other effects of long-duration deep-space travel. Two realistic high-thrust options exist that could be available for missions in the next 20 years: LOX/H_2 and nuclear thermal rockets. The engines must be capable of multiple restarts following prolonged periods of inactivity and must be extremely high reliability systems. There are currently no engines of either type that meet the requirements of performance, reliability, and re-start capability.

TA02 High-Priority Technologies

Technology 2.2.1, Electric Propulsion

Electric propulsion (EP) uses electrical power produced on the spacecraft to accelerate propellant to extremely high speeds. Solar electric propulsion (SEP), including arcjet, Hall thruster, and ion thruster systems, is routinely used today for spacecraft maneuvers. Modern laboratory-model ion thrusters and Hall thrusters have been demonstrated on the ground and flight versions of these thrusters may be developed in the mid-term timeframe. Further in the future, multi-MW systems enabled by nuclear power systems could use flight versions of various thrusters currently in early laboratory testing. The development of high-power SEP systems (~100 kW to ~1 MW) could enable larger-scale or faster missions, more efficient in-space transportation systems in Earth orbit, more affordable sample return missions, and pre-positioning of cargo and ISRU facilities for human exploration missions.

Technology 2.4.2, Propellant Storage and Transfer

Propellant Storage and Transfer in space includes both the long-term storage of cryogens and the transfer of these fluids between refueling stations and the propulsion systems on spacecraft, upper stages, and Moon/Mars landing and ascent vehicles. This technology has only been validated at the component level for cryogenic fluids in laboratory environments, although "storable" propellant storage and transfer has been demonstrated in space. Propellant storage and transfer is a game-changing technology for a wide range of applications because it enables long-duration, high-thrust, high-ΔV missions for large payloads and crew and can be implemented within the next three decades.

Technology 2.2.3, (Nuclear) Thermal Propulsion

The technology includes both solar and nuclear thermal sources that heat hydrogen propellant to achieve high specific impulse. Of these two, only nuclear thermal propulsion is rated as a high-priority technology. Nuclear thermal rockets (NTRs) are high-thrust propulsion systems with the potential for twice the specific impulse of the best liquid hydrogen/oxygen chemical rockets. Critical NTR technologies include the nuclear fuel, reactor and system controls, and long-life hydrogen pumps, and technology development will also require advances in ground test capabilities, as the open-air approach previously used is no longer environmentally acceptable.

Technology 2.1.7, Micro-propulsion

Micro-propulsion technology addresses all propulsion, chemical and non-chemical, that fulfills the needs for high mobility micro-satellites (<100 kg) and extremely fine pointing and positioning for certain astrophysics missions. Small satellites, either individually or flying in formation, are being considered for increasingly complex missions, driven by low costs, fast development times, and the potential to perform tasks previously limited to large systems. Many technologies have been proposed, including miniaturization of existing systems and innovative concepts, and several promising technologies have emerged. Micro-propulsion technology development properly includes a broad range of technologies, current and future applications, and NASA, DOD, and commercial users.

Additional Information

In an unconstrained funding environment, the TA02 roadmap presents a reasonable approach, particularly when focus is placed on the high-priority technologies listed above. However, in a constrained funding environment it is unlikely that all the level 3 technologies shown on the schedule will be affordable.

The planetary decadal survey identifies Mars ascent propulsion and precision landing as key capabilities. (NRC, 2011, p. 311) Current entry, descent, and landing technologies are near their limits for the martian atmosphere, and some improvements in propulsion systems for descent and landing will be required. While new engineering developments are certainly required, the propulsion challenges are more in system implementation than technology development.

TA03 Space Power and Energy Storage

TA03 is divided into four technology areas: power generation, energy storage, power management and distribution, and crosscutting technologies. NASA has many unique needs for space power and energy storage technologies that require special technology solutions due to extreme environmental conditions. These missions would all benefit from advanced technologies that provide more robust power systems with lower mass.

Before prioritizing the technologies included in TA03, several were renamed, deleted, or moved, and two additional approaches to energy storage were added: (1) electric and magnetic field storage and (2) thermal storage.

TA03 Top Technical Challenges

1. Power Availability: Eliminate the constraint of power availability in planning and executing NASA missions.

Power is a critical limitation for space science and exploration and the availability of more power opens up new paradigms for how NASA operates and what can be accomplished. For example, increased power availability for human exploration missions can support more astronauts at larger outposts with more capabilities, and for robotic science missions, power availability can determine the scope and duration of the mission.

2. High Power for Electric Propulsion: Provide enabling power system technologies for high power electric propulsion for large payloads and planetary surfaces.

Advances in solar and nuclear technologies during the last decade offer the potential of developing power generation systems that can deliver tens to hundreds of kilowatts. Various designs have been utilized to enhance power efficiency, using proven fuels, power conversion technologies, and reactor materials to reduce the development and operations risk to acceptable levels. Other aspects of fission systems require technology development including heat exchangers, fluid management, scaling of power conversion devices, heat rejection components, radiation shielding, and aspects of system integration and testing.

3. Reduced Mass: Reduce the mass and stowed launch volume of space power systems.

Power systems typically constitute one third of the mass of a spacecraft at launch, and the volume available in the launch vehicle fairing can limit the size of solar arrays that can be packaged on the vehicle. Further development of new power generation, energy storage, and power delivery technologies can potentially reduce the mass and volume of these systems, enabling missions to include more science instruments, use smaller and less expensive launch vehicles, and/or provide higher power levels.

4. Power System Options: Provide reliable power system options to survive the wide range of environments unique to NASA missions.

NASA missions require power systems and components to survive many different types of extreme environments. Technology developments to meet these challenges will enable NASA to plan and execute a wide array of missions.

TA03 High-Priority Technologies

Technology 3.1.3, Solar Power Generation (Photovoltaic and Thermal)

Photovoltaic (PV) space power systems have been the workhorse of NASA science missions as well as the foundation for commercial and military systems. Solar cells directly convert sunlight into electricity, and today's solar cells operate with 30 percent efficiency. Current emphasis is on the development of high efficiency cells as well as cells that can effectively operate in extreme environments. Nearly all spacecraft flown to date have been powered by solar arrays, and NASA has a vital interest in photovoltaic power system developments for higher power electric propulsion missions. Of particular interest are advanced array technologies that offer high specific mass and high power density. Solar power generation applies to all NASA mission areas plus DOD, as well as commercial and other civil or national applications.

Technology 3.1.5, Fission Power Generation

Space fission power systems use heat generated by fission of a nuclear fuel to power a thermal to electric conversion device to generate electric power. Key subsystems include the reactor, heat exchanger, power converter, heat rejection, and radiation shield. Space fission power systems would overcome mission infrastructure limitations associated with low power level availability, and can potentially provide a power rich environment to planetary surface exploration missions and enable high power electric propulsion systems for deep-space exploration and science missions.

Technology 3.3.3, Power Distribution and Transmission

As science and human exploration missions of the future are examined, the need for significant increases in electrical power on spacecraft becomes a clearer and higher priority. With these higher power levels, an extrapolation of the current technologies for the distribution and transmission (D&T) of power would result in unacceptably high mass and complexity; therefore more efficient D&T methods are considered high-priority. Proposed research would increase the D&T voltage, develop high frequency alternating current distribution options for space systems, and identify alternate materials to replace copper conductors.

Technology 3.3.5, Power Conversion and Regulation

The available power on any particular spacecraft will be in a form dictated by the power source and distribution architecture, and the various payloads will then likely require the power in a different form. The purpose of conversion and regulation is to provide the necessary bridge between the power source and payloads, and to regulate this power to within the tolerances required by the payloads. A current issue is the need to space-qualify existing terrestrial high voltage components to replace space qualified components that lag behind the commercial state of the art. Important parameters for improving power conversion and regulation devices include increasing conversion efficiency, operating temperature range, and radiation tolerance.

Technology 3.2.1, Batteries

Batteries are electrochemical energy storage devices that have been flown in space from the beginning. In space batteries must survive a variety of environments and load profiles more demanding than for most terrestrial applications. Many batteries are already proven in space, but a variety of advanced chemistry alternatives have yet to be developed and qualified for spaceflight. NASA missions would benefit from new electrochemical power technologies that provide higher specific energy and/or higher specific power.

Technology 3.1.4, Radioisotope Power Generation

Radioisotope power systems (RPS) have enabled many unique deep-space and planetary exploration missions, making scientific discovery possible. RPSs are based on plutonium-238 and have used thermoelectric converters to provide reliable power for many missions throughout the solar system, with operating lifetimes exceeding 30 years. Future RPSs could be developed to deliver both lower and higher power levels. While RPSs are well-established, there are significant technology issues due to the lack of available plutonium-238. Stirling engines, which require less plutonium-238, are being developed to replace thermoelectric converters. Establishing a reliable, recurring source of plutonium-238 and maturing Stirling engine technology are both critically important for NASA's future science and exploration programs. The planetary science decadal survey committee cited as "its highest priority for near-term multi-mission technology investment . . . the completion and validation of the Advanced Stirling Radioisotope Generator" (NRC, 2011, p. 307).

Additional Information

Schedules for space power and energy storage technologies are highly dependent on the level of funding. The schedules are possible if sufficient resources are applied to each item in the roadmap.

TA04 Robotics, Tele-Robotics, and Autonomous Systems

The roadmap for TA04 consists of seven technology subareas: sensing and perception; mobility; manipulation; human-systems integration; autonomy; autonomous rendezvous and docking (AR&D); and robotics, tele-robotics, and autonomous systems engineering. TA04 supports NASA space missions with the development of new capabilities, and can extend the reach of human and robotic exploration through a combination of dexterous robotics, better human/robotic interfaces, improved mobility systems, and greater sensing and perception. The TA04 roadmap focuses on several key issues for the future of robotics and autonomy: enhancing or exceeding human performance in sensing, piloting, driving, manipulating, and rendezvous and docking; development of cooperative and safe human interfaces to form human-robot teams; and improvements in autonomy to make human crews independent from Earth and make robotic missions more capable.

For the TA04 roadmap to describe and provide supporting text for each of the level 3 technologies (like the other roadmaps) it would have to be largely rewritten, and the panel made a number of suggestions for changes to TA04 for it to parallel the other roadmaps. As a result, the steering committee and responsible panel did not have a list of well-defined technologies originally identified in the draft roadmaps, and have recommended a new set of level 3 technologies.

TA04 Top Technical Challenges

1. Rendezvous: Develop the capability for highly reliable, autonomous rendezvous, proximity operations, and capture/attachment to (cooperative and non-cooperative) free-flying space objects.

The ability to perform autonomous rendezvous and safe proximity operations and docking/grappling are central to the future of diverse mission concepts. Major challenges include improving the robustness of the rendezvous and capture process to ensure successful capture.

2. Maneuvering: Enable robotic systems to maneuver in a wide range of NASA-relevant environmental, gravitational, and surface and subsurface conditions.

Current rovers cannot access extreme lunar or martian terrain, eliminating the possibility of robotic access and requiring humans to park and travel on foot in suits. In microgravity, locomotion techniques on or near asteroids and comets are undeveloped and untested. Challenges include developing robotics to travel into these otherwise denied areas, developing techniques to grapple and anchor with asteroids and non-cooperative objects, or building crew mobility systems to move humans into these challenging locations.

3. In Situ Analysis and Sample Return: Develop subsurface sampling and analysis exploration technologies to support in situ and sample return science missions.

A top astrobiological goal and a fundamental NASA exploration driver is the search for life or signs of previous life in our solar system. A significant planetary science driver exists to obtain unaltered samples (with volatiles intact) for either in situ analysis or return to Earth from planetary bodies. Terrestrial drilling technologies have limited applicability to these missions and robotic planetary drilling and sample handling is a new and different capability.

4. Hazard Avoidance: Develop the capabilities to enable mobile robotic systems to autonomously and verifiably navigate and avoid hazards.

Due to the large computational throughput requirements needed to quickly assess subtle terrain geometric and non-geometric properties fast enough to maintain speeds near vehicle limits, robotic systems lag behind the ability of human drivers to perceive terrain hazards at long range.

5. Time-Delayed Human-Robotic Interactions: Achieve more effective and safe human interaction with robotic systems (whether in proximity or remotely) that accommodates time-delay effects.

More effective and safe human interaction with robotic systems has a number of different focuses which range from the potential dangers of proxemic interactions to remote supervision with or without time delays. Remote interactions with robotic systems do not pose the same immediate potential level of danger to humans as close proximity interactions; however, it is often significantly more difficult for a remote human to fully understand the context of the environment in which the robotic system functions and the status of the system.

6. Object Recognition and Manipulation: Develop means for object recognition and dexterous manipulation that supports engineering and science objectives.

Object recognition requires sensing, and requires a perception function that can associate the sensed object with an object that is understood a priori. Sensing approaches to date have combined machine vision, stereo vision, lidar, structured light, and radar, while perception approaches often start with CAD models or models created by a scan with the same sensors that will later be used to identify the object. Major challenges include the ability to work with a large library of known objects, identifying objects that are partially occluded, sensing in poor lighting, estimating the pose of quickly tumbling objects, and working with objects at near and far range. Robotic hands with equivalent or superior grasping ability to human hands would avoid the added complexity of robot interfaces on objects and provide a sensate tool change-out capability for specialized tasks.

TA04 High-Priority Technologies

Technology 4.6.2, Relative Guidance Algorithms

Relative guidance technologies encompass algorithms that determine the desired trajectories to be followed between vehicles performing rendezvous, proximity operations, and/or docking and capture. These algorithms must anticipate applicable environmental effects, the nature of the trajectory change/attitude control effectors in use, and the inertial and relative navigation state data available to the guidance algorithms. The new level 3 technologies of interest provide real-time, onboard algorithmic functionality that can calculate and manage spacecraft maneuvers to achieve specific trajectory change objectives. Relative guidance aligns well with NASA's needs because it impacts crewed deep-space exploration, sample return, servicing, and orbital debris mitigation.

Technology 4.6.3, Docking and Capture Mechanisms/Interfaces

Docking and capture mechanisms enable the physical capture and attachment, as well as subsequent safe release, of two bodies in space that achieve part of their mission objectives when operating while joined. Development of a physical docking and capture interface for AR&D operations would greatly simplify the control demands for a working AR&D system. This technology will improve the reliability of AR&D and enable new interfaces that can be employed. Varieties of docking and capture mechanisms enable transfer of crew between delivery and destination vehicles, provide means for attachment of added equipment modules, facilitate execution of robotic servicing missions, and potentially enable grapple/capture of inactive, possibly tumbling spacecraft.

Technology 4.5.1, Vehicle Systems Management and FDIR

The panel combined the related and overlapping topics of integrated systems health management (ISHM), fault detection and isolation and recovery (FDIR), and vehicle systems management (VSM), which together provide the crucial capability for an autonomous spacecraft to operate safely and reliably. ISHM/FDIR/VSM will improve the reliability of future missions by providing a diagnostic capability that helps ground or crew failure assessment and an automated capability to fix/overcome faults; increase robotic mission flexibility in response to failures; and increase crew safety in the event of a detected need for crew escape and abort. This technology is highly aligned to NASA's needs because it will impact many missions, such as deep-space exploration, robotic science missions, planetary landers, and rovers.

Technology 4.3.2, Dexterous Manipulation

Dexterous manipulation is a system-level technology that encompasses multiple stand-alone technology areas and has high relevance for several current and future NASA applications including servicing and maintenance of the ISS, remote satellite servicing, on-orbit assembly of larger structures, and applications to remote exploration. Since 1997, NASA has focused on the development of Robonaut, which is now being evaluated on the ISS and approaches the dexterity of a suited astronaut. Development activities to date have focused primarily on human-in-the-loop teleoperation, and limitations of this system do exist because of high bandwidth, low latency communications requirements. NASA could explore options for extending Robonaut technologies and capabilities for operations in large latency and low bandwidth environments. Additionally, the size and weight of Robonaut preclude its use for exploration activities and NASA could benefit from the development of novel actuation technologies that dramatically increase the strength to weight ratio.

Technology 4.4.2, Supervisory Control

Supervisory control is defined as incorporating the techniques necessary for controlling robotic behaviors using higher-level goals instead of low-level commands, thus requiring robots to have semi-autonomous or autonomous behaviors. This increases the number of robots a single human can simultaneously supervise and also incorporates time-delayed supervision. Key components to be addressed include the development of robust high-level autonomous behaviors and control, multi-sensor fusion, clearly understood and usable presentations of information from multiple robots for human understanding, time-delayed interpretation and presentation of robot provided information, haptic feedback, and means for a supervisory control system to handle communication outages. This technology is highly aligned to NASA's needs due to the impact of reducing the number of personnel required to supervise robotic missions and the number of science and exploration missions to which the technology can be applied.

Technology 4.3.6, Robotic Drilling and Sample Processing

Robotic drilling and sampling processing (RDSP) technologies will improve the science return of robotic science missions to small bodies, moons, and planets, and will also benefit in situ resource utilization for human spaceflight to the moon and small bodies. The development of new robotic drilling, drill-like, and coring technologies coupled with sample processors will have a major beneficial impact on the quality of planetary science returned by future missions due to the relatively uncontaminated, unaltered, and volatile-rich nature of the samples acquired by the next generation of RDSP technology.

Technology 4.2.1, Extreme Terrain Mobility

Extreme mobility encompasses all ground or surface-level mobility. Extremely mobile platforms will be a critical component to both the success and diversity of extraterrestrial body exploration and determining the terrain that will be traversed. In addition, higher degrees of mobility serve to complement autonomy. This technology provides NASA with the capability to maneuver its surface vehicles in extreme terrain in order to "follow the water"—a high-priority science focus for Mars and lunar science missions, and is applicable to any exploration mission, human or robotic, to a planetary (or lunar) surface.

Technology 4.2.4, Small Body/Microgravity Mobility

Operating robots in microgravity poses many challenges and is particularly difficult without fixing or tethering to grounded structures. Even simple tasks such as turning a screw can be extreme challenges to mobile platforms that are not attached to other structures. The development of adaptive mobility systems with complementary perception and autonomy are key elements to performing exploration and sample return missions in tight spaces and microgravity environments. Variable or dynamic CG capabilities can greatly enhance the ability of platforms to move around and perform meaningful work by dynamically shifting the CG in conjunction with the motion of vehicles. This technology is well aligned with NASA's goals related to the exploration of small bodies both robotic and with crew, making this a critical technology for future missions; therefore, the panel designated this as a high-priority technology because the NASA 2010 Authorization Act (P.L. 111-267) has indicated that small body missions (to near-Earth asteroids) should be an objective for NASA human spaceflight beyond Earth orbit. If this goal is pursued as a high NASA priority, it would likely also require precursor robotic missions to small body surfaces with applicable mobility capability.

TA05 Communication and Navigation

TA05, Communication and Navigation, consists of six technology subareas: optical communication and navigation; radio frequency communication; internetworking; position, navigation, and timing; integrated technologies; and revolutionary concepts. Communication links are the lifelines to spacecraft, providing commanding, telemetry, and science data transfers as well as navigation support. Therefore, the Communication and Navigation technology area supports all NASA space missions. Advancement in communication and navigation technology will allow future missions to implement new and more capable science instruments, greatly enhance human missions beyond Earth orbit, and enable entirely new mission concepts.

Before prioritizing the level 3 technologies in TA05, several changes were made to the TA05 roadmap: Technologies 5.4.1 Time-keeping and 5.4.2 Time Distribution have been merged, and Technology 5.6.7 Reconfigurable Large Apertures has been renamed "Reconfigurable Large Apertures Using Nanosat Constellations."

TA05 Top Technical Challenges

1. Autonomous and Accurate Navigation: Meet the navigation needs of projected NASA missions by developing means for more autonomous and accurate absolute and relative navigation.

NASA's future missions include diverse navigational challenges that cannot be supported with current methods, such as precision position knowledge, trajectory determination, cooperative flight, trajectory traverse, and rendezvous with small bodies. Additionally, NASA spacecraft will need to perform these tasks farther from Earth and more autonomously.

2. Communications Constraint Mitigation: Minimize communication data rate and range constraints that impact planning and execution of future NASA space missions.

A recent analysis of NASA's likely future mission set indicates that communications performance will need to grow by about a factor of ten every ~15 years in order to keep up with projected robotic mission requirements and missions will additionally continue to be constrained by the legally internationally allocated spectral bandwidth. Many of the complex tasks of future missions are hampered by keeping Earth in the real-time decision loop, which can be mitigated by making decisions closer to the platform, minimizing reliance on Earth operations. Advancements in communications and navigation infrastructure will allow information to be gathered locally and computation to be performed either in the spacecraft or shared with nearby nodes.

3. Information Delivery: Provide integrity and assurance of information delivery across the solar system.

Future missions will include international partnerships and increased public interaction, which implies increased vulnerability to information compromise. As internetworking extends throughout the solar system, the communications architecture needs to operate in a safe and secure manner.

TA05 High-Priority Technologies

Technology 5.4.3, Onboard Autonomous Navigation and Maneuvering Systems

Onboard autonomous navigation and maneuvering (OANM) techniques are critical for improving the capabilities and reducing the support requirements for many future space missions, and will reduce the dependence on routine position fixes from Earth, freeing the communication network for other tasks. The onboard maneuver planning and execution monitoring will increase the vehicle agility, enabling new mission capabilities and reducing costs by eliminating the large work force required to support routine spacecraft operations. The alignment of this technology to NASA's needs is high because it will impact deep-space exploration with crew, robotic science missions, planetary landers, and rovers.

Technology 5.4.1, Timekeeping and Time Distribution

Underlying NASA's communications and navigation infrastructure are atomic clocks and time transfer hardware and software. New, more precise atomic clocks operating in space, as well as new and more accurate means of time distribution and synchronization of time among such atomic clocks, will enable the infrastructure improvements and expansion NASA requires in the coming decades. Advances in timekeeping and distribution of several orders of magnitude were judged to provide major benefits, since increased precision of timekeeping and transfer leads to increased precision of relative and absolute position and velocity which in turn provides better starting solutions to enable autonomous rendezvous, docking, landing, and formation flying remote from Earth. Alignment with NASA's needs is considered high due to the substantial impact of the technologies on multiple missions in multiple mission areas including human and robotic spaceflight involving rendezvous, relative station keeping, and landing missions.

Technology 5.3.2, Adaptive Network Topology

Adaptive network topology (ANT) is the capability for a network to change its topology in response to either changes or delays in the network, or additional knowledge about the relationship between the communication paths. ANT includes technologies to improve mission communications, methods of channel access, and techniques to maintain the quality of signal across dynamic networks to assure successful exchange of information needed to accommodate increased mission complexity and achieve great mission robustness. The benefit of this technology to NASA is due to the future multi-element missions that will require advanced network topologies, which will need to be adaptive to remain robust for their applications.

Technology 5.5.1, Radio Systems

Radio systems technology focuses on exploiting technology advances in RF communications, PNT, and space internetworking to develop advanced, integrated space and ground systems that increase performance and efficiency while reducing cost. While this technology can benefit from individual advances in many of the other level 3 technologies in TA05, this entry focuses on the challenges associated with integration of these advancements into operational systems. Advancements in radio systems integration focus on one of the highest-priority technical challenges within TA05: Minimize communications constraints on data rate and range that impact planning and execution of future NASA space missions. The steering committee assessed the benefit of radio systems technologies as resulting in major mission performance improvements due to the potential to improve throughput, versatility, and reliability with lower SWAP impact on the host spacecraft. The alignment to NASA needs is high because improvements in communication systems will impact nearly every NASA spacecraft, including near-Earth, deep-space, and human exploration missions.

Additional Information

All NASA missions require communication and navigation to some degree, so the priorities developed in this section are mostly independent of the mission mix, and in most cases, the prioritization of communication and navigation technologies is not impacted by specific missions in the mission model.

TA06 Human Health, Life Support, and Habitation Systems

TA06 includes technologies necessary for supporting human health and survival during space exploration missions and consists of five technology subareas: environmental control and life support systems and habitation systems; extravehicular activity systems; human health and performance; environmental monitoring, safety, and emergency response; and radiation. These missions can be short suborbital missions, extended microgravity missions, or missions to various destinations, and they experience what can generally be referred to as "extreme environments" including reduced gravity, high radiation and UV exposure, reduced pressures, and micrometeoroids and/or orbital debris.

The panel noted that unlike some of the roadmaps which contained multiple technologies for a design solution, TA06 was broader in scope and often vague with respect to technology descriptions. The five level 2 technology areas should be considered enabling systems rather than discrete technologies, as can the level 3 areas, and the panel was therefore required to review level 4. Despite the lack of technical detail, the panel concurred with NASA on the level 3 technical topics with the exception of topic 6.5.4 Space Weather. It concluded that space weather should be removed from this roadmap and possibly identified as a separate interagency roadmap. 6.5.4 was then restructured and renamed "Human Radiation Prediction."

TA06 Top Technical Challenges

1. Space Radiation Effects on Humans: Improve the understanding of space radiation effects on humans and develop radiation protection technologies to enable long-duration human missions.

Missions beyond low Earth orbit (LEO) present an expanded set of health hazards for a crew and lifetime radiation exposure is already a limiting flight assignment factor for career astronauts on the ISS. Human health radiation models for predicting health risks are currently hampered by large uncertainties based on the lack of appropriate in situ data. Without the collection of in situ biological data to support the development of appropriate models, as well as the development of new sensors, solar event predictions, and radiation mitigating designs, extended human missions beyond LEO may be beyond acceptable risk limits for both human health and mission success. An integrated approach is needed to develop systems and materials to protect crewmembers, and space weather technologies must be upgraded so that the radiation environment is well characterized and solar events can be forecasted from at least Earth to Mars.

2. Environmental Control and Life Support Closed Loop Systems: Develop reliable, closed-loop environmental control and life support systems (ECLSS) to enable long-duration human missions beyond low Earth orbit.

ECLSS for spacesuits, spacecraft, and surface habitats beyond Earth orbit is critical for safety and mission success. In missions without early return capability or remote safety "depots," the ECLSS system must be 100 percent reliable or easily repairable. Current ISS experience with both the U.S. and Russian segments shows significant rates of ECLSS hardware failures, and the systems should undergo further assessment prior to implementation. Additionally, new propulsion capabilities that reduce mission duration would have a positive impact on system design by reducing exposure to impacts.

3. Long-Duration Health Effects: Minimize long-duration crew health effects.

The accumulated international experience with long-duration missions to date reveals that physical and behavioral health effects and adverse events will occur, and are likely to be life threatening in the absence of correct diagnosis and effective treatment, not all of which can be predicted. Thus, autonomous, flexible, and adaptive technologies and systems to promote long-duration health and effectively restore it when accident or illness occurs are judged to be of high priority. Areas of interest include adverse effects of reduced gravity (such as bone loss, muscular and cardiovascular deconditioning, and neurovestibular disorders), in-flight surgery capability in microgravity environments, autonomous medical decision support and procedures management, and in-flight medical diagnosis enabled by biomedical sensors and "laboratory on a chip" technologies.

4. Fire Safety: Assure fire safety (detection and suppression) in human-rated vehicles and habitats in reduced gravity.

Research and testing are needed to understand why current fire detecting sensors have failed to detect smoldering electrical fires, and to develop more efficient and less hazardous fire suppression systems and remediation capabilities that do not impair ECLSS components and/or processes.

5. EVA Surface Mobility: Improve human mobility during extravehicular activity in reduced gravity environments in order to assure mission success and safety.

Since the Apollo lunar missions, relatively little supported research has taken place on space suits for environments other than microgravity. Differences in Apollo and future planetary suits will include the effects of long-term exposure to microgravity en route, prior to reduced gravity EVA operations for significant surface durations. Critical issues for research in this area include the effects of various reduced gravity levels on gait, posture, and suited biomechanics, and the use of advanced materials and techniques for extending life, enabling ease of maintenance, and reducing the effect of surface dust on bearings, seals, and closure mechanisms. Benefits exist from thorough integration of rovers, pressurized habitats, and robotic assist vehicles in extended surface operations. Innovative technologies providing sensory, data management, and actuation assistance to the suit wearer must be developed and assessed.

TA06 High-Priority Technologies

A total of 14 high-priority technologies were identified for TA06, which have been grouped into five theme areas: Radiation (5), ECLSS/Habitation (4), Human Health/Performance (1), EVA Systems (2), and Environmental Monitoring/Safety (2).

Radiation

NASA-supported research and a number of NRC studies over the past decade have confirmed radiation as responsible for many of the unsolved health issues of exploration missions. Therefore the highest-priority technologies for TA06 relate to radiation and are as follows:

Technology 6.5.5, Radiation Monitoring Technology

The ability to monitor the radiation environment will be critical to ensure the safety of astronauts and mission success. Measuring the local radiation environment, including the secondary particles generated in the shielding, is necessary to ensure that astronauts keep their total exposure "as low as reasonably achievable." Established technologies are not sensitive to the full range of threat radiation, nor do they give details about the types of particles contributing to the dose. Advances are needed for smaller, lower power dosimeters with active readout, and sensitive to a broad range of radiation.

Technology 6.5.3, Radiation Protection Systems

Radiation protection systems include materials and other approaches to limit astronauts' radiation exposure. Shielding is a critical design criterion for many elements of human exploration. It is generally considered that shielding alone will not eliminate galactic cosmic ray exposure, but a well-shielded vehicle or habitat could substantially reduce the exposure from solar particle events. The challenge is in finding the optimum approach that reduces radiation exposure while meeting overall mission mass, cost, and other design considerations.

Technology 6.5.1, Radiation Risk Assessment Modeling

Radiation risk is consistently ranked as one of the highest risks to long duration human exploration missions, and risk limits based on current risk assessment models focusing on cancer incidence would be exceeded after only 4 to 6 months in deep space. There are several layers of risk limits included in NASA's permissible exposure limits, and quantification of certain aspects is dominated by a significant uncertainty. Reducing the biological uncertainties would have significant benefits in reducing the cancer uncertainty, quantifying the value of alternative shielding, and quantifying the efficacy of possible radiation mitigation countermeasures.

Technology 6.5.4, Human Radiation Prediction

The ability to forecast the radiation environment, particularly solar particle events and periods of intense ionizing radiation associated with solar storms, is critical to ensuring the safety of astronauts and mission success. The implementation of improved forecasting would improve mission effectiveness and enable more cost effective mitigation strategies by increasing the time to respond, reducing the time spent under shelter, and avoiding false alarms.

Technology 6.5.2, Radiation Mitigation

It is generally considered that shielding alone will not eliminate galactic cosmic ray exposure; therefore there is a need to explore biological/pharmacological countermeasures to mitigate the effect of continuous radiation exposure, as well as to limit the severity of acute radiation effects should an astronaut be exposed during a solar particle event to a significant dose of radiation.

ECLSS/Habitation

Technology 6.1.4, ECLSS Habitation

The habitation technology area focuses on functions that closely interface with life support systems, including food production, food preparation/processing, crew hygiene, metabolic waste collection and stabilization, clothing/laundry, and re-use/recycling of logistics trash. These technologies provide food, sanitation, comfort, and protection for space-faring crew.

Technology 6.1.3, ECLSS Waste Management

Waste management technology safeguards crew health, increases safety and performance, recovers resources, and protects planetary surfaces. Key areas of concern for this technology include volume reduction, stabilization, odor control, and recovery of water, oxygen and other gases, and minerals.

Technology 6.1.2, ECLSS Water Recovery and Management

This technology provides a safe and reliable supply of potable water to meet crew consumption and operational needs. Due to the tremendous launch mass of water for an entire transit mission and impracticality of resupply from Earth, water recovery from waste-water is essential for long-duration transit missions.

Technology 6.1.1, ECLSS Air Revitalization

Air revitalization is essential for long-duration missions and includes carbon dioxide removal, carbon dioxide reduction, oxygen supply, gaseous trace contaminant removal, particulate removal, temperature control, humidity removal, and ventilation.

Human Health/Performance[3]

Technology 6.3.2, Long-Duration Crew Health

The accumulated international experience with long-duration missions to date reveals and predicts a simple truth, that physical and behavioral health effects and adverse events will occur. Autonomous, flexible, and adaptive technologies and systems to promote long-duration health, and effectively restore it when accident or illness

[3] During the execution of this study, the NRC completed its report titled *Recapturing a Future for Space Exploration: Life and Physical Sciences Research for a New Era* (April 2011), which represents a more in-depth review of subject matter covered in TA06.3, Human Health and Performance.

occurs, are judged by the panel to be of high priority. Among the long-duration health-related technologies, the panel identified artificial gravity evaluation/implementation as a game-changing technology with the potential to mitigate bone loss, muscular and cardiovascular deconditioning, and neurovestibular disorders. The highest-priority technologies within this category include in-flight surgery capability in microgravity environments; autonomous medical records, informatics, and procedures management; and in-flight medical diagnosis.

EVA Systems

Technology 6.2.1, EVA Pressure Garment

Pressure garments constitute the anthropomorphic articulated spacecraft in which each EVA crew member works and survives. Ideally pressure garments should be easy to don and doff, highly articulated and readily adjustable to the kinematics of the wearer's body, and able to minimize additional forces and torques which the wearer must overcome to accomplish all tasks. The current operational technology for pressure garments represents incremental changes to those developed more than 30 years ago; thus, significant potential exists for substantial increases in performance and operational capabilities.

Technology 6.2.2, EVA Portable Life Support Systems

Although they are not critical to basic functionality, and all portable life support system functions are limited in non-spaceflight applications, both thermal control and carbon dioxide capture were assigned high priority for special attention. Increasing the capacity, reliability, and maintainability of a personal life support system, while extending duration and reducing on-back weight for the user, are important but difficult goals.

Environmental Monitoring/Safety

Technology 6.4.2, Fire Detection and Suppression

This technology is concerned with ensuring crew health and safety by reducing the likelihood of a fire and, if one occurs, minimizing risk to crew, mission, and/or systems. Areas of research include fire prevention, fire detection, fire suppression, and a proposed free-flying fire test bed.

Technology 6.4.4, Fire Remediation

The panel elevated this technology to high priority status based on Mir, ISS, and space shuttle experiences with fire and post-fire remediation. The issues behind the failures which have occurred need to be thoroughly understood and corrected before long-duration mission are conducted. When vehicle abandonment is not an option, systems must operate throughout the mission, and situational awareness is critical to survival, not just mission success.

Additional Information

Additional comments are detailed in Appendix I.

TA07 Human Exploration Destination Systems

The roadmap for TA07, Human Exploration Destination Systems, includes six technology subareas: in situ resource utilization, sustainability and supportability, advanced human mobility systems, advanced habitat systems, missions operations and safety, and cross cutting technologies. The technologies included in TA07 are necessary for supporting human operations and scientific research during space exploration missions, both in transit and on surfaces. Roadmap TA07 is much broader in scope than other roadmaps, and the six level 2 technology areas of TA07 should be considered enabling systems, rather than competing discrete technologies, all of which are required for mission success. Before prioritizing the level 3 technologies, the steering committee made a number of substantial changes to the TA07 roadmap, which have been enumerated in more detail in the related appendix (Appendix J).

TA07 Top Technical Challenges

1. In Situ Resource Utilization (ISRU) Demonstration: Develop and demonstrate reliable and cost beneficial ISRU technologies for likely destinations to reduce cost and enhance and/or enable productive long-duration human or robotic missions into the solar system.

ISRU capabilities directly impact the deployment and success of exploration missions, having the potential to greatly reduce the cost while increasing the human safety margin and likelihood of mission success and extending mission lifetimes for robotic missions. Key technical challenges are the in situ characterization of the raw resources, demonstration of resource recovery and beneficiation, establishment of the optimum processes under the relevant gravity environment, and production of the strategic products necessary to support future explorations missions.

2. Dust: Characterize and minimize the impact that dust in destination environments will have on extravehicular activity (EVA), rover, and habitat systems.

Dust is a critical environmental hazard. Although dust samples from the Apollo landing sites have been well characterized, little is known about the composition and particle size of unexplored areas of the Moon and Mars. This information is needed in order to develop dust-mitigating technologies for EVA, design requirements for rover treads, and simulants for ISRU.

3. Supportability: Invest in autonomous logistics management, maintenance, and repair strategies in order to reduce mission costs and improve probabilities of mission success.

Improving supportability for long duration missions requires a "launch to end of mission" concept of operations that incorporates highly reliable, maintainable, and repairable systems with fully integrated autonomous logistics management. Reuse and recycling also will be required to reduce the logistics burden of resupply. Supportability systems should be integrated into the design of the systems themselves at the outset, ensuring vehicle systems can be easily maintained with a minimum of crew. Without significant supportability, requirements for future missions to distant destinations should specify a high level of reliability.

4. Food Production, Preservation, and Processing: Develop a food subsystem, as part of a closed-loop life support system, to provide fresh food and oxygen and to remove atmospheric carbon-dioxide during long duration missions.

Food systems for long duration missions are required in order to reduce the costs of up-mass and resupply, habitat volume, and consumables storage requirements at exploration sites. Human spaceflight to distant destinations requires that the nutritional needs of the crew be met for long periods of time.

5. Habitats: Develop space and surface habitats that protect the crew, implement self-monitoring capabilities, and minimize crew maintenance time.

Future human missions to distant destinations will almost certainly involve mission durations equal to or beyond those attempted to date and mass will likely be much more highly constrained. Practically nothing is currently known about humans living, working, and being productive for long periods of time in reduced gravity environments such as the Moon and Mars. Future habitats will need to provide radiation shielding, accommodate long-term exposure to dust, provide a highly reliable habitable volume for months or perhaps years, while additionally accommodating serious medical and surgical intervention, provisioning for world-class research equipment, and still providing a comfortable and sustainable living environment.

6. Surface Mobility (Rovers and EVA): Develop advanced rovers and EVA systems for large-scale surface exploration.

In the case of much longer missions to the Moon than previously attempted, and ultimately Mars, enhanced surface mobility at all levels will improve the science return of exploration missions. A comprehensive program of geological exploration needs access to high slopes, loose and unstable surfaces, and the subsurface access via drilling or excavation. Technology issues such as wheel-soil interactions, optimum mobility platform design, and high-reliability mechanisms with high tolerance for dust and exposure to extreme environments must be addressed.

TA07 High-Priority Technologies

The panel identified 11 high priority technologies in TA07. These technologies have been grouped into five theme areas: ISRU (3), Cross Cutting Systems (2), Sustainability and Supportability (3), Advanced Human Mobility (1), and Advanced Habitat Systems (2).

ISRU

Technology 7.1.3, In Situ Resource Utilization (ISRU) Products/Production

ISRU potentially carries huge economic benefits if destination resources can be utilized to produce key products for exploration, including return propellants, oxygen, water, fuel, metals, concrete, glasses and ceramics, fabrics/textiles/fiber, volatile gases, and plastics and other hydrocarbons. This technology is considered game-changing because it would significantly reduce the cost of and enhance the productivity of long-duration human or robotic missions. The production of oxygen, water, fuel, metals, and building/construction materials would be particularly beneficial, and these capabilities would be in strong alignment with NASA's human exploration program needs. Development of system components and autonomous plant operations also ranks high in benefits and alignment.

Technology 7.1.4, ISRU Manufacturing/Infrastructure

This area encompasses a number of technologies, including in situ infrastructure, in situ manufacturing, in situ derived structures, regolith deep excavation for infrastructure, spare parts manufacturing, and regolith stabilization. This area offers high benefit and alignment to NASA's needs due to the potential for reducing launch costs through reduction of up mass volume and mass.

Technology 7.1.2, ISRU Resource Acquisition

This ISRU element pertains to collecting and acquiring the raw materials to be used and/or processed into the appropriate product or use, and involves a number of subcategories, including regolith and rock acquisition, atmospheric acquisition, material scavenging and resource pre-processing, cold-trap technologies, shallow excavation of dry regolith, and excavation of icy regolith. These technologies will benefit NASA due to their contribution to the reduction in launch costs through reduced up mass and volume.

Crosscutting Technologies

Technology 7.6.3, Dust Prevention and Mitigation

Dust prevention and mitigation is an exceptional challenge and potential health risk for planetary missions. The development of technologies that mitigate the deleterious effects of dust will require knowledge of the chemistry and particle size distribution of the dust. For missions that entail longer stays and/or increased numbers of EVAs, or that involve dust properties that humans have not yet personally encountered (e.g., Mars), the imperative to preclude dust intrusion into the habitation areas, including the EVA suit, is essential.

Technology 7.6.2, Construction and Assembly

This category covers techniques and technologies for assembling structures anywhere in space which are too large, too heavy, or both to be launched in a single mission. Other than large module berthing performed routinely in the construction of ISS, most of the functionality of this technology area is readily available on the Earth but has not been adapted to spaceflight. It allows moving beyond deployable structures or modular assembly to erectable structures, including possible use of structural components obtained and fabricated in situ. There are also particular technologies of relevance to reduced gravity situations. All hardware developed for construction and assembly will have to be long-term suitable for the relevant environments and use alternative modes of achieving robustness and accuracy other than the use of massive body components.

Sustainability and Supportability

Technology 7.2.1, Autonomous Logistics Management

Autonomous logistics management includes the integrated tracking of location, availability, and status of mission hardware and software to facilitate decision making by the team with respect to consumables usage, spares availability, and the overall health and capability of the vehicle and subsystems. This system would automatically update the location of hardware items as they were moved around the vehicle or habitat, track life cycle times and conditions of equipment, and inform the mission team of resupply needs based upon the same. The potentially long duration of future missions coupled with long response times for resupply makes it imperative not only that the health of the vehicle and habitat be known, but also that the mission team know the failure tolerance of the integrated system.

Technology 7.2.4, Food Production/Processing/Preservation

The ability to reduce the volume, waste, and mass associated with the mission food supply must be a priority for the development team, as it will be one of the limiting consumables in any long endurance trip. In addition to the need to simply provide caloric intake for the crew, the food supply must provide the proper nutritional balance to ensure crew health during long duration missions.

Technology 7.2.2, Maintenance Systems

The inability to return faulty equipment to Earth before end of mission, coupled with potentially long resupply times, enhances the value of equipment designs that facilitate servicing by the crew—or eliminate the need for crew servicing. Intelligent/smart systems that autonomously determine and report their status, display graceful degradation, and are self-repairing will be valuable to habitat and vehicle development.

Advanced Human Mobility Systems

Technology 7.3.2, Surface Mobility

Surface mobility technologies are of high priority to the Moon and Mars because they enable scientific research over a large area from a single landing site and because they make dispersed landing areas acceptable. The ability to travel great distances over the lunar or martian surface is imperative for conducting large-scale scientific investigations in these environments.

Advanced Habitat Systems

Technology 7.4.2, Habitat Evolution

Advanced conceptual habitat systems would advance the state of the art, provide a higher level of safety and reliability, and mitigate the long-term effects of microgravity and/or radiation exposure to crew on prolonged transits to and from remote destinations. Habitat evolution was rated of critical importance and includes integrated systems, self-repairing materials, inflatable structures, and "cyclers" (solutions that allow the establishment and long-term utilization of transfer habitats between space destinations). These could also allow the use of substantial in situ resources to provide sufficient mass shielding.

Technology 7.4.3, Smart Habitats

This area involves the development of advanced avionics, knowledge-based systems, and potential robotic servicing capabilities to create long-term habitats with significantly reduced demands on human occupants for diagnosis, maintenance, and repair. While studies of three-person crews for ISS showed that about 2.5 crew was required to maintain space systems, this task envisions advanced habitation systems that augment the crew by providing many of the functions currently performed by mission control, and ultimately by the crew itself.

Additional Information

Additional information on development and schedule changes appears in Appendix J.

TA08 Science Instruments, Observatories, and Sensor Systems

The TA08 roadmap addresses technologies that are primarily of interest for missions sponsored by NASA's Science Mission Directorate and are primarily relevant to space research in Earth science, heliophysics, planetary science, and astrophysics. TA08 consists of three level 2 technology subareas: remote sensing instruments/sensors, observatories, and in situ instruments/sensors. Before prioritizing the level 3 technologies, a number of changes were made to the TA08 roadmap, which are further detailed in the relevant appendix (Appendix K). NASA's science program technology development priorities are generally driven by science goals and future mission priorities recommended in NRC decadal survey strategy reports; therefore, those priorities were considered in evaluating TA08 level 3 technologies (NRC, 2003, 2007, 2010, 2011).

TA08 Top Technical Challenges

1. Rapid Time Scale Development: Enable the exploration of innovative scientific ideas on short time scales by investing in a range of technologies that have been taken to sufficiently high TRLs and that cover a broad class of applications so that they can be utilized on small (e.g., Explorer and Discovery-class) missions.

Innovative ideas need to be tested and evaluated on a rapid time scale in order to be brought to maturity, but to accomplish this, there needs to be inexpensive and routine access to space for technology demonstration.

2. Low-Cost, High-Performance Telescopes: Enhance and expand searches for the first stars, galaxies, and black holes, and advance understanding of the fundamental physics of the universe by developing a new generation of lower-cost, higher-performance astronomical telescopes.

Cosmologically important astronomical objects are very distant and produce faint signals at Earth, the measurement of which requires much larger effective telescope collecting areas and more efficient detector systems, spanning the range of the electromagnetic spectrum. This goal requires new, ultra-stable, normal and grazing incidence mirrors with low mass-to-collecting area ratios.

3. High-Contrast Imaging and Spectroscopy: Enable discovery of habitable planets, facilitate advances in solar physics, and enable the study of faint structures around bright objects by developing high-contrast imaging and spectroscopic technologies to provide unprecedented sensitivity, field of view, and spectroscopy of faint objects.

Among the highest-priority and highest-visibility goals of the space science program is the search for habitable planets and life upon them. Only technologies that are fully developed and demonstrated to a high level will facilitate the large, expensive missions needed to achieve this goal.

4. Sample Returns and In Situ Analysis: Determine if synthesis of organic matter may exist today, whether there is evidence that life ever emerged, and whether there are habitats with the necessary conditions to sustain life on other planetary bodies, by developing improved sensors for planetary sample returns and in-situ analysis.

The needed technologies include integrated and miniaturized sensor suites, sub-surface sample gathering and handling, unconsolidated-material handling in microgravity, temperature control of frozen samples, portable geochronology, and instrument operations and sample handling in extreme environments.

5. Wireless Systems: Enhance effectiveness of spacecraft design, testing, and operations, and reduce spacecraft schedule risk and mass by incorporating wireless systems technology into spacecraft avionics and instrumentation.

Current ground-based network technologies will need to be adapted and improved to accommodate very high data rates, provide high throughput and low latency wireless protocols, support a myriad of avionics interfaces, and be immune to interference.

6. Synthetic Aperture Radar: Enable the active measurement from space of planetary surfaces and of solid-Earth and cryosphere surface deformation and monitoring of natural hazards by developing an affordable, lightweight, deployable synthetic aperture radar antenna.

Synthetic aperture radar can provide unique information regarding such natural phenomena as earthquakes, volcanoes, and glacier surges. In addition, synthetic aperture radar can enable measurements of planetary surfaces, such as geologic features on the cloud-shrouded surfaces of Venus or Titan. Major advances can come either via a large single structure or apertures distributed across two or more spacecraft and will additionally depend on advances in high-performance computing in space.

TA08 High-Priority Technologies

Technology 8.2.4, High-Contrast Imaging and Spectroscopic Technologies

Development of these technologies would enhance high-dynamic-range imaging and support exoplanet imaging, enabling the discovery of potentially habitable planets, facilitating advances in solar physics, and enabling the study of faint structures around bright objects. This technology would provide substantially increased sensitivity, field of view, and spectroscopy of exoplanetary systems, with many subsidiary applications such as solar physics and the study of faint structures around bright objects.

Technology 8.1.3, Optical Systems (Instruments and Sensors)

Two optical systems technologies are of particular interest: active wavefront control and grazing-incidence optical systems. Active wavefront control enables the modification of mirror figure and alignment in response to external disturbances, allowing automated on-orbit alignment of optical systems and the use of lightweight mirrors and telescopes. This technology closely aligns with NASA's need to develop the next generation of large-aperture astronomical telescopes, lightweight laser communication systems, and high-performance orbiting observatories for planetary missions. Further development in grazing-incidence optical systems to improve spatial resolution by at least a factor of ten, without increasing mass per unit area, is critical for future x-ray astronomy missions. These are game-changing technologies that would enable direct imaging of stars and detailed imaging of energetic objects such as active galactic nuclei.

Technology 8.1.1, Detectors and Focal Planes

Sub-kelvin coolers and high-sensitivity detectors are very high priority for future space astronomy missions and are strongly linked to the top technical challenge of developing a new generation of lower-cost astronomical telescopes. The availability of capable sub-kelvin refrigerators could enable long-duration space missions and could also enable new categories of devices with enormous commercial and social impact, such as superconducting and quantum computing and superconducting electronics. The increased sensitivity of detectors would improve detection magnitude in numerous wavelengths and thereby enable new missions.

Technology 8.3.3 In Situ Instruments and Sensors

In situ instruments and sensors would help determine if synthesis of organic matter may exist today, whether there is evidence that life ever emerged, and whether there are habitats with the necessary conditions to sustain life on other planetary bodies. Geological, geophysical, and geochemical sensors and instrumentation would need to be designed to survive in extreme environments, such as high atmospheric pressure, high or low temperature, and adverse chemistry. This technology is game-changing because it would enable missions to the surface and atmosphere of Venus and the surface and sub-surface of outer planet satellites such as the jovian and saturnian moons.

Technology 8.2.5, Wireless Spacecraft Technology

The use of wireless systems in spacecraft avionics and instrumentation can usher in a new, game-changing methodology in the way spacecraft and space missions will be designed and implemented. To make wireless systems ready for application in spacecraft, current ground based network technologies would need to be adapted and improved to accommodate very high as well as low data rates, provide high throughput and low latency wireless protocols, support a myriad of avionics interfaces, and be immune to interferences including multi-path self-interference. The panel designated this as a high-priority technology because it directly relates to meeting the top technical challenge to enhance effectiveness of spacecraft design, testing, and operations, and reduce spacecraft schedule risk and mass, by incorporating wireless systems architecture into spacecraft avionics and instrumentation.

Technology 8.1.5, Lasers

Lasers are fundamental components of topographic lidars, atmospheric composition probes, and Doppler wind instruments, and advances in laser efficiency and lengthening lifespan are critical to enabling space studies. The panel designated this as a high-priority technology due to its applications value. NASA would be well served by evaluating and encouraging emerging laser technologies as needed to support the ongoing needs of space missions identified in decadal survey reports and by focusing on approaches for qualifying laser systems for space.

Technology 8.1.2, Electronics for Instruments and Sensors

The design of future readout integrated circuitry to support larger detector sizes will require appropriate design, layout, simulation tools, and fabrication, making use of state-of-the-art ASIC technology. This technology is broadly applicable to many categories of NASA missions and there is a strong linkage between these technologies and making progress on the top technical challenge of this roadmap, regarding development and maturation of technologies for small missions in short time scales.

TA09 Entry, Descent, and Landing Systems

The roadmap for TA09, Entry, Descent, and Landing Systems, consists of four sub-technology areas: aeroassist and entry, descent, landing, and vehicle systems technology. Entry, descent, and landing (EDL) is a critical technology that enables many of NASA's landmark missions, including Earth reentry, Moon landings, and robotic landings on Mars. NASA's draft EDL roadmap defines entry as the phase from arrival through hypersonic flight, with descent being defined as hypersonic flight to the terminal phase of landing, and landing being from terminal descent to the final touchdown. EDL technologies can involve all three of these mission phases, or just one or two of them. Before prioritizing the level 3 technologies of TA09, a number of changes were made to the roadmap, which have been detailed in the corresponding appendix (Appendix L).

TA09 Top Technical Challenges

EDL has commonly been one of the more challenging areas of NASA missions and has been characterized by significant failures as well as many near misses. Additionally, the panel observed that NASA's draft EDL roadmap may be too narrow because it is focused on the development of human class, large payload delivery to Mars as the primary emphasis even though before such a mission is undertaken, many more robotic missions requiring EDL advances will be planned and executed.

1. Mass to Surface: Develop the ability to deliver more payload to the destination.

NASA's future missions will require ever greater mass delivery capability in order to place scientifically significant instrument packages on distant bodies of interest, to facilitate sample returns from bodies of interest, and to enable human exploration of planets such as Mars. As the maximum mass that can be delivered to an entry interface is fixed for a given launch system and trajectory design, the mass delivered to the surface will require reductions in spacecraft structural mass; more efficient, lighter thermal protection systems; more efficient lighter propulsion systems; and lighter, more efficient deceleration systems.

2. Surface Access: Increase the ability to land at a variety of planetary locales and at a variety of times.

Access to specific sites can be achieved via landing at a specific location(s) or transit from a single designated landing location, but it is currently infeasible to transit long distances and through extremely rugged terrain, requiring landing close to the site of interest. The entry environment is not always guaranteed with a direct entry, and improving the entry system's robustness to a variety of environmental conditions could aid in reaching more varied landing sites.

3. Precision Landing: Increase the ability to land space vehicles more precisely.

A precision landing capability allows a vehicle to land closer to the intended position, and the level of precision achievable at touchdown is a function of the closed loop guidance, navigation, and control (GN&C) design, control authority of the actual vehicle, and the subject environment. Motivations for highly precise landings include targets of interest and safe landing concerns.

4. Surface Hazard Detection and Avoidance: Increase the robustness of landing systems to surface hazards.

One does not know what hazards a landing surface brings until one has actually landed there, and reliance on passive systems alone to land the vehicle safely can be problematic. Active hazard detection methods can quickly optimize safe sites and reduce fuel costs while directly characterizing the landing surface in real time, but require technology development. A practical system for planetary landing must represent a logical compromise among such factors as landing site conditions, pre-mission landing site knowledge, trajectories, and sensors in order to support an overall landing vehicle solution that is simple, reliable, robust, and efficient for safe and robust exploration.

5. Safety and Mission Assurance: Increase the safety, robustness, and reliability of EDL.

Loss-of-mission events during EDL for NASA and the international community have been unacceptably high for Earth-entry and especially planetary entry missions. These events are painful for high-profile robotic missions and can result in tragedy for crewed missions. Adequate safety and mission assurance can be considered a necessary constraint in the mission and vehicle design process. Risk cannot be eliminated entirely from planetary exploration missions; however, this challenge seeks to improve safety and mission assurance while achieving important mission objectives in an affordable manner.

6. Affordability: Improve the affordability of EDL systems.

Improving EDL technology affordability will allow more missions to be flown within fixed and predictable budgets and also will allow new missions previously deemed unaffordable. Affordability needs to be improved either by making it less expensive to transport the same mass or by achieving the same mission objectives with lower mass and therefore lower costs. Affordability also must be balanced with risk so that a mission does not become too expensive in order that it not fail, nor experience so much cost-cutting that failure is likely.

TA09 High-Priority Technologies

Technology 9.4.7, Guidance, Navigation, and Control (GN&C) Sensors and Systems (EDL)

The ability to accurately hit entry corridors, to control the vehicle during entry and descent, to navigate the vehicle during all phases of EDL, and to safely and precisely land a vehicle in hazardous terrain are examples of a high-performing EDL GN&C system. The ability of the GN&C system to achieve its mission objectives is a function of GN&C sensor performance, vehicle actuator ability, and the designer's ability to craft them sensibly together onboard a capable, real time, computing platform. GN&C sensors and systems are common to all of the foreseen EDL generic reference missions and align extremely well with NASA's expertise, capabilities, and facilities. This technology is game-changing because it significantly enhances the ability to increase mass to the surface, the ability to land anywhere, and the ability to land at any time.

Technology 9.1.1, Rigid Thermal Protection Systems

Thermal protection systems (TPS) are used to protect the payload of the entry vehicle from the high temperature and high shear flow environment experienced during the hypersonic entry phase. Most NASA flight experience has been with rigid thermal protection systems, where the TPS are installed onto a rigid aeroshell/structure which can handle both high velocity and high heat fluxes but can also account for a large percentage of the entry vehicle mass. Recent research has been focused on the development of lower density ablators which can reduce the overall vehicle mass fraction. For higher-speed entries into the outer planets or their moons that have atmospheres, new materials would need to be developed that can also handle extreme environments that include both high convective and radiative components. This technology is game-changing because advances in this area would enable new missions in extreme thermal environment or reduced mass to increase vehicle payload and performance, far beyond what has been previously achieved.

Technology 9.1.2, Flexible Thermal Protection Systems

Like rigid TPS, flexible TPS can be reusable or ablative, or some combination thereof. Because of their flexible nature, these TPS systems could be packaged into tighter volumes, applied to irregular surfaces, and deployed when necessary, and, in addition to thermal protection, these systems can also be expected to carry significant aerodynamics loads. Because of their flexibility, it might be possible to tailor the shape of the TPS to improve the aerodynamic performance during the hypersonic entry phase to provide lifting and cross range capability, and these materials could also be used to control local boundary layer state and ultimately heating loads. This technology is game-changing because advances in the flexible area could manifest themselves in reductions in both TPS size and weight.

Technology 9.1.4, Deployable Hypersonic Decelerators

Current entry systems employ traditional rigid decelerator architectures to provide thermal protection and deceleration following entry interfaces. The shape and size of rigid devices define aerodynamic performance and in order to improve performance, size becomes the first order driving parameter. Deployable decelerators enhance the drag area of the spacecraft during the early phase of EDL and advancing these technologies could enable the safe landing of larger objects from sub-orbital terrestrial trajectories and enable heavier payloads to successfully arrive at planetary destinations. There are a number of technologies that must be pursued to enable successful deployment of decelerators, and various advantages exist to using rigid or inflatable decelerators. This technology is game-changing because it provides the ability to utilize much larger drag areas and novel vehicle shapes relative to rigid devices, both of which can enhance thermal protection and deceleration following entry interface and thus enable a whole class of new missions.

Technology 9.4.5, EDL Modeling and Simulation

EDL modeling and simulation (M&S) technology provides the ability to conduct computational predictions necessary for robust and efficient design in all phases of EDL missions. This technology includes computational fluid dynamics analysis, finite element modeling, fluid-structural interaction analysis, aerothermodynamics modeling, coupled stability and 6DOF (degrees of freedom) trajectory analysis, multi-disciplinary analysis tools, and other high-fidelity analysis. This technology also includes development and application of experimental validation including flight tests. This technology is widely applicable to all EDL missions and to the successful development and implementation of the other high-priority technologies in this roadmap.

Technology 9.4.6, Instrumentation and Health Monitoring

Complete simulation of the entry environment is impossible in ground-based test facilities. Although ground-based test facilities are indispensable in developing thermal protection systems, the complete rigorous validation of TPS design algorithms can be achieved only through comparison of predictions with flight data. Also, health monitoring instrumentation can provide system performance data as well as evidence that vehicle systems are operating properly prior to entry. This technology has wide applicability to and would improve the safety and reliability of EDL missions.

Technology 9.4.4, Atmosphere and Surface Characterization

The goal of this technology is to provide a description of the atmosphere and surface of a planet in sufficient detail to facilitate the planning and execution of planetary missions. In the case of planetary atmospheres, a predictive model is required that will define the spatial and temporal atmospheric characteristics on global, zonal, and local scales, including annual, seasonal, and daily variations. Such models exist for the Moon, Mars, and Venus, but they do not provide the needed level of detail. For other planets, the models that exist provide only gross descriptions with very little detail. Atmosphere models are of critical importance for entry missions that involve aeromaneuvering for increased landing accuracy and aerocapture to increase landed mass. Research and technology development topics of particular interest include distributed weather measurements on Mars, the development of a standard, low-impact measurement package for all Mars landed missions, the development of orbiter instruments for wind and atmospheric property characterization, and the development of higher fidelity atmospheric models. Both basic science investigations and the development of predictive engineering models are elements critical to this technology.

Technology 9.4.3, Systems Integration and Analyses

The design of EDL systems is a highly coupled and interdependent set of capabilities consisting of software and hardware components as well as multiple disciplines. The nature of this problem lends itself to technologies that develop improved methods of performing systems integration and analysis, such as multidisciplinary design optimization. Effective system integration and optimization involves incorporating the various disciplines involved in an EDL system while also capturing the multiple phases of flight of entry, descent, and landing. While systems integration and analyses technology is not expected to be game-changing, it is considered a high-priority technology because it supports the complete mission set, provides low risk and reasonableness, requires minimal time and effort, and is applicable to achieving all six of the EDL top technical challenges.

Additional Information

Facilities and continuity are two subjects that are not within the purview of the Office of the Chief Technologist but are critical to the success of EDL developments and therefore forefront to discussions by the panel and also by numerous participants in the EDL workshop and in the open survey. Therefore further comments about these elements have been included in the relevant appendix for this technology area (Appendix L).

TA10 Nanotechnology

The roadmap for TA10, Nanotechnology, addresses four subareas: engineered materials and structures, energy generation and storage, propulsion, and sensors, electronics, and devices. Nanotechnology describes the manipulation of matter and forces at the atomic and molecular levels and includes materials or devices that possess at least one dimension within a size range of 1-100 nm. At this scale, quantum mechanical forces become important in that the properties of nano-sized materials or devices can be substantially different than the properties of the same material at the macro scale. Nanotechnology can provide great enhancement in properties, and materials engineered at the nano-scale will shift the paradigm in space exploration, sensors, propulsion, and overall systems design. Before prioritizing the level 3 technologies in this technology area, several changes were made to the roadmap, which are illustrated in the relevant appendix (Appendix M).

TA10 Top Technical Challenges

1. ***Nano-Enhanced Materials:*** Reduce spacecraft and launch vehicle mass through the development of lightweight and/or multifunctional materials and structures enhanced by nanotechnologies.

Development of advanced materials using nanotechnologies can improve performance in numerous areas. Nano-enhanced composites have the capability to enhance mission performance by increasing the strength and stiffness of materials and reducing structural weight. Multi-scale models valid over nano- to macro-scales are needed to understand nano-enhanced composite materials' failure mechanisms and interfaces. Multi-physics models are needed to address fabrication processes, operations in extreme environments, and designing with active materials. Additionally, new production methodologies are required to manufacture the raw nanomaterials and to controllably incorporate them into other materials.

2. ***Increased Power:*** Increase power for future space missions by developing higher efficiency, lower mass, and smaller energy systems using nanotechnologies.

Energy generation and energy storage will remain a top technical challenge for all future space-related missions. Nanotechnology can improve performance for energy generation, energy storage, and energy distribution, and it will enable sensors to be self-powered and allow for distributed sensing in a networked fashion. Newer technologies such as nano-structured metamaterials and photonic or phononic crystals with spectral compression will improve collection efficiencies and provide new capabilities.

3. ***Propulsion Systems:*** Improve launch and in-space propulsion systems by using nanotechnologies.

Advances in nanotechnology will enable new propellants, potentially by providing higher combustion efficiency and enabling alternative fuel materials. More-energetic propellants will reduce fuel mass in solid motors and provide tailorable ignition and reaction rates. Higher-temperature and lower-erosion structural materials based on nanomaterials could reduce the weight of engine nozzles and propulsion structures.

4. ***Sensors and Instrumentation:*** Develop sensors and instrumentation with unique capabilities and better performance using nanotechnologies.

The success of NASA space missions relies heavily on a variety of sensing methods and sensor technologies for numerous environments in addition to scientific data collection. Nano-sensor technology allows the incorporation of sensors in structures and systems that are smaller, more energy efficient, and more sensitive, allowing for more complete and accurate health assessments. Nanotechnology also permits targeted sensor applications that improve functional efficiency and allows miniaturization of instruments with enhanced performance.

5. ***Thermal Management:*** Improve the performance of thermal management systems by using nanotechnology.

Thermal management can reduce overall system cost and weight, with the direct benefit of reducing overall launch vehicle weight. Thermal control is often required at the system level as well as at the subsystem and component level. Nanotechnology can be used to tailor the thermal conductivity of materials, making them more efficient conductors or insulators.

TA10 High-Priority Technologies

Technology 10.1.1, (Nano) Lightweight Materials and Structures

Nano-sized materials have the promise of substantially improving the thermal, electrical, and/or mechanical properties of components and structures while reducing weight, allowing for the development of multifunctional, lightweight materials and structures that will revolutionize aerospace system design and capability. This technology is game-changing because reductions in the structural and payload weight of a space vehicle allow for higher efficiency launches with increased payload capacity, allowing NASA greater flexibility in mission design. Lack of research into fabrication methodologies related to scale will slow development of lightweight materials and structures. Additionally, strength and performance gains may not be achieved if control of nanoparticle dispersion, ordering, and interface properties is not addressed.

Technology 10.2.1, (Nano) Energy Generation

Nanotechnology impacts energy generation by improving the material systems of existing energy storage and generation systems. This technology is game-changing because lighter, stronger materials and structures allow for more payload devoted to energy generation and power storage, and more efficient energy generation allows for lighter payloads at launch.

Technology 10.3.1, Nanopropellants

Nanopropellants include the use of nano-sized materials as a component of the propellant and as gelling agents for liquid fuels. The nano-size provides a material with enormous reactive surface areas. The use of nano-sized materials as a component of the propellant can solve several problems including potentially the toxicity and environmental hazards of hypergolic and solid propellants and the handling requirements for cryogenics, while also heightening combustion efficiency and potentially impacting the controllability of ignition and reaction rates. The use of nanopropellants can provide a 15-40 percent increase in efficiency and potentially provide multi-functionality.

Technology 10.4.1, (Nano) Sensors and Actuators

Nano-scale sensors and actuators allow for improvements in sensitivity and detection capability while operating at substantially lower power levels. Nanosensors are smaller, more energy efficient, and more sensitive, allowing for more complete and accurate health assessments as well as targeted sensor applications. The panel designated this as a high-priority technology because of the overall benefit offered to all missions.

Additional Information

Future NASA missions depend highly on advances such as lighter and stronger materials, increased reliability, and reduced manufacturing and operating costs, all of which will be impacted by the incorporation of nanotechnology. Major challenges to the broad use and incorporation of nano-engineered materials into useful products are the limited availability of certain raw nanomaterials and their variable quality. Nanotechnology is a very broad area of research and is crosscutting with and impacts every other roadmap. Furthermore, recognizing that much work sponsored by NSF and other agencies on a national R&D effort in nanotechnology is underway in government labs, universities, and industry, the NASA research for space applications should be well coordinated with this national effort. Nanotechnology research at NASA does not seem to be centrally coordinated, and thus the potential exists for substantial duplication of effort. The panel suggests that there be substantial coordination among the nanotechnology researchers at the various NASA centers, the national R&D effort, and specific NASA mission end users.

TA11 Modeling, Simulation, and Information Technology and Processing

The roadmap for TA11 consists of four technology subareas, including computing, modeling, simulation, and information processing. NASA's ability to make engineering breakthroughs and scientific discoveries is limited not only by human, robotic, and remotely sensed observation, but also by the ability to transport data and transform the data into scientific and engineering knowledge to meet sophisticated needs. With data volumes exponentially increasing into the petabyte and exabyte ranges, modeling, simulation, and information technology and processing requirements demand advanced supercomputing capabilities. Before prioritizing the level 3 technologies of TA11, 11.2.4, Science and Engineering Modeling, was divided into two parts: 11.2.4a, Science Modeling and Simulation, and 11.2.4b, Aerospace Engineering Modeling and Simulation.

TA11 Top Technical Challenges

1. Flight-Capable Devices and Software: Develop advanced flight-capable devices and system software for flight computing.

Space applications require devices that are immune to, or at least tolerant of, radiation-induced effects, within tightly constrained resources of mass and power. The software design that runs on these advanced devices also requires new approaches. The critical and complex software needed for these demanding applications requires further development to manage this complexity at low risk.

2. New Software Tools: Develop new flight and ground computing software tools to take advantage of new computing technologies by keeping pace with computing hardware evolution, eliminating the multi-core "programmability gap," and permitting the porting of legacy codes.

Since about 2004, the increase in computer power has come about because of increases in the number of cores per chip and use of very fast vector graphical processor units rather than increases in processor speed. NASA has not yet addressed the challenge of developing efficient codes for these new computer architectures. NASA's vast inventory of legacy engineering and scientific codes will need to be re-engineered to make effective use of rapidly changing advanced computational systems.

3. Testing: Improve the reliability and effectiveness of hardware and software testing and enhance mission robustness via new generations of affordable simulation software tools.

New software tools that allow insight into the design of complex systems will support the development of systems with well understood, predictable behavior while minimizing or eliminating undesirable responses.

4. Simulation Tools: Develop scientific simulation and modeling software tools to fully utilize the capabilities of new generations of scientific computers.

Supercomputers have become increasingly powerful, often enabling realistic multi-resolution simulations of complex astrophysical, geophysical, and aerodynamic phenomena, including the evolution of circumstellar disks into planetary systems, the formation of stars in giant molecular clouds in galaxies, and the evolution of entire galaxies, including the feedback from supernovas and supermassive black holes. However, efficient new codes that use the full capabilities of these new computer architectures are still under development.

TA11 High-Priority Technologies

Technology 11.1.1, Flight Computing

Low-power, radiation-hardened, high-performance processors will continue to be in demand for general application in the space community. Processors with the desired performance are readily available for terrestrial applications; however, radiation-hardened versions of these are not. A major concern is ensuring the continued availability of radiation-hardened integrated circuits for space. Action may be required if NASA and other government organizations wish to maintain domestic sources for these devices, or a technology development effort may be required to determine how to apply commercial devices in the space environment. This technology can have significant impact because multi-core/accelerated flight processors can yield major performance improvements in on-board computing throughput, fault management, and intelligent decision making and science data acquisition, and will enable autonomous landing and hazard avoidance. Its use is anticipated across all classes of NASA missions.

Technology 11.1.2, Ground Computing

Ground computing technology consists of programmability for multi-core/hybrid/accelerated computer architectures, including developing tools to help port existing codes to these new architectures. The vast library of legacy engineering and scientific codes does not run efficiently on the new computer architectures in use, and technology development is needed to create software tools to help programmers convert legacy codes and new algorithms so that they run efficiently on these new computer systems. Continuous technology improvements will be required as computer system architectures steadily change.

Technology 11.2.4a, Science Modeling and Simulation

This technology consists of multi-scale modeling, which is required to deal with complex astrophysical and geophysical systems with a wide range of length scales or other physical variables. Better methods also need to be developed to compare simulations with observations in order to improve physical understanding of the implications of rapidly growing NASA data sets. This technology can have significant impact because it optimizes the value of observations by elucidating the physical principles involved, and could impact many NASA missions.

Technology 11.3.1, Distributed Simulation

Distributed simulation technologies create the ability to share simulations between software developers, scientists, and data analysts. There is a need for large-scale, shared, secure, distributed environments with sufficient interconnect bandwidth and display capabilities to enable distributed analysis and visualization of observations and complex simulations. This technology could provide major efficiency improvements supporting collaborations, particularly interdisciplinary studies that would benefit numerous NASA missions in multiple areas.

TA12: Materials, Structures, Mechanical Systems, and Manufacturing

The TA12 portfolio is extremely broad, including five technology areas: materials, structures, mechanical systems, manufacturing, and crosscutting technologies. TA12 consists of enabling core disciplines and encompasses fundamental new capabilities that directly impact the increasingly stringent demands of NASA science and exploration missions. NASA identified human radiation protection and reliability technologies as two critical areas upon which the technologies in TA12 should be focused.

TA12 Top Technical Challenges

1. Multifunctional Structures: Conceive and develop multifunctional structures, including shielding, to enable new mission capabilities such as long-duration human spaceflight and to reduce mass.

Structures carry load and maintain shape. To the extent that a structure can simultaneously perform additional functions, mission capability can be increased with decreased mass. Such multifunctional materials and structures will require new design analysis tools and might exhibit new failure modes; these should be understood for use in systems design and space systems operations.

2. Reduced Mass: Reduce the mass of launch vehicle, spacecraft, and propulsion structures to increase payload mass fraction, improve mission performance, and reduce cost.

Lightweight materials and structures are required to enhance mission performance and enable new mission opportunities. Advanced composites, revolutionary structural concepts, more energetic propellants, and materials with higher temperature tolerance and lower erosion potential represent some of the possible strategies to reducing mass.

3. Computational Modeling: Advance new validated computational design, analysis, and simulation methods for materials and structural design, certification, and reliability.

First-principle physics models offer the game-changing potential to guide tailored computational materials design. A validated computational modeling methodology could provide the basis for certification by analysis, with experimental evidence, as available, used to verify and improve confidence in the suitability of a design.

4. Large-Aperture Systems: Develop reliable mechanisms and structures for large-aperture systems. These must be stowed compactly for launch, yet achieve high-precision final shapes.

Numerous NASA missions employ mechanical systems and structures that must deploy reliably in extreme environments, often to achieve a desired shape with high precision. These can be deployed, assembled, or manufactured in space and may involve flexible materials. Modularity and scalability are desirable features of such concepts and may require development of autonomous adaptive control systems and technology to address critical functional elements and materials.

5. Structural Health Monitoring: Enable structural health monitoring and sustainability for long-duration missions, including integration of unobtrusive sensors and responsive on-board systems.

Mission assurance would be enhanced by an integrated structural health monitoring system that could detect and assess the criticality of in-service damage or fault, and then define an amelioration process or trigger a repair in self-healing structures. An autonomous integrated on-board systems capability would be game-changing for long-duration, remote missions.

6. Manufacturing: Enable cost-effective manufacturing for reliable high-performance structures and mechanisms made in low-unit production, including in-space manufacturing.

Advanced NASA space missions need affordable structures, electronics systems, and optical payloads, requiring advances in manufacturing technologies. In-space manufacturing offers the potential for game-changing weight savings and new mission opportunities.

TA12 High-Priority Technologies

Nine high-priority technologies were identified in TA12, some of which connected directly to other technologies in TA12 and to other technology roadmaps in support of a common technical challenge.

Technology 12.2.5, Innovative, Multifunctional Concepts (Structures)

Structures that perform functions in addition to carrying load and maintaining shape can increase mission capability while decreasing mass and volume. Multifunctional structural concepts involve increasing levels of system integration and provide a foundation for increased autonomy. Examples of multifunctional structural concepts include habitat structures with integral shielding to reduce radiation exposure and micrometeoroid and orbital debris risk for long-duration human spaceflight missions. The human spaceflight applications of multifunctional structures technology are unique to NASA and dictate that NASA lead associated technology development. Other multifunctional structures concepts, such as those involving thermal-structural and electrical-structural functionality, are likely to find broader applications. Therefore NASA would benefit from partnerships in the development of these technology concepts.

Technology 12.2.1, Lightweight Concepts (Structures)

Lightweight structural concepts could significantly enhance future exploration and science missions and enable new missions. For example, lightweight cryo-tank concepts could improve launch vehicle performance and enable on-orbit fuel storage depots, and lightweight concepts for deployable solar sails, precision space structures, and inflatable, deployable heat shields could provide opportunities for new missions or significantly benefit planned science missions. Lightweight structural concepts developed by NASA and the aerospace industry have found extensive applications in transportation, commercial aircraft, and military systems.

Technology 12.1.1, Lightweight Structure (Materials)

Advanced composite, metallic, and ceramic materials, as well as cost-effective processing and manufacturing methods, are required to develop lightweight structures for future space systems. Lightweight structural materials developed by NASA and other government agencies, academia, and the aerospace industry have found extensive applications in transportation, commercial aircraft, and military systems. Continued NASA leadership in materials development for space applications could result in new materials systems with significant benefits in weight reduction and cost savings. This technology has the potential to significantly reduce the mass of virtually all launch vehicles and payloads, creating opportunities for new missions, improved performance, and reduced cost.

Technology 12.2.2, Design and Certification Methods (Structures)

Current structural certification approaches rely on a conservative combination of statistics-based material qualification and experience-based load factors and factors of safety, followed by design development and qualification testing. Verification testing and mission history indicate that structures tend to be over-designed and thus heavier than necessary. A model-based virtual design certification methodology could be developed to design and certify space structures more cost-effectively. This technology provides another path to lighter and more affordable space structures while assuring adequate reliability, and is applicable to all NASA space vehicles including uncrewed, robotic, and human-rated vehicles for use in science missions, and human exploration over extended periods of time.

Technology 12.5.1, Nondestructive Evaluation and Sensors (Crosscutting)

Non-destructive evaluation (NDE) has evolved from its early uses for quality control product acceptance, and periodic inspection to include continuous health monitoring and autonomous inspection. Early detection, localization, and mitigation of critical conditions will enhance mission safety and reliability. NASA has proposed an integrated NDE and sensor technology capability in a virtual digital flight leader (VDFL) that would include a digital representation of a vehicle with real-time assessment of vehicle structural health to predict performance and identify operational actions necessary to address vehicle performance. NDE and sensor technologies are likely to impact multiple areas and multiple missions, especially as mission durations continue to increase.

Technology 12.3.4, Design and Analysis Tools and Methods (Mechanical Systems)

High-fidelity kinematics and dynamics design and analysis tools and methods are essential for modeling, designing, and certifying advanced space structures and mechanical systems. A mechanism interrelation/correlation analysis methodology would enable creation of a single model of spacecraft mechanical systems and would reduce the stack-up of margins across disciplines. Such models could be integrated into a health-management system for diagnosis, prognosis, and performance assessment and in a VDFL system. This technology is applicable to all NASA space vehicles including uncrewed, robotic, and human-rated vehicles for use in science missions, and human exploration over extended periods of time.

Technology 12.3.1, Deployables, Docking and Interfaces (Mechanical Systems)

Many future science missions involving imaging and scientific data collection will benefit from the combination of a large aperture and precision geometry, the achievement of which will most likely involve deployment, possibly including flexible materials or other approaches such as assembly or in-space manufacturing. Docking and the associated interfaces provide another approach to building up larger platforms from smaller ones. These mechanical systems and structures must deploy reliably in extreme environments and achieve a desired shape with high precision; some systems may require the use of a control system to maintain a precise shape under operational disturbances. Large precise aperture systems are critical to some NASA science missions as well as some DOD surveillance missions, enabling advanced mission performance. This suggests that NASA lead associated technology development, finding partners when feasible. Space missions have not infrequently failed as the result of failure of a separation, release, or deployment system, and clearly technology development to pursue improvements in the reliability of such systems is needed.

Technology 12.3.5, Reliability/Life Assessment/Health Monitoring (Mechanical Systems)

In recent experience, the reliability of mechanical systems has been a more significant contributor to the failure of space missions than the reliability of structures designed to meet current certification standards. An integrated sensor system would provide a basis for determining the current state of a mechanical system, as well as prediction of future behavior. To be most effective in assuring mission reliability, the ability to take corrective action must also be designed into the system. This technology is closely linked with the area of deployables, docking, and interfaces, and could enable a dramatic increase in the reliability of mechanical systems and structures, especially for long-duration space missions.

Technology 12.4.2, Intelligent Integrated Manufacturing and Cyber Physical Systems (Manufacturing)

The fielding of high-performance materials, structures, and mechanisms for space applications requires specialized manufacturing capabilities. Through advances in technology, particularly IT-based, more general but flexible manufacturing methods can be adapted to produce specialized components and systems. There are existing industrial capabilities in such technologies, and investments in similar technologies from the Air Force Research Laboratory have contributed significantly and are expected to continue because of the potential impacts on affordability. Manufacturing is an area in which NASA can benefit from monitoring developments in hardware, software, and supply chain management, and there is potential to form government, university, and industry consortia to pursue these ends. This technology would enable physical components to be manufactured in space, on long-duration human missions if necessary. This could reduce the mass that must be carried into space for some exploration missions, and furthermore this technology promises improved affordability of one-off structures made from high-performance materials. This technology is applicable to all NASA space vehicles including uncrewed, robotic, and human-rated vehicles for use in science missions, and human exploration over extended periods of time.

Additional Comments

Perhaps as a result of the need to address such a broad range of technologies in a summary document, the TA12 roadmap devotes little space to discussion of the assumed mission model, or to the inter-dependence of technology development, and to some degree can be read as a catalog of technology items as much as a plan. Detailed interpretation of TA12 is left to the reader, making it challenging to suggest specific modifications to the schedule.

Additionally, the TA12 roadmap addresses neither improved understanding of the intense vibroacoustic environment of launch nor novel approaches that could reduce structural dynamic response, which frequently drive the structural design of spacecraft.

TA13 Ground and Launch Systems Processing

The goal of TA13 is to provide a flexible and sustainable U.S. capability for ground processing as well as launch, mission, and recovery operations to significantly increase safe access to space. The TA13 roadmap consists of four technology subareas, including technologies to optimize the operational life-cycle, environmental and green technologies, technologies to increase reliability and mission availability, and technologies to improve mission safety/mission risk. The primary benefit derived from advances in this technology area is reduced cost, freeing funds for other investments. Before prioritizing the technologies of TA13, the panel considered the TA13 breakdown structure but did not recommend any changes.

TA13 Top Technical Challenges

Although advanced technology can contribute to solving the major challenges of advances in ground and launch systems (for example, cost and safety concerns), they are most effectively addressed through improvements in management practices, engineering, and design. Therefore the panel did not identify any technical challenges related to TA13 on the level of those associated with the other roadmaps.

TA13 High-Priority Technologies

The panel did not identify any high-priority level 3 technologies for TA13.

Additional Information

The panel does not have any recommendations with regard to development and schedule changes.

TA14 Thermal Management Systems

Thermal management systems are systems and technologies that are capable of handling high thermal loads with excellent temperature control, with a goal of decreasing the mass of existing systems. The roadmap for TA14, Thermal Management Systems, consists of three technology subareas: cryogenic systems, thermal control systems, and thermal protection systems. Before prioritizing the technologies of TA14, the panel considered the TA14 breakdown structure but did not recommend any changes.

TA14 Top Technical Challenges

1. Thermal Protection Systems: Develop a range of rigid ablative and inflatable/flexible/deployable thermal protection systems (TPS) for both human and robotic advanced high-velocity return missions, either novel or reconstituted legacy systems.

TPS is mission-critical for all future human and robotic missions that require planetary entry or reentry. The current availability of high-TRL rigid ablative TPS is adequate for LEO re-entry but is inadequate for high-energy re-entries to Earth or planetary missions. Ablative materials are enabling for all NASA, military, and commercial missions that require high-mach-number re-entry, such as near-Earth asteroid visits and Mars missions, whether human or robotic.

2. Zero Boil-Off Storage: Accelerate research on advanced active and passive systems to approach near-zero boil-off in long-term cryogenic storage.

Long-term missions that require cryogenic life-support supplies, cryogenic propellants, or very low temperature for scientific instrument support will require near-zero boil-off rates. Multiple technologies in TA14 support this, and emphasis should be placed on reliable, repairable, supportable active and passive systems that can be integrated into many missions.

*3. **Radiators:*** Develop improved space radiators with reduced mass.

Radiators are used for energy removal from spacecraft and planetary base systems and are mission-critical for many proposed missions. To reduce radiator mass, area, and pumping power, research is needed on variable emissivity, very low absorptivity-to-emissivity ratio, self-cleaning, and high-temperature coatings, as well as on lightweight radiators or compact storage systems for extending extravehicular activity capability.

*4. **Multifunctional Materials:*** Develop high-temperature multifunctional materials that combine structural strength, good insulating ability, and possibly other functions.

Multifunctional systems can provide significant mass savings, allowing increased payload weight. Multifunctional TPS and multi-layer insulation (MLI) systems that combine thermal, structural, or micrometeoroid and orbital debris (MMOD) and crew radiation protection could provide significant weight savings and enable long-duration missions, and can also be used for planetary habitat thermal and multifunctional protection.

*5. **Verification and Validation:*** Develop, verify, validate, and quantify uncertainty analysis requirements for new or improved comprehensive computer codes for thermal analysis.

Upgrades to predictive codes for ablation during re-entry heating are needed to include closely coupled multi-phase ablation and radiative heating into the flow simulations, with careful attention given to verification, validation, and uncertainty quantification.

*6. **Repair Capability:*** Develop in-space thermal protection system repair capability.

Repair capability is especially important for long-duration missions. TPS repair developed for Space Shuttle Orbiter TPS should be continued and expanded to provide a repair method for future spacecraft.

*7. **Thermal Sensors:*** Enhance thermal sensor systems and measurement technologies.

Operational instrumentation is necessary to understand anomalies, material or performance degradation and performance enhancements, and advanced science mission measurements.

TA14 High-Priority Technologies

Technology 14.3.1, Ascent/Entry TPS

Effective heat shields and thermal insulation during ascent and atmospheric entry are mission-critical for all robotic and human missions that require entry into a planetary atmosphere. Ascent/entry TPS is game-changing because it is necessary for every planetary atmospheric ascent and/or entry mission, including every mission for return to Earth. Particularly critical level 4 technology items are rigid ablative TPS, obsolescence-driven TPS materials and process development, multi-functional TPS, and flexible TPS.

Technology 14.1.2, Active Thermal Control of Cryogenic Systems

Low to zero boil-off of cryogenic fluids will be mission-critical for long-duration missions, and cannot be achieved with present technology. A goal of this technology is to develop an overall cryogenic system design that integrates active and passive technologies into an optimal system, as well as instrumentation and sensors to monitor fluid mass. Minimization of active system capacity through effective use of passive control should help increase overall system reliability. This technology can enable a wide variety of long-duration missions.

Additional Information

Software validation and the use of ground test facilities are two overarching crosscutting issues pertinent to TA14 that are addressed in detail in Chapter 4.

NASA recognizes that budgetary and staffing constraints make it impossible to carry out all of the tasks proposed in the roadmap. It will be necessary to coordinate and cooperate with other organizations for funding research and portions of these technologies. Many of the tasks could be combined; for example, the draft roadmap breaks Technology 14.1.1 Passive Thermal Control into eight items, all of which deal with minimizing heat leaks, and the research should be attacked as an overall system rather than technology by technology.

REFERENCES

NRC (National Research Council). 2003. *The Sun to the Earth—and Beyond: A Decadal Research Strategy in Solar and Space Physics*. The National Academies Press, Washington, D.C.

NRC. 2007. *Earth Science and Applications from Space: National Imperatives for the Next Decade and Beyond*. The National Academies Press, Washington, D.C.

NRC. 2010. *New Worlds, New Horizons in Astronomy and Astrophysics*. The National Academies Press, Washington, D.C.

NRC. 2011. *Vision and Voyages for Planetary Science in the Decade 2013-2022*. The National Academies Press, Washington, D.C.

3

Integrated Ranking of Top Technical Challenges and High-Priority Technologies

As explained in Chapter 2, the panels' assessment of the level 3 technologies in each individual roadmap considered a wide variety of factors.[1] In prioritizing the 83 technologies evaluated as high-priority by the panels across all 14 draft roadmaps, the steering committee established an organizing framework that addressed balance across NASA mission areas; relevance in meeting the highest-priority technical challenges; and expectations that significant progress could be made in the next 5 years of the 30-year window of the roadmaps. Furthermore, the steering committee constrained the number of highest-priority technologies recommended in the final list in the belief that in the face of probable scarce resources, focusing initially on a small number of the highest-priority technologies offers the best chance to make the greatest impact, especially while agency mission areas, particularly in exploration, are being refined and can be shaped by technology options. Within this organizing framework, technology objectives were defined by the steering committee to address the breadth of NASA missions and group related technologies.

TECHNOLOGY OBJECTIVES

The 2011 NASA Strategic Plan (NASA, 2011, p. 4) states:

> New in this 2011 Strategic Plan is a strategic goal that emphasizes the importance of supporting the underlying capabilities that enable NASA's missions.

The steering committee interpreted this formulation of NASA's strategic vision as the need to assess the technologies by the measure of how well they supported NASA's various missions.

The question became one of identifying the totality of NASA's missions that were all-inclusive of the agency's responsibilities and yet easily distinguished by the type of technologies needed to support them. The steering committee defined the following technology objectives to serve as an organizing framework for prioritization of technical challenges and roadmap technologies.

[1] The draft space technology roadmaps are available at http://www.nasa.gov/offices/oct/strategic_integration/technology_roadmap.html.

Technology Objective A: Extend and sustain human activities beyond low Earth orbit.

Supporting technologies would enable humans to survive long voyages throughout the solar system, get to their chosen destination, work effectively, and return safely.

This objective includes a major part of NASA's mission to send humans beyond the protection of the Van Allen belts, mitigate the effects of space radiation and long exposure to the microgravity environment, enable the crew to accomplish the goals of the mission (contained in Technology Objective B), and then return to Earth safely. This objective includes using the International Space Station (ISS) for technology advancement to support future human space exploration, providing opportunities for commercial companies to offer services to low Earth orbit and beyond, and developing the launch capability required for safe access to locations beyond low Earth orbit.

Technology Objective B: Explore the evolution of the solar system and the potential for life elsewhere.

Supporting technologies would enable humans and robots to perform in situ measurements on Earth (astrobiology) and on other planetary bodies.

This objective is concerned with the in situ analysis of planetary bodies in the solar system. It includes the detailed analysis of the physical and chemical properties and processes that shape planetary environments and the study of the geologic and biological processes that explain how life evolved on Earth and whether it exists elsewhere. It involves development of instruments for in situ measurements and the associated data analysis. This objective includes all the in situ aspects of planetary science; measurement of interior properties, atmospheres, particles, and fields of planets, moons, and small bodies; and methods of planetary protection.

Technology Objective C: Expand our understanding of Earth and the universe in which we live.

Supporting technologies would enable remote measurements from platforms that orbit or fly by Earth and other planetary bodies, and from other in-space and ground-based observatories.

This objective includes astrophysics research; stellar, planetary, galactic, and extra-galactic astronomy; particle astrophysics and fundamental physics related to astronomical objects; solar and heliospheric physics; and magnetospheric physics and solar-planetary interactions. This objective also includes space-based observational Earth-system science and applications aimed at improving our understanding of Earth and its responses to natural and human-induced changes. This objective includes all space science activities that rely on measurements obtained remotely from various observational platforms.

These objectives are not independent and are often shared by a single mission (e.g., humans to explore planetary bodies or to service observatories, as was the case with the Hubble Space Telescope), and there are technologies that support more than one of these objectives (e.g., multifunctional structures, electric propulsion, GN&C). Yet this taxonomy is a useful way to categorize NASA's responsibilities as described in its strategic plan and serves to prioritize the various technologies and technical challenges identified in this study.

Balance

One of the steering committee's basic assumptions was that NASA would continue to pursue a balanced space program across its mission areas of human exploration, space science, space operations, space technology, and aeronautics. Indeed, this balance is emphasized in the 2011 NASA Strategic Plan (NASA, 2011) and addressed in the NRC report *America's Future in Space: Aligning the Civil Space Program with National Needs* (NRC, 2009), where breadth across NASA's mission areas contributes to making the U.S. a leader in space. Therefore, since the technology program of the Office of the Chief Technologist (OCT) should broadly support the breadth of the agency's missions and serve to open up options for future missions, the steering committee established priorities

TABLE 3.1 Relationships Among NASA's Mission Areas and the Three Technology Objectives

NASA Mission Areas	Technology Objective A: Extend and Sustain Human Activities Beyond Low Earth Orbit	Technology Objective B: Explore the Evolution of the Solar System and the Potential for Life Elsewhere (In Situ Measurements)	Technology Objective C: Expand Understanding of Earth and the Universe (Remote Measurements)
Planetary Science	X	X	X
Astrophysics			X
Earth Science		X	X
Heliophysics			X
Human Exploration	X	X	X[a]
Operations	X	X	X

[a] If telescopes and observatories are serviceable by astronauts.

for each of the three technology objectives independently. No one technology objective area was given priority over another.

Table 3.1 relates the three technology objectives with NASA's mission areas and illustrates the balance of using the adopted organizing framework. As mentioned previously, aeronautics was not part of the roadmap study.

TECHNICAL CHALLENGES

With the three technology objectives defined, the steering committee evaluated the top technical challenges from the panels' prioritized rankings for each roadmap TA01-14.[2] In some cases, the steering committee combined similar challenges, particularly across roadmaps. The steering committee members began the process by voting on the highest-priority technical challenges in multiple iterations, first using a 1-10 ranking to rate their top 10 challenges for each objective, and then using a weighted scale: 0 = Not relevant; 1 = Minor importance; 3 = Significant; 9 = Essential. The steering committee then discussed any significant scoring variations by different members and the relative priority of each challenge implied by the average and mean scores of the members' scores. This discussion continued until a final group consensus was reached on a prioritized list of the final ten technical challenges for each objective. The top 10 technical challenges for each of the three technology objectives are described below.

Top Technical Challenges for Technology Objective A: Extend and sustain human activities beyond low Earth orbit.

A1. Improved Access to Space: Dramatically reduce the total cost and increase the reliability and safety of access to space.

A2. Space Radiation Health Effects: Improve understanding of space radiation effects on humans and develop radiation protection technologies to enable long-duration space missions.

A3. Long-Duration Health Effects: Minimize the crew health effects of long-duration space missions (other than space radiation).

A4. Long-Duration ECLSS: Achieve reliable, closed-loop environmental control and life support systems (ECLSS) to enable long-duration human missions beyond low Earth orbit.

[2] See the sections titled "Top Technical Challenges" in Appendixes D-Q for the panels' prioritized technical challenge rankings.

A5. Rapid Crew Transit: Establish propulsion capability for rapid crew transit to and from Mars or other distant targets.

A6. Lightweight Space Structures: Develop innovative lightweight materials and structures to reduce the mass and improve the performance of space systems such as (1) launch vehicle and payload systems; (2) space and surface habitats that protect the crew, including multifunctional structures that enable lightweight radiation shielding, implement self-monitoring capability, and require minimum crew maintenance time; and (3) lightweight, deployable synthetic aperture radar antennas, including reliable mechanisms and structures for large-aperture space systems that can be stowed compactly for launch and yet achieve high-precision final shapes.

A7. Increase Available Power: Eliminate the constraint of power availability for space missions by improving energy generation and storage with reliable power systems that can survive the wide range of environments unique to NASA missions.

A8. Mass to Surface: Deliver more payload to destinations in the solar system.

A9. Precision Landing: Increase the ability to land more safely and precisely at a variety of planetary locales and at a variety of times.

A10. Autonomous Rendezvous and Dock: Achieve highly reliable, autonomous rendezvous, proximity operations, and capture of free-flying space objects.

Top Technical Challenges for Technology Objective B: Explore the evolution of the solar system and the potential for life elsewhere (in situ measurements).

B1. Improved Access to Space: Dramatically reduce the total cost and increase the reliability and safety of access to space.

B2. Precision Landing: Increase the ability to land more safely and precisely at a variety of planetary locales and at a variety of times.

B3. Robotic Maneuvering: Enable mobile robotic systems to autonomously and verifiably navigate and avoid hazards and increase the robustness of landing systems to surface hazards.

B4. Life Detection: Improve sensors for in situ analysis to determine if synthesis of organic matter may exist today, whether there is evidence that life ever emerged, and whether there are habitats with the necessary conditions to sustain life on other planetary bodies.

B5. High-Power Electric Propulsion: Develop high-power electric propulsion systems along with the enabling power system technology.

B6. Autonomous Rendezvous and Dock: Achieve highly reliable, autonomous rendezvous, proximity operations, and capture of free-flying space objects.

B7. Increase Available Power: Eliminate the constraint of power availability for space missions by improving energy generation and storage with reliable power systems that can survive the wide range of environments unique to NASA missions.

B8. Mass to Surface: Deliver more payload to destinations in the solar system.

B9. Lightweight Space Structures: Develop innovative lightweight materials and structures to reduce the mass and improve the performance of space systems such as (1) launch vehicle and payload systems; (2) space and surface habitats that protect the crew, including multifunctional structures that enable lightweight radiation shielding, implement self-monitoring capability, and require minimum crew maintenance time; and (3) lightweight, deployable synthetic aperture radar antennas, including reliable mechanisms and structures for large-aperture space systems that can be stowed compactly for launch and yet achieve high-precision final shapes.

B10. Higher Data Rates: Minimize constraints imposed by communication data rate and range.

Top Technical Challenges for Technology Objective C: Expand understanding of Earth and the universe in which we live (remote measurements).

C1. Improved Access to Space: Dramatically reduce the total cost and increase the reliability and safety of access to space.

C2. New Astronomical Telescopes: Develop a new generation of astronomical telescopes that enable discovery of habitable planets, facilitate advances in solar physics, and enable the study of faint structures around bright objects by developing high-contrast imaging and spectroscopic technologies to provide unprecedented sensitivity, field of view, and spectroscopy of faint objects.

C3. Lightweight Space Structures: Develop innovative lightweight materials and structures to reduce the mass and improve the performance of space systems such as (1) launch vehicle and payload systems; (2) space and surface habitats that protect the crew, including multifunctional structures that enable lightweight radiation shielding, implement self-monitoring capability, and require minimum crew maintenance time; and (3) lightweight, deployable synthetic aperture radar antennas, including reliable mechanisms and structures for large-aperture space systems that can be stowed compactly for launch and yet achieve high-precision final shapes.

C4. Increase Available Power: Eliminate the constraint of power availability for space missions by improving energy generation and storage with reliable power systems that can survive the wide range of environments unique to NASA missions.

C5. Higher Data Rates: Minimize constraints imposed by communication data rate and range.

C6. High-Power Electric Propulsion: Develop high-power electric propulsion systems along with the enabling power system technology.

C7. Design Software: Advance new validated computational design, analysis, and simulation methods for design, certification, and reliability of materials, structures, and thermal, EDL, and other systems.

C8. Structural Monitoring: Develop means for monitoring structural health and sustainability for long duration missions, including integration of unobtrusive sensors and responsive on-board systems.

C9. Improved Flight Computers: Develop advanced flight-capable devices and system software for real-time flight computing with low-power, radiation-hard, and fault-tolerant hardware.

C10. Cryogenic Storage and Transfer: Develop long-term storage and transfer of cryogens in space using systems that approach near-zero boil-off.

HIGHEST-PRIORITY LEVEL 3 TECHNOLOGIES ACROSS ALL ROADMAPS

Process for Prioritizing Technologies Across Roadmaps

Utilizing the panel results, which established a high degree of correlation between high-priority level 3 technologies and the respective technical challenges for each roadmap (see correlation matrices in Appendixes D-Q), the steering committee was able to relate high-priority technologies that aligned with each of the three technology objectives. This organizing principle in turn helped categorize similar technologies with similar drivers (i.e., technologies driven by keeping humans alive, able to be productive, and transported; in situ measurements; and remote measurements) and enabled prioritization among them on a meaningful basis.

The process followed by the steering committee was as follows: First, the steering committee considered only the 83 high-priority level 3 technology as selected by the panels. These 83 technologies are listed in Table 3.2. Next, following the correlation procedure used by the panels, the steering committee mapped those technologies against the top technical challenges for each of the three objectives. In many cases the correlation matrix was sparsely filled; for example, technologies from roadmaps relating to human exploration or life support would have little correlation with Technology Objective C, which is focused primarily on remote measurements from observational platforms, except if servicing is done by astronauts. The correlation information was then used by the steering committee as it voted on the priority of technologies against the three objectives. Each steering committee member voted on the importance of each technology to each objective using a weighted scale:

0 = Not relevant;
1 = Minor importance;
3 = Significant;
9 = Essential.

The total of the members' scores assigned to each technology was then summed to create a rank-ordered list of technologies for each technology objective. There were several iterations of voting and discussion first to develop an interim list of 11 to 15 technologies per objective, followed by another iteration of voting and discussion to obtain a consensus on the final list of 7 or 8 technologies per objective.

The robustness of the final results was tested by the steering committee in numerous ways. The steering committee used other weighting schemes (such as voting on top five technologies rather than using a 0-1-3-9 weighting factor) and other voting schemes (such as voting to remove technologies rather than voting to include them). Initially the steering committee had removed from the voting any technologies that were uncorrelated to any technical challenge; to make certain all technologies were properly considered, that constraint was relaxed and all 83 technologies were voted upon. In all cases, however, the changes to the methods had little or no impact on the final outcome.

Tables 3.3, 3.4, and 3.5 show the correlation matrices for the high-priority technologies and the top technical challenges for Objectives A, B, and C, respectively. All empty rows (i.e., all technologies that do not correlate to any challenges for that objective) have been removed from the matrix.

The steering committee determined that, in several instances, technologies on the original list of 83 high-priority technologies were highly coupled or addressed the same technology pedigree. During the prioritization process, these highly coupled technologies were grouped together and considered as one unit. Table 3.6 shows the mapping of the technology area breakdown structure (TABS) technologies to each unified technology that was considered during the final prioritization process.

Results and Recommendations for Prioritized Technologies Across Roadmaps

Table 3.7 represents the steering committee's consensus viewpoint following the first iteration of voting, discussion, and prioritization, with the technologies listed by objective in ranked order. To obtain as short a list as is reasonable in the face of anticipated constrained budgets, a second iteration of prioritization was conducted

TABLE 3.2 The 83 High-Priority Level 3 Technologies Selected by the Panels

TA01 Launch Propulsion Systems

- 1.3.1 Turbine Based Combined Cycle (TBCC)
- 1.3.2 Rocket Based Combined Cycle (RBCC)

TA02 In-Space Propulsion Technologies

- 2.2.1 Electric Propulsion
- 2.4.2 Propellant Storage and Transfer
- 2.2.3 (Nuclear) Thermal Propulsion
- 2.1.7 Micro-Propulsion

TA03 Space Power and Energy Storage

- 3.1.3 Solar Power Generation (Photovoltaic and Thermal)
- 3.1.5 Fission Power Generation
- 3.3.3 Power Distribution and Transmission
- 3.3.5 Power Conversion and Regulation
- 3.2.1 Batteries
- 3.1.4 Radioisotope Power Generation

TA04 Robotics, TeleRobotics, and Autonomous Systems

- 4.6.2 Relative Guidance Algorithms
- 4.6.3 Docking and Capture Mechanisms/Interfaces
- 4.5.1 Vehicle System Management and FDIR
- 4.3.2 Dexterous Manipulation
- 4.4.2 Supervisory Control
- 4.2.1 Extreme Terrain Mobility
- 4.3.6 Robotic Drilling and Sample Processing
- 4.2.4 Small Body/Microgravity Mobility

TA05 Communication and Navigation

- 5.4.3 Onboard Autonomous Navigation and Maneuvering
- 5.4.1 Timekeeping and Time Distribution
- 5.3.2 Adaptive Network Topology
- 5.5.1 Radio Systems

TA06 Human Health, Life Support, and Habitation Systems

- 6.5.5 Radiation Monitoring Technology
- 6.5.3 Radiation Protection Systems
- 6.5.1 Radiation Risk Assessment Modeling
- 6.1.4 Habitation
- 6.1.3 Environmental Control and Life Support System (ECLSS) Waste Management
- 6.3.2 Long-Duration Crew Health
- 6.1.2 ECLSS Water Recovery and Management
- 6.2.1 Extravehicular Activity (EVA) Pressure Garment
- 6.5.4 Radiation Prediction
- 6.5.2 Radiation Mitigation
- 6.4.2 Fire Detection and Suppression
- 6.1.1 Air Revitalization
- 6.2.2 EVA Portable Life Support System
- 6.4.4 Fire Remediation

TA07 Human Exploration Destination Systems

- 7.1.3 In Situ Resource Utilization (ISRU) Products/Production
- 7.2.1 Autonomous Logistics Management
- 7.6.2 Construction and Assembly
- 7.6.3 Dust Prevention and Mitigation
- 7.1.4 ISRU Manufacturing/Infrastructure etc.
- 7.1.2 ISRU Resource Acquisition
- 7.3.2 Surface Mobility
- 7.2.4 Food Production, Processing, and Preservation
- 7.4.2 Habitation Evolution
- 7.4.3 Smart Habitats
- 7.2.2 Maintenance Systems

TA08 Science Instruments, Observatories, and Sensor Systems

- 8.2.4 High-Contrast Imaging and Spectroscopy Technologies
- 8.1.3 Optical Systems (Instruments and Sensors)
- 8.1.1 Detectors and Focal Planes
- 8.3.3 In Situ Instruments and Sensors
- 8.2.5 Wireless Spacecraft Technology
- 8.1.5 Lasers for Instruments and Sensors
- 8.1.2 Electronics for Instruments and Sensors

TA09 Entry, Descent, and Landing (EDL) Systems

- 9.4.7 GN&C Sensors and Systems (EDL)
- 9.1.1 Rigid Thermal Protection Systems
- 9.1.2 Flexible Thermal Protection Systems
- 9.1.4 Deployment Hypersonic Decelerators
- 9.4.5 EDL Modeling and Simulation
- 9.4.6 EDL Instrumentation and Health Monitoring
- 9.4.4 Atmospheric and Surface Characterization
- 9.4.3 EDL System Integration and Analysis

TA10 Nanotechnology

- 10.1.1 (Nano) Lightweight Materials and Structures
- 10.2.1 (Nano) Energy Generation
- 10.3.1 Nanopropellants
- 10.4.1 (Nano) Sensors and Actuators

TA11 Modeling, Simulation, and Information Technology and Processing

- 11.1.1 Flight Computing
- 11.1.2 Ground Computing
- 11.2.4a Science Modeling and Simulation
- 11.3.1 Distributed Simulation

TA12 Materials, Structures, Mechanical Systems, and Manufacturing

- 12.2.5 Structures: Innovative, Multifunctional Concepts
- 12.2.1 Structures: Lightweight Concepts
- 12.1.1 Materials: Lightweight Structure
- 12.2.2 Structures: Design and Certification Methods
- 12.5.1 Nondestructive Evaluation and Sensors
- 12.3.4 Mechanisms: Design and Analysis Tools and Methods
- 12.3.1 Deployables, Docking, and Interfaces
- 12.3.5 Mechanisms: Reliability/Life Assessment/Health Monitoring
- 12.4.2 Intelligent Integrated Manufacturing and Cyber Physical Systems

TA14 Thermal Management Systems

- 14.3.1 Ascent/Entry Thermal Protection Systems
- 14.1.2 Active Thermal Control of Cryogenic Systems

NOTE: Technologies are listed by roadmap/technology area (TA01 through TA14; there are no high-priority technologies in TA13). Within each technology area, technologies are listed by the QFD score assigned by the panels, in descending order. This sequencing may be considered a rough approximation of the relative priority of the technologies within a given technology area.

TABLE 3.3 Correlation Matrix Between High-Priority Technologies Selected by Panels and Top Technical Challenges for Technology Objective A. Rows with all blanks are not shown.

Technology Objective A: Extend and sustain human activities beyond low Earth orbit	Improved Access to Space	Space Radiation Health Effects	Long Duration Health Effects	Long Duration ECLSS	Rapid Crew Transit	Lightweight Space Structures	Increase Available Power	Mass to Surface	Precision Landing	Autonomous Rendezvous and Dock
1.3.1 TBCC	●									
1.3.2 RBCC	●									
2.2.3 (Nuclear) Thermal Propulsion					●					
3.1.3 Solar Power Generation (Photovoltaic and Thermal)							●			
3.1.5 Fission Power Generation							●			
3.2.1 Batteries							●			
3.3.3 Power Distribution and Transmission							●			
3.3.5 Power Conversion and Regulation							●			
4.6.2 Relative Guidance Algorithms										●
4.6.3 Docking and Capture Mechanisms/Interfaces										●
5.4.1 Timekeeping and Time Distribution										●
5.4.3 Onboard Autonomous Navigation & Maneuvering									●	●
6.1.1 Air Revitalization				●						
6.1.2 ECLSS Water Recovery and Management				●						
6.1.3 ECLSS Waste Management				●						
6.1.4 Habitation				●						
6.2.2 EVA Portable Life Support System				●						
6.3.2 Long-Duration Crew Health			●							
6.5.1 Radiation Risk Assessment Modeling		●								
6.5.2 Radiation Mitigation		●								
6.5.3 Radiation Protection Systems		●								
6.5.4 Radiation Prediction		●								
6.5.5 Radiation Monitoring Technology		●								
9.1.1 Rigid Thermal Protection Systems								●	●	
9.1.2 Flexible Thermal Protection Systems								●	●	
9.1.4 Deployment Hypersonic Decelerators								●	●	
9.4.7 GN&C Sensors and Systems (EDL)									●	
10.1.1 (Nano) Lightweight Materials and Structures	●									
12.1.1 Materials: Lightweight Structure	●							●		
12.2.1 Structures: Lightweight Concepts	●					●		●		
12.2.2 Structures: Design and Certification Methods						●				
12.2.5 Structures: Innovative, Multifunctional Concepts						●				
12.3.1 Deployables, Docking and Interfaces										●
14.3.1 Ascent/Entry TPS								●	●	

TABLE 3.4 Correlation Matrix Between High-Priority Technologies Selected by Panels and Top Technical Challenges for Technology Objective B. Rows with all blanks are not shown.

Technology Objective B: Explore the evolution of the solar system and the potential for life elsewhere (in situ measurements)	Improved Access to Space	Precision Landing	Robotic Surface Maneuvering	Life Detection	High-Power Electric Propulsion	Autonomous Rendezvous and Dock	Increase Available Power	Mass to Surface	Lightweight Space Structures	Higher Data Rates
1.3.1 TBCC	●									
1.3.2 RBCC	●									
2.2.1 Electric Propulsion					●					
3.1.3 Solar Power Generation (Photovoltaic and Thermal)							●			
3.1.5 Fission Power Generation							●			
3.2.1 Batteries							●			
3.3.3 Power Distribution and Transmission							●			
3.3.5 Power Conversion and Regulation							●			
4.2.1 Extreme Terrain Mobility			●							
4.6.2 Relative Guidance Algorithms			●			●				
4.6.3 Docking & Capture Mechanisms/Interfaces						●				
5.3.2 Adaptive Network Topology										●
5.4.1 Timekeeping and Time Distribution						●				●
5.4.3 Onboard Autonomous Navigation and Maneuvering		●	●			●				
5.5.1 Radio Systems										●
8.3.3 In Situ Instruments				●						
9.1.1 Rigid Thermal Protection Systems		●						●		
9.1.2 Flexible Thermal Protection Systems		●						●		
9.1.4 Deployment Hypersonic Decelerators		●						●		
9.4.7 EDL GN&C Sensors and Systems		●								
10.1.1 (Nano) Lightweight Materials and Structures	●									
12.1.1 Lightweight Structure	●							●		
12.2.1 Structures: Lightweight Concepts	●							●	●	
12.2.2 Structures: Design and Certification Methods									●	
12.2.5 Structures: Innovative, Multifunctional Concepts									●	
12.3.1 Deployables, Docking and Interfaces						●				
14.3.1 Ascent/Entry TPS		●						●		

TABLE 3.5 Correlation Matrix Between High-Priority Technologies Selected by Panels and Top Technical Challenges for Technology Objective C. Rows with all blanks are not shown.

Technology Objective C: Expand understanding of Earth and the universe (remote measurements)	Improved Access to Space	New Astronomical Telescopes	Lightweight Space Structures	Increase Available Power	Higher Data Rates	High-Power Electric Propulsion	Design Software	Structural Monitoring	Improved Flight Computers	Cryogenic Propellant Storage
1.3.1 TBCC	●									
1.3.2 RBCC	●									
2.2.1 Electric Propulsion						●				
2.4.2 Propellant Storage and Transfer										●
3.1.3 Solar Power Generation (Photovoltaic and Thermal)				●						
3.1.5 Fission Power Generation				●						
3.2.1 Batteries				●						
3.3.3 Power Distribution and Transmission				●						
3.3.5 Power Conversion and Regulation				●						
5.3.2 Adaptive Network Topology					●					
5.4.1 Timekeeping and Time Distribution					●					
5.5.1 Radio Systems					●					
8.1.1 Detectors & Focal Planes		●								
8.1.3 Optical Systems		●								
8.2.4 High-Contrast Imaging and Spectroscopy		●								
9.4.5 EDL Modeling and Simulation							●			
10.1.1 (Nano) Lightweight Materials and Structures	●									
11.1.1 Flight Computing									●	
12.1.1 Lightweight Structure	●									
12.2.1 Structures: Lightweight Concepts	●		●							
12.2.2 Structures: Design and Certification Methods			●				●			
12.2.5 Structure Innovative, Multifunctional Concepts			●							
12.3.4 Design and Analysis Tools and Methods							●			
12.3.5 Mechanisms Reliability / Life Assessment / Health Monitoring								●		
12.5.1 Nondestructive Evaluation & Sensors								●		
14.1.2 Active Thermal Control of Cryogenic Systems										●

TABLE 3.6 Technologies That Represent Multiple Highly Coupled Technologies from the Technology Area Breakdown Structure

Unified Technology	Technology Area Breakdown Structure Technologies
X.1 Radiation Mitigation for Human Spaceflight	6.5.1 Radiation Risk Assessment Modeling 6.5.2 Radiation Mitigation 6.5.3 Radiation Protection Systems 6.5.4 Radiation Prediction 6.5.5 Monitoring Technology
X.2 Lightweight and Multifunctional Materials and Structures	10.1.1 (Nano) Lightweight Materials and Structures 12.1.1 Lightweight Structures 12.2.1 Structures: Lightweight Concepts 12.2.2 Structures: Design and Certification Methods 12.2.5 Structures: Innovative, Multifunctional Concepts
X.3 ECLSS	6.1.1 Air Revitalization 6.1.2 Water Recovery and Management 6.1.3 Waste Management 6.1.4 Habitation
X.4 GN&C	4.6.2 Relative Guidance Algorithms 5.4.3 Onboard Autonomous Navigation and Maneuvering 9.4.7 EDL GN&C Sensors and Systems
X.5 EDL TPS	9.1.1 Rigid Thermal Protection Systems 9.1.2 Flexible Thermal Protection Systems 14.3.1 Ascent/Entry TPS

TABLE 3.7 Initial Prioritization of Top Technologies, Categorized by Technology Objective

Technology Objective A: Extend and sustain human activities beyond low Earth orbit	Technology Objective B: Explore the evolution of the solar system and the potential for life elsewhere (in situ measurements)	Technology Objective C: Expand understanding of Earth and the universe (remote measurements)
Radiation Mitigation for Human Spaceflight (X.1)	GN&C (X.4)	(Instrument and Sensor) Optical Systems (8.1.3)
Long-Duration (Crew) Health (6.3.2)	Electric Propulsion (2.2.1)	High-Contrast Imaging and Spectroscopy Technologies (8.2.4)
ECLSS (X.3)	Solar Power Generation (Photovoltaic and Thermal) (3.1.3)	Detectors & Focal Planes (8.1.1)
GN&C (X.4)	In Situ (Instruments and Sensor) (8.3.3)	Lightweight and Multifunctional Materials and Structures (X.2)
Thermal Propulsion (2.2.3)	Fission (Power) (3.1.5)	Radioisotope (Power) (3.1.4)
Fission (Power) (3.1.5)	Extreme Terrain Mobility (4.2.1)	Electric Propulsion (2.2.1)
Lightweight and Multifunctional Materials and Structures (X.2)	Lightweight and Multifunctional Materials and Structures (X.2)	Solar Power Generation (Photovoltaic and Thermal) (3.1.3)
EDL TPS (X.5)	Radioisotope (Power) (3.1.4)	Science Modeling and Simulation (11.2.4a)
Atmosphere and Surface Characterization (9.4.4)	Robotic Drilling and Sample Handling (4.3.6)	Batteries (3.2.1)
Propellant Storage and Transfer (2.4.2)	EDL TPS (X.5)	(Instrument and Sensor) Electronics (8.1.2)
Pressure Garment (6.2.1)	Docking and Capture Mechanisms/ Interfaces (4.6.3)	Active Thermal Control of Cryogenic Systems (14.1.2)
		(Mechanisms) Reliability / Life Assessment / Health Monitoring (12.3.5)
		Vehicle System Management and FDIR (4.5.1)

TABLE 3.8 Final Prioritization of Top Technologies, Categorized by Technology Objective

Technology Objective A: Extend and sustain human activities beyond low Earth orbit	Technology Objective B: Explore the evolution of the solar system and the potential for life elsewhere (in situ measurements)	Technology Objective C: Expand understanding of Earth and the universe (remote measurements)
Radiation Mitigation for Human Spaceflight (X.1)	GN&C (X.4)	(Instrument and Sensor) Optical Systems (8.1.3)
Long-Duration (Crew) Health (6.3.2)	Solar Power Generation (Photovoltaic and Thermal) (3.1.3)	High-Contrast Imaging and Spectroscopy Technologies (8.2.4)
ECLSS (X.3)	Electric Propulsion (2.2.1)	Detectors and Focal Planes (8.1.1)
GN&C (X.4)	Fission (Power)(3.1.5)	Lightweight and Multifunctional Materials and Structures (X.2)
Thermal Propulsion (2.2.3)	EDL TPS (X.5)	Active Thermal Control of Cryogenic Systems (14.1.2)
Lightweight and Multifunctional Materials and Structures (X.2)	In Situ (Instruments and Sensor) (8.3.3)	Electric Propulsion (2.2.1)
Fission (Power) (3.1.5)	Lightweight and Multifunctional Materials and Structures (X.2)	Solar Power Generation (Photovoltaic and Thermal) (3.1.3)
EDL TPS (X.5)	Extreme Terrain Mobility (4.2.1)	

to determine the highest-priority technologies to emphasize over the next 5 years. It is not that other technology development is unimportant, but rather that some technology development can wait, some depends on obtaining prior results before progress can be made, and some is best served by low-level funding of exploratory concept development before proceeding. Alternatively, some technologies are so game-changing that early results are needed to define and shape possible paths to future missions (e.g., radiation protection). Table 3.8 shows the final technology prioritization for each technology objective, listed in ranked order.

It should be noted that the prioritizations in Tables 3.7 and 3.8 may differ from the prioritizations determined by the panels in Appendixes D through Q for two principal reasons. First, the steering committee organized and prioritized its technologies against the three different technology objectives, which the panels did not do; a technology's priority can change significantly depending upon the objective. Second, the steering committee emphasized technology development in the next 5 years, a specific timing constraint that was not given to the panels.

The steering committee's consensus viewpoint on a short list of the highest-priority technologies to emphasize over the next 5 years is shown in Table 3.8 (three columns with 16 different technologies). Table 3.9 provides a single list of the 16 technologies and shows which technology objectives each one supports. The relationship of these technologies to the top technical challenges is shown in Tables 3.10, 3.11, and 3.12.

The steering committee assumes NASA will pursue all three objectives in a balanced approach, each according to the approved resources and mission plans allocated. The steering committee does not recommend or advocate support for one objective over the others. The steering committee noted that Technology Objective B has many common technology needs with Objectives A and C. Objectives A and C each have dominant technologies to enable NASA's strategic goals; i.e., radiation protection, long-duration-mission crew health, and ECLS are mostly unique to Objective A, while optical systems, high-contrast imaging and spectrometry, and detectors are mostly unique to Objective C. However, GN&C, lightweight and multifunctional materials and structures, and solar power are primary to Objective B but are common to all three objectives.

The steering committee reasoned that this intentionally limited set of recommended high-priority technologies constituted a scope that could reasonably be accommodated within the most likely expected funding level available for technology development by OCT (in the range of $500 million to $1 billion annually). Also considered within the scope of a balanced technology development program is the importance of low technology readiness level (TRL; 1 and 2) exploratory concept development and high-TRL flight demonstrations. The steering committee

TABLE 3.9 The 16 Technologies That Appear in the Final Prioritization, Showing the Priority Assigned for Each Technology Objective

Technologies included in the final prioritization, listed by TABS number		Technology Objective A	Technology Objective B	Technology Objective C
2.2.1	Electric Propulsion		3	6
2.2.3	Thermal Propulsion	5		
3.1.3	Solar Power Generation (Photovoltaic and Thermal)		2	7
3.1.5	Fission (Power)	7	4	
4.2.1	Extreme Terrain Mobility		8	
6.3.2	Long-Duration (Crew) Health	2		
8.1.1	Detectors & Focal Planes			3
8.1.3	(Instrument and Sensor) Optical Systems			1
8.2.4	High-Contrast Imaging and Spectroscopy Technologies			2
8.3.3	In Situ (Instruments and Sensors)		6	
14.1.2	Active Thermal Control of Cryogenic Systems			5
X.1	Radiation Mitigation for Human Spaceflight	1		
X.2	Lightweight and Multifunctional Materials and Structures	6	7	4
X.3	ECLSS	3		
X.4	GN&C	4	1	
X.5	EDL TPS	8	5	

NOTE: The content of technologies X.1 through X.5 is shown in Table 3.6.

TABLE 3.10 Linkages Between Top Technologies and Technical Challenges for Technology Objective A

Technology Objective A: Extend and sustain human activities beyond low Earth orbit		Radiation Mitigation for Human Spaceflight (X.1)	Long-Duration (Crew) Health (6.3.2)	ECLSS (X. 3)	GN&C (X.4)	Thermal Propulsion (2.2.3)	Lightweight and Multifunctional Materials and Structures (X.2)	Fission (Power) (3.1.5)	EDL TPS (X.5)
1	Improved Access to Space						●		
2	Space Radiation Health Effects	●							
3	Long Duration Health Effects		●						
4	Long Duration ECLSS			●					
5	Rapid Crew Transit					●			
6	Lightweight Space Structures	●					●		
7	Increase Available Power							●	
8	Mass to Surface						●		●
9	Precision Landing				●				●
10	Autonomous Rendezvous and Dock				●				

TABLE 3.11 Linkages Between Top Technologies and Technical Challenges for Technology Objective B

Technology Objective B: Explore the evolution of the solar system and the potential for life elsewhere (in situ measurements)		GN&C (X.4)	Solar Power Generation (Photovoltaic and Thermal) (3.1.3)	Electric Propulsion (2.2.1)	Fission (Power) (3.1.5)	EDL TPS (X.5)	In Situ (Instruments & Sensor) (8.3.3)	Lightweight and Multifunctional Materials and Structures (X.2)	Extreme Terrain Mobility (4.2.1)
1	Improved Access to Space							●	
2	Precision Landing	●				●			
3	Robotic Surface Maneuvering	●							●
4	Life Detection						●		
5	High-Power Electric Propulsion			●					
6	Autonomous Rendezvous and Dock	●							
7	Increase Available Power		●		●				
8	Mass to Surface					●			
9	Lightweight Space Structures							●	
10	Higher Data Rates							●	

consensus is that low-TRL, NASA Innovative Advanced Concepts-like funding should be on the order of 10 percent of the total, and that the research should quickly weed out the least competitive concepts, focusing on those that show the greatest promise in addressing the top technical challenges. At the high-TRL end of the spectrum, flight demonstrations, while expensive, are sometimes essential to reach a readiness level required for transition of a technology to an operational system. Such technology flight demonstrations are considered on a case-by-case basis when there is ample "pull" from the user organization, including a reasonable level of cost sharing.

At some point, the scale of technology development for nuclear thermal propulsion and fission power technologies in Table 3.8 will grow to a level where large-scale efforts may need to be deferred if the OCT space technology research budget is substantially below the expected level. Even in such a case, technology development should still proceed at a low level in these high-priority areas because the technology will take years to advance and they represent game-changing approaches to NASA's mission.[3]

Recommendation. *Technology Development Priorities.* During the next 5 years, NASA technology development efforts should focus on (1) the 16 identified high-priority technologies and associated top technical challenges, (2) a modest but significant investment in low-TRL technology (on the order of 10 percent of NASA's technology development budget), and (3) flight demonstrations for technologies that are at a high-TRL when there is sufficient interest and shared cost by the intended user.

[3] By statute, DOE must take the lead in the development of reactor components for a NASA fission power or nuclear thermal rocket propulsion system.

TABLE 3.12 Linkages Between Top Technologies and Technical Challenges for Technology Objective C

Technology Objective C: Expand understanding of Earth and the universe (remote measurements)		(Instrument and Sensor) Optical Systems (8.1.3)	High-Contrast Imaging and Spectroscopy (8.2.4)	Detectors and Focal Planes (8.1.1)	Lightweight and Multifunctional Materials and Structures (X.2)	Active Thermal Control of Cryogenic Systems (14.1.2)	Electric Propulsion (2.2.1)	Solar Power Generation (Photovoltaic and Thermal) (3.1.3)
1	Improved Access to Space				●			
2	New Astronomical Telescopes	●	●	●				
3	Lightweight Space Structures				●			
4	Increase Available Power							●
5	Higher Data Rates				●			
6	High-Power Electric Propulsion						●	
7	Design Software							
8	Structural Monitoring				●			
9	Improved Flight Computers							
10	Cryogenic Storage and Transfer				●	●		

The Importance of Improved Access to Space

In most cases, the steering committee and the panels have identified technologies that will make substantial progress in achieving the top technical challenges at the steering committee level and at the panel level. However, the importance of a challenge is not diminished simply because technologies to achieve the challenge are not readily available. For example, improving access to space (by dramatically reducing the total cost and increasing the reliability and safety of access to space) was the number one technical challenge for each of the three technology objectives cited. Despite this, only one of the top 16 technologies selected by the steering committee that was relevant to this challenge, Lightweight and Multifunctional Materials and Structures (X.2), made it to the final short list of technologies for emphasis over the next 5 years. While low-cost access to space is critically important, technologies to achieve that particular challenge are few, and some of the high-leverage factors affecting launch cost tend to be operational rather than technological. Some non-technological approaches to solving this problem may exist, such as on-orbit assembly or ground operations. In addition, for any given set of launch vehicles, advanced technologies that reduce payload mass and volume could reduce launch costs on a per mission basis if they allow missions to be launched with smaller, less expensive launch vehicles.

In Appendix D, the Propulsion and Power Panel describes possible architecture changes that would increase launch rates and potentially reduce costs. In Appendix P, this panel addresses operational efficiencies associated with ground operations (TA13, Ground and Launch Systems Processing) that are not technology issues. The panel also identified two high-priority technologies to align with this challenge: Turbine Based Combined Cycle (TBCC) and Rocket Based Combined Cycle (RBCC) engines. RBCC and TBCC would provide a revolutionary new, next-generation reusable launch system. Although the steering committee acknowledges the potential benefits of TBCC/RBCC technologies toward the challenge of low-cost access to space, it did not recommend these for highest emphasis for the following reasons. The development of either an RBCC or a TBCC propulsion system would require a national-level effort that includes partnerships with the DOD and other organizations. Before a national-level program could be started, vehicle system design trades would need to clearly show the benefits of chosen configurations and designs. Because combined cycle propulsion is so integral to the design of the airframe, configuration and design are critical. To date, one of the main deterrents to RBCC and TBCC is their development cost. Also, since these systems are targeted for reusable configurations, high flight rates are required to attain promised cost savings.

Technologies Near a Tipping Point

A "tipping point" is defined as a point in the research process such that a small increase in the research effort could produce a large advance in its technology readiness. The steering committee identified two such technologies nearing a tipping point: ASRGs and cryogenic storage and transfer. Both of these technologies are ready for near-term flight demonstrations.

Advanced Stirling Radioisotope Generator

Radioisotope power systems provide electrical power for spacecraft and systems that are unable to use solar power. Radioisotope power systems, in the form of Radioisotope Thermoelectric Generators (RTGs), have been used reliably for 50 years to enable NASA's solar system exploration missions. Plutonium-238 (Pu-238) is the only isotope suitable as the heat source for RPSs, but no Pu-238 has been produced in the United States since the late 1980s. Currently, Pu-238 is not being produced anywhere in the world, and the stockpile of Pu-238 available to NASA is almost depleted (NRC, 2006, 2010).

Because of the limited supply of Plutonium-238, NASA and the Department of Energy have begun research and development in higher-efficiency technologies. The Advanced Stirling Radioisotope Generator (ASRG) is a new type of radioisotope power system still in development. An ASRG uses a Stirling engine coupled to linear alternators to convert heat to electricity. ASRG Stirling converters have efficiencies several times greater than the thermoelectric converters of traditional RTGs, and thus they require much less Pu-238 for the same electric power output (NRC, 2010). The demonstration of the long-duration reliability and flight readiness of ASRGs is still to be achieved, however. The planetary science decadal survey committee determined that the completion and validation of the Advanced Stirling Radioisotope Generator is its highest priority for near-term technology investment (NRC, 2011b, p. 307).

In 2011, NASA selected three candidate Discovery missions for potential downselect for launch in 2016. Two of the candidates would utilize ASRG flight units and would demonstrate their utility on long-duration, deep-space missions. NASA is on a good course to bring this critical technology at a "tipping point" to full demonstration.

> **Recommendation. *Advanced Stirling Radioisotope Generators*.** The NASA Office of the Chief Technologist should work with the Science Mission Directorate and the Department of Energy to help bring Advanced Stirling Radioisotope Generator technology hardware to flight demonstration on a suitable space mission beyond low Earth orbit.

> **Finding: *Plutonium-238.*** Consistent with findings of previous NRC reports on the subject of plutonium-238 (NRC, 2010, 2011b), restarting the fuel supply is urgently needed. Even with the successful development of Advanced Stirling Radioisotope Generators, if the funds to restart the fuel supply are not authorized and appropriated, it will be impossible for the United States to conduct certain planned critical deep-space missions after this decade.

Reduced Gravity Cryogenic Storage and Transfer

The storage and handling of cryogenic fluids will be needed to support missions beyond low Earth orbit. Technology to effectively store, manage, and transfer propellants over long periods in space would improve mission feasibility and affordability.

Flight experiments are needed to test and validate key capabilities and technologies required for the storage and transfer of cryogenic propellants to and from advanced propulsion stages and propellant depots. The ISS could play an important role in validating long-term storage and handling of cryogenic propellants. Technologies to be demonstrated include:

- Cryogenic fluid instrumentation and sensors
- Passive thermal control
- Active thermal control

Instrumentation and sensors are needed to ascertain and monitor fluid mass and location in reduced gravity tanks. Cryogenic systems include passive techniques (such as multilayer insulation and vapor-cooled shields), as well as active thermal control techniques (such as cryocoolers) to manage remaining heat leaks after passive techniques are applied. In addition to supporting cryogenic propellant storage and transfer, active thermal control technology can enable long-term storage of consumables such as LOX for human missions and support scientific instruments that require cryogenic conditions. The 2011 NRC decadal survey *Recapturing a Future for Space Exploration: Life and Physical Sciences Research for a New Era* recommended near-term research and technology development in zero-boil-off propellant storage (both passive and active techniques) and cryogenic handling and gauging (NRC, 2011a). These technologies are approaching a high level of technical maturity but remain to be tested and demonstrated in a reduced-gravity environment.

> **Recommendation**. ***Cryogenic Storage and Handling.*** Reduced-gravity cryogenic storage and handling technology is close to a "tipping point," and NASA should perform on-orbit flight testing and flight demonstrations to establish technology readiness.

Relevance of High-Priority Technologies to National and Commercial Space Needs

When pursuing the 16 technologies recommended by the steering committee as high-priority efforts in the next 5 years, it is useful to simultaneously consider the value of those technologies to the interests of others outside NASA, specifically those that address broader national needs as well as the needs of the commercial space industry. Alignment with national and commercial needs outside NASA (both aerospace needs and non-aerospace needs) was a scoring category used by the panels as they made their initial assessments of all the level 3 technologies, although the weighting factor given to this category was not high relative to other categories such as benefit and risk. The steering committee identified those technologies that would either be essential or could make a significant contribution to national and commercial space interests outside NASA (shown in Table 3.13). In the case of national needs—for example, dual-use technology of interest to the Department of Defense—this information shows those technologies that offer the best chance to partner with other government institutions through sharing of information and resources.

The technologies shown in Table 3.13 were selected based on NASA's most critical needs and highest priorities. NASA is the first and primary user, although there is relevance to other national interests. The strong importance of commercial space activities to NASA was recognized by the steering committee, and this relationship is discussed further in Chapter 4.

TABLE 3.13 Relevance of High-Priority Technologies to National and Commercial Space Needs

Technologies included in the final prioritization, listed by TABS number		National Needs	Commercial Needs
2.2.1	Electric Propulsion	⊙	⊙
2.2.3	Thermal Propulsion	⊙	○
3.1.3	Solar Power Generation (Photovoltaic and Thermal)	●	●
3.1.5	Fission (Power)	○	○
4.2.1	Extreme Terrain Mobility	○	○
6.3.2	Long-Duration (Crew) Health	⊙	○
8.1.1	Detectors & Focal Planes	⊙	⊙
8.1.3	(Instrument and Sensor) Optical Systems	⊙	⊙
8.2.4	High-Contrast Imaging and Spectroscopy Technologies	⊙	⊙
8.3.3	In Situ (Instruments and Sensors)	○	○
14.1.2	Active Thermal Control of Cryogenic Systems	⊙	⊙
X.1	Radiation Mitigation for Human Spaceflight	○	○
X.2	Lightweight and Multifunctional Materials and Structures	⊙	⊙
X.3	ECLSS	○	⊙
X.4	GN&C	⊙	⊙
X.5	EDL TPS	⊙	⊙
Key		Essential	●
		Significant	⊙
		Minor	○

REFERENCES

NASA (National Aeronautics and Space Administration). 2011. *2011 Strategic Plan.* NASA, Washington, D.C.

NRC (National Research Council). 2006. *Priorities in Space Science Enabled by Nuclear Power and Propulsion.* The National Academies Press, Washington, D.C.

NRC. 2009. *America's Future in Space: Aligning the Civil Space Program with National Needs.* The National Academies Press, Washington, D.C.

NRC. 2010. *Radioisotope Power Systems: An Imperative for Maintaining U.S. Leadership in Space Exploration.* The National Academies Press, Washington, D.C.

NRC. 2011a. *Recapturing a Future for Space Exploration: Life and Physical Sciences Research for a New Era.* The National Academies Press, Washington, D.C.

NRC. 2011b. *Vision and Voyages for Planetary Science in the Decade 2013-2022.* The National Academies Press, Washington, D.C.

4

Observations and General Themes

INTRODUCTION

In reviewing and evaluating the draft roadmaps and considering the purpose and strategic goals for the advanced technology development program managed by NASA's Office of the Chief Technologist (OCT), the steering committee formed some general observations concerning the program as a whole and reached some conclusions on how the effectiveness of the program can be maintained or enhanced.[1] Those observations and conclusions are described in this chapter. The topics dealt with tend to address multiple roadmaps.

SYSTEMS ANALYSIS

Effective management of NASA's space technology program requires careful consideration of technology priorities, trade-offs, and decision points for down-selecting from competing options, as well as the changing needs of future missions. Technology relationships and planning can be complex. Some focused technologies must be developed early to support the development of higher-level technologies. Conversely, other technologies cannot effectively move forward until substantial progress is made with more foundational technologies. In other cases, technology advances are only realized when a suite of multidisciplinary technologies are combined in a subsystem or system. An effective management process that is guided by systems analysis trade studies and includes systems engineering considerations, as appropriate, is necessary to establish and maintain a coherent and effectively phased technology program that coordinates and integrates the research conducted across multiple roadmaps, as necessary.

There will always be multiple technology options available to address a given technical challenge. Establishing milestones with well-defined performance criteria coinciding with down-select points in project plans provides a structured approach to selecting technologies that show the most promise, while terminating those that are less relevant and unlikely to contribute to emerging capabilities. Down-selecting too soon can limit options, but in a constrained budgetary environment pursuing too many competing parallel technical approaches is unaffordable. These competing pressures—keeping options open and down-selecting early to the most promising technology—highlight the importance of emphasizing technologies that can meet a range of likely future needs. For example, the draft Entry, Descent, and Landing Systems roadmap is focused largely on meeting the needs of a human mission

[1]The draft space technology roadmaps are available online at http://www.nasa.gov/offices/oct/strategic_integration/technology_roadmap.html.

to Mars. While this mission beneficially stresses and challenges the technology envelope in entry, descent, and landing (EDL), it would be prudent to ensure that the EDL technology under development is not tied too closely to a specific mission or destination. Technologies that enable a broad spectrum of future missions by accommodating a wide range of destination and schedule options were more highly valued in the roadmap evaluation.

Some technologies in different roadmaps have connections that are not delineated in the draft roadmaps. For example, both a better understanding of the effects of space radiation on humans and development of technologies to mitigate those effects more effectively are a high priority for future human deep-space missions. The risk posed by space radiation is closely linked to mission duration and thus to advances in in-space propulsion that could shorten the duration of long missions, such as a human mission to Mars. Systems analysis could help address the multifaceted challenge of reducing the health risk posed by space radiation. More generally, systems analysis could be used throughout the technology development process to guide technology selection, refinement, redirection, and down-selection in the dynamic environment that shapes NASA's current and future research and mission priorities.

Recommendation. ***Systems Analysis.*** NASA's Office of the Chief Technologist (OCT) should use disciplined systems analysis for the ongoing management and decision support of the space technology portfolio, particularly with regard to understanding technology alternatives, relationships, priorities, timing, availability, down-selection, maturation, investment needs, system engineering considerations, and cost-to-benefit ratios; to examine "what-if" scenarios; and to facilitate multidisciplinary assessment, coordination, and integration of the roadmaps as a whole. OCT should give early attention to improving systems analysis and modeling tools, if necessary to accomplish this recommendation.

Recommendation. ***Managing the Progression* of *Technologies to Higher Technology Readiness Levels (TRLs).*** OCT should establish a rigorous process to down-select among competing technologies at appropriate milestones and TRLs to ensure that only the most promising technologies proceed to the next TRL.

FOUNDATIONAL TECHNOLOGY BASE AND DEVELOPMENT OF LOW-TRL TECHNOLOGY

The successful development of game-changing technologies that lead to revolutionary capabilities applicable to a wide range of potential missions is a priority for NASA's space technology program and was treated as such in the roadmap evaluation process. Not to be overlooked, however, is the fact that many of the game-changing breakthroughs emerge only from a sustained level of disciplinary research and its contribution to a foundational technology base. As such, maintaining (or reestablishing) a technology base that produces a pipeline of evolutionary improvements over time is an important element of the OCT program. There are many areas where continuity of effort is all-important, such as research on aerothermodynamics and hypersonic flow, advanced lightweight materials, fault-tolerant guidance and control, and human factors, to name just a few. These are areas that address problems that will never be completely solved, yet advancing the level of understanding in these disciplines is critical for many of NASA's missions. Also, in executing challenging missions, it is often necessary to have access to individuals with in-depth competence who reside in the government, at an academic institution, or perhaps in industry, and who are experts in a given subject. These experts are invaluable in providing advice to help solve problems of a critical nature as they arise during a program. Furthermore, hiring the very brightest people as they graduate from universities will help NASA maintain and improve its workforce over the long term. These individuals are usually best placed initially in a disciplinary research organization where they can continue to grow their expertise in their fields. They can later transition to other parts of the organization as they progress in their chosen careers. The steering committee recognizes that many of these core-competency and workforce issues do not rest principally with OCT, but they are mentioned here because the technology development program can influence how NASA addresses these issues.

In 2011, NASA OCT reestablished the NASA Innovative Advanced Concepts (NIAC) program to fund visionary technologies at TRLs 1 to 3. The NIAC program, which will investigate individual technology concepts at a relatively low level of effort, should not be limited to those technologies identified as high priority by the steering committee or the panels. Rather, research supported by the NIAC program will complement the more substantial efforts that are necessary to investigate high-priority technologies at a higher TRL. The NIAC approach is also appropriate to meet the policy objectives of enhancing the education of future scientists and engineers and facilitating international collaboration in the development of low-TRL technology.

> **Recommendation.** ***Foundational Technology Base.*** OCT should reestablish a discipline-oriented technology base program that pursues both evolutionary and revolutionary advances in technological capabilities and that draws upon the expertise of NASA centers and laboratories, other federal laboratories, industry, and academia.

COOPERATIVE DEVELOPMENT OF NEW TECHNOLOGIES

Programs such as the International Space Station (ISS) demonstrate the benefit of interagency and international cooperation at the mission level. The development of many technologies relevant to NASA is supported and/or conducted by other federal agencies, foreign governments, industry, and academic institutions. In many cases NASA is already cooperating with other organizations to develop critical new technologies and/or adapt the results of work by others to meet NASA's needs. NASA's 2011 Strategic Plan confirms NASA's intent to continue such cooperation, noting that NASA should "facilitate the transfer of NASA technology and engage in partnerships with other government agencies, industry, and international entities to generate U.S. commercial activity and other public benefits" while "expanding partnerships with international, intergovernmental, academic, industrial, and entrepreneurial communities and recognizing their role as important contributors of skill and creativity to our missions and for the propagation of results" (NASA, 2011, p. 3-5). Department of Defense research laboratories have space technology development efforts that the workshops identified as areas where collaboration with OCT would be mutually beneficial. Europe has made significant long-term investments in basic and industrial research to advance and sustain its space program. Similarly, the space programs of Japan and other Asian countries are also advancing rapidly. NASA's technology roadmaps would be more valuable and actionable if they provided more detail about how various goals might be accomplished through partnerships with outside organizations.

With some technologies, international partnerships are hampered by limitations imposed by U.S. International Traffic in Arms Regulations (ITAR). Even so, technology development efforts by OCT provide a new opportunity for NASA to engage in cooperative development of new technologies. This cooperation could enable NASA to achieve more of its technological goals with available funding, in part by drawing on the available specialized expertise and prior investments made elsewhere.

NASA recognizes that resource constraints of funding and staffing will always be a limiting factor to carrying out all technology development recommended by the roadmaps. Accordingly, cooperative development of applicable high-priority technologies with other organizations will expand the scope of advanced technologies that will be available to future missions.

> **Recommendation.** ***Cooperative Development of New Technologies.*** OCT should pursue cooperative development of high-priority technologies with other federal agencies, foreign governments, industry, and academic institutions to leverage resources available for technology development.

FLIGHT TESTING AND DEMONSTRATIONS AND TECHNOLOGY TRANSITION

Testing and demonstrating new space technologies under realistic flight conditions are always desirable. The steering committee makes a distinction between flight *testing* and flight *demonstrations:* flight testing deals with acquiring performance data at any TRL below 6 that happens be in flight, and flight demonstration deals with the TRL 6 validation of a system or subsystem performance to confirm technology readiness and level of risk to the satisfaction of those who will decide to incorporate the technology in a mission.

Flight testing is needed to validate the maturity of technologies when ground-based testing and/or modeling and simulation (M&S) are inadequate. It can also (1) increase the visibility of new technologies with mission offices regarding the potential of the technology to meet mission needs in terms of performance and reliability in a way that M&S and ground tests might not convey; (2) lay the groundwork for dialog between technology developers and mission offices to define a rigorous approach to achieving TRL 6; and (3) provide opportunities to train new members of the workforce and give systems engineers and instrument scientists hands-on experience with a new technology across the full span of space mission phases (design, development, fabrication, testing, data analysis, and so on) over relatively short time spans and in risk-tolerant environments.

Flight demonstration would be the final phase of a NASA technology development program. Flight demonstrations should be conducted only if there is sufficient "pull" (and typically cost sharing) from the user. Such co-funding between OCT and the using Mission Directorate in the case of a NASA application is a mechanism for bridging the "valley of death" that often impedes or prevents the transition of advanced technologies from technology development offices and/or organizations to mission development offices and/or organizations.

Various platforms are available to support flight testing and demonstrations, depending on the technology and application in question. Possibilities include high-altitude airborne flights, sub-orbital space flights, and orbital flights on dedicated spacecraft, government or commercial spacecraft (as a secondary payload), and on the ISS.

Recommendation. *Flight Demonstrations and Technology Transition.* OCT should collaborate with other NASA mission offices and outside partners in defining, advocating, and where necessary co-funding flight demonstrations of technologies. OCT should document this collaborative arrangement using a technology transition plan or similar agreement that specifies success criteria for flight demonstrations as well as budget commitments by all involved parties.

FACILITIES

Although facility capability is outside OCT's direct line of responsibility and is not explicitly addressed in the study's statement of task, the health and availability of facilities are closely linked to development of advanced technology.

State-of-the-art facilities for aerospace research and development are often large, complex, and expensive. As a result, many aerospace research facilities have historically been built and operated by government laboratories. This tradition was first established in Europe at the beginning of the 20th century (e.g., the National Physical Laboratory in the United Kingdom, which began aeronautics research and testing in 1908). This was followed by the creation of the National Advisory Committee for Aeronautics (NACA) in the United States in 1915 and the opening of the NACA Langley Memorial Laboratory in 1920. The need for such government-run facilities continues today, as underscored by a number of NRC reports, most recently an assessment of NASA laboratories for basic research (NRC, 2010).

Adequate ground test facilities are required to validate analytical models, to benchmark complex computer simulations such as computational fluid dynamics models, and to examine new designs and concepts. Testing is a critical element in material development, such as new TPS materials. Such testing is normally carried out in arcjet facilities that can produce convective heating rates and accommodate test articles in sizes of interest to simulate entry from LEO, NEO, and Mars missions. Large thermal vacuum chambers are needed to perform thermal response testing at or near vacuum or low pressure. As old facilities become obsolete, some may need to be replaced with modern facilities to support the development of new technology.

The ISS is a unique research and test facility that is critical for the development of space technologies. It provides a platform for testing in microgravity and the harsh environment of space (cosmic rays, solar coronal ejecta, micrometeorites, etc.) for long durations. Low-TRL initiatives will develop many technologies that may or may not survive the space environment, and testing in simulated space environments on the ground may not provide credible results. Thus, testing on the ISS is an important step in moving a technology from TRL 3 to TRL 5 or 6. Testing of materials, components, and/or subsystems is mentioned in all but two of the roadmaps (TA01, Launch Propulsion Systems and TA13, Ground and Launch Systems Processing). Examples of level 3 technologies from the roadmaps that would benefit from the testing on the ISS include 2.4.2 Propellant Storage and Transfer; 3.2.1 Energy Storage: Batteries; 4.6.3 Docking and Capture Mechanisms/Interfaces; 5.5.1 Radio Systems; 10.4.1 Sensors and Actuators; 12.1.1 Lightweight Materials and Structures; and 14.3.1 Ascent/Entry TPS. In addition, there are at least four level 3 technologies in TA06 related to space radiation prediction, monitoring, and protection that would benefit from the ability to use the ISS as a testing facility.

Astronauts and machines are inevitably exposed to foreign environments during space exploration. Therefore, there is a continued need for exploration surface environment chambers, consisting of both small and large ground-based facilities that simulate space environments in terms of vacuum, CO_2 dust, and solar radiation (but not reduced gravity). Such facilities are vital to the future development of EVA suits, rovers, and habitats.

There are only a few locations where synergistic effects of reduced gravity and high radiation can be studied on biological and physical systems prior to committing to a 500+ day mission to Mars. A centrifuge in high Earth orbit or on the ISS would enable testing at all gravity levels of interest, from 0 to 1 *g*, but there are no plans to build such facilities, and they would not accommodate human subjects. If NASA human exploration returns to the surface of the Moon, testing on the moon would provide the opportunity to conduct long-term research and testing in 1/6 *g*. Although such data would not be taken in the microgravity environment experienced during a transit to and from Mars or the 3/8 *g* experienced on the Mars surface, this data would provide much needed information that is not available from current testing in the microgravity environment of the ISS or the 1-*g* environment on Earth.

Finding. ***Facilities.*** Adequate research and testing facilities are essential to the timely development of many space technologies. In some cases, critical facilities do not exist or no longer exist, but defining facility requirements and then meeting those requirements fall outside the scope of NASA's Office of the Chief Technologist (and this study).

PROGRAM STABILITY

The productivity and the effectiveness of technology development programs are diminished when the direction, content, and/or funding of those programs abruptly change from year to year. Some redirection of effort based on program progress, results, and new understanding is appropriate, but when substantial changes occur repeatedly and unexpectedly, those changes can be extremely disruptive, especially to university research programs. Reconstituting lost capabilities or recovering from major changes in program direction can take years. Stability is important in the short term to avoid disrupting individual programs and in the long term to ensure that other federal agencies, industry, academia, and foreign organizations recognize NASA as a reliable partner.

Maintaining a stable research and technology development program can be particularly difficult when that program is too closely tied to near-term mission priorities. For example, after the Apollo program, Project Viking and other planetary probes capitalized on the ablative heat shield technologies developed for the Apollo spacecraft. However, in more recent years, the focus has been more on the reusable thermal protection systems used by the space shuttle for return from low Earth orbit. During this era, much momentum was lost in the ablative material development and supply chain, and there is a concern that reusable thermal protection systems (TPS) will suffer the same fate in the coming years. In fact, key materials suppliers are already terminating production, and technology development in this area is faltering (Grantz, 2011).

Disruptions caused by reduced budgets and changing goals of space technology programs within NASA and other federal agencies can cascade from one agency to another. Reduced support by one agency can threaten the viability and the continuation of multiagency efforts. In some cases, the resulting disruptions have led to a loss of experienced technology specialists. These losses impact NASA as well as the national aerospace community (NRC, 2009, 2010). Consequently, the need to restore these capabilities across NASA, industry, and academia and to preserve stability and continuity in a core space technology program has become a national issue.

Program stability has been a long-standing concern to the EDL community. Their concern for maintaining core capabilities, skills, and knowledge raises the issue of the role NASA should play in maintaining knowledge so it is not lost (e.g., losing TPS capabilities after Apollo) between periods of peak demand from major mission programs. Ideally, EDL research projects, technology demonstrations, and interim technology goals in the roadmap would smooth out these peak demands while building on past work to meet future requirements. A successful technology program will preserve test capabilities and advance key technologies at a steady pace that does not depend solely on flight mission initiatives. By ensuring knowledge capture, NASA will not have to relearn lessons from the past. Struggles with Avcoat are a good example of loss of knowledge, experience, and lessons learned. Such an approach is similar to that employed with great success by NACA.

> **Finding. *Program Stability.*** Repeated, unexpected changes in the direction, content, and/or level of effort of technology development programs have diminished their productivity and effectiveness. In the absence of a sustained commitment to address this issue, the pursuit of OCT's mission to advance key technologies at a steady pace will be threatened.

COMMERCIAL SPACE

The draft roadmaps could be improved by explicitly addressing the needs of the commercial space sector. The National Space Policy affirms the importance of commercial space activities, stating that "a robust and competitive commercial space sector is vital to continued progress in space. The United States is committed to encouraging and facilitating the growth of a U.S. commercial space sector"[2] (White House, 2010, p. 3). Further, The National Aeronautics and Space Act declares that "the general welfare of the United States requires that the Administration seek and encourage, to the maximum extent possible, the fullest commercial use of space." In addition, NASA is directed to "encourage and provide for federal government use of commercially provided space services and hardware, consistent with the requirements of the federal government" (P.L. 111-314, sec. 20102). NASA's contribution to accomplishing these important objectives would be enhanced by a technology development program that:

- Identifies how the commercial space sector could benefit from advanced technology.
- Makes appropriate efforts to develop pre-competitive technology relevant to the needs of the commercial space sector, in much the same way that NASA supports pre-competitive technology development in support of the aeronautics industry.
- Transfers advanced technologies to U.S. industry to help satisfy the needs of the commercial space sector as well as NASA's own mission needs.

Meeting these objectives requires a proactive and sustained partnership between NASA and industry that goes beyond treating the private sector as a contractor, which is typically the case when NASA funds industry to achieve NASA goals.

The U.S. aerospace industry has developed and matured as a result of the government's civil and military missions in space. It seems ready to exploit emerging commercial opportunities (beyond traditional services such as commercial communications and imagery), often by selling commercial space products and services where earlier the government would have purchased the space system itself. Promising non-governmental commercial opportunities include orbital human habitats and satellite servicing. Current U.S. space policies are intended to take advantage of the strengths of the United States with its free-market, entrepreneurial business culture. The transition to a more robust commercial space industry would be facilitated if NASA made new and existing research and development data more accessible to U.S. industry (especially industry that is working on its own commercial goals apart from NASA missions). The active collaboration of NASA with industry on precompetitive technologies of interest to industry would also be helpful. It is not up to NASA to predict or select viable commercial endeavors. Industry will initiate relationships with NASA in the technology development areas of interest to them, informed by improved access to archived and ongoing technology program data mentioned, and be prepared to finance their own participation. Such relationships would be confined to pre-competitive technologies.

---

[2] As used in the National Space Policy, the term *commercial* refers to "space goods, services, or activities provided by private sector enterprises that bear a reasonable portion of the investment risk and responsibility for the activity, operate in accordance with typical market-based incentives for controlling cost and optimizing return on investment, and have the legal capacity to offer these goods or services to existing or potential nongovernmental customers" (White House, 2010, p. 10).

It can be difficult for U.S. industry to access some NASA research results, especially for companies not under contract to NASA. Improving this situation requires addressing multiple issues that constrain data transfer to U.S. commercial enterprises. Dissemination of technical data held by NASA to commercial entities is sometimes limited by the International Traffic in Arms Regulations (ITAR) and by intellectual property rights associated with a given research project. These factors are complicated by the multi-national nature of many aerospace firms and their involvement in the space programs of foreign nations. NASA prime contractors typically have good access to NASA's technical data for projects in which they participate, but the impact of NASA technology development would be enhanced by more effectively disseminating technology data—past, present, and future—to companies that are not under contract to NASA. For example, NASA has considerable experimental information about human adaptation to the microgravity environment of LEO and the design requirements for the various life support systems needed to sustain life and human operations in a closed environment. This information would be of particular interest to commercial companies developing crewed systems not only for NASA but also for purely commercial missions. (See for instance Appendix I (TA06), which references robust human factors data going back to the earliest days of human spaceflight.) In addition, new commercial space orbital and suborbital vehicles most likely could take advantage of NASA data on the performance of EDL technologies. Currently, the Life Sciences Data Archive at Johnson Space Center provides a positive example of effective data archiving, sharing, and transparency (see http://lsda.jsc.nasa.gov/lsda_home1.cfm).

In addition, new NASA programs could implement data plans that target specific governmental and commercial markets. A good (non-NASA) example of this practice is found in the National Science Foundation (NSF); the NSF Grant Proposal Guide requires a data management plan which is reviewed as an integral part of every grant proposal submitted (see http://www.nsf.gov/bfa/dias/policy/dmpfaqs.jsp).

Recommendation. ***Industry Access to NASA Data.*** OCT should make the engineering, scientific, and technical data that NASA has acquired from past and present space missions and technology development more readily available to U.S. industry, including companies that do not have an ongoing working relationship with NASA and which are pursuing their own commercial goals apart from NASA's science and exploration missions. To facilitate this process in the future, OCT should propose changes to NASA procedures so that programs are required to archive data in a readily accessible format.

Recommendation. ***NASA Investments in Commercial Space Technology.*** While OCT should focus primarily on developing advanced technologies of high value to NASA's own mission needs, OCT should also collaborate with the U.S. commercial space industry in the development of precompetitive technologies of interest to and sought by the commercial space industry.

CROSSCUTTING TECHNOLOGIES

OCT's draft technology roadmaps identify many crosscutting technologies that have the potential for broad and significant advances. In fact, all but one of the roadmaps (TA09, EDL Systems) has a section on interdependencies with the other roadmaps, and TA09 still addresses many technologies related to other roadmaps. For example, many of the level 3 technologies in the roadmaps for TA10 (Nanotechnology), TA11 (Modeling, Simulation, Information Technology, and Processing), and TA12 (Materials, Structures, Mechanical Systems, and Manufacturing) support technology advances in other technology areas. The current set of draft roadmaps would be improved if they explicitly and systematically addressed two additional crosscutting technologies: avionics and space weather beyond radiation effects.

Space weather refers to the dynamic state of the space environment. It includes space radiation as well as other phenomena, such as solar electromagnetic flux, magnetic fields, charged and neutral components of the solar wind, and energetic particles superimposed on the solar wind from solar and galactic sources. The space environment extends from the Sun throughout the solar system, and it includes the magnetospheres and ionospheres of planets and moons. The space environment changes over time scales ranging from seconds to millennia, but the most common time scales of interest to NASA mission operations range from minutes to hours or days. For mission planning and design the relevant time scales range from days to years or decades.

Space weather affects NASA operations through multiple phenomena, including spacecraft charging and discharging from plasma effects; single event effects (SEEs) in electronics; thermal and material degradation from exposure to ultraviolet radiation and atomic oxygen; communications and navigation disruption from x-rays and geomagnetic storms; and enhanced orbital drag from atmospheric heating. Advanced technologies are needed to improve space situation awareness, to provide dynamic models of the space environment, and to develop innovative approaches for mitigating the varied effects of space weather and to resolve operational failures and anomalies.

Currently, space weather and the space environment beyond radiation seem to be touched upon in just one of the draft roadmaps: TA08, Science Instruments, Observatories, and Sensor Systems (see technology 8.3.1, which comprises sensors for particles, fields, and waves, including charged and neutral particles, magnetic fields, and electric fields).

Avionics systems are critical to the success of a wide range of space vehicle operations. Avionics systems include processors, software, data buses, and other electronic components that assess overall system health. Avionics systems require some share of available vehicle volume, mass, power, and thermal management capacity, and they operate properly throughout the environmental envelope within which a given space vehicle operates. Technologies related to avionics appear in nine of the draft roadmaps, as shown in Table 4.1, though avionics per se as a technology is not specifically mentioned. Because no single roadmap is responsible for presenting a

TABLE 4.1 Existing Level 3 Technologies at Least Partly Applicable to Avionics

Technology Area (TA)	Technology Number	Technology Name	Identified Objectives
01: Launch Propulsion Systems	1.4.5	Health Management and Sensors	Fault management
03: Space Power and Energy Storage	3.3.1	Fault Detection, Isolation, and Recovery (FDIR)	Fault management
04: Robotics, Tele-Robotics, and Autonomous Systems	4.1.6	Multi-Sensor Data Fusion	Processing speed and data throughput
	4.5.1	Vehicle System Management and FDIR	Fault management
	4.7.3	On-Board Computing	Processing speed and data throughput
08: Science Instruments, Observatories, and Sensor Systems	8.1.2	Electronics	– Volume, mass, and power reduction – Integrated capabilities
	8.2.5	Wireless Spacecraft Technologies	Data throughput
09: Entry, Descent, and Landing	9.4.6	Instrumentation and Health Monitoring	Fault management
10: Nanotechnology	10.4.2	Electronics	– Volume, mass, and power reduction – Radiation tolerance
11: Modeling, Simulation, and Information Technology and Processing	11.11.1	Flight Computing	– Radiation tolerance – Fault-tolerant processing
12: Materials, Structures, Mechanical Systems, and Manufacturing	12.3.5	Reliability/Life Assessment/Health Monitoring	Fault management
13: Ground and Launch Systems Processing	13.3.3	Inspection, Anomaly Detection, and Identification	Fault management
	13.3.4	Fault Isolation and Diagnostics	Fault management
	13.4.5	Safety Systems	Fault management

NOTE: Excludes technologies specific to GN&C or scientific sensors.

TABLE 4.2 Space Vehicle Avionics Technology Gaps Across the NASA Roadmaps

Avionics Technology Gap Area	Limits of Draft Roadmaps
Processing Speed and Data Throughput	High-performance computing is mentioned in TA 11 as a flight computing technology, but only based on use of multi-cores as a technical approach. Discussion of data bus technology is absent except for (1) TA08 under Wireless Spacecraft Technologies and (2) wireless and optical networks addressed in TA 13 under Safety Systems technology.
Radiation Tolerance	In TA10, there is passing reference to possible radiation-tolerant benefits of nanomaterial-based electronics. In TA11, only integrated circuits are identified as needing radiation-hard capabilities (as Flight Computing technology).
Reliable, Fault-Tolerant Processing	This area of interest is superficially mentioned in the TA11 under Flight Computing technology where it is addressed solely in the context of multi-core processing.
Fault Management	This area is not consistently identified across the draft roadmaps where these technologies are applicable.
Integrated Capabilities	This area is only superficially mentioned in TA08 under Electronics technology.

comprehensive and coordinated approach for developing avionics technologies, important gaps in avionics technologies remain, as detailed in Table 4.2.

Finding. *Crosscutting Technologies.* Many technologies, such as those related to avionics and space weather beyond radiation effects, cut across many of the existing draft roadmaps, but the level 3 technologies in the draft roadmaps provide an uneven and incomplete list of the technologies needed to address these topics comprehensively.

Recommendation. *Crosscutting Technologies.* OCT should review and, as necessary, expand the sections of each roadmap that address crosscutting level 3 technologies, especially with regard to avionics and space weather beyond radiation effects. OCT should assure effective ownership responsibility for crosscutting technologies in each of the roadmaps where they appear and establish a comprehensive, systematic approach for synergistic, coordinated development of high-priority crosscutting technologies.

REFERENCES

Grantz, A.C., Experimental Systems Group, Boeing Company. 2011. "TA-09 Entry, Descent, and Landing Panel Discussion," presentation at the National Research Council NASA Technology Roadmaps Panel 6 Workshop in Irvine, Calif., on March 23. National Research Council, Washington, D.C.

NASA (National Aeronautics and Space Administration). 2011. *2011 NASA Strategic Plan.* NASA Headquarters, Washington, D.C.

NRC (National Research Council). 2009. *America's Future in Space: Aligning the Civil Space Program with National Needs.* The National Academies Press, Washington, D.C.

NRC. 2010. *Capabilities for the Future: An Assessment of NASA Laboratories for Basic Research.* The National Academies Press, Washington, D.C. Available at http://www.nap.edu/catalog/12903.html.

White House. 2010. *National Space Policy of the United States of America.* White House, Washington, D.C. June 28. Available at http://www.whitehouse.gov/sites/default/files/national_space_policy_6-28-10.pdf.

Appendixes

A

Statement of Task

The NRC will appoint a steering committee and [six] panels to solicit external inputs to and evaluate the 14 draft technology roadmaps that NASA has developed as a point of departure. The steering committee will also provide recommendations that identify and prioritize key technologies. The scope of the technologies to be considered includes those that address the needs of NASA's exploration systems, Earth and space science, and space operations mission areas, as well as those that contribute to critical national and commercial needs in space technology. (This study will not consider aeronautics technologies except to the extent that they are needed to achieve NASA and national needs in space; guidance on the development of core aeronautics technologies is already available in the National Aeronautics Research and Development Plan.)

The steering committee and panels will prepare two reports, as follows:

- The steering committee will establish a set of criteria to enable prioritization of technologies within each and among all of the technology areas that the NASA technology roadmaps should satisfy.
- Each panel will conduct a workshop focused on one or more roadmaps, as assigned, to solicit feedback and commentary from industry, academia, and government on the 14 draft roadmaps provided by NASA at the initiation of the study.
- Based on the results of the community input and its own deliberations, the steering committee will prepare a brief interim report that addresses high-level issues associated with the roadmaps, such as the advisability of modifying the number or technical focus of the draft NASA roadmaps.
- Each panel will meet individually to suggest improvements to the roadmaps in areas such as:
  — The identification of technology gaps,
  — The identification of technologies not covered in the draft roadmaps,
  — Development and schedule changes of the technologies covered,
  — A sense of the value (such as potential to reduce mass and/or volume, number of missions it could support, new science enabled, facility to operate, terrestrial benefit) for key technologies,
  — The risk, or reasonableness, of the technology line items in the NASA technology roadmaps, and
  — The prioritization of the technologies within each roadmap by groups such as high, medium, or low priority; this prioritization should be accomplished, in part, via application of relevant criteria described above in a uniform manner across panels.

- Each panel will prepare a written summary of the above for the steering committee.
- The steering committee will subsequently develop a comprehensive final report that
  - — Summarizes findings and recommendations for each of the 14 roadmaps,
  - — Integrates the outputs from the workshops and panels to identify key common threads and issues, and
  - — Prioritizes, by group, the highest-priority technologies from all 14 roadmaps.

B

Revised Technology Area Breakdown Structure

The revised technology area breakdown structure (TABS) that appears below reflects all of the changes described in the introductions in Appendixes D through Q.[1] The names of level 3 technologies that have been added, moved, or renamed relative to technologies as they appear in the roadmaps appear in bold. To avoid confusion, technologies that have not been changed have the same numerical designation in both the steering committee's revised TABS and the TABS generated by NASA. As a result, the numbering scheme for the level 3 technologies is discontinuous where technologies have been deleted or moved.

TA01 Launch Propulsion Systems

- 1.1 Solid Rocket Propulsion Systems
  - 1.1.1 Propellants
  - 1.1.2 Case Materials
  - 1.1.3 Nozzle Systems
  - 1.1.4 Hybrid Rocket Propulsion Systems
  - 1.1.5 Fundamental Solid Propulsion Technologies
- 1.2 Liquid Rocket Propulsion Systems
  - 1.2.1 LH2/LOX Based
  - 1.2.2 RP/LOX Based
  - 1.2.3 CH4/LOX Based
  - 1.2.4 Detonation Wave Engines (Closed Cycle)
  - 1.2.5 Propellants
  - 1.2.6 Fundamental Liquid Propulsion Technologies
- 1.3 Air Breathing Propulsion Systems
  - 1.3.1 Turbine Based Combined Cycle (TBCC)
  - 1.3.2 Rocket Based Combined Cycle (RBCC)
  - 1.3.3 Detonation Wave Engines (Open Cycle)

[1] The original TABS (for the draft space technology roadmaps) is available at http://www.nasa.gov/offices/oct/strategic_integration/technology_roadmap.html.

- 1.3.4 Turbine Based Jet Engines (Flyback Boosters)
- 1.3.5 Ramjet/Scramjet Engines (Accelerators)
- 1.3.6 Deeply Cooled Air Cycles
- 1.3.7 Air Collection and Enrichment System
- 1.3.8 Fundamental Air Breathing Propulsion Technologies

1.4 Ancillary Propulsion Systems
- 1.4.1 Auxiliary Control Systems
- 1.4.2 Main Propulsion Systems (Excluding Engines)
- 1.4.3 Launch Abort Systems
- 1.4.4 Thrust Vector Control Systems
- 1.4.5 Health Management and Sensors
- 1.4.6 Pyro and Separation Systems
- 1.4.7 Fundamental Ancillary Propulsion Technologies

1.5 Unconventional / Other Propulsion Systems
- 1.5.1 Ground Launch Assist
- 1.5.2 Air Launch / Drop Systems
- 1.5.3 Space Tether Assist
- 1.5.4 Beamed Energy / Energy Addition
- 1.5.5 Nuclear
- 1.5.6 High Energy Density Materials/Propellants

TA02 In-Space Propulsion Technologies

2.1 Chemical Propulsion
- 2.1.1 Liquid Storable
- 2.1.2 Liquid Cryogenic
- 2.1.3 Gels
- 2.1.4 Solid
- 2.1.5 Hybrid
- 2.1.6 Cold Gas/Warm Gas
- 2.1.7 Micro-Propulsion

2.2 Non-Chemical Propulsion
- 2.2.1 Electric Propulsion
- 2.2.2 Solar Sail Propulsion
- 2.2.3 Thermal Propulsion
- 2.2.4 Tether Propulsion

2.3 Advanced (TRL <3) Propulsion Technologies
- 2.3.1 Beamed Energy Propulsion
- 2.3.2 Electric Sail Propulsion
- 2.3.3 Fusion Propulsion
- 2.3.4 High Energy Density Materials
- 2.3.5 Antimatter Propulsion
- 2.3.6 Advanced Fission
- 2.3.7 Breakthrough Propulsion

2.4 Supporting Technologies
- 2.4.2 Propellant Storage and Transfer

TA03 Space Power and Energy Storage

- 3.1 Power Generation
  - 3.1.1 Energy Harvesting
  - 3.1.2 Chemical (Fuel Cells, Heat Engines)
  - 3.1.3 Solar (Photovoltaic and Thermal)
  - 3.1.4 Radioisotope
  - 3.1.5 Fission
  - 3.1.6 Fusion
- 3.2 Energy Storage
  - 3.2.1 Batteries
  - 3.2.2 Flywheels
  - 3.2.3 Regenerative Fuel Cells
  - **3.2.4 Electric and Magnetic Field Storage**
  - **3.2.5 Thermal Storage**
- 3.3 Power Management and Distribution
  - 3.3.1 Fault Detection, Isolation, and Recovery (FDIR)
  - 3.3.2 Management and Control
  - 3.3.3 Distribution and Transmission
  - 3.3.4 Wireless Power Transmission
  - 3.3.5 [Power] Conversion and Regulation
- 3.4 Crosscutting Technology
  - 3.4.1 Analytical Tools
  - 3.4.2 Green Energy Impact
  - 3.4.3 Multi-functional Structures
  - 3.4.4 Alternative Fuels

TA04 Robotics, Tele-Robotics, and Autonomous Systems

- 4.1 Sensing and Perception
  - **4.1.1 Vision**
  - **4.1.2 Tactile Sensing**
  - **4.1.3 Natural Feature Image Recognition**
  - **4.1.4 Localization and Mapping**
  - **4.1.5 Pose Estimation**
  - **4.1.6 Multi-Sensor Data Fusion**
  - **4.1.7 Mobile Feature Tracking and Discrimination**
  - **4.1.8 Terrain Classification and Characterization**
- 4.2 Mobility
  - **4.2.1 Extreme Terrain Mobility**
  - **4.2.2 Below-Surface Mobility**
  - **4.2.3 Above-Surface Mobility**
  - **4.2.4 Small Body/Microgravity Mobility**
- 4.3 Manipulation
  - **4.3.1 Robot Arms**
  - **4.3.2 Dexterous Manipulators**
  - **4.3.3 Modeling of Contact Dynamics**
  - **4.3.4 Mobile Manipulation**
  - **4.3.5 Collaborative Manipulation**
  - **4.3.6 Robotic Drilling and Sample Processing**

4.4 Human-Systems Integration
- **4.4.1 Multi-Modal Human-Systems Interaction**
- **4.4.2 Supervisory Control**
- **4.4.3 Robot-to-Suit Interfaces**
- **4.4.4 Intent Recognition and Reaction**
- **4.4.5 Distributed Collaboration**
- **4.4.6 Common Human-Systems Interfaces**
- **4.4.7 Safety, Trust, and Interfacing of Robotic/Human Proximity Operations**

4.5 Autonomy
- **4.5.1 Vehicle System Management and FDIR**
- **4.5.2 Dynamic Planning and Sequencing Tools**
- **4.5.3 Autonomous Guidance and Control**
- **4.5.4 Multi-Agent Coordination**
- **4.5.5 Adjustable Autonomy**
- **4.5.6 Terrain Relative Navigation**
- **4.5.7 Path and Motion Planning with Uncertainty**

4.6 Autonomous Rendezvous and Docking
- **4.6.1 Relative Navigation Sensors (long, mid, and near range)**
- **4.6.2 Relative Guidance Algorithms**
- **4.6.3 Docking and Capture Mechanisms/Interfaces**

4.7 RTA Systems Engineering
- **4.7.1 Modularity/Commonality**
- **4.7.2 Verification and Validation of Complex Adaptive Systems**
- **4.7.3 Onboard Computing**

TA05 Communication and Navigation

5.1 Optical Communication and Navigation
- 5.1.1 Detector Development
- 5.1.2 Large Apertures
- 5.1.3 Lasers
- 5.1.4 Acquisition and Tracking
- 5.1.5 Atmospheric Mitigation

5.2 Radio Frequency Communications
- 5.2.1 Spectrum Efficient Technologies
- 5.2.2 Power Efficient Technologies
- 5.2.3 Propagation
- 5.2.4 Flight and Ground Systems
- 5.2.5 Earth Launch and Reentry Communication
- 5.2.6 Antennas

5.3 Internetworking
- 5.3.1 Disruptive Tolerant Networking
- 5.3.2 Adaptive Network Topology
- 5.3.3 Information Assurance
- 5.3.4 Integrated Network Management

5.4 Position, Navigation, and Timing
- **5.4.1 Timekeeping and Time Distribution**
- 5.4.3 Onboard Autonomous Navigation and Maneuver

- 5.4.4 Sensors and Vision Processing Systems
- 5.4.5 Relative and Proximity Navigation
- 5.4.6 Auto Precision Formation Flying
- 5.4.7 Auto Approach and Landing

5.5 Integrated Technologies
- 5.5.1 Radio Systems
- 5.5.2 Ultra Wideband
- 5.5.3 Cognitive Networks
- 5.5.4 Science from the Communication System
- 5.5.5 Hybrid Optical Communication and Navigation Sensors
- 5.5.6 RF/Optical Hybrid Technology

5.6 Revolutionary Concepts
- 5.6.1 X-Ray Navigation
- 5.6.2 X-Ray Communications
- 5.6.3 Neutrino-Based Navigation and Tracking
- 5.6.4 Quantum Key Distribution
- 5.6.5 Quantum Communications
- 5.6.6 SQIF Microwave Amplifier
- **5.6.7 Reconfigurable Large Apertures Using Nanosat Constellations**

TA06 Human Health, Life Support, and Habitation Systems

6.1 Environmental Control, Life Support Systems, and Habitation Systems
- 6.1.1 Air Revitalization
- 6.1.2 Water Recovery and Management
- 6.1.3 Waste Management
- 6.1.4 Habitation

6.2 Extravehicular Activity Systems
- 6.2.1 Pressure Garment
- 6.2.2 Portable Life Support System
- 6.2.3 Power, Avionics, and Software

6.3 Human Health and Performance
- 6.3.1 Medical Diagnosis/Prognosis
- 6.3.2 Long-Duration Health
- 6.3.3 Behavioral Health and Performance
- 6.3.4 Human Factors and Performance

6.4 Environmental Monitoring, Safety, and Emergency Response
- 6.4.1 Sensors: Air, Water, Microbial, etc.
- 6.4.2 Fire: Detection, Suppression
- 6.4.3 Protective Clothing/Breathing
- 6.4.4 Remediation

6.5 Radiation
- 6.5.1 Risk Assessment Modeling
- 6.5.2 Radiation Mitigation
- 6.5.3 Protection Systems
- **6.5.4 Radiation Prediction**
- 6.5.5 Monitoring Technology

TA07 Human Exploration Destination Systems

- 7.1 In Situ Resource Utilization
  - 7.1.1 Destination Reconnaissance, Prospecting, and Mapping
  - 7.1.2 Resource Acquisition
  - **7.1.3 ISRU Products/Production**
  - 7.1.4 Manufacturing and Infrastructure Emplacement
- 7.2 Sustainability and Supportability
  - **7.2.1 Autonomous Logistics Management**
  - 7.2.2 Maintenance Systems
  - 7.2.3 Repair Systems
  - **7.2.4 Food Production, Processing, and Preservation**
- 7.3 Advanced Human Mobility Systems
  - 7.3.1 EVA Mobility
  - 7.3.2 Surface Mobility
  - 7.3.3 Off-Surface Mobility
- 7.4 Advanced Habitat Systems
  - 7.4.1 Integrated Habitat Systems
  - 7.4.2 Habitat Evolution
  - **7.4.3 Smart Habitats**
- 7.5 Mission Operations and Safety
  - 7.5.1 Crew Training
  - **7.5.5 Integrated Flight Operations Systems**
  - **7.5.6 Integrated Risk Assessment Tools**
- 7.6 Crosscutting Systems
  - 7.6.2 Construction and Assembly
  - 7.6.3 Dust Prevention and Mitigation

TA08 Science Instruments, Observatories, and Sensor Systems

- 8.1 Remote Sensing Instruments/Sensors
  - 8.1.1 Detectors and Focal Planes
  - 8.1.2 Electronics
  - **8.1.3 Optical Systems**
  - 8.1.4 Microwave/Radio
  - 8.1.5 Lasers
  - 8.1.6 Cryogenic/Thermal
  - **8.1.7 Space Atomic Interferometry**
- 8.2 Observatories
  - 8.2.2 Structures and Antennas
  - 8.2.3 Distributed Aperture
  - **8.2.4 High Contrast Imaging and Spectroscopy Technologies**
  - **8.2.5 Wireless Spacecraft Technologies**
- 8.3 In Situ Instruments/Sensors
  - **8.3.1 Particles, Fields, and Waves: Charged and Neutral Particles, Magnetic and Electric Fields**
  - 8.3.3 In Situ (Instruments and Sensors)
  - **8.3.4 Surface Biology and Chemistry Sensors: Sensors to Detect and Analyze Biotic and Prebiotic Substances**

TA09 Entry, Descent, and Landing Systems

- 9.1 Aeroassist and Atmospheric Entry
  - 9.1.1 Rigid Thermal Protection Systems
  - 9.1.2 Flexible Thermal Protection Systems
  - 9.1.3 Rigid Hypersonic Decelerators
  - 9.1.4 Deployable Hypersonic Decelerators
- 9.2 Descent
  - 9.2.1 Attached Deployable Decelerators
  - 9.2.2 Trailing Deployable Decelerators
  - 9.2.3 Supersonic Retropropulsion
- 9.3 Landing
  - 9.3.1 Touchdown Systems
  - 9.3.2 Egress and Deployment Systems
  - 9.3.3 Propulsion Systems
  - 9.3.5 Small Body Systems
- 9.4 Vehicle Systems Technology
  - 9.4.2 Separation Systems
  - 9.4.3 System Integration and Analyses
  - 9.4.4 Atmosphere and Surface Characterization
  - **9.4.5 EDL Modeling and Simulation**
  - **9.4.6 Instrumentation and Health Monitoring**
  - **9.4.7 GN&C Sensors and Systems**

TA10 Nanotechnology

- 10.1 Engineered Materials and Structures
  - 10.1.1 Lightweight Materials and Structures
  - 10.1.2 Damage Tolerant Systems
  - 10.1.3 Coatings
  - 10.1.4 Adhesives
  - 10.1.5 Thermal Protection and Control
- 10.2 Energy Generation and Storage
  - 10.2.1 Energy Generation
  - 10.2.2 Energy Storage
  - 10.2.3 Energy Distribution
- 10.3 Propulsion
  - 10.3.1 Nanopropellants
  - 10.3.2 Propulsion Systems
  - 10.3.3 In-Space Propulsion
- 10.4 Sensors, Electronics, and Devices
  - 10.4.1 Sensors and Actuators
  - 10.4.2 Electronics
  - 10.4.3 Miniature Instrumentation

TA11 Modeling, Simulation, and Information Technology and Processing

- 11.1 Computing
  - 11.1.1 Flight Computing
  - 11.1.2 Ground Computing

11.2 Modeling
- 11.2.1 Software Modeling and Model-Checking
- 11.2.2 Integrated Hardware and Software Modeling
- 11.2.3 Human-System Performance Modeling
- **11.2.4a Science Modeling and Simulation**
- **11.2.4b Aerospace Engineering Modeling and Simulation**
- 11.2.5 Frameworks, Languages, Tools, and Standards

11.3 Simulation
- 11.3.1 Distributed Simulation
- 11.3.2 Integrated System Lifecycle Simulation
- 11.3.3 Simulation-Based Systems Engineering
- 11.3.4 Simulation-Based Training and Decision Support Systems

11.4 Information Processing
- 11.4.1 Science, Engineering, and Mission Data Lifecycle
- 11.4.2 Intelligent Data Understanding
- 11.4.3 Semantic Technologies
- 11.4.4 Collaborative Science and Engineering
- 11.4.5 Advanced Mission Systems

TA12 Materials, Structures, Mechanical Systems, and Manufacturing

12.1 Materials
- 12.1.1 Lightweight Structure
- 12.1.2 Computational Design
- 12.1.3 Flexible Material Systems
- 12.1.4 Environment
- 12.1.5 Special Materials

12.2 Structures
- 12.2.1 Lightweight Concepts
- 12.2.2 Design and Certification Methods
- 12.2.3 Reliability and Sustainment
- 12.2.4 Test Tools and Methods
- 12.2.5 Innovative, Multifunctional Concepts

12.3 Mechanical Systems
- 12.3.1 Deployables, Docking, and Interfaces
- 12.3.2 Mechanism Life Extension Systems
- 12.3.3 Electro-mechanical, Mechanical, and Micromechanisms
- 12.3.4 Design and Analysis Tools and Methods
- 12.3.5 Reliability/Life Assessment/Health Monitoring
- 12.3.6 Certification Methods

12.4 Manufacturing
- 12.4.1 Manufacturing Processes
- 12.4.2 Intelligent Integrated Manufacturing and Cyber Physical Systems
- 12.4.3 Electronics and Optics Manufacturing Process
- 12.4.4 Sustainable Manufacturing

12.5 Crosscutting
- 12.5.1 Nondestructive Evaluation and Sensors
- 12.5.2 Model-Based Certification and Sustainment Methods
- 12.5.3 Loads and Environments

TA13 Ground and Launch Systems Processing

- 13.1 Technologies to Optimize the Operational Life-Cycle
  - 13.1.1 Storage, Distribution, and Conservation of Fluids
  - 13.1.2 Automated Alignment, Coupling, and Assembly Systems
  - 13.1.3 Autonomous Command and Control for Ground and Integrated Vehicle/Ground Systems
- 13.2 Environmental and Green Technologies
  - 13.2.1 Corrosion Prevention, Detection, and Mitigation
  - 13.2.2 Environmental Remediation and Site Restoration
  - 13.2.3 Preservation of Natural Ecosystems
  - 13.2.4 Alternate Energy Prototypes
- 13.3 Technologies to Increase Reliability and Mission Availability
  - 13.3.1 Advanced Launch Technologies
  - 13.3.2 Environment-Hardened Materials and Structures
  - 13.3.3 Inspection, Anomaly Detection, and Identification
  - 13.3.4 Fault Isolation and Diagnostics
  - 13.3.5 Prognostics Technologies
  - 13.3.6 Repair, Mitigation, and Recovery Technologies
  - 13.3.7 Communications, Networking, Timing, and Telemetry
- 13.4 Technologies to Improve Mission Safety/Mission Risk
  - 13.4.1 Range Tracking, Surveillance, and Flight Safety Technologies
  - 13.4.2 Landing and Recovery Systems and Components
  - 13.4.3 Weather Prediction and Mitigation
  - 13.4.4 Robotics/TeleRobotics
  - 13.4.5 Safety Systems

TA14 Thermal Management Systems

- 14.1 Cryogenic Systems
  - 14.1.1 Passive Thermal Control
  - 14.1.2 Active Thermal Control
  - 14.1.3 Systems Integration
- 14.2 Thermal Control Systems
  - 14.2.1 Heat Acquisition
  - 14.2.2 Heat Transfer
  - 14.2.3 Heat Rejection and Energy Storage
- 14.3 Thermal Protection Systems
  - 14.3.1 Ascent/Entry TPS
  - 14.3.2 Plume Shielding (Convective and Radiative)
  - 14.3.3 Sensor Systems and Measurement Technologies

C

Acronyms

AC	alternating current
AFRL	Air Force Research Laboratory
ALM	autonomous logistics management
ANT	adaptive network topology
AR&D	autonomous rendezvous and docking
ASRAT	Astrophysics Sounding Rocket Assessment Team
ASRG	Advanced Stirling Radioisotope Generator
ATA	Allen Telescope Array
CCD	charge-coupled device
CG	center of gravity
CMOS	complementary metal oxide semiconductor
CNT	carbon nanotube
CPS	cyber-physical system
CTE	coefficient of thermal expansion
DC	direct current
DDM	direct digital manufacturing
DOD	Department of Defense
DOE	Department of Energy
DOF	degrees of freedom
D&T	distribution and transmission
E3	environment, economy, and energy
ECLSS	environmental control and life support system
EDL	entry, descent, and landing
EELV	Evolved Expendable Launch Vehicle
EP	electric propulsion
ETDP	Exploration Technology Development Program
EVA	extravehicular activity

FAST	Fast Access Testbed Spacecraft
FDIR	fault detection, isolation, and recovery
GCR	galactic cosmic radiation
GEO	geosynchronous Earth orbit
GN&C	guidance, navigation, and control
GPU	graphical processor units
GRM	generic reference mission
H_2	hydrogen
HDWL	hybrid Doppler wind lidar
HEFT	Human Exploration Framework Team
IBIS	Integrated Blanket/Interconnect System
ICMSE	integrated computational materials science and engineering
IHPRPT	Integrated High Payoff Rocket Propulsion Technology
IMM	inverted metamorphic
IMU	inertial measurement unit
IR	infrared
ISHM	integrated systems health management
Isp	specific impulse
ISRU	in situ resource utilization
ISS	International Space Station
ITAR	International Traffic in Arms Regulations
JAXA	Japan Aerospace Exploration Agency
JIMO	Jupiter Icy Moons Orbiter
kW	kilowatt
LEO	low Earth orbit
LH2	liquid hydrogen
LILT	low-intensity/low-temperature
LOX	liquid oxygen
LTD	low-temperature detector
M	mach number
M&S	modeling and simulation
MCP	microchannel plate
MEMS	micro-electro-mechanical system
MEO	mid-Earth orbit
MLI	multi-layer insulation
MMOD	micrometeoroid and orbital debris
NACA	National Advisory Committee for Aeronautics
NASA	National Aeronautics and Space Administration
NDE	non-destructive evaluation
NEA	near-Earth asteroid
NEO	near-Earth object
NERVA	Nuclear Engine for Rocket Vehicle Application

NIAC	NASA Innovative Advanced Concepts (program); NASA Institute for Advanced Concepts
NIST	National Institute of Standards and Technology
NNI	National Nanotechnology Initiative
NRC	National Research Council
NSF	National Science Foundation
NTR	nuclear thermal rocket
OANM	onboard autonomous navigation and maneuvering
OCT	Office of the Chief Technologist
OSR	Orbital Sounding Rocket
PLSS	personal life support system
PNT	position, navigation, and timing
PV	photovoltaic
QE	quantum efficiency
QFD	quality function deployment
R&D	research and development
RBCC	rocket-based combined cycle
RDSP	robotic drilling and sampling processing
RF	radio frequency
RFID	radio frequency identification
RLCAS	routine low cost access to space
ROIC	readout integrated circuit
RP	rocket propellant (hydrocarbon-based)
RPS	radioisotope power system(s)
RTG	radioisotope thermoelectric generator
SDR	software defined radio
SEE	single event effects
SEP	solar electric propulsion
SPE	solar particle event
SWAP	size, weight, and power
SWNT	single wall nanotube
TA	technology area
TA01	Launch Propulsion Systems Technology Area
TA02	In-Space Propulsion Technologies Technology Area
TA03	Space Power and Energy Storage Technology Area
TA04	Robotics, TeleRobotics, and Autonomous Systems Technology Area
TA05	Communication and Navigation Technology Area
TA06	Human Health, Life Support, and Habitation Systems Technology Area
TA07	Human Exploration Destination Systems Technology Area
TA08	Science Instruments, Observatories, and Sensor Systems Technology Area
TA09	Entry, Descent, and Landing Systems Technology Area
TA10	Nanotechnology Technology Area
TA11	Modeling, Simulation, and Information Technology and Processing Technology Area
TA12	Materials, Structures, Mechanical Systems, and Manufacturing Technology Area

TA13	Ground and Launch Systems Processing Technology Area
TA14	Thermal Management Systems Technology Area
TABS	technology area breakdown structure
TBCC	turbine-based combined cycle
TDRSS	Tracking and Data Relay Satellite System
TPS	thermal protection system(s)
TRL	technology readiness level
USAF	U.S. Air Force
UV	ultraviolet
V&V	verification and validation
VASIMR	Variable Specific Impulse Magnetoplasma Rocket
VDFL	virtual digital flight leader
VSM	vehicle systems management
ZBO	zero boil-off

D

TA01 Launch Propulsion Systems

INTRODUCTION

The draft roadmap for technology area (TA) 01, Launch Propulsion Systems, consists of five level 2 technology subareas:[1]

- 1.1 Solid Rocket Propulsion Systems
- 1.2 Liquid Rocket Propulsion Systems
- 1.3 Air Breathing Propulsion Systems
- 1.4 Ancillary Propulsion Systems
- 1.5 Unconventional/Other Propulsion Systems

TA01 includes all propulsion technologies required to deliver space missions from the surface of Earth to Earth orbit or Earth escape. The Earth to orbit launch industry includes mature technologies, proven designs, and well-established companies, as well as innovative technologies and designs and some relatively new companies. For launch propulsion, in particular, the fundamental technologies are based on chemical propulsion and are decades old. Only small incremental improvements are possible in these technology areas. Breakthrough or game changing technologies in launch are not on the near-term horizon, although many ideas exist and were included in the roadmap.

The challenge for the panel was to prioritize these technologies in light of 50 years of spaceflight development experience, the current status of all the technologies, an assessment of the likely benefits that would result from successfully developing new technology, and a general understanding of NASA's mission objectives. The main challenge in launch is cost, measured by the cost per kilogram to low Earth orbit (LEO).

Prior to prioritizing the level 3 technologies included in TA01, the panel considered whether to rename, delete, or move technologies in the technology area breakdown structure (TABS). No changes were recommended for TA01. The TABS for TA01 is shown in Table D.1, and the complete, revised TABS for all 14 TAs is shown in Appendix B.

[1] The draft space technology roadmaps are available at http://www.nasa.gov/offices/oct/strategic_integration/technology_roadmap.html.

TABLE D.1 Technology Area Breakdown Structure for TA01, Launch Propulsion Systems

NASA Draft Roadmap (Revision 10)	Steering Committee-Recommended Changes
TA01 Launch Propulsion Systems	The structure of this roadmap remains unchanged.
1.1. Solid Rocket Propulsion Systems	
1.1.1. Propellants	
1.1.2. Case Materials	
1.1.3. Nozzle Systems	
1.1.4. Hybrid Rocket Propulsion Systems	
1.1.5. Fundamental Solid Propulsion Technologies	
1.2. Liquid Rocket Propulsion Systems	
1.2.1. LH2/LOX Based	
1.2.2. RP/LOX Based	
1.2.3. CH4/LOX Based	
1.2.4. Detonation Wave Engines (Closed Cycle)	
1.2.5. Propellants	
1.2.6. Fundamental Liquid Propulsion Technologies	
1.3. Air Breathing Propulsion Systems	
1.3.1. Turbine Based Combined Cycle (TBCC)	
1.3.2. Rocket Based Combined Cycle RBCC)	
1.3.3. Detonation Wave Engines (Open Cycle)	
1.3.4. Turbine Based Jet Engines (Flyback Boosters)	
1.3.5. Ramjet/Scramjet Engines (Accelerators)	
1.3.6. Deeply Cooled Air Cycles	
1.3.7. Air Collection & Enrichment System	
1.3.8. Fundamental Air Breathing Propulsion Technologies	
1.4. Ancillary Propulsion Systems	
1.4.1. Auxiliary Control Systems	
1.4.2. Main Propulsion Systems (Excluding Engines)	
1.4.3. Launch Abort Systems	
1.4.4. Thrust Vector Control Systems	
1.4.5. Health Management & Sensors	
1.4.6. Pyro & Separation Systems	
1.4.7. Fundamental Ancillary Propulsion Technologies	
1.5. Unconventional / Other Propulsion Systems	
1.5.1. Ground Launch Assist	
1.5.2. Air Launch / Drop Systems	
1.5.3. Space Tether Assist	
1.5.4. Beamed Energy / Energy Addition	
1.5.5. Nuclear	
1.5.6. High Energy Density Materials/Propellants	

TOP TECHNICAL CHALLENGES

The panel has identified two top technical challenges for launch propulsion, listed below in priority order.

1. Reduced Cost: Develop propulsion technologies that have the potential to dramatically reduce the total cost and to increase reliability and safety of access to space.

One major barrier to any space mission is the high cost of access to space. In spite of billions of dollars in investment over the last several decades, the cost of launch has not decreased. In fact, with the end of the Space Shuttle Program and uncertainty in the future direction in human spaceflight, launch costs for NASA science missions are actually increasing. This is because without the space shuttle or a human spaceflight program, the propulsion industrial base is at significant overcapacity. The resulting high costs limit both the number and scope of NASA's space missions. Finding technologies that dramatically reduce launch cost is a tremendous challenge given the past lack of success.

Reliability and safety continue to be major concerns in the launch business. For NASA space missions, the cost of failure is extreme. Finding ways to improve reliability and safety without dramatically increasing cost is a major technology challenge.

2. Upper Stage Engines: Develop technologies to enable lower cost, high specific impulse upper stage engines suitable for NASA, DOD, and commercial needs, applicable to both Earth-to-orbit and in-space applications.

The venerable RL-10 engine is the current upper stage engine for both the Atlas V and Delta IV launch vehicles, but it is based on 50-year-old technology, and it has become expensive and difficult to produce. There are alternative engine cycles and designs that have the promise to reduce cost and improve reliability, and the opportunity exists for a joint NASA-Air Force technology development effort. Also, as discussed below, high-rate production can substantially reduce unit costs. To maximize production rates, new technologies should be applicable to both upper stage and in-space applications.

QFD MATRIX AND NUMERICAL RESULTS FOR TA01

The results of the panel's quality function deployment (QFD) scoring for the level 3 launch propulsion technologies are shown in Figures D.1 and D.2. Two technologies were assessed to be high priority based on their QFD scores:

- Air Breathing Propulsion Systems: Rocket Based Combined Cycle (RBCC)
- Air Breathing Propulsion Systems: Turbine Based Combined Cycle (TBCC)

These technologies, which received identical QFD scores, both burn oxygen extracted from the atmosphere (during the atmospheric portion of flight) giving some promise for increased efficiency and reduced cost. As discussed below, however, the greatest potential to reduce launch cost actually comes from high-priority technologies in other roadmaps.

Two medium-priority technologies deserve some mention. RP/LOX propulsion offers potential benefit for booster stages for all NASA space missions. However, this technology is already at a very mature state of development and application in Russia, and it is available commercially through products such as the RD-180 and AJ-26 engines. Therefore any decision for NASA to invest in this technology should primarily be made for programmatic and political reasons (e.g., the desire to create a domestic production capability), not technological reasons. These non-technological reasons could be important, even compelling, but the priorities in this report are based on technical—not political—considerations.

LH2/LOX propulsion is used for both upper stage and in-space applications. This basic technology area appears here in TA01 and in TA02, In-Space Propulsion (see technology 2.1.2, Liquid Cryogenic). LH2/LOX propulsion scored medium in both of these TAs, though it might have ranked higher if these two areas had been ranked together.

CHALLENGES VERSUS TECHNOLOGIES

A matrix showing the linkage between technology rankings and top technical challenges is shown in Figure D.3. The highest ranked launch propulsion technologies are strongly correlated to the first technology challenge. The various air breathing technologies offer some prospects for reducing the cost of launch, but the correlation with launch propulsion technologies is diluted by the fact that these breakthrough technologies are somewhat speculative. The launch industry has searched for a breakthrough to lower launch costs for decades and, unfortunately, it has yet to materialize. The greatest potential for reduction in launch costs may reside in technologies included in other roadmaps, as discussed below.

There is a very strong correlation between the second challenge and the third ranked technology area.

	Benefit	Alignment with NASA Needs	Alignment with Non-NASA Aero Tech Needs	Alignment with Non-Aerospace National Goals	Technical Risk and Reasonableness	Sequencing and Timing	Time and Effort	QFD Score (Weighted)	Panel Priority
Multiplier	27	5	2	2	10	4	4		
	0/1/3/9	0/1/3/9	0/1/3/9	0/1/3/9	1/3/9	-9/-3/-1/1	-9/-3/-1/0		
Technology Name	Benefit	Alignment			Risk/Difficulty				
1.1.1. (Solid Rocket) Propellants	1	3	3	0	3	-1	-1	70	L
1.1.2. (Solid Rocket) Case Materials	1	3	3	1	3	-1	-1	72	L
1.1.3. (Solid Rocket) Nozzle Systems	1	3	3	0	3	-3	-1	62	L
1.1.4. Hybrid Rocket Propulsion Systems	1	3	3	0	3	-3	-3	54	L
1.1.5. Fundamental Solid Propulsion Technologies	1	9	3	0	3	-3	-1	92	M
1.2.1. LH2/LOX Based	1	9	9	0	3	1	-3	112	M
1.2.2. RP/LOX Based	1	9	9	0	3	1	-3	112	M
1.2.3. CH4/LOX Based	1	3	3	0	3	-3	-3	54	L
1.2.4. Detonation Wave Engines (Closed Cycle)	1	3	3	0	3	-3	-3	54	L
1.2.5. (Liquid Rocket) Propellants	1	9	3	1	3	-3	-1	94	M
1.2.6. Fundamental Liquid Propulsion Technologies	1	9	3	1	3	-3	-1	94	M
1.3.1. TBCC	3	9	9	0	3	-3	-3	150	H
1.3.2. RBCC	3	9	9	0	3	-3	-3	150	H
1.3.3. Detonation Wave Engines (Open Cycle)	1	3	3	0	3	-3	-3	54	L
1.3.4. Turbine Based Jet Engines (Flyback Boosters)	1	3	1	0	3	-3	-3	50	L
1.3.5. Ramjet/Scramjet Engines (Accelerators)	1	0	3	0	3	-3	-3	39	L
1.3.6. Deeply Cooled Air Cycles	1	3	3	0	3	-3	-1	62	L
1.3.7. Air Collection and Enrichment System	1	3	1	0	3	-3	-1	58	L
1.3.8. Fundamental Air Breathing Propulsion Technologies	1	3	3	1	3	-1	-3	64	L
1.4.1. Auxiliary Control Systems	1	9	3	0	3	-1	-1	100	M
1.4.2. Main Propulsion Systems (Excluding Engines)	1	9	3	0	3	-1	-1	100	M
1.4.3. Launch Abort Systems	3	3	1	0	3	-1	-3	112	M
1.4.4. Thrust Vector Control Systems	1	9	3	0	3	-1	-1	100	M
1.4.5. Health Management & Sensors	1	9	3	1	3	-1	-1	102	M
1.4.6. Pyro and Separation Systems	1	9	3	0	3	-1	-1	100	M
1.4.7. Fundamental Ancillary Propulsion Technologies	1	9	3	0	3	-3	-1	92	M
1.5.1. Ground Launch Assist	1	3	3	1	3	-3	-3	56	L
1.5.2. Air Launch / Drop Systems	1	3	3	0	3	-3	-3	54	L
1.5.3. Space Tether Assist (for launch)	0	3	1	0	1	-3	-3	3	L
1.5.4. Beamed Energy / Energy Addition	1	3	1	1	1	-3	-3	32	L
1.5.5. Nuclear (Launch Engines)	0	0	0	0	1	-3	-9	-38	L
1.5.6. High Energy Density Materials/ Propellants	1	3	3	1	1	-3	-1	44	L

FIGURE D.1 Quality function deployment (QFD) summary matrix for TA01 Launch Propulsion Systems. The justification for the high-priority designation of all high-priority technologies appears in the section "High-Priority Level 3 Technologies." H = High Priority; M = Medium Priority; L = Low Priority.

Some of the medium-ranked technologies and all of the low-ranked technologies are judged to have a weak linkage because of the limited benefit of investing in these technologies regardless of how closely they may overlap with various challenges in terms of subject matter.

HIGH-PRIORITY LEVEL 3 TECHNOLOGIES

Panel 1 identified two high-priority technologies in TA01: Turbine Based Combined Cycle (TBCC) and Rocket Based Combined Cycle (RBCC). The justification for ranking each of these technologies as a high priority is discussed below. TBCC and RBCC would benefit other users, such as the DOD, which also has the ability to advance these technologies. However, they are ranked as a high priority for NASA because they would provide a

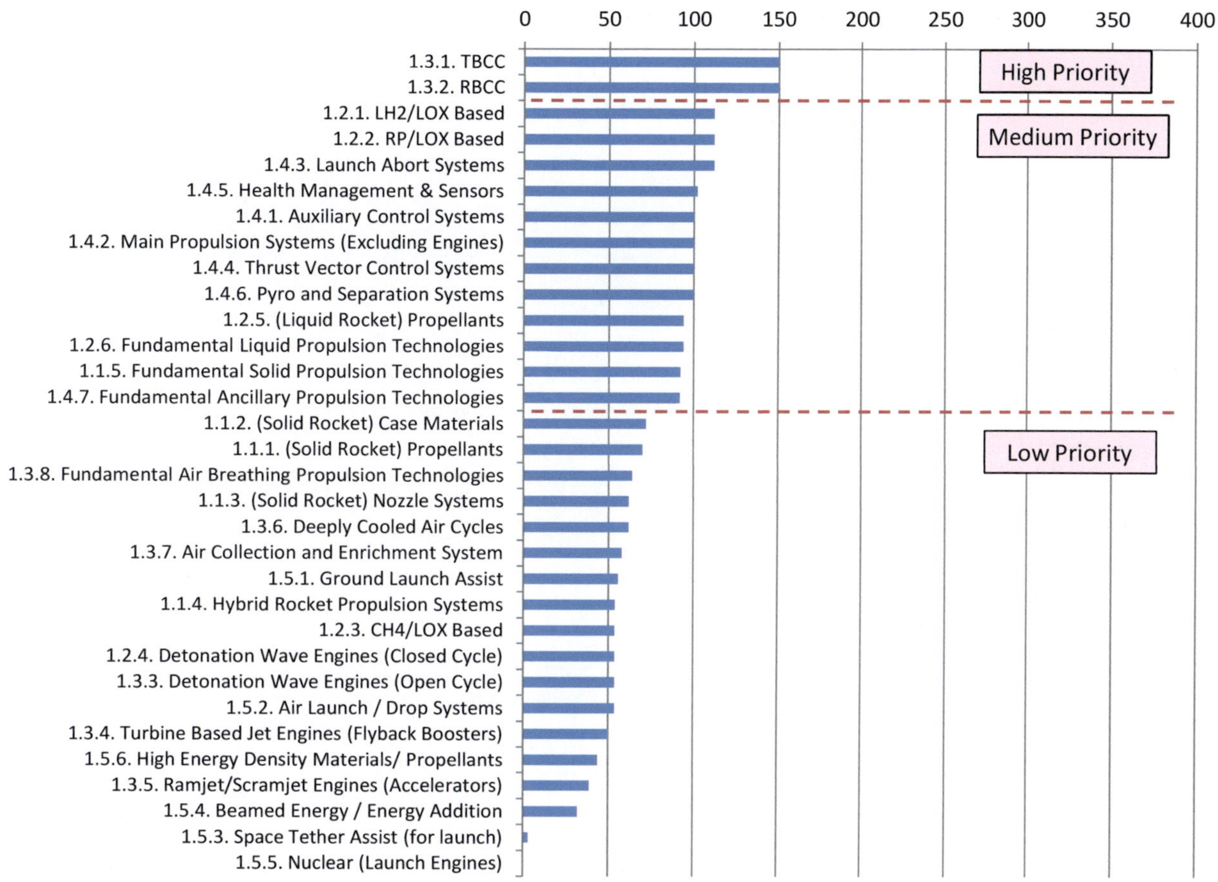

FIGURE D.2 Quality function deployment rankings for TA01 Launch Propulsion Systems.

large benefit to NASA and because they are a good match with NASA's mission and expertise. In fact, the current state of the art in TBCC and RBCC technology has benefited from past research supported by NASA's aeronautics research and technology program.

The International Space Station is not an appropriate test bed for any launch propulsion technologies.

Technology 1.3.2, Rocket Based Combined Cycle

Rocket Based Combined Cycle (RBCC) propulsion systems combine the high specific impulse of air breathing ramjet and scramjet engines with the high thrust-to-weight ratio of a chemical rocket. They promise to deliver launch systems with much lower costs than present launch systems. A vehicle using an RBCC propels itself from the ground using a rocket with secondary air to increase its thrust (ejector ramjet). At high enough Mach numbers (M ~ 3) for ramjet operation, the rockets turn off and air breathing propulsion is used. The ramjet mode transitions to scramjet mode at even higher Mach numbers. After the altitude is high enough to make scramjet operation impractical due to lack of oxygen, the vehicle reverts to a pure rocket mode. This type of propulsion system usually has a single flow path for the entire operating range. RBCC system components are at TRL 3 to 4.

NASA has been investigating rocket-air breathing cycles for many years and has been at the helm of experimental and numerical studies. The experimental X-43 program exemplifies NASA's commitment to and expertise in hypersonic air breathing cycles. There is also considerable Air Force expertise in air breathing hypersonic flight as demonstrated by the recent X-51 flight of a hydrocarbon scramjet. Based on the common need within NASA

Priority	TA01 Technologies, Listed by Priority	Top Technology Challenges: 1. Reduced Cost: Develop propulsion technologies that have the potential to dramatically reduce the total cost and to increase the reliability and safety of access to space.	Top Technology Challenges: 2. Upper Stage Engines: Develop technologies to enable lower cost, high specific impulse upper stage engines suitable for NASA, DOD, and commercial needs, applicable to both Earth-to-orbit and in-space applications.
H	1.3.1. TBCC	●	
H	1.3.2. RBCC	●	
M	1.2.1. LH2/LOX Based	○	●
M	1.2.2. RP/LOX Based	○	
M	1.4.3. Launch Abort Systems	○	
M	1.4.5. Health Management and Sensors	○	○
M	1.4.1. Auxiliary Control Systems		○
M	1.4.2. Main Propulsion Systems (Excluding Engines)		
M	1.4.4. Thrust Vector Control Systems		
M	1.4.6. Pyro and Separation Systems		
M	1.2.5. (Liquid Rocket) Propellants		
M	1.2.6. Fundamental Liquid Propulsion Technologies		
M	1.1.5. Fundamental Solid Propulsion Technologies		
M	1.4.7. Fundamental Ancillary Propulsion Technologies		
L	1.1.2. (Solid Rocket) Case Materials		
L	1.1.1. (Solid Rocket) Propellants		
L	1.3.8. Fundamental Air Breathing Propulsion Technologies		
L	1.1.3 (Solid Rocket) Nozzle Systems		
L	1.3.6. Deeply Cooled Air Cycles		
L	1.3.7. Air Collection and Enrichment System		
L	1.5.1. Ground Launch Assist		
L	1.1.4. Hybrid Rocket Propulsion Systems		
L	1.2.3. CH4/LOX Based		
L	1.2.4. Detonation Wave Engines (Closed Cycle)		
L	1.3.3. Detonation Wave Engines (Open Cycle)		
L	1.5.2. Air Launch / Drop Systems		
L	1.3.4. Turbine Based Jet Engines (Flyback Boosters)		
L	1.5.6. High Energy Density Materials/Propellants		
L	1.3.5. Ramjet/Scramjet Engines (Accelerators)		
L	1.5.4. Beamed Energy / Energy Addition		
L	1.5.3. Space Tether Assist (for launch)		
L	1.5.5. Nuclear (Launch Engines)		

●	Strong Linkage: Investments by NASA in this technology would likely have a major impact in addressing this challenge.
○	Moderate Linkage: Investments by NASA in this technology would likely have a moderate impact in addressing this challenge.
[blank]	Weak/No Linkage: Investments by NASA in this technology would likely have little or no impact in addressing the challenge.

FIGURE D.3 Level of support that the technologies provide to the top technical challenges for TA01 Launch Propulsion Systems.

and DOD for lower launch costs, it would be appropriate for NASA to embark on a joint RBCC development effort with DOD.

RBCC technology is potentially game changing because it could enable revolutionary new launch systems that could be used for a broad spectrum of missions. The performance of RBCC engines is projected to be higher than that of separate rocket and ramjet/scramjet systems, with an average specific impulse at least twice that of a rocket (Bulman and Siebenhaar, 2011; Hampsten and Hickman, 2010). RBCCs have also been considered as part of reusable launch systems and as candidates for operationally flexible and cost-effective launch systems for the Air Force (Hampsten and Hickman, 2010). A reusable booster combined with a reusable RBCC orbiter is projected to offer significant launch cost savings (Hampsten and Hickman, 2010). Compared with a Turbine Based Combined Cycle (TBCC) systems, an RBCC system would be lighter due to the lack of turbine engines and additional ducting (Bulman and Siebenhaar, 2011). However, with state-of-the-art technology, an RBCC system would be heavier than traditional rockets. This is a key design trade that technology development should address.

Some of the challenges associated with RBCCs include high-temperature materials, thermal management, airframe integration, the air-breathing engines, nozzle design, ejector-ramjet optimization, and the smooth transition between modes. The panel believes it will take decades of research and development and a large and sustained financial investment to make this technology feasible.

Technology 1.3.1, Turbine Based Combined Cycle

Turbine Based Combined Cycle (TBCC) propulsion systems have the potential to combine the advantages of gas turbines and rockets in order to enable lower launch costs and more responsive operations. A TBCC-equipped vehicle, which could be configured as a two-stage reusable vehicle to improve payload capacity while reducing life cycle costs, would propel itself using a gas turbine engine. At high enough Mach numbers (M ~ 3) the engine would shift modes and operate as a ramjet. The engine would then transition to a scramjet mode at even higher Mach numbers. The vehicle would then transition to a pure rocket mode when high altitude makes scramjet operation impractical due to lack of oxygen. For most TBCC concepts, the turbine engines are mounted in separate ducts to protect them from damage during hypersonic flight conditions (Bulman and Siebenhaar, 2011). TBCC system components are at TRL 3 to 4.

As noted in the discussion of RBCC technology, above, NASA and the U.S. Air Force have been investigating rocket-air breathing cycles for many years, and it would be appropriate for NASA to embark on a joint TBCC development effort with the DOD.

TBCC technology is potentially game changing because it could enable revolutionary new launch systems that could be used for a broad spectrum of missions. Because of the air-breathing operation from take-off to scramjet, TBCCs offer loiter, fly-out, and abort capabilities (Eklund et al., 2005). Also, they provide horizontal take-off and powered landing. If hydrocarbon fuels are used for all propulsion modes, then the turnaround times and launch responsiveness could resemble that of aircraft (i.e., the launch turnaround time could be hours instead of days or weeks) (Bulman and Siebenhaar, 2011; Eklund et al., 2005). TBCCs have been considered as candidates for operationally flexible and cost-effective launch systems by the Air Force (Eklund et al., 2005).

Some of the challenges associated with TBCCs include high-temperature materials, thermal management, airframe integration, high-speed air-breathing engines, and the smooth transition between propulsion modes. TBCCs may have poor transonic acceleration, and so rockets might be needed for additional thrust (Bulman and Siebenhaar, 2011). Although gas turbines have very high specific impulse, they are heavy, and the overall system weight could be heavier than conventional launch vehicles (Bulman and Siebenhaar, 2011; Hampsten and Hickman, 2010). TBCCs are expected to be heavier than RBCCs due to the use of turbine engines and the need for additional ducting. The committee believes it will take decades of research and development and a sustained and large financial investment to make this technology feasible.

MEDIUM- AND LOW-PRIORITY TECHNOLOGIES

The assessment of the TA01 roadmap technologies identified 30 level 3 technologies as medium or low priority. Two medium-ranked technologies (RP/LOX and LH2/LOX) are so widely used that they are particularly important to the overall launch industry and future NASA programs and missions.

RP/LOX Based Propulsion

RP/LOX based propulsion systems are a good choice for main propulsion stages of expendable launch vehicles. The combination of high-density fuel, allowing for smaller volume tanks, and high thrust and reasonably high Isp are all desirable attributes for booster stages. The technology for RP/LOX main engines is quite mature and many RP/LOX engines are employed in expendable launch vehicles around the world. These include the RD-170 used in the Russian Zenit rocket, the RD-180 which powers the U.S. Atlas V vehicle, and the AJ-26 (formally the Russian NK-33) which will be the booster engine for the U.S. Taurus II vehicle. Thrust from these engines ranges from approximately 400,000 lb for the AJ-26 to 1,500,000 lb for the RD-170. Unfortunately the nexus of this technology resides within Russia. The high-performance engines described above use staged combustion, a process that can produce very high combustion chamber pressures, which results in high specific impulse. Staged compression, however, requires specialized materials, coatings, and combustion chamber design for engine parts to resist these high temperatures and pressures. Nozzle designs also need to be carefully considered to ensure proper propellant-oxidizer mixing and to prevent coking.

The technology for staged combustion RP/LOX engines can be imported from Russia or developed independently within the United States. Significant progress on each approach has been made over the last decade. U.S. companies Pratt & Whitney Rocketdyne and Aerojet have made progress in being able to establish a U.S. production capability for the RD-180 and AJ-26 engines, respectively. AFRL and NASA have both invested significant funds in establishing an independent U.S. RP/LOX technology base.

If a U.S. capability to produce RP/LOX main engines is deemed necessary, a national strategy should be developed that considers the interests of NASA, DOD, and industry. For example, the U.S. Air Force and NASA could jointly invest in the development of a modular family of RP/LOX engines to meet a wide range of mission requirements (medium lift through super heavylift) in partnership with a team from the U.S. propulsion and launch industry. The cost of such an endeavor is likely to be on the order of $1 billion to $3 billion.

Because RP/LOX technology is applicability to such a wide range of missions, this technology received the highest possible score for both NASA mission needs and non-NASA aerospace needs. However, technology investment in this technology would provide little additional benefit in terms of launch vehicle performance given that the technology is available commercially. U.S. capabilities are at TRL 4 to 5, but Russian technology is at TRL 9.

LH2/LOX Based Propulsion

LH2/LOX based propulsion systems are especially useful for launch system upper stages and in-space stages where thrust and volume are less important but high specific impulse is critical. The technology for LH2/LOX engines is quite mature. The 25,000 lb thrust RL-10 engine has been used for decades in many different variants to power virtually every NASA mission beyond Earth orbit. It is also the workhorse for DOD launches. The RL-10, however, has become increasingly expensive and difficult to produce. NASA is developing the J-2X engine with roughly 250,000 lb of thrust. This engine is appropriate for very large upper stages but is too big and heavy for in-space applications.

A low-cost, producible engine is needed to replace the RL-10 for upper stages, to power an in-space cryogenic propulsive stages for exploration missions, and for other in-space applications. Several options exist, including turbine-based, piston-pump-based, and staged-combustion-based configurations.

As with RP/LOX, if the decision is made to develop a new LH2/LOX engine, it may be prudent for NASA to partner with DOD and industry. One key to achieving low cost is high production rate, so a new engine should be designed to meet the needs of as many launch users as possible (in the upper stage configuration) and the maximum number of in-space applications.

LH2/LOX received the highest possible score in terms of NASA needs because it is applicable to nearly every NASA mission. Currently the U.S. Air Force and industry are investing in this technology for upper stage applications. For those applications, additional NASA technology investment would have little impact in terms of overall cost and performance of the launch system. However, for in-space applications, there are unique requirements that may not be addressed without a NASA technology investment.

Other Medium- and Low-Priority Technologies

The panel assessed 12 technologies in TA01 as medium priority and 18 as low priority. Two of the low-priority technologies, nuclear propulsion and tethers, were deemed non-credible for launch propulsion applications. One medium-priority technology, launch abort systems, has the potential for a major improvement in mission performance. All of the other medium- and low-priority technologies were determined to have the potential for only a minor improvement in mission performance, life cycle cost, or reliability.

The major discriminator between medium- and low-priority technologies in TA01 was alignment with NASA needs. With one exception (launch abort systems), all the medium-priority technologies scored higher in this area than all of the low-priority technologies.

DEVELOPMENT AND SCHEDULE CHANGES FOR THE TECHNOLOGIES COVERED BY THE ROADMAP

The development timeline for launch propulsion technologies will be critically dependent on the overall strategy and architecture chosen for exploration, and the funding available. Until these factors are known, it makes little sense to define a timeline.

OTHER GENERAL COMMENTS ON THE ROADMAP

The economics of an operational launch system are described by the following equation:

$$\$/\text{kg} = ((\text{fixed cost}) + N * (\text{variable cost})) / (N * (\text{kg/launch})),$$

where fixed cost = annual cost of the fixed infrastructure and critical skill base, variable cost = cost to build and launch one unit, N = launch rate (number of launches per year), and kg/launch = payload mass delivered by one launch.

The fixed cost for a launch vehicle program is typically very high. Rockets for orbital launches are large, complex objects and require large factories, large and specialized transportation and handling equipment, and extensive infrastructure at the launch site. For example, the fixed cost of the Space Shuttle Program was historically \$3 billion to \$4 billion per year. The fixed cost of the Evolved Expendable Launch Vehicle (EELV) program exceeds \$1 billion per year. Both the fixed cost and the variable cost are non-linear increasing functions of the size of the rocket. In general, the fixed cost is many times the variable cost of a single launch.

Given the fundamentals of launch economics, it is clear that one way to significantly reduce launch cost per kilogram is to increase N, the launch rate. The launch rate is largely determined by the market demand, but for complex missions that require very large payload mass, there is an architecture choice between one large launch carrying all the payload mass and two or more launches, each of which delivers a smaller payload mass. All else being equal, the economics of launch would prefer the latter option. Of course, launch economics is only one consideration, albeit a very important one. This consideration has to be balanced with the difficulty and complexity of breaking payloads into smaller pieces and the logistics of multiple launches and assembly in space.

Some of the technologies in other TAs, especially TA02 (in-space propulsion) and TA04 (robotics) could open the trade space to architecture options that use smaller rockets to increase launch rate. For example, many of NASA's most challenging space missions require large quantities of propellant be delivered to LEO. Technologies that enable the storage and transfer of propellants (especially cryogenic LOX and LH2) would allow propellants to be launched in smaller quantities. These technologies could be more effective in reducing launch costs than specific launch vehicle technologies. In fact, one could imagine a commodity market being established for propellant launches to LEO, where market forces come to bear to reduce cost.

PUBLIC WORKSHOP SUMMARY

The workshop for the Launch Propulsion Systems technology area was conducted by the Propulsion and Power Panel on March 23, 2011, on the campus of the California Institute of Technology in Pasadena, California. The discussion was led by panel member George Sowers, who started the day by giving a general overview of the roadmaps and the NRC's task to evaluate them. He also provided some direction for what topics the invited speakers should cover in their presentations. Experts from industry, academia, and government were invited to lead a 25 minute presentation and discussion of their perspective on the draft NASA roadmap for TA01. At the end of the session, there was a short open discussion by the workshop attendees that focused on the recent session. At the end of the day, there was a concluding discussion led by Sowers summarizing the key points observed during the day's discussion.

Session 1: Academia

Bill Anderson (Purdue University) started the session with academia by emphasizing the need for the NASA roadmap to reduce the number of options and focus on a few of the most promising options. He suggested that an objective, rigorous, and transparent study of launch missions and requirements is necessary to determine the proper focus. At the present time when there is no clear and compelling mission, he urged NASA to systematically investigate foundational engineering challenges such as variable-fidelity modeling of advanced and new propulsion systems and their components, whereas incremental development and implementation of heritage launch systems should be left to industry. He also discussed the need to maintain a skilled workforce, and NASA's important role of inspiring and developing new scientists and engineers by identifying and providing new and challenging problems, including actual flight.

Bob Santoro (Penn State) noted that most NASA personnel who worked on the development of earlier generations of launch propulsion systems have or soon will retire. He stated that the biggest factor in lowering the cost of launch is increasing the flight rate. (This point was made throughout the workshop by multiple presenters.) He suggested that in the near term, the most promising launch propulsion technology is a hydrocarbon-based liquid engine. Over the long term, he said NASA should invest in technologies to support a two-stage, air-breathing, combined cycle launch vehicle. He believes at the moment there is no need to down-select between TBCC and RBCC systems because of their many commonalities. He also thought that it might be beneficial to invest in pulse detonation engines because of their game-changing potential. Finally he observed that the current roadmap is too broad and needs focus.

Bill Saylor (Air Force Academy) remarked that his role as an educator at the Air Force Academy is to make his students smart buyers of commercial systems. He suggested that NASA's main role in technology development should be basic research, and that such investments promote science, technology, engineering, and mathematics education and improve the excising workforce. He acknowledged that industry partners are probably most useful for making near-term improvements. He suggested that perhaps that could be encourage by higher launch rates. He also agreed that the most promising long-term technology is air-breathing propulsion systems such as TBCC and RBCC.

In the discussion period that followed many speakers endorsed the development of air-breathing technologies as potentially the biggest game changers in TA01. However, there also was some skepticism that there would be a large enough market to support a reusable launch system that leveraged air-breathing technology.

Session 2: U.S. Air Force

Toby Cavallari (USAF / SMC/LR) started the U.S. Air Force session with a presentation that focused on the history of the Air Force launch vehicle program, its current status, and near-term plans. He noted some of the near-term challenges that face the Air Force include parts obsolescence, increasing costs, dependence on foreign suppliers, and a declining U.S. industrial base. He said that a new affordable upper stage engine is needed, and that such an engine should leverage both Air Force and NASA technology investments. He concluded by saying the Air Force and NASA should pursue joint development programs with interagency cooperation and commercial partnerships, particularly for liquid rocket engines, noting that neither agency can afford standalone programs.

Greg Rudderman (Air Force Research Laboratory; AFRL) presented charts generated by Richard Cohn (AFRL) who was not able to attend the workshop. This presentation started with a review of relevant research, past and present, by AFRL, including the Integrated High Payoff Rocket Propulsion Technology (IHPRPT) program. IHPRPT is a joint DOD-NASA-industry program to develop technologies that will lead to more capable rockets. Military applications of interest include tactical missiles, strategic missiles, and spacecraft. The IHPRPT goals are similar in nature to many of the goals laid out in NASA's draft roadmap for TA01. Propulsion Directorate is interested and actively working on both solid and liquid motor technologies as well as improved modeling while other parts of AFRL pursue air-breathing engine concepts. In reviewing the draft TA01 roadmap Cohn noted that it includes some technologies that have been shown in the past to lack promise and agreed that greater focus on a smaller number of promising technologies would be beneficial.

Randy Kendall (The Aerospace Corporation) said that modern launch options have plenty of performance and reliability, but are very expensive. He also stated that increased flight rates were key to reducing launch costs. He described current Air Force plans to build a reusable booster system and said that a combustion engine suitable for a reusable hydrocarbon stage would be the most promising NASA technology to support a reusable booster system. Over the mid- to long-term, he said that the highest priorities should be air-breathing propulsion technologies (RBCC and TBCC), pulse detonation engines, and an air collection and enrichment system.

Tim Lawrence (Air Force Institute of Technology) included comments on in-space propulsion and more advanced concepts. He suggested that nuclear-based propulsion technologies are good options for solving NASA's transportation needs, but there are several challenges to be overcome before they can be implemented. He strongly encouraged development in the field of green propulsion technologies (that do not include hazardous materials) because they are compatible with small scale and student-run spacecraft. In addition, revolutionary advances in propulsion technology would enable missions that are currently inconceivable.

In the discussion period many speakers discussed how the U.S. Air Force and NASA should cooperate. It was noted that, although the two agencies are trying to cooperate, it is difficult to execute joint programs because of the potential for redirection by either participant. It was also mentioned that the nation may have too many underused test facilities because of overlap between the Air Force and NASA. One speaker said the Air Force launch rates would probably remain unchanged unless a revolutionary system is developed that leads to higher launch rates.

Session 3: Propulsion System Manufacturers

Stan Graves (ATK) started the session with propulsion system manufacturers by noting that current launch systems all use a combination of liquid and solid propulsion systems. He expects that trend to continue due to the physics, economics, and programmatics of the launch vehicle industry. Having reviewed the draft NASA roadmaps for TA01, he observed that many of the technologies would benefit both commercial and NASA heavy lift launch systems and both liquid and solid propulsion systems. He also suggested that two technologies, electrical-hydrostatic and electrical-mechanical thrust vector control, should be high priorities. Graves also asserted that investments should be made in developing a low-cost, safe, and green system.

Jeff Greason (XCOR) stated improving rocket performance is not likely to be a cost effective approach for NASA to improve the economics, reliability, and safety of access to space. Instead, he advocated increasing flight rates and reducing production costs. He declared that one of the best approaches for increasing flight rate is for commonality in performance requirements established by launch customers, and NASA could contribute to this approach by investing in technologies that allow large exploration missions to be broken into smaller pieces for launch. With a higher flight rate, reusable launch systems become more advantageous, especially if maximum payload mass per launch is contained. He identified two other high priorities: thermal protection systems for reusable launch vehicles and low-cost engines with adequate performance.

Russ Joyner (Pratt Whitney Rocketdyne) said that the current roadmaps are too broad and should be focused, but not before NASA establishes mission priorities. In the meantime, he urged NASA to invest in crosscutting technologies, such as manufacturing. He also called for technology investments to focus on reducing cost rather than increasing performance. He also supported providing a steady level of funding for small-scale efforts to improve capabilities over the long term, with periodic reviews.

Todd Neill (Aerojet) provided very specific suggestions for the full list of launch propulsion technologies. In particular, he mentioned that NASA should move toward HTPB (hydroxy-terminated polybutadiene) propellants for solids, develop a nozzle extension for hydrogen engines and develop a new hydrocarbon boost engine. He saw little benefit to either hybrid propulsion systems or advanced propellants (besides the previously mentioned HTPB).

In the discussion session a few members of the audience suggested that hybrid propulsion systems should be high priority. They argued that hybrid systems have improved significantly in recent years, suggesting that they have higher efficiency than solid propulsion system, that they are less complex than liquid engines, and they are easy to manufacture and operate. There was also a discussion on the IHPRPT program as a model for propulsion technology development, with some suggesting that is a good program for attracting bright talent and developing new tools. Others criticized IHPRPT for starting with too much of a focus on improving performance, with not enough attention to cost reduction. One speaker disagreed that increasing launch rates is the best solution for reducing costs, suggesting that mission payloads could generally be repackaged in such a way to significantly increase the national launch rate. When asked what technologies would help improve affordability, various speakers mentioned improved materials, manufacturing, and health monitoring, and they cautioned that industry's ability to invest in these technologies as they pertain to launch vehicles is constrained by high costs and low production rates.

Session 4: Launch Vehicle Manufacturers

Bernard Kutter (United Launch Alliance) started the session with launch vehicle manufacturers by emphasizing some of the points made earlier in the day. These include increased flight rates as a key to reducing cost and investing in cost reduction and operability instead of performance. Kutter noted that numerous attempts in the past 30 years to develop revolutionary systems had failed. He also said that it is unclear if reusability will show economic benefit. One technology he supported was integrated vehicle fluids, which would use primary engine propellants to serve the needs of auxiliary vehicle systems that currently use other fluids. Given the uncertainty in the future optimum vehicle configuration, he favored making technology investments in crosscutting technologies with broad applicability.

Gwynne Shotwell (SpaceX) reviewed the history of SpaceX. She believes that the highest priority propulsion technology would be a hydrocarbon boost engine with a thrust on the order of 1.5 million pounds or greater. Such an engine could support a NASA super heavy lift vehicle as well as smaller commercial launch systems. She suggested that this engine should be developed through a public-private partnership using a fixed price competition similar to the one NASA used for its Commercial Orbital Transportation Services program. This approach gives industry the flexibility and the incentives to produce optimum solutions.

John Steinmeyer (Orbital) agreed that a new high-thrust hydrocarbon boost engine should be the highest priority launch development. He suggested that the current Russian engines could be used as starting points, with the goal of a developing a propulsion system that could support the proposed NASA super heavy lift vehicle, smaller commercial launchers, and the proposed Air Force RBS. He asserted that the recent U.S. industrial space policy has hampered emerging technology through lack of focus and constancy. He said that new efforts should be properly funded and coordinated programs that capitalize on past developments and strategic, focused investments.

In the discussion session the hydrocarbon engine was further discussed with several speakers endorsing it as the best path forward for a super heavy lift system, especially if it were also used in other launch vehicles to reduce costs. Some speakers said that two competing engines should be developed to foster competition, but others countered that the market might be too small to support two vendors.

REFERENCES

Bulman, M.J., and Siebenhaar, A. 2011. Combined cycle propulsion: Aerojet innovations for practical hypersonic vehicles. AIAA Paper 2011-2397. American Institute of Aeronautics and Astronautics, Reston, Va.

Eklund, D.R., Boudreau, A.H., and Bradford, J.E. 2005. A turbine-based combined cycle solution for responsive space access. AIAA Paper 2005-4186. American Institute of Aeronautics and Astronautics, Reston, Va.

Hampsten, K.R., and Hickman, R.A. 2010. Next Generation Air Force Spacelift. AIAA Paper 2010-8723. American Institute of Aeronautics and Astronautics, Reston, Va

E

TA02 In-Space Propulsion Technologies

INTRODUCTION

The draft roadmap for technology area (TA) 02, In-Space Propulsion Technologies, consists of four level 2 technology subareas:[1]

- 2.1 Chemical Propulsion
- 2.2 Non-Chemical Propulsion
- 2.3 Advanced (TRL<3) Propulsion Technologies
- 2.4 Supporting Technologies

TA02 includes all propulsion-related technologies required by space missions after the spacecraft leaves the launch vehicle from Earth. The technology area includes propulsion for such diverse applications as fine pointing of an astrophysics satellite in low Earth orbit (LEO), robotic science and Earth observation missions, high-thrust Earth orbit departure for crewed vehicles, low-thrust cargo transfer for human exploration, and planetary descent, landing and ascent propulsion. This wide range of applications results in a very diverse set of technologies, including traditional space-storable chemical, cryogenic chemical, various forms of EP, various forms of nuclear propulsion, chemical and electric micropropulsion, solar sails, and space tethers. The challenge for the panel was to prioritize these technologies in light of 50 years of spaceflight development experience, the current status of all the technologies, an assessment of the likely benefits which would result from successfully developing each technology, and a general understanding of NASA's mission objectives.

Prior to prioritizing the level 3 technologies included in TA02, several technologies were deleted. The changes are explained below and illustrated in Table E.1. The complete, revised technology area breakdown structure (TABS) for all 14 TAs is shown in Appendix B.

The steering committee deleted the following level 3 technologies

- 2.4.1. Engine Health Monitoring & Safety,
- 2.4.3. Materials & Manufacturing Technologies,
- 2.4.4. Heat Rejection, and
- 2.4.5. Power.

[1] The draft space technology roadmaps are available at http://www.nasa.gov/offices/oct/strategic_integration/technology_roadmap.html.

TABLE E.1 Technology Area Breakdown Structure for TA02, In-Space Propulsion Systems

NASA Draft Roadmap (Revision 10)	Steering Committee-Recommended Changes
TA02 In-Space Propulsion Technologies	Four technologies have been deleted.
2.1. Chemical Propulsion	
2.1.1. Liquid Storable	
2.1.2. Liquid Cryogenic	
2.1.3. Gels	
2.1.4. Solid	
2.1.5. Hybrid	
2.1.6. Cold Gas/Warm Gas	
2.1.7. Micro-Propulsion	
2.2. Non-Chemical Propulsion	
2.2.1. Electric Propulsion	
2.2.2. Solar Sail Propulsion	
2.2.3. Thermal Propulsion	
2.2.4. Tether Propulsion	
2.3. Advanced (TRL <3) Propulsion Technologies	
2.3.1. Beamed Energy Propulsion	
2.3.2. Electric Sail Propulsion	
2.3.3. Fusion Propulsion	
2.3.4. High Energy Density Materials	
2.3.5. Antimatter Propulsion	
2.3.6. Advanced Fission	
2.3.7. Breakthrough Propulsion	
2.4. Supporting Technologies	
2.4.1. Engine Health Monitoring & Safety	Delete: 2.4.1. Engine Health Monitoring & Safety
2.4.2. Propellant Storage & Transfer	
2.4.3. Materials & Manufacturing Technologies	Delete: 2.4.3. Materials & Manufacturing Technologies
2.4.4. Heat Rejection	Delete: 2.4.4. Heat Rejection
2.4.5. Power	Delete: 2.4.5. Power

The scope of each of these technologies actually falls outside the scope of TA02, and NASA's draft roadmap for TA02 does not suggest that any of them should be developed as part of TA02. Except for item 2.4.2, this section of the roadmap is used to highlight level 1 or level 2 topics in other roadmaps that are important to the TA02 roadmap—but that belong to other roadmaps. For example, with regard to 2.4.5. Power, the roadmap says:

> Power systems play an integral role in all in-space propulsion systems for both human and robotic missions. The reader is referred to the Technology Area 3, Space Power and Energy Storage Systems.

Similarly, with regard to technologies 2.4.1, 2.4.3, and 2.4.4, roadmap TA02 refers readers to roadmaps TA04, TA12, and TA14, respectively, to learn the details of what should be done in these areas.

TOP TECHNICAL CHALLENGES

The panel identified four top technical challenges for TA02, all of which are related to the provision of safe, reliable, and affordable in-space transportation consistent with NASA's mission needs. The challenges are listed below in priority order.

1. High-Power Electric Propulsion (EP) Systems: Develop high-power EP system technologies to enable high-ΔV missions with heavy payloads.

EP systems have a higher propellant efficiency than other in-space propulsion technologies that will be available in the foreseeable future, with applications to all NASA, Department of Defense (DOD), and commercial space mission areas. Specifically, low-power EP systems are currently used for small robotic interplanetary missions (e.g., Hayabusa and Dawn), for post-launch circularization of the orbits of large geosynchronous communications satellites (e.g., Advanced Extremely High Frequency satellite), and stationkeeping for a wide range of spacecraft (e.g., GOES-R and commercial communications satellites). Development of high-power EP systems (30 kW to 600 kW) will enable larger scale missions with heavy payloads, including development of a more efficient in-space transportation system in Earth-space, sample returns from near-Earth objects (NEOs), the martian moons, other deep space destinations (including extensions of the JUNO mission to Jupiter), precursor demonstrations of in situ resource utilization (ISRU) facilities, and pre-placement of cargo for human exploration missions. In addition to these specific propulsion and power system technologies, demonstration of large scale EP vehicles is required to ensure adequate control during autonomous rendezvous and docking operations necessary for either cargo or small body proximity operations.

2. Cryogenic Storage and Transfer. Enable long-term storage and transfer of cryogens in space and reliable cryogenic engine operation after long dormant periods in space.

Deep space exploration missions will require high-performance propulsion for all mission phases, including Earth departure, destination arrival, destination departure, and Earth return, occurring over the entire mission duration. Both high-thrust propulsion options, LOX/H_2 chemical propulsion and LH2 nuclear thermal rocket (NTR), will require storage of cryogens for well over a year to support all mission phases. Chemical and NTR engines must also operate reliably after being dormant for the same period. While LOX can currently be stored for extended periods, LH2 boil-off rates using state-of-the-art technology are far too high for deep-space missions, allowing only a few days of storage. Additionally, cryogenic fluid transfer technology would enable other exploration architectures, including propellant aggregation and the use of propellants produced using ISRU facilities. This technical challenge is enabling for the most plausible transportation architectures for human exploration beyond the Moon.

3. Microsatellites: Develop high-performance propulsion technologies for high-mobility microsatellites (<100 kg).

The broader impact of small satellites is hindered by the lack of propulsion systems with performance levels similar to those utilized in larger satellites (high ΔV, high Isp, low mass fractions, etc.). Most existing propulsion systems are not amenable for miniaturization and work is needed to develop concepts that scale and perform favorably. In addition to small satellites, high-performing miniature propulsion would also provide functionality in different applications, for example in distributed propulsion for controlling large, flexible structures and address missions requiring fine thrust for precise station keeping, formation flight, accurate pointing and cancellation of orbital perturbations. A moderate investment in many of these technologies (including chemical, electric, and advanced propulsion concepts, such as tethers and solar sails) could validate their applicability to small satellites.

4. Rapid Crew Transit: Establish propulsion capability for rapid crew transit to/from Mars.

Trip times for crewed missions to NEOs, Phobos, and the surface of Mars should be minimized to limit impacts to crew health from radiation (galactic and solar), exposure to reduced gravity, and other effects of long-duration deep space travel. Developing high-performance, high-thrust propulsion systems to reduce transit times for crewed missions would mitigate these concerns. Two realistic high-thrust options exist that could be available for missions in the next 20 years: LOX/H_2 and NTR. Engines used for rapid crew transport must be capable of multiple restarts following prolonged periods of inactivity, and they must demonstrate extremely high reliability. There are no engines of either type currently available that meet the requirements of performance, reliability, and restart capability. The two LOX/H_2 engines that come closest are the J2X, with about ~250,000 pounds of thrust and the RL-10, with about ~25,000 pounds of thrust. Both are high-performance engines and both have some restart capability, but neither has demonstrated the ability to accomplish multiple restarts following prolonged dormancy. Also, NTRs have never been tested in space, and the last ground test was conducted more than 40 years ago. There is also considerable uncertainty regarding the effort it would take to reconstitute the state of the art as it existed 40 years ago or to define test and operational requirements, and the environmental issues are substantial.

QFD MATRIX AND NUMERICAL RESULTS FOR TA02

The panel evaluated 23 In-Space Propulsion level 3 technologies. The results of the panel's QFD scoring for the level 3 in-space propulsion technologies are shown in Figures E.1 and E.2. As noted above, four technologies in the draft roadmap for TA02 were eliminated from consideration (2.4.1. Engine Health Monitoring & Safety, 2.4.3. Materials & Manufacturing Technologies, 2.4.4. Heat Rejection, and 2.4.5. Power) because they are properly addressed in other roadmaps. The results of the QFD scoring are shown in Figure E.3. The seven technologies in the Advanced Propulsion Technologies subarea received the same score and they are listed as a single (low priority) line item. Four technologies were assessed as high-priority technologies:

- Electric propulsion
- Propellant storage and transfer
- Thermal propulsion
- Micropropulsion systems

The first three technologies were designated as high-priority technologies because they received the highest QFD scores based on the panel's initial assessment. The panel subsequently decided to override the QFD scoring results to designate micropropulsion systems as a high-priority technology to highlight the importance of developing propulsion systems that can support the rapidly developing micro-satellite market, as well as certain large astrophysics spacecraft.

CHALLENGES VERSUS TECHNOLOGIES

Figure E.3 shows how each of the TA02 level 3 technology supports the top technical challenges described above. This shows that the high-priority technologies, which are discussed in the next section, provide potential

	Benefit	Alignment with NASA Needs	Alignment with Non-NASA Aero Tech Needs	Alignment with Non-Aerospace National Goals	Technical Risk and Reasonableness	Sequencing and Timing	Time and Effort	QFD Score (Weighted)	Panel Priority
Multiplier	27	5	2	2	10	4	4		
	0/1/3/9	0/1/3/9	0/1/3/9	0/1/3/9	1/3/9	-9/-3/-1/1	-9/-3/-1/0		
Technology Name	Benefit	Alignment			Risk/Difficulty				
2.1.1. Liquid (Non-Cryogenic) Storable Propulsion	1	9	3	0	3	-1	-1	100	M
2.1.2. Liquid Cryogenic (Propulsion)	3	3	1	0	3	1	-3	120	M
2.1.3. Gels (Propulsion)	1	3	1	0	3	-3	-1	58	L
2.1.4. Solid (Propulsion)	1	3	1	0	3	-3	-1	58	L
2.1.5. Hybrid (Propulsion: Solid and Liquid)	1	3	1	0	3	-3	-1	58	L
2.1.6. Cold Gas/Warm Gas (Propulsion)	1	3	1	0	1	-3	-1	38	L
2.1.7. Micro-propulsion	3	3	3	0	3	1	-1	132	H*
2.2.1. Electric Propulsion	9	9	9	0	9	1	-3	388	H
2.2.2. Solar Sail Propulsion	1	1	1	0	1	-3	-3	20	L
2.2.3. (Nuclear) Thermal Propulsion	9	3	1	0	3	-1	-3	274	H
2.2.4. Tether Propulsion	3	3	1	0	3	-3	-3	104	M
2.3. Advanced (TRL <3) Propulsion Technologies	0	0	0	0	0	0	0	0	L
2.4.2. Propellant Storage & Transfer	9	3	1	1	9	1	-3	344	H

FIGURE E.1 Quality function deployment (QFD) summary matrix for TA02 In-Space Propulsion Systems. The justification for the high-priority designation of all high-priority technologies appears in the section "High-Priority Level 3 Technologies." H = High Priority; H* = High Priority, QFD score override; M = Medium Priority; L = Low Priority.

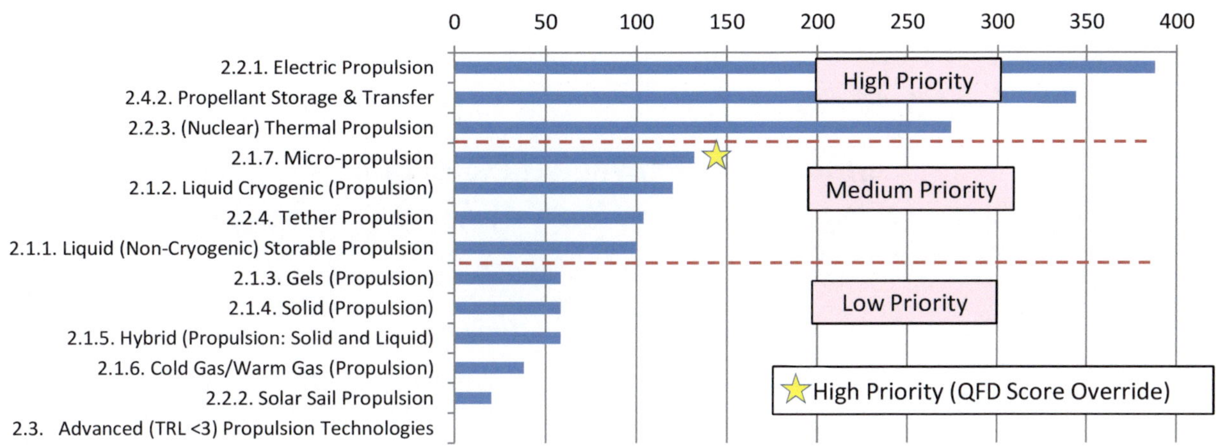

FIGURE E.2 Quality function deployment results for TA02 In-Space Propulsion Systems.

Priority	TA 02 Technologies, Listed by Priority	Top Technology Challenges			
		1. High-Power Electric Propulsion Systems: Develop high power electric propulsion system technologies to enable high ΔV missions with heavy payloads.	2. Cryogenic Storage and Transfer: Enable long-term storage and transfer of cryogens in space and reliable cryogenic engine operation after long dormant periods in space.	3. Microsatellites: Develop high performance propulsion technologies for high-mobility micro-satellites (<100 kg).	4. Rapid Crew Transit: Establish propulsion capability for rapid crew transit to/from Mars.
H	2.2.1. Electric Propulsion	●		●	
H	2.4.2. Propellant Storage and Transfer		●		●
H	2.2.3. (Nuclear) Thermal Propulsion		○		●
H	2.1.7. Micro-propulsion			●	
M	2.1.2. Liquid Cryogenic Propulsion		●		○
M	2.2.4. Tether Propulsion	○		○	
M	2.1.1. Liquid (Non-Cryogenic) Storable (Propulsion)			○	
L	2.1.3. Gels (Propulsion)				
L	2.1.4. Solid (Propulsion)				
L	2.1.5. Hybrid (Propulsion: Solid and Liquid)				
L	2.1.6. Cold Gas/Warm Gas (Propulsion)				
L	2.2.2. Solar Sail Propulsion				
L	2.3. Advanced (TRL<3) Propulsion Technologies				

●	Strong Linkage: Investments by NASA in this technology would likely have a major impact in addressing this challenge.
○	Moderate Linkage: Investments by NASA in this technology would likely have a moderate impact in addressing this challenge.
[blank]	Weak/No Linkage: Investments by NASA in this technology would likely have little or no impact in addressing the challenge.

FIGURE E.3 Level of support that the technologies provide to the top technical challenges for TA02 In-Space Propulsion Systems.

solutions that will meet these challenges. The low-ranked technologies are judged to have a weak linkage because of the limited benefit of investing in these technologies regardless of how closely they may overlap with various challenges in terms of subject matter.

HIGH-PRIORITY LEVEL 3 TECHNOLOGIES

Panel 1 identified four high-priority technologies in TA02. The justification for ranking each of these technologies as a high priority is discussed below.

Technology 2.2.1, Electric Propulsion

Technology 2.2.1, electric propulsion (EP) uses electrical power produced onboard a spacecraft to accelerate propellant to extremely high speeds. Solar electric propulsion (SEP), including arcjet, Hall thruster, and ion thruster systems, is routinely used today on more than 230 space vehicles for spacecraft maneuvers, mostly north-south station-keeping and orbit-raising. A handful of U.S. and international lunar and interplanetary probes (SMART-1, Hayabusa, Dawn) use or have used SEP for primary propulsion. SEP has also been used for drag make-up in LEO (GOCE) and for orbit-raising and station-keeping of large geosynchronous communications satellites. These systems are at TRL 9.

Modern laboratory-model ion thrusters and Hall thrusters have been demonstrated on the ground by NASA at 30 kW and ~100 kW, respectively. These systems are at TRL 3. Laboratory-model Hall thrusters at the 100 to 250 kW power level are currently being developed by NASA and the U.S. Air Force. Flight versions of these thrusters may be developed in the mid-term (2017 to 2022) timeframe. Over the longer term, multi-MW systems enabled by space nuclear power systems could use flight versions of the lithium magnetoplasmadynamic thrusters, pulsed inductive thrusters, field reversed configuration thrusters, and VASIMR thrusters that are in early laboratory testing today.

NASA has the expertise and ground facilities to lead the critical EP technology developments in cooperation with the U.S. Air Force, industry, and academia. There is also potential for international cooperation as Europe, Japan, and Russia have very productive EP programs. In addition to thruster development, advances in high-power EP systems will require:

- Developing the components and architectures needed for high-capacity power processing units;
- Gaining a better understanding of thruster wear mechanisms so full-length life tests are not always necessary;
- Characterizing EP/spacecraft interactions more completely;
- Developing the infrastructure needed to test high-power EP systems on the ground; and
- Demonstrating autonomous operation and control of high-power, large-scale EP systems in space.

The International Space Station is not well suited as a test platform for high-power EP. In its current configuration, the ISS provides no benefit to high-power EP testing in space given the limited power available (~5 kW) and the requirement to validate vehicle system operation in rendezvous and docking scenarios

The primary benefit of EP is its high specific impulse, which is typically an order of magnitude greater than those of chemical propulsion systems: 10^3 to 10^4 s for EP versus 500 s or less for chemical propulsion systems. As a result, EP systems are the most propellant efficient in-space propulsion technology available for the foreseeable future, with applications to all NASA, DOD, and commercial space mission areas. While EP's large specific impulse enables a host of space missions that are not possible or affordable with conventional propulsion, its thrust is low, which results in long trip times for many missions. This characteristic of EP places constraints on departure orbits and travel through deep gravity wells. The development of high-power SEP systems (from ~100 kW to ~1 MW) would enable missions with heavier spacecraft and/or shorter transit time, resulting in more efficient in-space transportation systems in Earth orbit; more affordable sample return missions from destinations such as the Moon, Mars, and the asteroid belt; pre-positioning of cargo and ISRU facilities for human exploration missions to Mars orbit; and efficient crew transfers to near-Earth objects (NEOs) and Phobos from departure points such as the Earth-Moon Lagrange points. The large benefits, broad applicability, and reasonable development timescales and challenges are the basis for the high priority placed on EP technology.

Technology 2.4.2, Propellant Storage and Transfer

Technology 2.4.2, propellant storage and transfer in space, includes both the long-term storage of cryogens (liquid hydrogen, oxygen, and potentially methane, as well as propellants for EP) and the transfer of these fluids between refueling stations (depots) and the propulsion systems on spacecraft, upper stages, and Moon/Mars landing and ascent vehicles. This technology has only been validated at the component level for cryogenic fluids in laboratory environments (TRL 4), although "storable" (non-cryogenic) propellant storage and transfer has been demonstrated in space (TRL 7).

NASA has the expertise and facilities to lead this development effort, with multiple ground test facilities and considerable experience in cryogenic fluid management at several NASA centers. The ISS could easily contribute to the development of this technology. Simple yet extremely beneficial experiments could be performed to validate long-term storage and handling of cryogenic propellants. Alternatively, expendable launch vehicles could carry large masses of residual cryogens into orbit for independent experiments, without introducing any risk to the ISS. This could lead to precursor demonstrations of the ISS as a deep space transportation node.

Propellant storage and transfer is a game changing technology that could provide big benefits for NASA exploration missions, and it may also benefit DOD and commercial missions. Propellant storage and transfer in space can reduce operational costs and enable affordable human exploration of the Moon and Mars, as follows:

- Human exploration of Mars: high-ΔV maneuvers will be required for all mission phases, including Earth departure, Mars arrival, and Earth return. The time-scales for these mission phases will require long-term storage of propellants. Additionally, it is likely that vehicles departing from Earth will need more cryogenic propellant than can be reasonably carried to orbit in a single launch, and therefore long-term storage of propellant is an absolute requirement for human missions to Mars.
- Refueling vehicles in lunar or Mars orbit with ISRU propellants has the potential to reduce exploration costs by perhaps an order of magnitude (compared to an exploration architecture that requires all fuel to be carried into space from Earth).
- Enabling launch of unfueled deep space vehicles, reducing the mass of deep space vehicles, and potentially reducing the maximum required launch mass per launch.

This technology requires in-space demonstrations to validate cryogenic fluid management in microgravity. Propellant storage and transfer is an interdisciplinary capability, which may overlap with other technology areas such as advanced thermal control to minimize boil-off and perhaps to provide active cooling. Propellant storage and transfer is a game changing technology for a wide range of applications because it enables long-duration, high-thrust, high-ΔV missions with large payloads and crew.

Technology 2.2.3, Thermal Propulsion

Technology 2.2.3, thermal propulsion, includes the option of using either solar and nuclear thermal sources to heat hydrogen propellant for high specific impulse. Of these two, only nuclear thermal propulsion is rated as a high-priority technology. Solar-thermal propulsion has limited benefit compared to other propulsion options and comes with a high degree of complexity and mission constraints.

Nuclear thermal rockets (NTRs) are high-thrust propulsion systems with the potential for twice the specific impulse of the best liquid hydrogen/oxygen chemical rockets. Multiple mission studies have shown that nuclear thermal rockets would enable rapid Mars crew transfer times with half the propellant and about 60 percent of the launch mass required by chemical rockets. Demonstrated NTRs use a solid-core nuclear reactor to heat hydrogen propellant, exhausting it through a standard nozzle to achieve a specific impulse of 800 s to 900 s. An extensive development program, Project Rover, was conducted between 1951 and 1971, during which 20 separate reactors and engines were tested at thrust levels between 7,500 and 250,000 pounds of thrust. The program culminated with the Nuclear Engine for Rocket Vehicle Applications (NERVA) system, which fired for almost 2 hours with 28 restarts. Since then intermittent demonstration efforts have been focused on advanced nuclear fuels development, non-nuclear validation of advanced engines such as the LOX-augmented NTR, assessments of ground test requirements, cost reduction studies, and mission studies.

Critical NTR technologies include the nuclear fuel, reactor and system controls, and long-life hydrogen pumps. Technology development will also require advances in ground test capabilities, as the open-air approach used during Project Rover is no longer environmentally acceptable. While NTR technology was close to TRL 6 in 1971, inactivity since then has resulted in the loss of experienced personnel and facilities, and the current TRL is probably at TRL 4 or less. The immediate challenge is to capture the engineering and technical knowledge base of the NERVA program. The next steps would be to validate nuclear fuels for long life at high temperature, to ensure no nuclear material would be released during ground tests, and to validate ground test site capability for handing NTR effluents. NERVA used graphite-based fuels, whereas modern fuels rely on cermets or tungsten to ensure a radiation-free exhaust. In parallel with the development of nuclear fuels, sub- and full-scale evaluations of ground testing NTRs using existing borehole testing would be needed to fully characterize effluent behavior. Initial studies for using existing boreholes at the Nevada Test Site have shown no major roadblocks to date, though considerable development and validation remains. Use of existing boreholes may minimize the cost of ground testing. NTR development could readily take a phased approach, with parallel efforts to develop nuclear fuels, validate ground test capabilities, and develop and demonstrate a low-thrust (5,000 pounds) NTR. This would be followed by full-scale development and flight of an NTR with 20,000 to 25,000 pounds of thrust. Such a system would have enough thrust for a crewed Mars mission. Growth options exist for follow-on systems such as the LOX-Augmented NTR, which can provide much higher thrust (at lower Isp) for operation in planetary gravity wells.

NTR technology development will require NASA to collaborate closely with DOE (Department of Energy) national laboratories and the Nevada Test Site. NASA has all the expertise required to develop an NTR except for the nuclear fuels and reactor, which by statute are the responsibility of the DOE. There is ample precedent for NASA-DOE collaboration in developing nuclear systems. There is no need for access to the ISS for NTR development.

The reduction in launch mass enabled by this technology could significantly reduce the cost and mission complexity of crewed missions to Mars. The panel could not identify credible non-NASA or non-aerospace applications of NTR technology. Although NTR development would be a major program, its benefits resulted in ranking NTRs as a high-priority technology.

Technology 2.1.7, Micro-propulsion

Technology 2.1.7, micro-propulsion, encompasses all propulsion options, both chemical and non-chemical, that could be used to fulfill the propulsion needs of (1) high-mobility micro-satellites (<100 kg) and (2) the extremely fine pointing and positioning requirements of certain astrophysics missions. Recent advances in the miniaturization of spacecraft subsystems have triggered a large growth in the field of micro-satellites (<100 kg), nano-satellites (~10 kg), pico-satellites (~1 kg), and femto-satellites (<1 kg). Small satellites, operating individually or flying in formation, are being considered for increasingly complex missions (e.g., flight testing and validation of new technologies, scientific missions, and commercial missions). Low costs, fast development times, and the potential to perform tasks so far limited to large systems have made small satellites an area of interest for NASA, DOD, other government agencies, and many research centers and educational institutions worldwide. The lack of micro-propulsion is currently a roadblock in the development of advanced high-mobility micro-satellites. Ideally, new and evolved micro-propulsion technologies would be characterized by:

- Low mass and low volume fractions, scalable to the smallest of satellites,
- Wide range of ΔV capability to provide hundreds or even thousands of m/s,
- Wide range of Isp capability, up to thousands of seconds,
- Precise thrust vectoring and low vibration for precision maneuvering,

- Efficient use of onboard resources (i.e., high power efficiency and simplified thermal and propellant management),
- Affordability, and
- Safety for users and primary payloads.

Many micro-propulsion technologies have been proposed, including miniaturization of existing systems and innovative concepts, but very few are beyond TRL 3. Nevertheless, several promising technologies based on chemical, electric, and other propulsion concepts have advanced to the point where modest investments in a low to moderate risk environment may be able to validate their operational principles in the laboratory, accelerate their engineering development, and enable their in-space demonstration. NASA has built expertise in the field over the last decade, and this trend could be improved by working with industry and research institutions to retain U.S. leadership in this globally growing area. The ISS could be used for in-space demonstrations of micro-propulsion technologies. In fact, microsatellites could be used to remotely inspect the ISS.

The benefits of developing micro-propulsion concepts are not confined to small satellites, to NASA, or to the aerospace industry. For instance, micro-propulsion could be used by larger satellites for missions requiring accurate thrust delivery to counteract orbital perturbations (e.g., LISA). They could also be used for precise formation flying of spacecraft clusters or as modular distributed propulsion for the control of large space structures. These and many other concepts are currently being explored by NASA, the Air Force, the National Reconnaissance Office, the Defense Advanced Research Projects Agency, and a growing community of commercial users. In addition, many micro-propulsion technologies are based on the use of non-conventional materials and micro/nano-fabrication processes that will likely find non-aerospace applications.

The ability to increase the value of space missions at a relatively modest costs as enabled by micro-propulsion would be "game-changing." However, the broad field of micro-propulsion appears as a level 3 technology in the draft TA03 roadmap as part of the chemical propulsion technology subarea. Limiting research in the technology to chemical propulsion alternatives would exclude many other promising alternatives. The panel ranked this technology as a high priority assuming that the scope of this technology would be broadened to include all applicable propulsion technologies.

MEDIUM- AND LOW-PRIORITY TECHNOLOGIES

A total of 11 technologies in TA02 were not ranked as high priority by the steering committee. A clear break in QFD score is observed between the six initially ranked as medium priority, and the other six ranked as low priority. (As noted above, Technology 2.1.7, micro-propulsion, which was initially ranked as medium priority, was subsequently designated by the panel as a high-priority technology despite its relatively modest QFD score.) The remaining technologies in the medium-priority group include Liquid Cryogenic and Liquid Storable, which the committee agreed were less focused on technology development than on engineering implementation. Tether Propulsion technology is believed to provide a low return on investment. In particular, electrodynamic and momentum exchange tethers have been investigated in the past with mixed results. Materials and Manufacturing technology is relevant to propulsion, but is already included in TA12. Likewise, Engine Health Monitoring and Safety is better served by investments in different technology areas (TA04).

The rest of the TA02 technologies were ranked low. This group includes Hybrid, Solid, Gelled, and Metalized-Gelled propellants. The benefit to NASA from the development of these technologies would be marginal, as they do not significantly improve the performance or reduce the cost of in-space propulsion systems. Solar Sail Propulsion is ranked low mostly because of the limited improvement to NASA or other agencies' capabilities that can be achieved in the near- or medium-term. Cold Gas/Warm Gas propulsion is already at a high TRL, and minimal gains would be expected from continued investment under a technology development program. Finally, the category of Advanced (TRL<3) Propulsion Technologies is ranked low because, even though success in developing any of these technologies would be "game-changing" in every possible sense, it is highly unlikely that any of the approaches described in the roadmaps will materialize in the next 20 to 30 years. However, this low ranking of such advanced concepts should not be interpreted as a recommendation to eliminate them from NASA's portfolio.

The panel recommends that the NASA Institute for Advanced Concepts provide a low level of funding for this category of low-TRL, very-high-risk technologies.

DEVELOPMENT AND SCHEDULE CHANGES FOR THE TECHNOLOGIES COVERED BY THE ROADMAP

The key to successful technology development is to establish a phased, evolutionary approach with reasonable decision gates for down-selecting options. For an unconstrained funding environment the TA02 roadmap presents a reasonable approach, particularly when focus is placed on the high-priority technologies listed above. However, in a funding constrained environment it is unlikely that all the level 3 technologies shown on the schedule will be affordable.

OTHER GENERAL COMMENTS ON THE ROADMAP

The planetary decadal survey identifies Mars ascent propulsion and precision landing as key capabilities. Mars ascent systems are unproven, but initial systems will likely rely on conventional solid rockets or storable liquids with strong environmental control systems to enable long-term storage in the martian environment. Additionally, Mars human exploration will require the delivery of large payloads to the surface of Mars. Current entry, descent, and landing technologies are near their limits for the martian atmosphere, and some improvement in propulsion systems for descent and landing will be required. While new engineering developments are certainly required, the propulsion challenges are more in system implementation than technology development.

PUBLIC WORKSHOP SUMMARY

The workshop for the In-Space Propulsion Systems technology area was conducted by the Power and Propulsion Panel on March 22, 2011, on the campus of the California Institute of Technology in Pasadena, California. The discussion was led by panel member Roger Myers, who started the day by giving a general overview of the roadmaps and the NRC's task to evaluate them. He also provided some direction for what topics the invited speakers should cover in their presentations. Experts from industry, academia, and government were invited to lead a 25 minute presentation and discussion of their perspective on the draft NASA roadmap for TA02. At the end of the session, there was a short open discussion by the workshop attendees that focused on the recent session. At the end of the day, there was a concluding discussion led by Myers summarizing the key points observed during the day's discussion.

Session 1: In-Space Low Thrust

Scott Miller (Aerojet) started the session on in-space low thrust technologies with a discussion of the need to meet future mission requirements rather than focus on performance improvements. He said that NASA should resist the urge to fund a little bit of everything, while at the same time providing a consistent level of low-TRL funding for future high-payoff technologies. He also suggested that NASA partner with other agencies and industry to increase advocacy, reduce costs, and develop common requirements. He said his top priority technologies for in-space propulsion would be high-power EP and in-space cryogenic propulsion. In the low thrust area he would focus on high-power Hall and ion thrusters, advanced storable bipropellant engines, and advanced monopropellant systems for attitude control.

Mike Micci (Penn State) focused his remarks on micropropulsion technology, which he said can be used both as a main propulsion system for small spacecraft and fine control of larger spacecraft. He believes that micropropulsion is a young technology that is poised to make a large impact on near term missions. Micci said that micropropulsion also has significant terrestrial applications as well as providing substantial benefits to small business and academic projects. In critiquing the NASA roadmap, Micci stated that there were significant gaps in micropropulsion technologies that are currently being developed outside of NASA, and the roadmap seems to place too much emphasis on low-performing technologies that show little promise.

Vlad Hruby (Busek) focused his presentation on micropropulsion and electrostatic and electromagnetic thrusters, which he believes are mission enabling and potentially game changing. He suggests that the United States should continue to develop EP systems at all power levels to cover a wide range of applications. In the near term he suggests deploying a small EP demonstration tug to gain operational experience. He identified Hall Effect Thrusters as the most promising approach for high-power EP systems. He also emphasized the need to improve the entire EP system, including power conditioning and power control components subsystems.

Lyon (Brad) King (Michigan Technological University) made a presentation on behalf of the American Institute for Aeronautics and Astronautics (AIAA) Electric Propulsion Technical Committee. He noted the large number of spacecraft currently in operation that employ EP systems. He believes that there are numerous technologies within the EP field that are near tipping points. He said that most of the work in advancing EP involves the often difficult challenge of scaling technologies to higher and higher power levels. He perceived two gaps in the draft roadmap for TA02:

- A lack of facilities, which will create technology development bottlenecks (there are only about one or two facilities in the United States capable of ground-testing 50-kW-class EP devices, and these are in government laboratories). In order for EP technology to grow, King asserted that universities, small businesses, and large corporations need to work in parallel, which would require additional facilities.
- The reliance on xenon propellant, which is typically the propellant of choice for EP systems. The United States may need to consider alternative propellants (to include condensables) that may have reduced performance, but which may mitigate other hurdles to development, such as the cost of fuel materials and the availability of facilities.

Dave Byers (consultant) discussed technology development strategies. He said that in the current environment, major economic forces and near-term payoffs dominate. He suggested prioritizing in-space technology investments to be responsive to multi-sector drivers that are relevant to the larger community and competitiveness. He would place a high priority on technologies that improve U.S. competitiveness in the global market place. He also suggested that technology developments must lead to practical solutions to be successful.

In the group discussion it was noted that power processing units for EP systems must be significantly improved to realize the advances made in EP thrusters. This led to a discussion on the importance of optimizing EP using a systems perspective that considers (1) all elements from power generation to thrust generation and (2) mission needs. Several speakers expressed agreement about the value of end-to-end in-space demonstrations of EP that scale to higher power over time. Several speakers also expressed shared concerns over the future availably of Xenon propellant as well as the quality and number of ground test facilities.

Session 2: In-Space High Thrust

Russ Joyner (Pratt & Whitney Rocketdyne) started the session on in-space high-thrust propulsion technologies with a discussion of the need to build on established technologies to achieve progress within affordable limits. He suggested that focusing on the current set of missions laid out by NASA is the best path forward, because propulsion technologies will most likely remain applicable even if the missions change. In his examination of the roadmap he recognized six technologies that he would consider a high priority for advancing high-thrust systems:

- Liquid cryogenic rockets,
- Liquid storable rockets,
- Nuclear thermal propulsion,
- Manufacturing/materials,
- Engine health monitoring, and
- Propellant storage.

Bruce Schnitzler (Idaho National Laboratory) focused his talk on nuclear thermal propulsion (NTP). He briefly reviewed the history of U.S. NTP efforts, which resulted in full scale engine ground testing in the 1960s and 1970s. He then noted the main advantage of NTP relative to conventional chemical propulsion systems is the much higher specific impulse that NTP can provide. This significantly reduces the launch mass requirements for human exploration missions. The challenges that must be overcome to develop a NTP system include identifying the best fuel, long development times, finding a safe and affordable means of testing, and overcoming the complications that typically arise with joint agency missions. (As with all nuclear systems developed for NASA, NTP would necessarily be a joint NASA-DOE effort.) Schnitzler suggested the best way to overcome these challenges would be to begin with a relatively low thrust system.

Joe Cassady (Aerojet) stressed the importance of a framework to guide the development of technologies and missions to prevent wasted efforts. He suggested basing this framework on an analysis of technical alternatives that examines relative merits and synergies between technologies and the flexibility of various technologies to support multiple missions and/or multiple destinations. He suggested that the technologies to be considered should be realistic for use within 20 years and have performance metrics from already demonstrated ground tests. The high-thrust options that meet this requirement are LOX/H_2 cryogenic engines, LOX/methane engines and NTP. He suggested that the keys to affordable missions is to maximize the use and reuse of common components and developing in-space propulsion technologies that allow for smaller launch vehicle. Finally he noted that technology selection has ripple effects. For example, ISRU capabilities that can produce a particular fuel would increase the usefulness of propulsion systems that can use that fuel.

Much of the group discussion focused on the challenges in pursuing NTP technology. The main challenges mentioned were developing the capability to work with specific fuels, down selecting to narrow the fuel options, the need for safe and affordable ground testing, the long time and high cost required to mature NTP technology, and anti-nuclear groups that could oppose development, testing, transportation, launch, and operation of a nuclear propulsion system. An architecture built around methane engines and ISRU production of methane fuel was suggested, but in response it was noted that development of methane engines would be difficult to develop and they would not satisfy near-term needs for advanced propulsion.

Session 3: In-Space Propulsion: Supporting Propulsion Technologies

Bernard Kutter (United Launch Alliance) started the session on supporting propulsion technologies by discussing his top priorities for in-space propulsion technologies:

- Developing an integrated vehicle fluid system so the main propellants could be used for attitude control, pressurization, and power generation.
- Improving in-space cryogenic storage.
- Improving in-space cryogenic fluid transfer.

These three technologies, when combined with efficient structural design, would lead toward an integrated cryogenic propulsion stage that would have much better performance than NASA is currently projecting. He supports a series of integrated ground tests and low-cost flight demonstrations to mature these technologies. He also supports small EP demonstrators to advance EP design and improve operating experience.

Tom Kessler (Boeing Phantom Works) focused his talk on high-power EP systems in the belief that they offer the greatest payoff in terms of affordability for a wide range of exploration and commercial space missions. He noted that EP systems are well proven at lower power levels and that a concerted effort is underway to increase EP power levels in the near term at a much faster rate than historical trends. He suggested that a 30 kW SEP flight demonstration could be conducted within the next 5 years and that a 200 to 400 kW reusable demonstrator could be flight tested by 2020. The technical challenges he identified for these EP systems concerned the availability of:

- Reliable, high-yield next generation solar cells,
- High-power, high-voltage power processing units,

- A 200 volt spacecraft power system, and
- A long-life/high-power thruster.

Al Herzl (Lockheed Martin) discussed how past technological advances are often made to directly support an identified mission need. He presented numerous examples, such as the advances in hydrogen propellant operations that were achieved during the development of the space shuttle external tank. He suggested that it is important to identify the technologies actually needed by future NASA missions, seek out adjacent commercial markets, and have those projects invest in the technology. At the same time, he indicated that projects should only be approved when required technologies are mature. He then reviewed his proposed list of near term technology needs:

- High-pressure, low-mass systems,
- Autonomous and integrated health management,
- Long-life cryogenics storage and refueling capabilities
- EP, and
- Development tools.

Jim Berry (Northrop Grumman) emphasized the need to prioritize technology investments based on benefit and cost impacts to reference missions for both exploration and science. He said NASA can only afford to invest in a limited number of technologies, and the selection of which technologies to fund should be based on approved missions. Once NASA has committed to conduct a particular mission, Berry suggested that NASA should stick with those decisions and carry them through. He also proposed choosing suites of technologies that work well together, and he urged NASA to establish technology backup options when a preferred technology is particularly risky. Berry also saw value in improving the operational lifetime of cryogenic engines and their supporting systems, including long-term in-space storage and propellant transfer. For EP systems he suggested that advances in power processing units and radiators will be keys for high-power systems. Berry also observed that small flight experiments can demonstrate the potential to operate at larger scales while validating the small systems.

The group discussion spent much of its time focused on cryogenic storage. Some speakers suggested that long-term in-space cryogenic storage should be addressed as a systems problem. One participant stated that a flight demonstrations should validate models before architectures with cryogenic storage move forward. Several speakers asserted that cryogenic fluid transfer will be required for almost any future human exploration beyond LEO. One participant questioned the value of high-power SEP in an era with low flight rates.

Session 4: In-Space Advanced Concepts

Terry Kammash (University of Michigan) started the session on advanced concepts with a discussion of fusion-based propulsion. He presented four paths that might be used to realize a fusion-based system, all of which involve technologies that are much less mature than the propulsion technologies discussed in the earlier sessions.

Rob Hoyt (Tethers Unlimited) presented an overview of three potential uses for space tethers:

- Electrodynamic tethers: a current is applied along the tether which interacts with Earth's magnetic field, imparting a force without using propellant.
- Momentum exchange tethers: enable the kinetic energy of one spacecraft to be transferred to another.
- Formation flying: tethers join multiple spacecraft that need to fly in a tight formation.

Kammash presented multiple operational scenarios using tethers. For example, a tug could boost payloads using momentum exchange tethers, and then the tug could restore its kinetic energy using an electrodynamic tether. Kammash also observed that tether flight tests had experienced a 70 percent success rate; all the failures were caused by engineering problems, not unexpected physics problems. He said that an operational demonstration of electrodynamic tethers is possible in the near term; demonstration of momentum exchange would be a long-term project.

Andrew Ketsdever (Air Force Research Laboratory) expressed support for NASA's new effort to support advanced, low-TRL concepts. He suggested that investments into advanced concepts should be agile and based on sound physics. He identified three technologies that he believes should be high priority:

- Micropulsion, which Ketsdever said was enabling for small satellites and could be developed at a low cost.
- Beamed energy, which would use ground-based power generation infrastructure to provide power to spacecraft, particularly for EP.
- Advanced high-power EP, including a concept that uses field reverse configurations and rotating magnetic fields.

Robert Frisbee (formerly of JPL) remarked that developing advanced concepts to the point where they can be used by operational systems can take a very long time and a lot of money. He noted the value in looking back at historical developments to avoid reinventing the wheel. He said that the draft roadmap for TA02 did a very good job of covering all the advanced concepts, but it should possibly add advanced tethers such as the "space elevator" concept. He noted that both fusion and advanced fission propulsion technologies would enable astronauts to complete a round trip to Mars in just 3 or 4 months, instead of the multiyear missions envisioned with near-term technology. He went on to state that, except for these advanced concepts, NTP is the only technology that allows for reasonably quick human missions to Mars. Frisbee also pointed out several crosscutting technologies from other roadmaps, such as advanced radiators and lightweight materials and structures, that are vital to advances in in-space power systems.

In the group discussion some speakers suggested making improvement in in-space infrastructure via tethers or beamed energy. Several speakers suggested that a large number of low-TRL technologies and less mature advanced concepts should be investigated at a low level of effort.

F

TA03 Space Power and Energy Storage

INTRODUCTION

The draft roadmap for technology area (TA) 03, Space Power and Energy Storage, is divided into four level 2 technology subareas:[1]

- 3.1 Power Generation
- 3.2 Energy Storage
- 3.3 Power Management and Distribution
- 3.4 Crosscutting Technology

NASA has many unique needs for space power and energy storage technologies that require special technology solutions due to extreme environmental conditions. For example,

- Venus surface operations require very high sustained temperatures (~460 °C),
- Surface penetrators must survive high-impact decelerations (hundreds of g's or more),
- Deep space planetary surface missions operate at very cold temperatures,
- Missions that travel very far from the Sun cannot rely on solar energy, and
- Missions to Jupiter and many other destinations must survive high-radiation environments.

These missions would all benefit from advanced technologies that provide more robust power systems with lower mass. In particular, this technology area encompasses pacing technology challenges for the volume, mass, and reliability critical space exploration systems. Even in the reduced gravity of the Moon or Mars, the large mass of EVA suits degrades crew operations. Advanced power and energy storage systems would directly improve the performance of EVA suits, rovers, surface habitats, and spacecraft.

The ability of space power and energy storage technologies to enable and enhance NASA's ability to learn about Earth and the solar system is illustrated by the following quotes from a recently completed decadal survey on planetary science (NRC, 2011):

---

[1] The draft space technology roadmaps are available at http://www.nasa.gov/offices/oct/strategic_integration/technology_roadmap.html.

> As future mission objectives evolve, meeting these challenges will require continued advances in several technology categories, including . . . more efficient power and propulsion for all phases of the missions.
>
> Of all the multi-mission technologies that support future missions, none is more critical than high-efficiency power systems for use throughout the solar system.
>
> Since more efficient use of the limited plutonium supply will help to ensure a robust and ongoing planetary program, the committee's highest priority for near-term multi-mission technology investment is for the completion and validation of the Advanced Stirling Radioisotope Generator.
>
> The committee recommends that NASA consider making equivalent systems investments in the advanced Ultraflex solar array technology that will provide higher power at greater efficiency. . .
>
> Investing in these system capabilities will yield a quantum leap in our ability to explore the planets and especially the outer solar system and small bodies.

Prior to prioritizing the level 3 technologies included in TA03, several technologies were renamed, deleted, or moved. The changes are explained below and illustrated in Table F.1. The complete, revised technology area breakdown structure (TABS) for all 14 TAs is shown in Appendix B.

Energy storage can be accomplished using many fundamentally different approaches. The current roadmap includes three: batteries, flywheels, and regenerative fuel cells. Two other approaches may also prove feasible for space applications: (1) electric and magnetic field storage and (2) thermal storage (especially for surface power applications). Accordingly, the structure for this roadmap has been modified by adding two new level 3 technologies:

- 3.2.4. Electric and Magnetic Field Storage
- 3.2.5. Thermal Storage

TABLE F.1 Technology Area Breakdown Structure for TA03, Space Power and Energy Storage

NASA Draft Roadmap (Revision 10)	Steering Committee-Recommended Changes
TA03 Space Power & Energy Storage	Two technologies have been added.
3.1. Power Generation	
3.1.1. Energy Harvesting	
3.1.2. Chemical (Fuel Cells, Heat Engines)	
3.1.3. Solar (Photovoltaic & Thermal)	
3.1.4. Radioisotope	
3.1.5. Fission	
3.1.6. Fusion	
3.2. Energy Storage	
3.2.1. Batteries	
3.2.2. Flywheels	
3.2.3. Regenerative Fuel Cells	
	Add: 3.2.4. Electric and Magnetic Field Storage
	Add: 3.2.5. Thermal Storage
3.3. Power Management & Distribution	
3.3.1. Fault Detection, Isolation, and Recovery (FDIR)	
3.3.2. Management & Control	
3.3.3. Distribution & Transmission	
3.3.4. Wireless Power Transmission	
3.3.5. [Power] Conversion & Regulation	
3.4. Crosscutting Technology	
3.4.1. Analytical Tools	
3.4.2. Green Energy Impact	
3.4.3. Multi-functional Structures	
3.4.4. Alternative Fuels	

TOP TECHNICAL CHALLENGES

The panel identified four top technical challenges for TA03, all of which are related to the provision of safe, reliable, and affordable in-space power systems consistent with NASA's current and potential mission needs. They are listed below in priority order.

1. Power Availability: Eliminate the constraint of power availability in planning and executing NASA missions.

Power is a critical limitation for space science and exploration. The availability of more power opens up new paradigms for how NASA operates and even what individual missions can accomplished. Increased power availability for human exploration missions translates into capabilities to support more astronauts at larger outposts with higher-capacity in situ resource utilization (ISRU) systems, higher data transmission rates, more capable mobility systems with shorter turnaround times, and higher capability science instruments. For robotic science missions, power availability has become critical in determining the scope of a mission that can be planned and how long it takes to reach mission destinations. This is due to the emergence of electric propulsion systems that enhance robotic mission design, so that the more power that is available, the shorter the trip time to any destination. Once at the destination, high power levels enable scientists to develop new approaches to scientific discovery and to communicate larger volumes of information more quickly back to Earth.

2. High-Power for Electric Propulsion: Provide enabling power system technologies for high-power electric propulsion for large payloads and planetary surface operations.

Advances in solar and nuclear technologies in the United States and elsewhere during the past decade offer the potential of developing power generation systems that can deliver tens to hundreds of kilowatts. For example, inverted metamorphic (IMM) solar cells are being developed to deliver 40 percent efficiency with very little mass due to removing the thick substrate used to grow the multi-layer photovoltaic semiconductor materials. New lightweight structures also greatly reduce the mass of solar arrays and enable higher power outputs. As solar arrays grow to large sizes (such as hundreds of kW for electric propulsion planetary missions), new technology will be needed for the control and pointing of the large arrays. Nuclear fission system concepts have been developed for lunar and Mars missions that provide pathways to reasonable mass reactors that can be placed and operated on a planetary surface to deliver 10 kW to 100 kW. These designs use proven fuels, power conversion technologies, and reactor materials to reduce the development and operations risk to acceptable levels. Other aspects of fission systems require technology development including heat exchangers, fluid management, scaling of power conversion devices, heat rejection components, radiation shielding, and aspects of system integration and testing.

3. Reduced Mass: Reduce the mass and stowed launch volume of space power systems.

Power systems typically comprise one third of the mass of a spacecraft at launch, and the volume available in the launch vehicle fairing can limit the size of solar arrays that can be packaged on the vehicle. New power generation, energy storage, and power delivery technologies have the potential to cut the mass and volume of these systems by a factor of two to three. Successfully developing these technologies would enable missions to include more science instruments, use smaller and less expensive launch vehicles, and/or provide higher power levels.

4. Power System Options: Provide reliable power system options to survive the wide range of environments unique to NASA missions.

NASA missions require power systems and components to survive many different types of extreme environments. This can include high radiation levels, very high or very low temperatures, very high impact forces (for planetary surface penetrators), highly corrosive environments, dusty atmospheres, and other unique extremes. In all of these challenging environments, the power system must operate predictably and reliably or the mission is lost. Continued advances in space power technology that improve NASA's ability to overcome these challenges will enable NASA to plan and execute a wider array of missions.

QFD MATRIX AND NUMERICAL RESULTS FOR TA03

The panel evaluated 20 level 3 technologies in TA03, Space Power and Energy Storage. Eighteen of these technologies appear in the draft roadmap for TA03; the other two were added by the panel, as explained above. The results of the QFD scoring are shown in Figure F.1. Figure F.2 shows the breakdown of the technologies into high, medium, and low priority categories. Six technologies were assessed to be high-priority technologies:

- Solar (photovoltaic and thermal),
- Fission,
- Distribution and transmission,
- Conversion and regulation,
- Batteries, and
- Radioisotope power systems.

The first five technologies were designated as high-priority technologies because they received the highest QFD scores based on the panel's initial assessment. The panel subsequently decided to override the QFD scoring results

	Benefit	Alignment with NASA Needs	Alignment with Non-NASA Aero Tech Needs	Alignment with Non-Aerospace National Goals	Technical Risk and Reasonableness	Sequencing and Timing	Time and Effort	QFD Score (Weighted)	Panel Priority
Multiplier	27	5	2	2	10	4	4		
	0/1/3/9	0/1/3/9	0/1/3/9	0/1/3/9	1/3/9	-9/-3/-1/1	-9/-3/-1/0		
Technology Name	**Benefit**	**Alignment**			**Risk/Difficulty**				
3.1.1. Energy Harvesting	1	3	1	1	3	-3	-1	60	L
3.1.2. Chemical (Fuel Cells, Heat Engines)	3	3	3	3	3	-1	-1	130	M
3.1.3. Solar (Photovoltaic and Thermal Power)	9	9	9	9	9	1	-3	406	H
3.1.4. Radioisotope (Power)	3	3	1	1	3	1	-3	122	H*
3.1.5. Fission (Power)	9	9	1	1	9	1	-3	374	H
3.1.6. Fusion	0	3	1	3	1	-9	-9	-39	L
3.2.1. Batteries	3	9	9	9	3	1	-1	192	H
3.2.2. Flywheels	1	3	3	3	3	-1	-1	76	L
3.2.3. Regenerative Fuel Cells	1	1	1	1	3	-1	-3	50	L
3.2.4. Electric and Magnetic Field Storage	1	3	3	1	3	-3	-1	64	L
3.2.5. Thermal Storage	3	3	1	3	3	-3	-1	118	M
3.3.1. (Power) Fault Detection Isolation and Recovery	1	9	9	3	3	-1	-1	118	M
3.3.2. Management and Control	1	9	9	3	3	-1	-1	118	M
3.3.3. Distribution and Transmission	3	9	9	3	9	-3	-3	216	H
3.3.4. Wireless Power Transmission	1	3	3	3	3	-3	-3	60	L
3.3.5. (Power) Conversion and Regulation	3	9	9	3	9	-3	-3	216	H
3.4.1. Analytical Tools	1	9	3	1	3	-1	0	106	M
3.4.2. Green Energy Impact	1	1	1	3	3	-3	-1	54	L
3.4.3. Multi-functional Structures	1	9	3	1	3	-1	-1	102	M
3.4.4. Alternative Fuels	1	3	1	1	3	-3	-3	52	L

FIGURE F.1 Quality function deployment (QFD) summary matrix for TA03 Space Power and Energy Storage. The justification for the high-priority designation of all high-priority technologies appears in the section "High-Priority Level 3 Technologies." H = High Priority; H* = High Priority, QFD score override; M = Medium Priority; L = Low Priority.

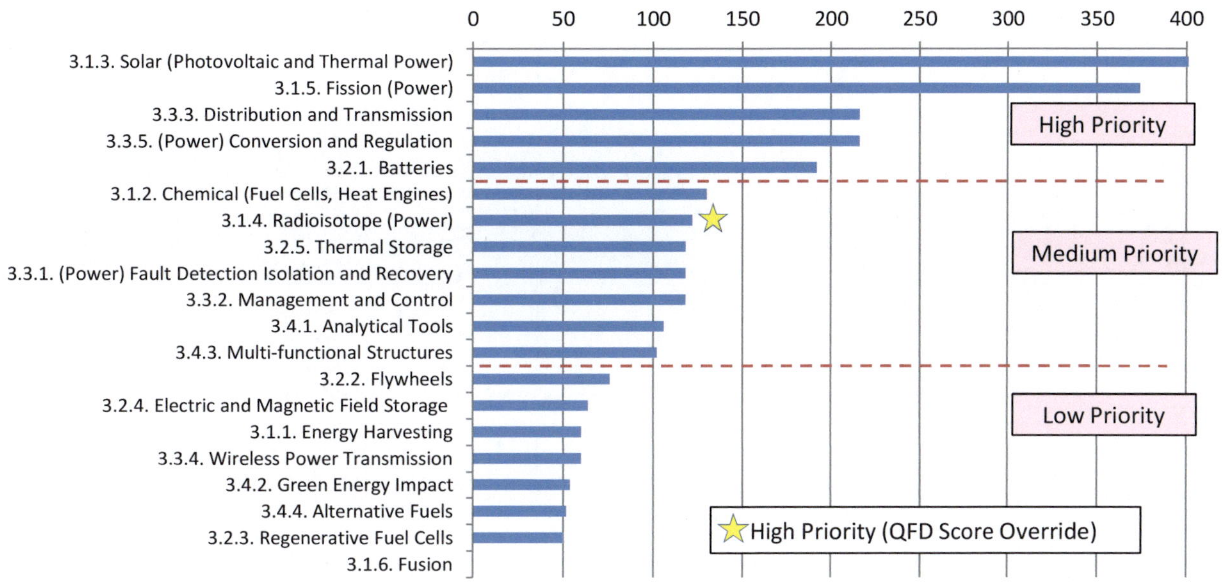

FIGURE F.2 Quality function deployment rankings for TA03 Space Power and Energy Storage.

to designate radioisotope power systems as a high-priority technology to highlight the critical importance of currently funded and planned programs for Pu-238 production and Stirling engine development and qualification.

Figure F.3 correlates the applicability of the top technical challenges, as described above, to each of the Space Power and Energy Storage level 3 technologies. This shows that the high-priority technologies, discussed in the next section, provide the potential solutions that will meet these challenges. The medium- and low-ranked technologies are judged to have a weak linkage because of the limited benefit of investing in these technologies regardless of how closely they may overlap with various challenges in terms of subject matter.

HIGH-PRIORITY LEVEL 3 TECHNOLOGIES

Panel 1 identified six high-priority technologies in TA03. The justification for ranking each of these technologies as a high priority is discussed below.

Technology 3.1.3, Solar (Photovoltaic and Thermal)

Photovoltaic space power systems have been the workhorse of NASA science missions as well as the foundation for commercial and military space systems. Solar cells directly convert sunlight into electricity. Today's solar cells are made from III-V materials[2] and are composed of multiple junctions of various band gaps to achieve solar conversion efficiency of 30 percent. Current emphasis is on the development of high-efficiency cells. NASA also needs

- Cells that can effectively operate in low-intensity/low-temperature (LILT) conditions (which is typical when spacecraft are more than three astronomical units from the Sun),
- Cells and arrays that can operate for long periods at high temperatures (>200°C),
- High specific power arrays (500 to 1000 W/kg), and
- Electrostatically clean, radiation tolerant, dust tolerant, and durable, re-stowable, and/or deployable arrays.

[2] III-V materials are semiconductor materials made of materials from Groups III and V of the periodic table, such as gallium-arsenide.

Priority	TA 03 Technologies, Listed by Priority	Top Technology Challenges: 1. Power Availability: Eliminate the constraint of power availability in planning and executing NASA missions.	2. High-Power for Electric Propulsion: Provide enabling power system technologies for high power electric propulsion for large payloads and planetary surface operations.	3. Reduced Mass: Reduce the mass and stowed launch volume of space power systems.	4. Power System Options: Provide reliable power system options to survive the wide range of environments unique to NASA missions.
H	3.1.3. Solar (Photovoltaic and Thermal Power)	●	●	●	●
H	3.1.5. Fission (Power)	●	●	●	●
H	3.3.3. (Power) Distribution and Transmission	○	●	○	
H	3.3.5. (Power) Conversion and Regulation	○	●	○	○
H	3.2.1. Batteries	○		●	●
H	3.1.4. Radioisotope (Power)	○			●
M	3.1.2. Chemical (Fuel Cells, Heat Engines)			○	
M	3.2.5. Thermal Storage				
M	3.3.1. (Power) Fault Detection Isolation and Recovery				
M	3.3.2. Management and Control				
M	3.4.1. Analytical Tools				
M	3.4.3. Multi-functional Structures				
L	3.2.2. Flywheels				
L	3.2.4. Electric and Magnetic Field Storage				
L	3.1.1. Energy Harvesting				
L	3.3.4. Wireless Power Transmission				
L	3.4.2. Green Energy Impact				
L	3.4.4. Alternative Fuels				
L	3.2.3. Regenerative Fuel Cells				
L	3.1.6. Fusion				

●	Strong Linkage: Investments by NASA in this technology would likely have a major impact in addressing this challenge.
○	Moderate Linkage: Investments by NASA in this technology would likely have a moderate impact in addressing this challenge.
[blank]	Weak/No Linkage: Investments by NASA in this technology would likely have little or no impact in addressing the challenge.

FIGURE F.3 Level of support that the technologies provide to the top technical challenges for TA03 Space Power and Energy Storage.

Because of the critical importance of photovoltaic power systems, the Department of Defense (DOD) is funding technology development at the cell level to raise that efficiency first to 33 percent and then to 39 percent. The most common usage is in planar solar arrays but certain concentrator arrays have flown successfully and two types are currently under development. Concentrator arrays offer cost reductions due to reduced solar cell material and efficiency gains.

Nearly all spacecraft flown to date have been powered by solar arrays. Photovoltaic power systems provide the energy for NASA science missions in low Earth orbit (LEO), including the International Space Station (ISS) and higher altitude communication systems such as the Tracking and Data Relay Satellite Systems (TDRSS). They have powered science missions to Mars (both in orbit and on the surface), Venus, and Mercury. Advanced photovoltaic power systems will be used on the Juno mission to Jupiter and the Solar Probe Plus near the Sun. Commercial communications satellites in geosynchronous Earth orbit (GEO) rely exclusively on photovoltaic power systems. NOAA polar orbiting weather satellites also rely on photovoltaic power systems. The DOD uses photovoltaic power systems for satellites in LEO, GEO and mid-Earth orbits (MEO) for observation, event detection, navigation (the Global Positioning System), and others applications. The high priority assigned to this level 3 technology is based on the benefit provide by photovoltaic research. The draft roadmap for TA03 also includes solar thermal systems in technology 3.1.3, but the panel does not perceive solar thermal power technology as a high priority because it has not been used in space and it has not proven to be cost competitive.

Many photovoltaic power systems have achieved a TRL of 9, including the III-V triple junction solar cells. Advances in new photovoltaic technology, such as IMM photovoltaic cells (which remove the substrate to leave a cell that is only a few microns thick and has a power conversion efficiency of more than 30 percent with exceedingly low mass and inherent radiation tolerance. IMM arrays will be storable and deployable. The increased efficiency of IMM cells will enable smaller arrays, and the lightweight cells may enable alternate array structures to further reduce mass. Lightweight solar array structures with specific mass beyond the 100 W/kg offered by the current state of the art would also be beneficial. These technologies are currently at TRL 3 or below.

As discussed in Appendix E (TA02, In-Space Propulsion Technologies), NASA has a vital interest in photovoltaic power system developments for high-power electric propulsion (EP) missions. Because power system mass reduction is critical to EP missions, advanced array technologies that offer high specific mass (>500 W/kg) and high power density (>300 W/m^2) are critical technology development areas. Increases in solar cell efficiencies offered by IMM photovoltaic cell technology and lightweight array technology, including both planar and concentrator approaches (now at TRL 2), is expected to achieve a specific mass of more than 600 W/kg and a power density of more than 400 W/m^2.

The NASA Glenn Research Center has been the primary NASA center developing photovoltaic technologies, with the Jet Propulsion Laboratory (JPL) and NASA Goddard Space Flight Center (GSFC) also providing significant capabilities and facilities to advance the state of the art. NASA is well qualified to lead development of high-power solar array technology because of its expertise and capabilities plus the diversity of its mission needs, while collaborating with DOD, DOE, commercial industry, and academia. While the DOD has a modest investment in IMM solar cell technology, NASA is highly motivated to invest in IMM technology to ensure timely development of next generation solar cells to meet its own mission needs. Continued interactions between NASA and other countries investing in space photovoltaic technology, including multiple European countries and Japan, would be beneficial. The ISS has been used to test and qualify new photovoltaic cell technologies in the past, and it remains available for this function.

Photovoltaic power technology is a high priority because of the game-changing impact that higher power, lighter weight solar arrays would have on future NASA missions. Solar power generation applies to virtually all NASA mission areas plus DOD, commercial, and civil space enterprises. Development efforts will therefore lead to widespread benefits across the user spectrum. The development risks for high-power solar arrays are moderate to high, which is appropriate for NASA technology investments. Space demonstration tests of lightweight arrays with power output of 30 to 50 kW level are warranted, and the risks at that size are relatively low. Development of larger arrays with power output as high as 1 MW will be required to support large-scale exploration missions. Solar arrays with such a high power output would be game changing. Structures large enough for such a large array have not been developed, however, and a focused development program for large arrays, that addresses control and pointing issues and the higher risk associated with higher power arrays, is warranted to support NASA exploration missions.

Technology 3.1.5, Fission

Space fission power systems use heat generated by fission of a nuclear fuel to power a thermal to electric conversion device to generate electric power. Key subsystems include the reactor, heat exchanger (to move the heat out of the reactor and into the power converter), power converter, heat rejection radiator, and radiation shield. As noted in an earlier study that focused on space applications of fission reactor systems (NRC, 2006, pp. 10-11):

> Nuclear reactor systems, which can provide relatively high power over long periods, make it possible to design missions with more numerous and more capable science instruments, high-bandwidth communications systems, shorter transit times, and greater flexibility to change the course and speed of spacecraft enough to conduct extended investigations (rather than brief flybys) of bodies of interest; visit multiple bodies much more easily; and significantly alter a spacecraft's trajectory in response to information collected during a particular mission. Nuclear reactors have the potential to overcome limitations associated with low energy and power. They do this by providing electricity and

propulsion over a wide range of power levels for extended periods (years to decades), including during both transit and surface operations, without regard to the availability of either solar energy or large quantities of chemical fuel. Nuclear reactor systems, however, are expensive to develop, and their potential will be realized only if key technology issues can be overcome.

Space fission technology is currently assessed to be at TRL 3. While some components are demonstrated at higher TRLs, many of the required elements require technology development to advance beyond TRL 3. Other components have reached higher TRLs in past programs such as the SP-100 and Prometheus programs, but technology capability has been lost and must be redeveloped. Key subsystems that must be addressed include the reactor (including instrumentation and control/safety), energy conversion, heat transfer, heat rejection, and radiation shields.

NASA has some of the expertise needed to develop space fission power system technologies, but it will need to work in collaboration with the DOE and private industry in order to advance the technology to TRL 6 and beyond. Because of their unique capabilities and statutory requirements, DOE must take the lead on the development reactor components and technologies, including the fuel. NASA centers including the Glenn Research Center and the Marshall Space Flight Center have the expertise and facilities to lead development of the energy conversion and heat rejection subsystems. NASA is also qualified to lead overall systems engineering efforts, with DOE assistance for nuclear subsystems. This is a technology that is primarily suited to space exploration needs, and therefore NASA is best positioned to take the lead in maturing the technology to TRL 6 or beyond. There may be opportunities to collaborate with some international partners that have capabilities and facilities in fast reactor technologies, such as Russia, Japan, and France. Use of the space station is not appropriate for developing this technology.

Space fission power systems would be game-changing due to the potential (1) to provide a power rich environment to planetary surface exploration missions, especially for crewed missions, and (2) to enable high-power electric propulsion systems for deep space exploration and science missions. The alignment of this technology with NASA's needs is high because of the game-changing impact it would make on both robotic science and human exploration capabilities. Alignment with other aerospace and national needs is considered to be low because space power reactors will be designed as fast neutron reactors, and there are no significant terrestrial applications for fast neutron reactor technology. The risk is assessed to be moderate to high, which is appropriate for NASA. (This risk level assumes that NASA will set realistic goals for the fission system to be developed; otherwise the risk could increase to the "very high" level.) The next space fission development program would hopefully adopt performance and life goals that are not as ambitious as the fission power system goals incorporated in the cancelled Jupiter Icy Moons Orbiter (JIMO) mission, so that entirely new materials or fuels will not need to be developed. A space reactor concept based on liquid metal cooling, stainless steel structures, and UO_2 fuel would have the strongest technology base, drawing on decades of development. The pursuit of higher temperature systems has been the primary source of technical problems and associated cost and schedule overruns in prior space fission power system development projects. The cost of the power system development is expected to be high, in the range of $1 billion to $2 billion over 10 to 12 years. This cost is easily justified, however, by the potential benefits that can be realized in both the human and robotic exploration of the outer solar system and beyond. Thus, fission is judged to be a high-priority technology.

Technology 3.3.3, Distribution and Transmission

Interest in the components of a spacecraft electrical power system often centers on the power generation and storage functions, and many resources have been devoted to exploring options, developing alternatives, and improving the specific power and energy of the systems that serve these two functions. As game-changing science and human exploration missions of the future are examined, the need for significant increases in electrical power on spacecraft becomes a clearer and higher priority. With these higher power levels, an extrapolation of the current technologies for the distribution and transmission (D&T) of power would result in unacceptably high mass and complexity. Thus, D&T is judged to be a high-priority technology area.

Distribution and transmission on a space system is comprised primarily of cables (copper conductors and insulation) and connectors. Currently, electrical power on spacecraft is distributed by direct current (DC) at fixed voltages ranging from 28V DC to 100V DC. The ISS is the exception to this rule, with a 120V DC distribution architecture. As higher power systems are developed, higher voltage distribution will be required to avoid very heavy wire bundles. This can be done with either DC or alternating current (AC) architectures. Paschen's law limits high-voltage DC systems to a maximum of about 270 V DC. Therefore, high power levels will require more attention to AC systems, probably at relatively high frequencies. The demand for lower mass will challenge the traditional copper materials for the bus, and the need to operate at higher temperatures (to increase the efficiency of the thermal management systems as well as electronics efficiencies) may appear counter to the need to reduce ohmic losses in transmission lines.

Proposed research under technology 3.3.3 would increase the voltage of D&T subsystems, develop high-frequency AC distribution options for space systems, and identify alternate materials to replace copper conductors. Copper wire has long been a conductor of choice for spacecraft, but as power levels increase, so too will current and voltage and with them the conductor mass will grow. With DC currents, to reduce the mass penalty of larger cables, alternate materials such as superconductors or nano-material conductors may need to be developed, along with lighter space-qualified insulating materials capable of protecting systems at high voltage. With AC power systems, advancing beyond the 116 V AC system in the space shuttle may require very high operating frequencies; for example, NASA funded development of a 440 volt, 20 kHz AC power system for Space Station Freedom until it was reconfigured to use a DC power system (Patel, 2005). Technical needs include keeping transmission losses to a minimum, reducing transformer masses, incorporating fault protection and smart telemetry into power distribution architectures, and developing new connectors.

Ultimately, the nature of future missions will dictate the architecture and technologies used for vehicle power systems, and that, in turn, will define the requirements for electrical power D&T. For example, the electrical power from a nuclear reactor-turboalternator system will likely be high-voltage AC, while power from photovoltaics is always generated as a DC voltage. If electric thrusters are needed for the mission, very-high-voltage DC power (kilovolts) will be required, perhaps from a nuclear prime-power source. Each of these options will impose D&T technology requirements that are, in most cases, not yet at TRL 3. The risks associated with developing the needed D&T technologies are high over the next two decades, but there may be no alternative to addressing those needs if the needed power is to be delivered to the load with acceptable mass, volume, and efficiency constraints.

NASA is well suited to lead development of advanced D&T technologies. Advances in this technology will likely require significant participation from both industry and academia for voltage selection, architecture technology option development, and advanced transmission and insulation materials. The ISS provides limited benefit to the development this technology, as testing can be accomplished on the ground using vacuum chambers and test fixtures. In addition, the introduction of high voltages on the ISS could pose safety risks, which would increase the cost and schedule of any ISS test program. However, in-space testing is ultimately required (on the ISS or some other platform) to validate new D&T technology in the space environment. For example, in-space testing is needed to address plasma interactions and micro-meteoroid impacts.

The panel determined that advanced D&T technology could provide significant benefits in terms of system mass reduction due to the major reductions in cable harness mass that could be achieved with high-voltage systems. Advanced technologies in this area would be broadly applicable to many classes of NASA and non-NASA space systems, and potentially to a broad range of terrestrial systems as well. The risk was judged to be medium to high, within the typical range of NASA technology programs.

Technology 3.3.5, Conversion and Regulation

The voltage and current of electrical power available on any particular spacecraft will be dictated by the power source and the power management and distribution architecture. Various payloads will then most likely require the power in a different form, such as higher voltage for electric propulsion. The purpose of electrical power conversion and regulation is therefore to provide the necessary bridge between the power source and payloads, and to regulate this power to within the tolerances required by the payloads.

Currently unresolved issues include the need to (1) space qualify existing terrestrial high-voltage components and (2) replace space qualified components that currently lag significantly behind the commercial state of the art. Important parameters for improving power conversion and regulation devices include increasing conversion efficiency, operating temperature range, and radiation tolerance. Advanced conversion and regulation technology with relatively near-term application is at TRL 4.

Future space power generation and distribution systems are likely to operate at high voltages than current systems to increase efficiency. Higher current ratings, lower switching and conduction losses, and higher junction temperature tolerances would improve the functionality of future system components. The need for high-voltage regulation is also associated with some electric propulsion technologies that require high voltages (kilovolts) to function. Development of high-voltage regulation capabilities would require a major project that includes the development of many new technologies and facilities. The next generation technology for power conversion and regulation is at TRL 2 to 3.

An example of advanced conversion and regulation technology is a higher band gap material such as silicon-carbide or gallium-nitride that would replace the traditional silicon materials in switching components, thereby increasing device operating temperature and efficiency while decreasing mass and volume. Another example is advanced magnetics for improved conversion and regulation devices.

NASA has the extensive capabilities, equipment, and facilities needed to lead research and technology development of advanced conversion and regulation technology. NASA will need to collaborate with industry and academia to leverage the power electronics advances being made for other applications, including commercial uses. Advancements that NASA makes in this area will have application to future electric vehicles and commercial smart-grid systems where power is generated and regulated close to its point of use. Some aspects of technology development can potentially be accomplished jointly with offices of the DOD and DOE. Qualification of the hardware in the space environment could be facilitated by attaching an experimental payload externally to the ISS, providing a low-risk, in-space demonstration.

Conversion and regulation of power was highly ranked because it was identified as potentially providing a major improvement in mission performance, as well as having broad applicability across several NASA mission areas. The main benefits of advanced power conversion and regulation on system performance come both directly in the form of lower subsystem mass, as well as indirectly in the form of better conversion efficiencies that provide higher power margin. These benefits are especially important at the higher power levels needed for electric propulsion systems or high-bandwidth communications. Increasing the efficiency of power conversion could potentially reduce the size of solar arrays, batteries, and thermal control systems by more than 10 percent on lower power systems, with a bigger impact for higher power systems. Conversion and regulation was assessed to have broad application across the general aerospace community, as all spacecraft use power conversion and regulation components. The technical risk associated with future advancement of this technology is moderate to high, which is a good fit to NASA's level of risk tolerance for technology development, and the likely cost to NASA and the timeframe to complete technology development is not expected to substantially exceed that of past efforts to develop comparable technologies.

Technology 3.2.1, Batteries

Batteries are electrochemical energy storage devices that have been flown in space since the beginning of the space age. Battery technology has advanced continuously, and further high-payoff improvements are possible through recent scientific discoveries. In space, batteries must survive a variety of harsh, sometimes unique environments and load profiles are more demanding than for most terrestrial applications. Many batteries are already proven in space, at TRL 9, but a variety of advanced chemistry alternatives have yet to be developed and qualified for spaceflight.

After relying on nickel-cadmium and nickel-hydrogen rechargeable batteries for decades, the aerospace industry is now moving to lithium-ion (Li-ion) batteries as the standard energy storage component for space systems. Li-ion technology provides a substantial improvement in specific energy, charge and discharge rates, and cycle life at high depths-of-discharge. Li-ion batteries are also being used as a primary battery (where the battery is used once and not recharged) in applications such as launch vehicle electric power systems.

NASA missions would benefit from new electrochemical power technologies that offer higher specific energy and/or higher specific power. NASA has world-class expertise in space battery technology, with capabilities at multiple centers including JPL, Glenn Research Center, Goddard Space Flight Center, and Ames Research Center. NASA is best qualified to lead development of advanced battery technology for their unique mission needs. Ideally, research in this technology would leverage commercial technology developments, as NASA did with the development of Li-ion batteries for space applications.

Rechargeable battery technology is advancing rapidly to meet the commercial needs of electric vehicles, cell phones, laptop computers/tablets, and renewable energy systems. NASA would therefore benefit from collaborations with DOE and commercial industry in developing advanced battery technology. DOE's Advanced Research Projects Agency-Energy (ARPA-E) is pursuing a wide range of advanced battery technologies, and the commercial automotive industry is investing in efficient batteries for plug-in hybrid/electric vehicles (where longevity, energy density, and reliability are quite important, as they are in space applications. The ISS can be an asset in testing some advanced batteries (such as those with liquid electrodes or electrolytes) in a relevant thermal, electrical, and microgravity environment.

The steering committee assessed the benefit of battery technology to be significant due to the potential to reduce mass for many space systems and to enable missions in extreme environment missions. The alignment of this technology with NASA's needs is high due to its potential impact on both robotic science and human exploration capabilities; its alignment with other aerospace and national needs is also high because batteries are so widely used. NASA can capitalize on the investments by other government and commercial organizations that are making substantial investments in advanced battery technologies. However, the unique requirements posed by NASA missions in extreme environments do require NASA-specific research and development with moderate risk.

Technology 3.1.4, Radioisotope

Radioisotope power systems (RPSs) provide power to scientific and human exploration missions over long periods almost anywhere in the solar system and beyond. RPSs have enabled many unique deep space and planetary exploration missions, making important scientific discoveries possible. RPSs used plutonium-238 (Pu-238) as a heat source, and they have used thermoelectric converters since 1961 to provide reliable electrical power for many missions throughout the solar system, including Pioneer, Viking, Viking landers, Galileo, Ulysses, Apollo 12-17, Cassini, and New Horizons. Demonstrated mission operating lifetimes have exceeded 30 years. Future RPSs could be developed to deliver both lower power levels (watts or fractions of a watt) and higher power levels (hundreds of watts up to 1 kW). The higher power systems would enable radioisotope electric propulsion for deep space missions, making several new classes of missions possible.

While RPSs have a well-established foundation, there are significant technology issues that must be overcome to maximize the effectiveness of the United States' dwindling supply of available Pu-238. DOE no longer has the ability to produce Pu-238 (except for very small amounts for research), and the United States has purchased all available Pu-238 from Russia. No other country, including Russia, is currently producing Pu-238, and multiple studies have shown that there is no other available radioisotope material that can meet even a significant fraction of NASA's RPS needs. Supported by recent NRC studies (NRC, 2010, 2011), NASA and DOE have been attempting to restart production of a limited annual quantity of Pu-238 for the past few years, but Congress has not yet provided funding to the DOE and/or NASA for this purpose.[3]

NASA and DOE have been developing advanced RPSs that would use Stirling engines to replace thermoelectric converters. Because the energy conversion efficiency of the Stirling engine under development is about 5 times that of thermoelectric converters, Stirling engines require significantly smaller quantities of Pu-238 to achieve similar power levels. Given the scarcity of Pu-238 (which will persist for years even after Pu-238 production is approved and funded), the much higher efficiency of Stirling engines is necessary if RPSs are to be available for NASA's

[3] By statute, the U.S. DOE is the only federal agency authorized to produce nuclear material. Congressional action through October 2011 indicates that the 2012 budget for DOE will include no funds to restart production of Pu-238. The NASA budget for 2012 may include up to $10 million for this purpose, but it remains to be seen when the DOE will be able to restart production of Pu-238.

planned science and exploration missions. As discussed in Chapter 3, establishing a reliable, recurring source of Pu-238 and maturing Stirling engine technology are both critically important to provide power for NASA's future science and exploration missions that cannot rely on solar power.

Radioisotope technology using Stirling engines is currently assessed to be at TRL 6. Although some components have been demonstrated at higher TRLs, a flight test is needed to advance beyond TRL 6. Using the ISS to demonstrate this technology is not an option due to restrictions regarding operation of nuclear power systems in LEO. RPS technology is somewhat unique to NASA, as interplanetary space missions are the only driving need that has been identified to justify restarting Pu-238 production. NASA and DOE have the unique capabilities and facilities necessary to develop RPSs. By statute, DOE must be responsible for the nuclear aspects of RPS technology development. NASA Glenn Research Center has led the development of Stirling engines and the Jet Propulsion Laboratory leads NASA efforts in RPS development and spacecraft integration.

In assigning the QFD scores for this technology, the panel assumed that Pu-238 production and Stirling technology development would continue as currently planned by NASA. As noted above, RPS technology was selected as a high-priority technology despite its relatively modest QFD score because this technology is critically important to the future of NASA's deep space missions. The committee assessed the benefit of additional investments in RPS technology to be low because there are few good options with the potential to improve on the performance of RPSs that couple Stirling engines with Pu-238 heat sources. However, as noted above, this rating would be much higher if those technologies were not already being developed. The alignment of this technology to NASA's needs is high due to the high impact on both robotic science and human exploration capabilities, while alignment with other aerospace and national needs is considered to be low. The risk is assessed to be moderate to high, within the bounds of NASA's acceptable risk levels for technology development. The cost of the power system development is expected to be moderate. Thus, RPS technology would be assessed as a medium-priority technology based on its QFD score, which is based on two assumptions: (1) the current program for Stirling engine development is continued and (2) domestic production of Pu-238 is restored in a timely fashion. Given that the second assumption remains in doubt, the panel overrode the QFD score to assign this technology a high priority.

MEDIUM- AND LOW-PRIORITY TECHNOLOGIES

TA03 includes 14 level 3 technologies that ranked low or medium priority. This includes two technologies that were added to the TABS for completeness. Thermal storage was added due to its potential to improve energy storage mass, as compared to advanced batteries, for many special purpose applications. For example, if a heat-engine-based power system is used, energy can be efficiently stored thermally instead of using an electrochemical system. In situ resource utilization (ISRU) is another application where thermal energy storage might be preferable, especially if thermally based processes are used. Electric and magnetic field storage was added as a technology to cover many advancing technologies such as super-capacitors, ultra-capacitors, and superconducting magnetic energy storage. In general, these technologies are useful options when storage times are very short such as peak load management for a high-power-radar instrument.

Seven of the eight technologies that were assessed to be low priority were judged to have marginal benefits to NASA missions within the next 20 to 30 years. These technologies included energy harvesting, flywheels, regenerative fuel cells, electric and magnetic field storage, green energy impact, alternative fuel storage, and wireless power transmissions. The marginal benefit (less than 10 percent improvement) evaluation was based on an assessment of the expected improvement, at the system level, in the primary parameter of interest for each technology. In most cases, this was improvement in spacecraft mass or reliability that the panel believed could be achieved given reasonable investments in that technology. While higher claims have been made for some of these technologies, such as flywheels or electric and magnetic field storage, the panel's review of available information did not produce any credible technology development paths that would achieve the ambitious performance levels specified in the draft roadmap with reasonable investments. Also, currently available approaches for advancing these technologies tended to have a lower risk level than is usually appropriate for NASA technology investments. The remaining low-priority technology, fusion, was judged to provide no likely value to NASA in the next 20 to 30 years due to a very low probability of success during that timeframe.

The six power and energy storage technologies that were ranked as medium priority included four that had marginal benefit scores and two that had more substantial benefits, but application of those benefits was limited to a small set of mission areas. The four with marginal benefits included fault detection, isolation, and recovery (FDIR), management and control, analytical tools, and multi-functional structures. While the benefit of these technologies was assessed to be similar to the low-priority technologies discussed previously, all four ranked higher because any advancement would apply to all or nearly all NASA and non-NASA space missions in all or most mission classes. In other words, success in these technologies helps everyone. The remaining two medium-priority technologies, thermal storage and chemical power generation, were judged to provide substantial benefit to a smaller set of mission opportunities. Thermal storage applications include heat engine power systems and ISRU, as described above. Chemical power generation, including fuel cells and heat engines, may be valuable in human exploration missions where large quantities of hydrogen and oxygen are being used for propulsion. In these cases, power can be generated essentially for free using the boil-off from the propellant tanks.

DEVELOPMENT AND SCHEDULE CHANGES FOR THE TECHNOLOGIES COVERED BY THE ROADMAP

Schedules for Space Power and Energy Storage technologies are highly dependent on the level of funding applied to the development programs. The schedules depicted in the roadmap are generally feasible if sufficient resources are applied to each item in the roadmap.

OTHER GENERAL COMMENTS ON THE ROADMAP

Space Power and Energy Storage is related to several other technical areas. Many challenging requirements arise from high-power electric propulsion applications discussed in TA02. Heat rejection from power and energy storage components relies on technologies from the thermal control systems covered by TA14. Advances in many power technologies are possible due to advancing materials technologies discussed under the materials and structures technologies in TA12 and nanomaterials covered by TA10.

PUBLIC WORKSHOP SUMMARY

The workshop for the Space Power and Energy Storage technology area was conducted by the Propulsion and Power Panel on March 21, 2011, on the campus of the California Institute of Technology in Pasadena, California. The discussion was led by panel member Douglas Allen, who began with a general overview of the draft roadmap and the NRC's task for this study. He also provided some direction for what topics the invited speakers should cover in their presentations. Experts from industry, academia, and government were invited to lead a 25 minute presentation and discussion of their perspective on the draft NASA roadmap for TA03. At the end of the session, there was a short open discussion by the workshop attendees that focused on the recent session. At the end of the day, a concluding discussion led by Allen summarized the key points observed during the day's discussion.

Session 1: Solar Arrays

Ed Gaddy (Applied Physics Laboratory) started the solar arrays session with a discussion on properly measuring the impact of technology improvements. His argument was that often the cost impact of weight saving technology improvement is significantly underestimated. This can lead to under-investment in advanced technologies. He also observed that solar array history has shown that slow and steady investments, which lead to slow and steady technology improvements can be highly successful and, over time, this approach can yield revolutionary improvements in capabilities. He advocated support for Inverted Metamorphic Multijunction (IMM) photovoltaic cells as the next evolutionary step forward in this technology. Finally, he stated operational missions need better insight into the benefits of adopting specific new technologies and/or incentives should be established to encourage mission managers to adopt advanced technologies that have achieved an appropriately high TRL.

Alan Jones (ATK) began his review of the TA03 roadmap by saying that its near-term goals were generally achievable, but the long-term goals would be much more challenging. He then reviewed the solar power array challenges that are specific to NASA missions, such as very low or very high environmental temperatures. He endorsed NASA investment in IMM photovoltaic cell technology, but added that support for advanced array technologies is also needed. In particular, he suggested that investing in wing-level structural platforms is needed to fully realize the benefits of the new cells.

Ted Stern (Vanguard Space Technologies) discussed integrated array manufacturing methods that could reduce array cost and weight and increase array reliability. He also discussed a concept of modular solar arrays that would simplify spacecraft design by making all arrays using the same method, which would allow for easier qualification. He also discussed the advantages and disadvantages of near-term technology for solar concentrators to improve solar array performance. For the long-term, he supported investments in spectrum converting and multi-photon enabled photovoltaics that may provide game-changing performance. Finally, he noted the value in showing the relevance of improved solar arrays to terrestrial spinoffs.

Paul Sharps (Emcore Photovoltaics) began by providing some information on Emcore and noting that it is one of two U.S. developers and manufacturers of high-efficiency multi-function solar cells and arrays for space applications. He noted that while solar cells have received continuous DOD funding for 25 years, there has been little improvements in cell integration and array technologies. He provided his perspective on IMM technology, stating that Emcore had achieved 34 percent energy conversion efficiency with this technology in the laboratory, and work is progressing with concepts that may achieve 37 percent efficiency. (State-of-the-art cells have less than 30 percent conversion efficiency.) Sharps believes that a 34 percent efficient, radiation hard IMM cell is 2 to 3 years away from being inserted into flight hardware. He suggested that NASA focus its photovoltaic cell investments on this technology, adapting it for NASA specific requirements such as very-high- and very-low-temperature environments and the very low light intensity experienced in deep space.

In the group discussion the speakers were asked to forecast where state-of-the-art array technology might be in the next 20 years. The speakers seemed to generally agree that a power density of 400 W/kg will be available in the near term, but that long-term forecasting is very difficult given that the next generation of technology that will be successful beyond IMM remains to be determined. The group discussion also covered the benefits that NASA is likely to receive from investing its own resources in solar array technology. The responses included arrays built for NASA specific environments (particularly missions to outer planets, the inner planets, and the Sun), quicker deployment of advanced technologies, and the ability to act as a smart buyer of array technology provided by industry.

Session 2: Power Storage

Joe Troutman (ABSL Space Products) started the session on power storage by noting that Li-ion technology has driven recent advances in battery performance. He said that basic Li-ion battery technology has been proven, but more work is needed to improving cell performance and safety, primarily through the use of advanced cathode materials and anode coatings. He also suggested that increasing cell voltage would lead to higher energy density and cycle life.

Michael Tomcsi (Quallion) said that Li-ion batteries have shown a steady improvement in energy density (W-hr/kg) of about 6 to 8 percent per year for the past 15 years. He then discussed the different materials that can be used to improve batteries in the future. He suggested that small improvements of a few percent per year are achievable. Tomcsi asserted that metallic lithium batteries may provide a leap in performance, although significant safety issues must be overcome.

In the group discussion both speakers agreed that cathode advancements will be the biggest challenge in future battery improvements. For advancements needed to support NASA specific needs, they suggested that low-temperature performance could be improved with some loss in performance, but the high-temperature requirements stated in the draft roadmap for TA03 are very challenging and may be achievable only with a completely new type of battery. There was also some discussion on better modeling for battery development and the challenge of developing good physics-based battery models to replace current empirical models.

Session 3: Power System Engineering

The purpose of this session was to obtain insights into how solar-based power systems work as a whole and the performance improvements that could arise from advanced technology.

Robert Francis (The Aerospace Corporation) began the session by discussing some of the general issues related to improving solar-based power systems. He noted that the DOD has a technology roadmap with specific performance numbers in terms of energy density and volumetric efficiency for solar arrays. This DOD roadmap projects improvements that are about half the level of improvement proposed in the draft NASA roadmap for TA03. Francis thinks that most of the mass savings in future solar power systems will come from improvements to solar array supporting structures—not from improvements in the solar cells mounted on the arrays. He noted some of the challenges in moving to very-high-power arrays include the need for higher voltages for the spacecraft electrical bus and difficulties in ground testing.

Azam Arastu (Boeing) focused most of his talk on the two advanced, lightweight, and compact, solar arrays currently under development at Boeing with the support of the Defense Advanced Research Projects Agency (DARPA) and the U.S. Air Force Research Laboratory. Both the Fast Access Testbed Spacecraft (FAST) and the Integrated Blanket/Interconnect System (IBIS) arrays are capable of generating very high power and employ highly efficient IMM solar cell technology and advanced deployment structural concepts to provide significant improvements in the specific power over the current state of the art. The FAST array uses linear solar concentrators and is less expensive as it uses fewer solar cells. It also is more tolerant to radiation and more effective in the low-light environments of deep space, but it has more precise pointing requirements. The IBIS array, which is constructed using thin, flexible, and planar solar modules, packs more compactly than FAST and has less precise pointing requirements. Both array systems are designed to be highly modular and scalable so they can eventually be assembled into systems providing solar arrays capable of generating upwards of 300 kW of power.

The group discussion spent some time on the issue of modularity with both speakers saying that the slight increase in weight caused by modular systems is outweighed by the savings due to simplicity and the ability to mass produce components. When asked why NASA needs to invest in arrays when there is already a robust technology development, the speakers suggested that NASA can focus its investment on spacecraft integration issues and the unique environmental requirements for operation in space. One of the speakers also said that even non-financial support from NASA could help ensure technological progress continues. Finally there was some discussion on bus voltage: 200 V DC was identified as a near term possibility; 300 V DC possible over the long term; and improving performance beyond that achievable with 300 V DC may require shifting from a DC power system to an AC power system.

Session 4: Nuclear Systems

Joseph Nainiger (Alphaport) began the session on nuclear systems by discussing the history of RPSs and fission power systems. He noted that while the United States has extensive flight experience with RPSs, it has only flown one fission power system (in 1965) and has not manufactured and ground tested a fission nuclear thermal rocket since 1972. Over the near term, RPSs will continue to fill the need for a reliable source of electrical power in environments that cannot use solar power (even with advances in solar power technology). Fission systems are enabling for missions with high power demands independent of sun proximity or illumination. Nainiger praised the current effort at NASA to develop and test the non-nuclear components of a fission power system, while adding that a system capable of producing at least 1 MW of electrical power will only be possible with a sustained and aggressive technology development effort. He noted that fission power systems will require a significant investment in infrastructure, and he suggested that efforts should be taken to capture technology developed in the past. Nainiger also asserted that restarting the production of Pu-238 is a critical national need.

Gary Bennett (formerly with NASA) concurred that NASA has a critical need for Pu-238. He postulated that more than half of the missions proposed in the recent planetary decadal survey (NRC, 2011) would benefit from an RPS, if enough Pu-238 were available. He discussed the new RPSs under development and urged NASA to focus on the Stirling engine power converter that is at the heart of the Advanced Stirling Radioisotope Generator

(ASRG). To advance fission power technology, he recommended starting small, using ideas that are known to work and evolving over time with an emphasis on safety and reliability. He said that NASA should acknowledge the high cost and complexity of fission space power systems, and he cautioned against chasing new fission power concepts that are unrealistic and unproven, with no technological foundation.

H. Sterling Bailey (formerly with General Electric) focused his remarks on fission power systems, which he supports as a game-changing technology that could dramatically expand NASA's science and exploration mission capabilities. He noted the programmatic challenges of fission power and the legal requirement that NASA partner with DOE in developing nuclear systems. He also said that the history of fissions space power systems has been characterized by bold efforts that are cancelled before development of operational systems is completed. As a result, confidence in this technology is low and the pool of personnel experienced with this technology is small and shrinking. He advocated the development of a fission space power system that could produce 1 kW of electricity, and he strongly supported ongoing work by NASA's Fission Surface Power System Technology Project. Bailey cautioned against adopting a megawatt-class system as a goal for a new fission space power program; it would be better to start small.

The group discussion period spent some time on what size fission system should be built. Most of the speakers supported development of the 40 kW system that NASA is currently funding. (The current effort, which is funded entirely by NASA, does not include the development of nuclear technology.) In the end, system performance requirements should be based on the requirements of whatever missions are expected to use the system. Several speakers noted that efforts to develop new nuclear power systems typically face both technological and political challenges. A member of the NRC panel noted that the latter is outside the scope of this study, which is focused on technology.

REFERENCES

NRC (National Research Council). 2006. *Priorities in Space Science Enabled by Nuclear Power and Propulsion.* The National Academies Press, Washington, D.C.

NRC. 2010. *Radioisotope Power Systems: An Imperative for Maintaining U.S. Leadership in Space Exploration.* The National Academies Press, Washington, D.C.

NRC. 2011. *Vision and Voyages for Planetary Science in the Decade 2013-2022.* The National Academies Press, Washington, D.C.

Patel, M.R. 2005. *Spacecraft Power Systems.* CRC Press, Boca Raton, Fla.

G

TA04 Robotics, Tele-Robotics, and Autonomous Systems

INTRODUCTION

The draft roadmap for Technology Area (TA) 04, Robotics, TeleRobotics, and Autonomous Systems, was unique in that it did not describe or provide supporting text for the level 3 technologies in its portion of the technology area breakdown structure (TABS). For this roadmap to be parallel with the other 13 roadmaps, it would need to be largely rewritten to do the following:

- Define a clear technology breakdown structure of the TA04 roadmap to level 3 that identifies technologies for development consideration without predetermination of technical solutions.
- Provide supporting roadmap text for each of the level 3 technologies, including an explanation of each technology's potential application(s) and motivating its development.
- Ensure that each of the included level 3 technologies connects to one or more of the top challenges that are also identified within the applicable roadmap document.

Although the existing roadmap for TA04 includes a version of a technology breakdown structure to level 3 in its included Figure 2, the document did not meet the criteria above for the following reasons:

- The breakdown structure in the existing TA04 roadmap (see Figure 2 in the draft roadmap) does not correlate to any of the document's supporting text below level 2.
- The level 3 categories included in the existing TA04 roadmap often identify candidate technical solutions rather than technology development categories.

As a result, the steering committee and the responsible panel did not have a list of well-defined level 3 technologies to execute the study statement of task. As detailed below, the panel responded by generating a substantially revised set of level 3 technologies. The content of some of the technologies in the revised TABS (the ones that cover important technology gaps in the draft roadmap) are described in the remainder of this introduction. All of the TA04 technologies that the panel assessed as high-priority are described in detail in the section titled "High-Priority Level 3 Technologies."

The panel did not make any changes to the set of seven level 2 technology subareas in the draft roadmap for TA04:[1]

- 4.1 Sensing and Perception
- 4.2 Mobility
- 4.3 Manipulation
- 4.4 Human-Systems Integration
- 4.5 Autonomy
- 4.6 Autonomous Rendezvous and Docking
- 4.7 Robotics, Tele-Robotics, and Autonomous Systems (RTA) Systems Engineering

Because the TA04 roadmap did not include a viable list of level 3 technologies, public comments for the TA04 roadmap were solicited using these level 2 technology subareas rather than at level 3 as was done for all the other roadmaps.

The TA04 draft roadmap supports NASA space missions with the development of new capabilities to enable new missions and enhance the efficiency and effectiveness of existing mission concepts. Through a combination of dexterous robotics, better human/robotic interfaces, improved mobility systems, and greater sensing and perception, TA04 technologies will extend the reach of human and robotic exploration. Improvements in autonomy will allow robotic and human systems to operate with greater efficiency and effectiveness, both for near-Earth missions and in deep space. TA04 technologies will enable robotic and crewed servicing missions, which have the potential to fundamentally change the way satellites are designed, built, and operated.

This roadmap describes the robotics, tele-robotics, and autonomy technology developments that are necessary to meet the needs of future missions, provide enhanced capabilities, or enable new mission concepts. This includes identification of representative future missions and key capabilities and investments that will enable or enhance these missions. The roadmap focuses on several key issues for the future of robotics and autonomy: enhancing or exceeding human performance in sensing, piloting, driving, manipulating, and rendezvous and docking; development of cooperative and safe human interfaces to form human-robot teams; and improvements in autonomy to make human crews independent from Earth and make robotic missions more capable.

As noted above, the panel created a new set of level 3 technologies for TA04 (see Table G.1). (The complete, revised TABS for all 14 TAs is shown in Appendix B.) This new set of level 3 technologies is consistent both with the relevant panel's understanding of applicable TA04 technology needs and with the intent of much of the existing roadmap text. Some new level 3 technologies address important technology gaps within the existing TA04 roadmap:

- *4.1.6. Multi-Sensor Data Fusion.* Since effective robotics operation often requires the integrated use of data collected simultaneously from a variety of sensors, the fusion of that data into more useful information is a critical capability. The original TA04 TABS included the narrower topic of Sensor Fusion for Grasping (4.5.3), which this technology treats more broadly.
- *4.1.7. Mobile Feature Tracking and Discrimination.* This area covers the unique challenges to object tracking and discrimination faced by a surface mobility vehicle while in motion.
- *4.2.1. Extreme Terrain Mobility.* Extreme terrain includes cliffs, crater walls, and very rugged surfaces. Means of mobility to safely reach and loiter at designated locations on such terrain poses unique challenges.
- *4.2.2. Below-Surface Mobility.* Vehicles that would transit under regolith, in caves, or immersed in bodies of liquid have mobility challenges not covered in other level 3 technology entries in this roadmap.
- *4.3.3. Modeling of Contact Dynamics.* Robotic vehicles that dock or that manipulate objects require detailed models of the contact dynamics to enable proper control of their interactions.

[1] The draft space technology roadmaps are available at http://www.nasa.gov/offices/oct/strategic_integration/technology_roadmap.html.

TABLE G.1 Technology Area Breakdown Structure for TA04, Robotics, Tele-Robotics, and Autonomous Systems

NASA Draft Roadmap (Revision 10)	Steering Committee-Recommended Changes
TA04 Robotics, Tele-Robotics and Autonomous (RTA) Systems	The technologies have been largely rewritten.
4.1. Sensing and Perception	4.1. Sensing and Perception
4.1.1. Stereo Vision	4.1.1. Vision
4.1.2. LIDAR	4.1.2. Tactile Sensing
4.1.3. Proximity Sensing	4.1.3. Natural Feature Image Recognition
4.1.4. Sensing Non-Geometric Terrain Properties	4.1.4. Localization and Mapping
4.1.5. Estimating Terrain Mechanical Properties	4.1.5. Pose Estimation
4.1.6. Tactile Sensing Arrays	4.1.6. Multi-Sensor Data Fusion
4.1.7. Gravity Sensors & Celestial Navigation	4.1.7. Mobile Feature Tracking and Discrimination
4.1.8. Terrain Relative Navigation	4.1.8. Terrain Classification and Characterization
4.1.9. Real-time Self-calibrating of Hand-eye Systems	
4.2. Mobility	4.2. Mobility
4.2.1. Simultaneous Localiz. and Mapping	4.2.1. Extreme Terrain Mobility
4.2.2. Hazard Detection Algorithms	4.2.2. Below-Surface Mobility
4.2.3. Active Illumination	4.2.3 Above-Surface Mobility
4.2.4. 3-D Path Planning w/Uncertainty	4.2.4. Small Body/Microgravity Mobility
4.2.5. Long-life Extr. Enviro. Mechanisms	
4.2.6. Robotic Jet Backpacks	
4.2.7. Smart Tethers	
4.2.8. Robot Swarms	
4.2.9. Walking in Micro-*g*	
4.3. Manipulation	4.3. Manipulation
4.3.1. Motion Planning Alg., High DOF	4.3.1. Robot Arms
4.3.2. Sensing and Control	4.3.2. Dexterous Manipulators
4.3.3. Robot Arms (light, high strength)	4.3.3. Modeling of Contact Dynamics
4.3.4. Dexterous Manipul., Robot Hands	4.3.4. Mobile Manipulation
4.3.5. Sensor Fusion for Grasping	4.3.5. Collaborative Manipulation
4.3.6. Grasp Planning Algorithms Robotic Drilling Mechanisms	4.3.6. Robotic Drilling and Sample Processing
4.3.7. Multi-arm/Finger Manipulation	
4.3.8. Planning with Uncertainty	
4.4. Human-Systems Integration	4.4. Human-Systems Integration
4.4.1. Crew Decision Support Systems	4.4.1. Multi-Modal Human-Systems Interaction
4.4.2. Immersive Visualization	4.4.2. Supervisory Control
4.4.3. Distributed Collaboration	4.4.3. Robot-to-Suit Interfaces
4.4.4. Multi Agent Coordination	4.4.4. Intent Recognition and Reaction
4.4.5. Haptic Displays	4.4.5. Distributed Collaboration
4.4.6. Displaying Range Data to Humans	4.4.6. Common Human-Systems Interfaces
	4.4.7. Safety, Trust, and Interfacing of Robotic/Human Proximity Operations
4.5. Autonomy	4.5. Autonomy
4.5.1. Spacecraft Control Systems	4.5.1. Vehicle System Management and FDIR
4.5.2. Vehicle Health, Prog/Diag Systems	4.5.2. Dynamic Planning and Sequencing Tools
4.5.3. Human Life Support Systems	4.5.3. Autonomous Guidance and Control
4.5.4. Planning/Scheduling Resources	4.5.4. Multi-Agent Coordination
4.5.5. Operations	4.5.5. Adjustable Autonomy
4.5.6. Integrated Systems Health Management	4.5.6. Terrain Relative Navigation
4.5.7. FDIR and Diagnosis	4.5.7. Path & Motion Planning with Uncertainty
4.5.8. System Monitoring and Prognosis	
4.5.9. V&V of Complex Adaptive Systems	
4.5.10. Automated Software Generation	
4.5.11. Software Reliability	
4.5.12. Semi Automatic Systems	

continues

TABLE G.1 Continued

NASA Draft Roadmap (Revision 10)	Steering Committee-Recommended Changes
4.6. Autonomous Rendezvous and Docking	4.6. Autonomous Rendezvous and Docking
4.6.1. Rendezvous and Capture	4.6.1. Relative Navigation Sensors (long, mid, and near range)
4.6.2. Low impact and Androgenous Docking Systems and Interfaces	4.6.2. Relative Guidance Algorithms
4.6.3. Relative Navigation Sensors	4.6.3. Docking & Capture Mechanisms/Interfaces
4.6.4. Robust AR&D GN&C Algorithms and FSW	
4.6.5. Onboard Mission Manager	
4.6.6. AR&D Integration and Standardization	
4.7. RTA Systems Engineering	4.7. RTA Systems Engineering
4.7.1. Human safety	4.7.1. Modularity/Commonality
4.7.2. Refueling Interfaces and Assoc. Tools	4.7.2. Verification and Validation of Complex Adaptive Systems
4.7.3. Modular/Serviceable Interfaces	4.7.3. Onboard Computing
4.7.4. High Perf., Low Power Onboard Computers	
4.7.5. Environment Tolerance	
4.7.6. Thermal Control	
4.7.7. Robot-to-Suit Interfaces	
4.7.8. Common Human-Robot Interfaces	
4.7.9. Crew Self Sufficiency	

- *4.3.4. Mobile Manipulation.* Performing any kind of manipulation with a surface vehicle while it is mobile adds complexity to the operation that will require dedicated technology to overcome.
- *4.4.4. Intent Recognition and Reaction.* More effective robotic aiding of humans will require means for the robots to reliably recognize human intent through speech, gesture, or facial expression to enable appropriate responses.
- *4.4.7. Safety, Trust, and Interfacing of Robotic/Human Proximity Operations.* Human reliance on proximate robots for critical support capabilities will necessitate a high level of confidence that the robot will respond usefully, safely, and predictably. This capability necessitates means for integrated interaction assessment and associated verification and validation. The original TA04 TABS included the topic of Human Safety (4.7.1), which has been broadened.
- *4.5.3. Autonomous Guidance and Control.* This capability requires the means to adjoin effective decision making algorithms regarding appropriate and efficient vehicle trajectory, path, and orientation management with suitably robust guidance and control implementations.
- *4.5.5. Adjustable Autonomy.* There are many levels of autonomy that can be designed into systems. An important capability would provide means to adapt the level of applied autonomy to the mission circumstances, either automatically or by providing a means for timely human command. The original TA04 TABS included the topic of Semi Automatic Systems (4.5.12), which has been broadened.

The other new level 3 technologies reorder, restate, or regroup the entries from the original TABS. If NASA accepts the recommended restructuring of this roadmap, it may be prudent for the NASA roadmap team to reconvene to revise the roadmap to include these new technologies and to provide descriptions of all of the TA04 level 3 technologies.

TOP TECHNICAL CHALLENGES

The panel identified six top technical challenges for TA04. These are listed below in priority order.

1. Rendezvous: Develop the capability for highly reliable, autonomous rendezvous, proximity operations, and capture/attachment to (cooperative and non-cooperative) free-flying space objects.

The ability to perform autonomous rendezvous and safe proximity operations and docking/grappling is central to the future of mission concepts for satellite servicing, Mars sample returns, active debris removal scenarios, and other cooperative space activities. Major challenges include improving the robustness of the rendezvous and capture process to ensure successful capture despite wide variations in lighting, target characteristics, and relative motion.

2. Maneuvering: Enable robotic systems to maneuver in a wide range of NASA-relevant environmental, gravitational, and surface and subsurface conditions.

Current crewed and robotic rovers cannot access extreme lunar or martian terrain, eliminating the possibility of robotic access and requiring humans to park and travel on foot in suits. Extreme terrain mobility would allow robotic rovers access to more scientifically interesting samples. In microgravity, locomotion techniques on or near asteroids and comets are undeveloped and untested. Challenges include developing robotics to travel into these otherwise denied areas, developing techniques to grapple and anchor with asteroids and non-cooperative objects, or building crew mobility systems to move humans into these challenging locations.

3. In Situ Analysis and Sample Return: Develop subsurface sampling and analysis exploration technologies to support in situ and sample return science missions.

A top astrobiological goal and a fundamental NASA exploration driver is the search for life or signs of previous life in our solar system, and NASA scientists and engineers have been told to "follow the water." A significant planetary science driver exists to obtain unaltered samples (with volatiles intact) for either in situ analysis, or return to Earth from planetary bodies both large and small. These pristine samples are found subsurface and are (mostly) acquired with robotic drilling devices. Due to the autonomy demands of robotic drilling/sampling along with very low mass and power constraints, terrestrial drilling technologies (including the National Science Foundation and U.S. Army ice drilling efforts) have limited applicability. Robotic planetary drilling and sample handling is a new and different capability.

4. Hazard Avoidance: Develop the capabilities to enable mobile robotic systems to autonomously and verifiably navigate and avoid hazards.

Human drivers have a remarkable ability to perceive terrain hazards at long range, but robotic systems lag behind due to the large computational throughput requirements needed to quickly assess subtle terrain geometric and non-geometric properties fast enough to maintain speeds near the vehicle limits.

5. Time-Delayed Human-Robotic Interactions: Achieve more effective and safe human interaction with robotic systems (whether in proximity or remotely) that accommodate any time-delay effects.

More effective and safe human interaction with robotic systems has a number of different focuses that range from the potential dangers of proxemic interactions to remote supervision with or without time delays. Proxemic interactions require that the robotic systems can safely operate and communicate their actions and intent to the humans working in close proximity, either in the case of peer-based interaction or independent actions. Similarly, humans interacting with robotic systems in close proximity must be able to provide direction and instruction to the robotic systems.

Remote interaction with robotic systems do not pose the same immediate potential level of danger to humans as close proximity interactions; however, it is often significantly more difficult for a remote human to fully understand the context of the environment in which the robotic system functions and the status of the system. This lack of understanding can be further complicated by the effects of communication time delays, which can range from fractions of a second to significant periods of time. An improper level of understanding often results in incorrect or improper instructions being provided by the human to the robotic system. Robotic systems for missions with long-duration communication time delays must be developed to support autonomous actions without immediate human interaction. Such systems must also ensure that the remote human is provided with tools to quickly understand the situation once communications are received from the robotic system.

6. Object Recognition and Manipulation: Develop means for object recognition and dexterous manipulation that supports engineering and science objectives.

Object recognition requires sensing, often fusing multiple sensing modalities, with a perception function that can associate the sensed object with an object that is understood a priori. Sensing approaches to date have combined machine vision, stereo vision, lidar, structured light, and radar, while perception approaches often start with CAD models or models created by a scan with the same sensors that will be used to identify the object later. Major challenges include the ability to work with a large library of known objects, identifying objects that are partially occluded, sensing in poor (high, low, and sharply contrasting) lighting, estimating the pose of quickly tumbling objects, and working with objects at near and far range. These challenges are important for object manipulation and in mobility for object following and avoidance.

While human hands are generally capable, a robotic hand with equivalent or superior grasping ability would avoid the added complexity of robot interfaces on objects, and provide a sensate tool change-out capability for specialized tasks. Dexterity can be measured by range of grasp types, scale, strength, and reliability. Challenges include fundamental actuation and sensing physics, discrimination, contact localization, extrinsic and intrinsic actuation, and hand/glove coverings that do not attenuate sensor or object motion but are rugged when handling rough and sharp objects.

QFD MATRIX AND NUMERICAL RESULTS FOR TA04

The process used to evaluate the level 3 technologies is described in detail in Chapter 2. The results of the evaluation are shown in Figures G.1 and G.2, which show the relative ranking of each technology. The panel assessed eight of the technologies as high priority. Seven of these were selected based on their QFD scores, which significantly exceeded the scores of lower ranked technologies. After careful consideration, the panel also designated 4.2.4 Small Body/Microgravity Mobility as a high-priority technology.[2]

CHALLENGES VERSUS TECHNOLOGIES

Figure G.3 shows the relationship between the individual level 3 TA04 technologies and the top technical challenges. Note that the lowest-priority technologies as determined by QFD rankings tend not to be strongly connected to the top technical challenges. All but one of the high-priority technologies have a strong connection to two or more of the top technical challenges. Just two of the medium-priority technologies have a strong connection to one of the top technical challenges. This shows a good level of consistency between the evaluations and the QFD rankings.

HIGH-PRIORITY LEVEL 3 TECHNOLOGIES

Panel 2 identified 8 high-priority technologies in TA04. The justification for ranking each of these technologies as a high priority is discussed below.

[2] In recognition that the QFD process could not accurately quantify all of the attributes of a given technology, after the QFD scores were compiled, the panels in some cases designated some technologies as high priority even if their scores were not comparable to the scores of other high-priority technologies. The justification for the high-priority designation of all the high-priority technologies for TA04 appears in the section "High-Priority Level 3 Technologies," below.

	Benefit	Alignment with NASA Needs	Alignment with Non-NASA Aero Tech Needs	Alignment with Non-Aerospace National Goals	Technical Risk and Reasonableness	Sequencing and Timing	Time and Effort	QFD Score (Weighted)	Panel Priority
Multiplier	27	5	2	2	10	4	4		
	0/1/3/9	0/1/3/9	0/1/3/9	0/1/3/9	1/3/9	-9/-3/-1/1	-9/-3/-1/0		
Technology Name	Benefit	Alignment			Risk/Difficulty				
4.1.1. Vision (including active illumination)	3	9	3	3	3	-3	-1	152	M
4.1.2. Tactile Sensing	3	3	1	1	3	-3	-1	114	M
4.1.3. Natural Feature Image Recognition	1	3	1	1	3	-3	-1	60	L
4.1.4. Localization and Mapping	3	3	3	1	3	-3	-1	118	M
4.1.5. Pose Estimation	1	3	1	0	3	-3	-1	58	L
4.1.6. Multi-Sensor Data Fusion	3	9	9	9	3	-3	-1	176	M
4.1.7. Mobile Feature Tracking and Discrimination	1	3	1	1	3	-3	-1	60	L
4.1.8. Terrain Classification and Characterization	3	3	1	1	3	-3	-3	106	M
4.2.1. Extreme Terrain Mobility	3	9	0	1	9	-3	-3	194	H
4.2.2. Below-Surface Mobility	3	3	0	1	3	-9	-3	80	L
4.2.3. Above-Surface Mobility	3	3	1	0	3	-3	-1	112	M
4.2.4. Small Body / Microgravity Mobility	3	3	1	0	3	-3	-1	112	H*
4.3.1. Robot Arms	1	3	1	1	3	-3	-1	60	L
4.3.2. Dexterous Manipulators (including robot hands)	3	9	1	3	9	-3	-1	208	H
4.3.3. Modeling of Contact Dynamics	1	3	1	1	3	-3	-1	60	L
4.3.4. Mobile Manipulation	1	3	1	1	9	-3	-1	120	M
4.3.5. Collaborative Manipulation	3	3	1	3	9	-3	-1	178	M
4.3.6. Robotic Drilling and Sample Handling	3	9	0	1	9	-3	-3	194	H
4.4.1. Multi-Modal Human-Systems Interaction	3	9	3	3	3	-3	-3	144	M
4.4.2. Supervisory Control (including time delay supervision)	3	9	3	3	9	-3	-3	204	H
4.4.3. Robot-to-Suit Interfaces	1	3	1	1	3	-3	-1	60	L
4.4.4. Intent Recognition and Reaction	1	3	1	3	3	-3	-3	56	L
4.4.5. Distributed Collaboration	3	9	3	3	3	-3	-3	144	M
4.4.6. Common Human-Systems Interfaces	1	3	1	3	3	-3	-1	64	L
4.4.7. Safety, Trust, and Interfacing of Robotic/Human Proximity Operations	3	9	3	9	3	-3	-3	156	M
4.5.1. Vehicle System Management & FDIR	3	9	3	9	9	-3	-3	216	H
4.5.2. Dynamic Planning & Sequencing Tools	3	9	3	3	3	-3	-1	152	M
4.5.3. Autonomous Guidance & Control	3	9	3	3	3	-1	-1	160	M
4.5.4. Multi-Agent Coordination	3	3	3	9	3	-3	-3	126	M
4.5.5. Adjustable Autonomy	3	9	3	9	3	-3	-1	164	M
4.5.6. Terrain Relative Navigation	3	3	3	3	3	-3	-1	122	M
4.5.7. Path & Motion Planning with Uncertainty	1	3	1	3	3	-3	-1	64	L
4.6.1. Relative Navigation Sensors	3	9	1	0	3	-3	-1	142	M
4.6.2. Relative Guidance Algorithms	9	9	1	0	3	-3	-1	304	H
4.6.3. Docking and Capture Mechanisms/Interfaces	9	9	1	0	3	-3	-1	304	H
4.7.1. Modularity / Commonality	3	9	1	1	3	-3	-1	144	M
4.7.2. V&V of Complex Adaptive Systems	3	9	9	9	3	-3	-3	168	M
4.7.3. Onboard Computing	3	9	9	1	3	1	-1	176	M

FIGURE G.1 Quality function deployment (QFD) summary matrix for TA04 Robotics, Tele-Robotics, and Autonomous Systems. The justification for the high-priority designation of all high-priority technologies appears in the section "High-Priority Level 3 Technologies." H = High Priority; H* = High Priority, QFD score override; M = Medium Priority; L = Low Priority.

Technology 4.6.2, Relative Guidance Algorithms

Relative guidance technologies encompass algorithms that determine the desired trajectories to be followed between vehicles performing rendezvous (at long range), proximity operations (at short range), and/or docking and capture. These algorithms must anticipate the applicable environmental field effects that drive the relative motion of the spacecraft with respect to the objects of interest, the nature of the trajectory change/attitude control effectors in use, as well as the quality of the inertial and relative navigation state data available to the guidance algorithms. To

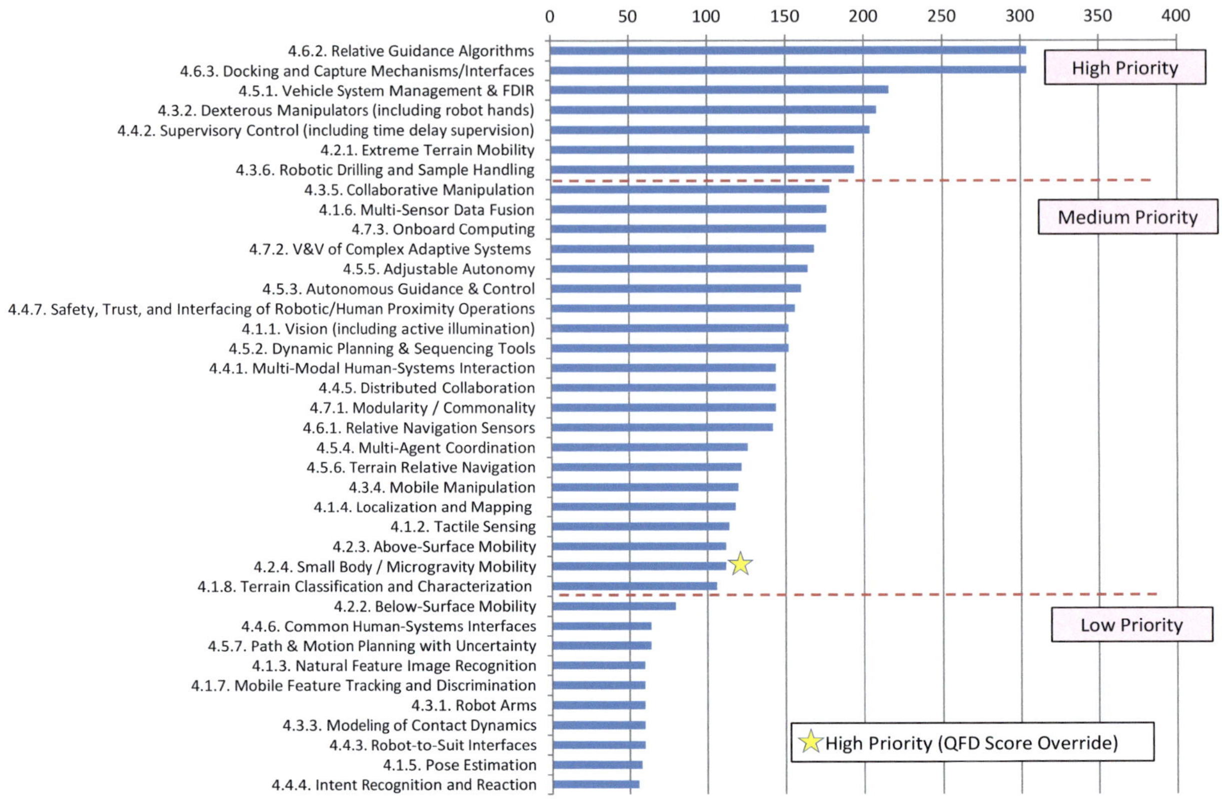

FIGURE G.2 Quality function deployment rankings for TA04 Robotics, Tele-Robotics, and Autonomous Systems.

date, most spacecraft relative guidance functionality has been accomplished on the ground with reliance on communication links to uplink mission maneuver target data. This technology provides real-time, onboard algorithmic functionality that can calculate and manage spacecraft maneuvers to achieve specific trajectory change objectives.

Ground-managed relative guidance for spacecraft with large impulsive thrusters is operational (TRL 9), but the TRL drops to 3-5 for systems applying alternative (e.g., electric, low-thrust, and/or variable-thrust) propulsion systems, when performing AR&D with tumbling bodies, or when trying to guarantee algorithmic convergence to enable onboard use in space without ground or crew intervention. The alignment to NASA's needs is high because relative guidance impacts crewed deep space exploration, sample return, servicing, and orbital debris mitigation. Some aspects of relative guidance verification and validation are dependent on evaluation in an actual space environment. This can be supported by experiments run parallel to baseline relative guidance operations on ISS visiting vehicles.

The committee assessed the relative guidance technologies as "game-changing" because they enable an array of NASA missions including deep space human exploration missions, Mars sample return, large orbital debris capture, and station keeping for closely coordinated spacecraft constellations. The safety margins of deep space missions (including human exploration missions) could be substantially improved by reducing dependence on ground management of relative guidance that would experience long (and risky) communication lags.

Many of the historical relative guidance capability developments are the direct result of NASA investments, and were achieved by NASA personnel, supporting contractors, and universities. That will likely continue to be the case for the foreseeable future. Non-NASA investment in this area is likely restricted to specialized military organizations and by organizations located in other space-faring nations. Continued relative guidance advances for the U.S. civil space sector will likely depend on NASA technology development investment.

●	Strong Linkage: Investments by NASA in this technology would likely have a major impact in addressing this challenge.
○	Moderate Linkage: Investments by NASA in this technology would likely have a moderate impact in addressing this challenge.
[blank]	Weak/No Linkage: Investments by NASA in this technology would likely have little or no impact in addressing the challenge.

		Top Technology Challenges					
Priority	TA 04 Technologies, Listed by Priority	1. Rendezvous: Develop the capability for highly reliable, autonomous rendezvous, proximity operations, and capture/attachment to (cooperative and non-cooperative) free-flying space objects.	2. Maneuvering: Enable robotic systems to maneuver in a wide range of NASA-relevant environmental, gravitational, and surface and subsurface conditions.	3. In Situ Analysis and Sample Return: Develop subsurface sampling and analysis exploration technologies to support in situ and sample return science missions.	4. Hazard Avoidance: Develop the capabilities to enable mobile robotic systems to autonomously and verifiably navigate and avoid hazards.	5. Time-Delayed Human-Robotic Interactions: Achieve more effective and safe human interaction with robotic systems (whether in proximity or remotely) that accommodate any time-delay effects.	6. Object Recognition and Manipulation: Develop means for object recognition and dexterous manipulation that supports engineering and science objectives.
H	4.6.2. Relative Guidance Algorithms	●	●	●	●		
H	4.6.3. Docking and Capture Mechanisms/Interfaces	●		●			
H	4.5.1. Vehicle System Management and FDIR	●	○	○	●	●	
H	4.3.2. Dexterous Manipulators	●		●		○	●
H	4.4.2. Supervisory Control (including time delay supervision)	●	○		○	●	
H	4.2.1. Extreme Terrain Mobility		●	○	●		
H	4.3.6. Robotic Drilling and Sample Processing			●			●
H	4.2.4. Small Body / Microgravity Mobility		●	○	○		
M	4.3.5. Collaborative Manipulation			○		○	●
M	4.1.6. Multi-Sensor Data Fusion	○	○		○		○
M	4.7.3. Onboard Computing	●	○		○		
M	4.7.2. V&V of Complex Adaptive Systems	○			○	○	
M	4.5.5. Adjustable Autonomy					○	
M	4.5.3. Autonomous Guidance and Control	○	○		○		
M	4.4.7. Safety, Trust, and Interfacing of Robotic/Human Proximity Ops.					○	
M	4.1.1. Vision (including active illumination)	○		○			○
M	4.5.2. Dynamic Planning and Sequencing Tools	○			○		
M	4.4.1. Multi-Modal Human-Systems Interaction					○	
M	4.4.5. Distributed Collaboration	○				○	
M	4.7.1. Modularity / Commonality						
M	4.6.1. Relative Navigation Sensors	○	○		○		
M	4.5.4. Multi-Agent Coordination	○				○	
M	4.5.6. Terrain Relative Navigation				○		
M	4.3.4. Mobile Manipulation						
M	4.1.4. Localization and Mapping		○		○		
M	4.1.2. Tactile Sensing						○
M	4.2.3. Above-Surface Mobility		○		○		
M	4.1.8. Terrain Classification and Characterization		○		○		
L	4.2.2. Below-Surface Mobility		○	○			
L	4.4.6. Common Human-Systems Interfaces					○	
L	4.5.7. Path and Motion Planning with Uncertainty				○		
L	4.1.3. Natural Feature Image Recognition				○		○
L	4.1.7. Mobile Feature Tracking and Discrimination	○			○		
L	4.3.1. Robot Arms						○
L	4.3.3. Modeling of Contact Dynamics				○		
L	4.4.3. Robot-to-Suit Interfaces					○	
L	4.1.5. Pose Estimation				○		○
L	4.4.4. Intent Recognition and Reaction					○	

FIGURE G.3 Level of support that the technologies provide to the top technical challenges for TA04 Robotics, Tele-Robotics, and Autonomous Systems.

Technology 4.6.3, Docking and Capture Mechanisms/Interfaces

Docking and capture mechanisms enable the physical capture and attachment, as well as subsequent safe release of two bodies in space that achieve part of their mission objectives when operating while joined. Docking and capture mechanisms/interfaces facilitate final relative alignment prior to docking/capture, enable initial soft capture/docking and then function to complete hard docking/mate. For some missions they also do the reverse when a flight scenario calls for separation of docked bodies. Docking interfaces to date have been designed for man-in-the-loop rendezvous/docking systems and operations; this is true even for the semi-automated Russian ISS cargo docking interface. Robotic or automated rendezvous and docking operations will not involve real-time human control. Development of a physical docking and capture interface for AR&D operations would greatly simplify the control demands for a working AR&D system.

Generic/modular AR&D docking and capture systems have not been tested in space and are presently only at TRL 2 or 3. While the DARPA Orbital Express Mission did perform robotic dockings, including some that were done autonomously, the design of the docking and capture mechanisms/interfaces were intended only to enable performing this demonstration. The alignment to NASA's needs is high because this technology will impact crewed deep space exploration, sample return, servicing, and orbital debris mitigation. The ISS could be an effective platform for evaluating the performance of new docking and capture mechanisms by enabling the repeated execution of experiments either inside or outside the ISS to test the reliability of these AR&D interfaces in a microgravity environment.

This technology was deemed a game changer due to the improvements in reliability of AR&D operations that will come from new interfaces that facilitate AR&D success and reduce the demands for AR&D control. Variations of docking and capture mechanisms enable transfer of crew between delivery and destination vehicles, provide means for attachment of added equipment modules, facilitate execution of robotic servicing missions, and could enable grapple/capture of inactive, possibly tumbling spacecraft for missions such as orbital debris mitigation. The development of a new spacecraft docking or capture interface (or multiple ones) designed to facilitate reliable AR&D operations will have a major beneficial impact on the overall development of a cost effective and practical AR&D architecture.

NASA has been a primary developer of the docking and capture capabilities to date. While Russia, Japan, and Europe have also developed applicable technology, it is important that the United States develop its own capability to ensure a means of accomplishing scientific and exploration missions of national priority and importance. National security agencies are leveraging applicable developments by NASA, and potential commercial users of the technology will rely heavily on technologies developed by NASA.

Technology 4.5.1, Vehicle Systems Management and FDIR

The panel combined the related and overlapping topics of integrated systems health management (ISHM), fault detection, isolation, and recovery (FDIR), and vehicle systems management (VSM). Together these algorithms provide the crucial capability for an autonomous spacecraft to operate safely and reliably, even in the face of changing mission objectives and/or vehicle failures. ISHM/FDIR/VSM will improve the reliability of future missions by providing a diagnostic capability that helps ground or crew failure assessment and an automated capability to fix/overcome faults; increasing robotic mission flexibility in response to failures; and increasing crew safety in the event of a detected need for crew escape and abort.

ISHM and FDIR are commonplace in many terrestrial applications, and basic FDIR capabilities have been flown on the space shuttle and the ISS, giving a TRL of 8-9 for the basic capabilities, but only a TRL of 3-5 for the advanced capabilities required for future, more autonomous spacecraft missions. Also, the integration of different algorithms to provide a consistent management function has not yet been accomplished. The required research is well aligned with the NASA experience and capabilities and consistent with NASA's past roles. Access to the ISS is not required for this technology development.

NASA might be able to draw from the significant other work being done in this area (e.g., commercial robotics, aircraft/UAVs, and commercial spacecraft), but given the length, variety, and complexity of the missions, the challenges faced by NASA are unique and it is very unlikely that they will be fully addressed by an external research organization. As such, NASA investment in this technology area is critical.

The steering committee assessed the ISHM/FDIR/VSM systems as providing a major benefit due to the potential to significantly improve the robustness and reliability of future missions. The alignment to NASA's needs is high because it will impact many missions, such as deep space exploration, robotic science missions, planetary landers and rovers. The alignment with other aerospace needs is considered to be medium, as this research has the potential to have a large impact on a subset of other aerospace sectors, such as commercial spacecraft. The alignment with national needs is considered to be higher, as improving reliability and confidence in the autonomy algorithms is one of the major hurdles to deployment/acceptance in many other safety critical sectors (e.g., ground and air transportation) and for autonomous operations in vicinity of humans. Thus autonomous platform operations in these other sectors could also benefit from NASA advances in these technology areas. The risk is assessed to be moderate to high, within the bounds of NASA's acceptable risk levels for technology development. In particular, a continued capability evolution of the ISHM/FDIR/VSM techniques that have been developed for many years can be applied to develop a flight proven foundation. Due to the potential for major mission improvements, strong alignment with NASA needs, and reasonable risk and development effort, ISHM/FDIR/VSM are rated as high-priority technologies.

Technology 4.3.2, Dexterous Manipulation

Dexterous Manipulation is a system-level technology that encompasses multiple stand-alone technology areas including: electro-mechanical devices for manipulators, end effectors, and sensor positioning; sensors for both tele-operated and autonomous control of joint, end point, and grasping functions; and control software for various levels of autonomy ranging from tele-operation to autonomous control. Dexterous Manipulation has great relevance for several current and future NASA applications:

- Use of dexterous manipulation for servicing and maintenance on the ISS can reduce the risk and cost of EVA operations. Moreover, control of operations from the ground would further unburden flight crews and leverage more extensive facilities not available on the ISS. In particular, the ISS is highly constrained by space, power, and thermal considerations. Ground facilities could incorporate larger displays, immersive user interface environments, and larger processing capacity to enable autonomous operations.
- Dexterous manipulation can enable remote satellite servicing and on-orbit assembly of larger structures.
- Applied to remote exploration, dexterous manipulation can provide the ability for scientists to experience and interact with an extra-terrestrial environment more naturally, ultimately improving the science return.

TRLs for dexterous manipulation range from 1 to 9. Since 1997, NASA has focused on the development of Robonaut, which has now been deployed to the ISS for evaluation. The result of this investment is a state-of-the-art system with dexterity approaching that of a suited astronaut. In addition, this system has passed the critical milestones associated with flight qualification and safety certification for use around astronauts. Development activities to date have focused primarily on human-in-the-loop tele-operation. Further, the high-bandwidth, low-latency communications requirement limits the utility of this system for remote operations where latency has adverse control stability effects.

To overcome these limitations, next-generation dexterous manipulation technologies must be able to operate with embedded control loops that can bypass the effects of large latencies and very low bandwidth associated with external operators. Two approaches have promise toward this objective. First, a combination of supervised autonomy coupled with autonomous sensing and control has the potential to endure very large latencies and can be implemented with limited bandwidth communication links. This approach relies on the development of autonomous skills that the remote manipulator can execute and then report the status thereof. Dexterous manipulation also relies heavily on developing human-system integration technologies (Subarea 4.4). Remote dexterous manipulation will benefit from the ability to autonomously recognize and control fundamental activities such as contact, grasping, and manipulation. If desired, NASA could explore options for extending Robonaut technologies and capabilities for operations in large latency and low-bandwidth environments.

While Robonaut is well suited for ISS operations, its size and weight preclude its use for exploration activities. Exploration missions could benefit significantly from a dexterous manipulation capability. However, significant advances in actuation technologies will reduce weight and power consumption while retaining end point forces, speed of response, and control bandwidth. NASA could benefit from the development of novel actuation technologies that dramatically increase the strength to weight ratio.

NASA has been at the forefront of developing dexterous manipulator systems that rely on tele-operation control. NASA has also worked well with DARPA under the MARS 2020 program to leverage common interests in manipulator technology. While NASA has focused on the electro-mechanical and human aspects of dexterous manipulation, DOD has placed a much greater emphasis on higher levels of autonomy and supervisory control necessary to achieve remote operations. NASA could either partner with these groups or leverage technologies currently under development to take advantage of complementary existing programs.

There is tremendous potential for dexterous manipulation beyond the confines of NASA applications including military operations (investigation and handling of improvised explosive devices), civilian hazardous materials operations and first responder operations, and manufacturing operations (increasing the level of automation of low-volume manufacturing).

The risk of dexterous manipulation capability development is varied. The deployment of Robonaut for use on the ISS in a strict tele-operation role has low risk. The development of means to tolerate higher remote operator latencies through the use of embedded automation increases the risk and necessary investment level. However, this is the area of dexterous manipulation that could have the largest payback for NASA, military, and civilian applications. Development of dexterous manipulation for remote exploration would require fundamental advancements in the areas of lightweight, highly efficient actuators and would entail greater risk and a longer technology development time horizon.

Technology 4.4.2, Supervisory Control

Supervisory Control was defined as incorporating the techniques necessary for controlling robotic behaviors using higher-level goals instead of low-level commands, thus requiring robots to have semi-autonomous or autonomous behaviors. This increases the number of robots a single human can simultaneously supervise, reducing the costs and impacts of time delays on remotely supported robotic teams, improving the synergy of combined human/robot teams, and facilitating distributed robot teams. Supervisory control for NASA also incorporates time-delayed supervision. Supervisory control provides major benefits because it will minimize the manpower impacts required to deploy robotic missions over long periods anywhere and anytime. This technology will support the design of game-changing science and exploration missions, such as new robotic missions at remote locations, and simultaneous robotic missions with reduced human oversight.

Limited supervisory control has been deployed for the Mars Rovers, thus basic capabilities have a high TRL (9), but the advanced capabilities have a relatively low TRL (2-3). Key components to be addressed include the development of robust high-level autonomous behaviors and control, multi-sensor fusion, clearly understood and usable presentations of information from multiple robots for human understanding, time-delayed interpretation and presentation of robot provided information, haptic feedback, and means for a supervisory control system to handle communication outages. This technology will not substantially benefit from access to the ISS.

The alignment to NASA's needs is high due to the impact of reducing the number of personnel required to supervise robotic missions and the number of science and exploration missions to which the technology can be applied. Supervisory control generally has applications across the government agencies, including the Departments of Defense, Energy, and Homeland Security. Additionally, submersible unmanned vehicles can encounter similar time delay effects while under water; however, the communications time-delayed component is primarily unique to space exploration. Thus, NASA is uniquely positioned to take the lead to mature the technology to TRL 6, but there may be opportunities to collaborate with international partners, such as Japan, France, and Germany.

The alignment with other aerospace and national needs is considered to be moderate, since the results can impact supervisory control for any robotic system. The risk is assessed as moderate to high, based on the fact that providing supervisory control is a systems engineering problem, thus development of the technology is highly dependent upon the development of underlying robotic and human-machine interaction capabilities. The program will need to leverage existing NASA and DOD capabilities in order to meet reasonable goals ensuring timely development of various related technologies, such as robust, autonomous behaviors. The concepts related to this technology are roughly sketched out and users should be clearly defined. This effort is easily justified when considering the potential benefits for both human and robotic exploration of the outer solar system and beyond. Thus, supervisory control is judged to be a high-priority technology.

Technology 4.3.6, Robotic Drilling and Sample Processing

Robotic Drilling and Sample Processing (RDSP) technologies refer to robotic drilling and drill-like technologies, as well as automated rock coring devices and new automated technologies that transfer, store, and process acquired samples such as sample caches, sample crushers, sample slicers, and sample-to-instrument carousels. In a game-changing manner, RDSP will improve the science return of robotic science missions to small bodies, moons, and planets. This is due to the almost uncontaminated, unaltered, and volatile rich nature of the samples acquired by the next-generation RDSP technology. RDSP technology developments will also benefit in situ resource utilization (ISRU) for human spaceflight to the Moon and small bodies through a better understanding of the raw materials available to human explorers in situ. For missions such as Mars Sample Return, the acquisition of uncontaminated samples is essential. Development of sample cache systems and sample transfer elements designed to facilitate reliable planetary protection implementations will be a significant benefit, as will the protection of samples from loss of volatiles during the return trip and the preservation of the samples physically from the re-entry environment upon reaching Earth.

The panel deemed it important to develop new RDSP technologies for various reasons. First, the search for past or present evidence of life off of Earth is a fundamental goal of NASA exploration. Analyses of surface samples from any planetary body are of great scientific interest, and future RDSP research and development will continue to produce more capable surface and near-surface samplers. But with respect to the search for life, targeting surface samples would likely only lead to a limit on the science yield. Analyses of subsurface rock and regolith samples from planetary bodies will provide planetary science with a high-quality gateway to the climate history of the target body and locations on the body that may be or might have been conducive to life. Direct life detection is likely to require instrument analysis of subsurface regions where water is found in the liquid phase. To support these searches, new RDSP technologies and systems would need to access, acquire, and process pristine and unaltered subsurface samples, as well as samples associated with liquid water from below the surface. These RDSP technology developments (including technology elements developed to achieve reliable forward and back planetary protection for sample return missions) scored very high in the panel valuations because, in part, no other agency is believed to be working on these technologies. Due to extreme mass and power constraints associated with planetary missions, there is also little or no technology transfer to be had from the power- and mass-rich terrestrial commercial drilling technology base. In addition, planetary drilling and sample processing systems possess a very high degree of autonomy that is not necessary for terrestrial drilling. RDSP is a singular technology set required to realize a critical NASA goal. Support for a very robust RDSP technology development program scored high because such a technology development push corresponds directly to one of the NASA strategic goals (exploring the origins and fate of life in the universe).

Beginning in the 1960s and continuing up to the present, surface oriented robotic sampling, shallow drilling, and sample preservation systems have achieved TRL 9. Many of these are well known, such as the Soviet Luna drills or the Mars Exploration Rover Rock Abrasion Tools. Recent systems include a shallow-drilling solid rotary drill integrated into the Mars Science Laboratory (Curiosity), which can deliver drill cuttings to instrumentation. RDSP systems supporting future explorations capable of penetrating deeper into a wide range of targets and capable also of acquiring unaltered cores for processing (or systems that transport instrumentation down the hole) are only at the TRL 2-5 range. With earlier basic sampling systems already developed to a high TRL and recent RDSP R&D producing limited but steady progress, the risk of developing the RDSP systems for future, more ambitious missions is perceived as moderate. Going deeper with subsurface robotic systems and aiming for ever higher quality samples and processing and protecting those samples will mean a corresponding increase in RDSP R&D to cover this increasing scope of RDSP technologies necessary for future missions.

Future RDSP systems will benefit from field testing on Earth in remote terrains such as Antarctica to approximate target planetary conditions. To support unmanned and manned sampling missions and surface and subsurface operations at comets and asteroids, the use of experiments conducted at the ISS might prove valuable. The microgravity ISS environment could be employed to address many engineering unknowns associated with robotic drilling such as cutting chip behavior in a microgravity environment and unknowns associated with processing and manipulation of powdered or sliced samples in microgravity. Elements of the planetary protection transfer break scheme could be tested at the ISS as well.

RDSP technology is applicable to the continued exploration of Mars for in situ missions and for Mars Sample Return. RDSP technology developments will also benefit manned and unmanned missions to small bodies and the Moon, and RDSP will benefit missions to the surfaces of the rocky planets and the moons of the outer planets.

Technology 4.2.1, Extreme Terrain Mobility

Extreme mobility encompasses all ground or surface level mobility. This involves moving across horizontal surfaces while navigating over, around, or through obstacles, vertical surfaces, and all terrain soil types. Additionally, it includes roving (wheeled, tracked, and walking), crawling, horizontal hopping, slithering, swimming through sand, rolling, climbing, and tethering. This category applies to the ability of a ground vehicle to negotiate extremely challenging, unstructured terrain. This category assumes a nominal gravitational field; mobility in a microgravity field is handled in the section addressing technology 4.2.4 Small Body/Microgravity Mobility.

Extremely mobile platforms will be a critical component to both the success and diversity of extraterrestrial body exploration. In addition, higher degrees of mobility serve to complement autonomy. While it is not universally true to say that more mobility will negate any deficiencies in autonomy, it is fair to say that the more capable a platform is in terms of mobility, the better it will perform across the spectrum of autonomy. The need for extremely mobile platforms is also evident in the terrain that will be traversed. Even structured obstacles such as stairs, walls, sewers, etc., can be challenging for vehicles designed specifically for those applications. However, in chaotic unstructured environments with random geometries and large variations on terrain composition, robotic platforms will be challenged more than ever to avoid getting stuck while still performing relevant and meaningful exploration.

The TRL for extreme mobility technologies covers a broad spectrum. Some of the vertical and liquid traversing technologies are at TRL 1 or 2, especially when considered for NASA applications. However, other ground and vertical surface traversing technologies vary from TRL 3 to 7 in terms of NASA mission applications. The DOD (the Army in particular) is investing heavily in this area. The Department of Energy and Department of Homeland Security are considering investments in this area given access mobility challenges resulting from recent worldwide disasters such as at the Fukushima power plant, and due to tornados and flooding in the United States. Therefore, NASA could benefit from partnering with other federal agencies in the development of extreme mobility technologies. While some of the environments for NASA's extreme mobility applications are unique, the work currently being done outside of NASA does have direct relevance to NASA's missions.

Development of this technology would not significantly benefit from access to the ISS. However, access to Moon/Mars surface base/exploration missions and access to Antarctic research sites and reduced-gravity flight test opportunities would significantly benefit the development of this technology.

This technology is game changing because it provides NASA with the capability to maneuver its surface vehicles in extreme terrain in order to "follow the water"—a high-priority science focus for Mars and lunar surface missions. The non-NASA aerospace needs received a low rating for this technology because industry does not do operations on the surface. Technical risk and reasonableness scored high because this technology category encompasses a wide range of specific-capability investment opportunities, some of which have applications scope and readiness levels very well suited to OCT investment. This technology area is applicable to any exploration mission, human or robotic, to a planetary (or moon) surface with appreciable gravity.

Technology 4.2.4, Small Body/Microgravity Mobility

Operating robots in microgravity poses many challenges. In particular, performing meaningful work in microgravity environments is difficult without fixturing or tethering to grounded structures. The small body/microgravity mobility area involves maneuvering on or above surface environments with very low gravitational forces. This includes walking or roving in microgravity and propulsive thrusting technologies. Examples include working near or on the surface of near-Earth asteroids or comets and all on-orbit EVA and IVA operations. Simple tasks such as turning a screw, drilling a hole, or pushing a button can be extreme challenges to mobile platforms that are not attached to other structures. For structured environments, work can often be performed by grasping parts of the structure to be worked on, such as grabbing a door handle and pushing off of the door frame. However, in unstructured environments (such as on the surface of an asteroid), it is often impractical or impossible to use this methodology to perform work. Therefore, adaptive mobility systems with complementary perception and autonomy are key to performing exploration and sample return missions in tight spaces and microgravity environments.

The TRL for microgravity mobility technologies ranges from 1 to 9. TRL 9 on-orbit operations include examples such as AERCam Sprint (EVA) and SPHERES (intra-vehicular activity). An example of a mission with high TRL was Japan Aerospace Exploration Agency (JAXA) Hayabusa spacecraft, which operated in the proximity of the asteroid Itokawa and successfully performed a sample acquisition operation. The JAXA Minerva asteroid rover was flown on the same Itokawa mission, but an error in deployment kept it from successfully landing on the surface. As a result there has only been a single attempt and no successful surface mobility operations on a small body. Surface mobility (as opposed to proximity operations) has had some technological prototypes such as the variable center of gravity (CG) technology implemented in Minerva and a hybrid momentum wheel. Variable or dynamic CG capabilities can greatly enhance the ability of platforms to move around and perform meaningful work by dynamically shifting the CG in conjunction with the motion of the vehicles. Techniques for operating on a surface with volatiles (e.g., comets) are only at preliminary conceptual stages. DOD and commercial satellite companies are the only other U.S. organizations that might have some interest in EVA operations (for the purposes of inspection and servicing/repair of orbital assets).

Development of this technology would significantly benefit from access to the ISS, especially with respect to EVA and IVA operations and testing. This technology is also well aligned with NASA's goals related to the exploration of small bodies both robotic and with crew as demonstrated by recent selections (e.g., selection of OSIRIS-Rex), making this a critical technology for future missions.

The panel overrode the QFD score for this technology to designate it as a high-priority technology because the QFD scores did not capture the value of this technology in terms of its relevance with NASA priorities. In particular it is recommended as a high priority because the NASA 2010 Authorization Act (P.L. 111-267) has indicated that small body missions (to near-Earth asteroids) should be an objective for NASA human spaceflight beyond Earth orbit. If this goal is pursued as a high NASA priority, it would likely also require precursor robotic missions to small body surfaces with applicable mobility capability. This item's overall score was low because the nature of development and testing of microgravity systems is expensive and it has little relevance to the non-aerospace community.

MEDIUM- AND LOW-PRIORITY TECHNOLOGIES

TA04 includes 30 level 3 technologies that ranked low or medium priority. This includes a substantial number of technologies that were added to the TABS in an attempt to align the TA04 TABS with that of the other 13 roadmaps.

In this roadmap, two technologies (4.3.5 Collaborative Manipulation and 4.3.4 Mobile Manipulation) were assessed to be medium priority because of limited alignment with NASA mission areas. While several potential missions (e.g., planetary missions with multiple robotic surface vehicles, on-orbit servicing and/or assembly) were identified, the technologies were not broadly applicable to multiple missions in multiple NASA mission areas.

Seven other medium-priority technologies were judged to provide major improvements in mission performance, but were assessed to have a lower risk level than is usually appropriate for NASA OCT technology investments. These technologies included 4.7.3 Onboard Computing, 4.5.3 Autonomous Guidance and Control, 4.1.1 Vision, 4.7.1 Modularity/Commonality, 4.6.1 Relative Navigation Sensors, 4.1.4 Localization and Mapping, and 4.1.2 Tactile

Sensing. For most of these technologies, the development risk is low because the underlying technology is well understood, and a gradual development program focused on advances in these areas could be formulated, tested, and applied incrementally. 4.7.3 Onboard Computing received a somewhat lower score because so many other agencies and commercial entities are working on it that NASA's contribution would be relatively small.

The remaining 11 medium-priority technologies and one low-priority technology were also judged to provide major improvements in mission performance with broad applicability to multiple NASA missions, but development cost and cost risk were a concern. These technologies included 4.1.6 Multi-Sensor Data Fusion, 4.7.2 V&V of Complex Adaptive Systems; 4.5.5 Adjustable Autonomy; 4.4.7 Safety, Trust, and Interfacing of Robotic/Human Proximity Operations; 4.5.2 Dynamic Planning and Sequencing Tools; 4.4.1 Multi-Modal Human Systems Interaction; 4.4.5 Distributed Collaboration; 4.5.4 Multi-Agent Coordination; 4.5.6 Terrain Relative Navigation; 4.2.3 Above-Surface Mobility; 4.1.8 Terrain Classification and Characterization; and 4.2.2 Below-Surface Mobility. Despite the potential for significant benefits, these technologies were assessed to have a high probability of encountering pitfalls which could complicate the effort and cause additional development problems, possibly leading to significant cost growth and schedule delays.

The final nine low-priority technologies were judged to have minor benefits to NASA missions within the next 20 to 30 years. These technologies included 4.4.6 Common Human-Systems Interfaces, 4.3.1 Robot Arms, 4.3.3 Modeling of Contact Dynamics, 4.1.5 Pose Estimation, 4.5.7 Path and Motion Planning w/ Uncertainty, 4.1.3 Natural Feature Image Recognition, 4.1.7 Mobile Feature Tracking and Discrimination, 4.4.3 Robot-to-Suit Interfaces, and 4.4.4 Intent Recognition and Reaction. Each of these technologies fall into at least one of three categories: (1) technologies that are fairly mature with limited room for additional improvements (e.g., 4.1.5 Pose Estimation); (2) technologies where even substantial improvements in component or subsystem technology will not easily translate into order-of-magnitude improvements in mission performance or mission risk (e.g., 4.5.7 Path and Motion Planning with Uncertainty); or (3) technologies still lack proven applicability of their underlying principles (e.g., 4.4.6 Common Human-Systems Interfaces).

The panel noted technologies where investment from organizations outside NASA would overshadow any potential NASA investments. When this is the case, NASA's limited funds could be better spent on areas where DOD, international space agencies, or commercial industry are not making substantial financial contributions. This impacts technologies such as 4.7.3 Onboard Computing, 4.4.4 Intent Recognition and Reaction, and 4.1.1 Vision. NASA's key work in these areas could focus on adapting advances made by other organizations to the specific requirements of NASA missions.

PUBLIC WORKSHOP SUMMARY

The workshop for the Robotics, Tele-Robotics, and Autonomous Systems technology area was conducted by the Robotics, Communications, and Navigation Panel on March 30, 2011, at the Keck Center of the National Academies, Washington, D.C. The discussion was led by panel chair Stephen Gorevan. He started the day by giving a general overview of the roadmaps and the NRC's task to evaluate them. He also provided some direction for what topics the invited speakers should cover in their presentations. After this introduction, the day started with an overview of the NASA roadmap by the NASA authors, followed by several sessions addressing the key areas of the roadmap. For each of these sessions, experts from industry, academia, and/or government were invited to provide a 30 to 45 minute presentation/discussion of their comments on the NASA roadmap. At the end of the day, there was approximately 1 hour for open discussion by the workshop attendees, followed by a concluding discussion by the panel chair summarizing the key points observed during the day's discussion.

Roadmap Overview by NASA

The presentation by the NASA roadmap development team described the seven level 2 subareas within TA04, including examples of past and current missions, previous technology development programs, and future applications for each technology subarea. These were described at a very high level, with little supporting information that described the level 3 technologies within each subarea. Key benefits of TA04 technologies were presented, but were only described as benefits of the entire robotics, tele-robotics, and autonomy fields rather than benefits tied to specific level 3 technologies. The briefing included an unprioritized list of 12 top technical challenges, though the briefing did not describe a direct one-to-one mapping between the challenges and the level 3 technologies, and did not focus the challenges on NASA-specific problems. The briefing also described a specific future mission manifest, and tried to identify technology push versus mission pull assessments against the mission manifest.

The panel described its struggle to evaluate technologies in TA04 whose endpoints were not defined in the roadmap. Without defining endpoints of a specific technology development, it is challenging to evaluate whether improvements in that technology will be sufficient to enable game-changing capabilities or fundamentally new types of missions. For example, mobility requirements are very different for long-distance traverses on a planetary body with substantial gravity than they are on small asteroids. Without a prioritized mission set, it will be hard to rank these challenges against each other. This was especially challenging because NASA had not provided a set of reference missions (consistent with NASA's current priorities and also consistent across all TAs) that could be used to filter these technologies, and the endpoints associated with different missions are very different. NASA stated that it recognized this was an issue during roadmap development, but each roadmap team was ultimately responsible for developing its own mission list.

Briefing 1: Satellite Servicing

David Kang (ATK Aerospace Systems) provided a presentation that evaluated the roadmap in terms of its applicability to in-space satellite servicing missions. Kang stated that on-orbit servicing is a game-changing capability that is very near a tipping point, but development of this capability requires a combination of technical advances (and good systems engineering/integration of less advanced technology) that cannot be tied to a specific level 3 technology. He also stated that satellite servicing offers a method to evaluate the utility of technology against a specific mission, by examining money spent on technology versus money earned from new capabilities (e.g., costs of communication and ground staff versus cost to develop and test onboard autonomy). Commercial servicing must have a business case that closes, so these types of trades are critically important; NASA may not have to make the same type of business case trades for high-priority assets.

Overall, Kang felt that the roadmap had a very good list of key technologies at a high level. He felt there was a need for additional specification for each technology, but this would require mission-oriented applications that define the specific performance level that is required. This is more challenging for TA04 than for any of the other 13 roadmaps, because TA04 deals with autonomy and robotic systems (e.g., the entire spacecraft system).

Discussion comments indicated that a primary impediment to on-orbit servicing may be the high cost of access to space, rather than the specific servicing technologies. When coupled with large fuel requirements for plane change maneuvers, servicing has only looked cost-effective for clusters of satellites in the same orbital plane. Kang recognized this as an issue, but felt that his organization has enough interest from commercial satellite operators in GEO to make a business case close.

Briefing 2: Orbital Debris Removal and Proximity Operations

Wade Pulliam (Logos Technology) presented observations that he generated along with other colleagues who have expertise in orbital debris or related technologies. He provided an overview of the current debris environment in LEO and GEO, projections of future debris environments, and operations concepts for large-scale debris removal. This presentation was relevant because techniques for capturing orbital debris have a similar technology set and requirements list as satellite servicing, with additional complications because the debris is non-cooperative with few things to hold on to. It is more analogous to landing on a small asteroid, including the need for characterization of the state of the target and potential grappling locations when the servicer arrives.

Pulliam described the utility of dexterous robotics for grappling a free-flying object, but also stated that harpoons with tethers rated highly for capturing inert debris objects. Orbital debris removal would also benefit from sensor fusion technologies (especially techniques for combining electro-optical / infrared imaging with LIDAR sensor data) and autonomous anomaly handling.

Overall, Pulliam felt that the roadmap was in pretty good shape, except that it did not sufficiently address grappling of non-man-made or non-intact objects. Pulliam noted that a significant difficulty is stopping and grappling non-cooperative rotating objects, and that technology development to achieve this capability. The NASA authors identified the section where these are briefly mentioned. In addition, Pulliam stated that advances should be tied to a mission and its needs, not included in a technology program for the technology's sake (i.e., let the mission requirements define which parts of the roadmap should take precedence).

During the discussion section, the panel raised the issue of potential roadblocks related to international policy and legal issues. Pulliam stated that 80 percent of the value in LEO is from the United States, and 80 percent of the debris is from Russian upper stages, leading to speculation that much of the risk could be reduced with a bipolar agreement between the United States and Russia.

Briefing 3: Mobility

William "Red" Whittaker (Carnegie Mellon University) followed with a presentation about future science missions that may look to explore "lava tubes" on planetary surfaces—enclosed subsurface areas that scientists have identified as a high priority for exploration. Mobility within these features drives requirements for autonomy (due to denial of line-of-sight communication links), mobility and access, power generation and storage, perception, modeling, and mobility modalities for extreme terrain.

Whittaker believes that future interplanetary missions with mobility should emphasize technologies for long-range traverses over technologies for precision landing. For investigations looking for volatiles, the descent and landing process will substantially impact the environment and prevent an authentic exploration of a pristine area. With traverse technologies, the lander/rover can move to the desired area and perform science without concerns about contamination.

The speaker stated that processing is no longer a constraint to robotic mobility, and sensing is getting very fast. The panel commented that this was truer for terrestrial robotics than for space missions due to the lag between terrestrial and radiation-hardened processors. Advancements in space-qualified processors will be required to close this gap.

Briefing 4: Planetary Surface Robotics

Edward Tunstel (Johns Hopkins University Applied Physics Laboratory) provided an assessment of the TA04 roadmap from a planetary surface robotics perspective. He stated that the roadmap has the same type of technologies as previous roadmaps because the technology had not progressed rapidly; the only change appeared to be the "landmark" missions that the roadmap focuses on. Tunstel stated that there were no glaring omissions in the technology list, but suggested that the roadmap should include more items that offer functional longevity for longer-duration (planetary surface robotic) missions.

Tunstel felt that there was not enough emphasis on software in the technology list, but the panel responded that there may not be enough emphasis on hardware. During discussions, it became apparent that the speaker was addressing his comments to NASA's original work breakdown structure, but between the kickoff and the workshop the panel had already made substantial alterations that addressed his original concern.

Tunstel identified two potential technology gaps: proprioception (both for robustly stable performance on challenging terrain and for manipulation) and low-risk learning/adaptation (i.e., learning by demonstration, maximizing functional capability in the face of degraded subsystems). Tunstel also prioritized the technologies from a planetary surface perspective, listing (1) mobility and manipulation for Mars Sample Return sample caching robotics; (2) multi-arm dexterous telerobotics (including immersive telepresence, haptics, and operation over delay-tolerant networks); (3) access to small body surfaces; (4) access to planetary subsurfaces; and (5) low-risk learning and adaptation.

Tunstel stated that the "open source" software community may have a lot to offer in this technology area, provided that the issues with ITAR sensitivity of flight code can be solved. The panel described a positive recent experience with SourceForge for a ground-based project, providing optimism that there are technical workarounds to these types of issues.

Briefing 5: Autonomous Rendezvous and Docking

Pierro Miotto (Draper Laboratory) focused his briefing on the Autonomous Rendezvous and Docking (AR&D) subarea, describing past missions that demonstrated manual and autonomous rendezvous and docking in space. Previous efforts in this area have been specialized for each application, with very little reuse or lessons learned. To prevent these mistakes in the future, Miotto argued that next-generation systems must emphasize common standards, open architectures, and non-proprietary solutions.

Miotto split the AR&D subarea into five categories: AR&D analysis, design, and test; autonomous GN&C; autonomous mission management; sensors; and mechanisms. Within the context of a spiral development process, he defined technology requirements and goals for early and future spirals for each of the five categories. He also described the importance of adequate trade space analysis for future AR&D missions, including ensuring that reliability, lighting constraints, and operational concepts do not overly constrain the technology space to require only the state of the art. As a result, he recommended that NASA explore development and demonstration of a range of AR&D technologies at different performance levels. He also recommended that NASA provide access for in-flight testing of systems developed independently of NASA, which would help promote an open architecture for commercial and noncommercial entities. Ideally, this would include creation of a dedicated AR&D facility to perform these kinds of tests in a closed-loop environment.

Briefing 6: Robotics

David Akin (University of Maryland) presented a top-level taxonomy of space robotics and described the current state of the art for mobility versus manipulation and microgravity versus planetary surfaces. He described a lower-level taxonomy of function-based manipulative robotics, and outlined key technology issues for mobility and for manipulation technologies. Each of these applications and technologies drives the required complexity, which can run the gamut from dexterous manipulation to simple actuation.

Akin also described future applications that include niches for a wide variety of space robots, each of which performs varied tasks at high efficiency. He does not see a need for a single generalist robot that can perform any task, but that robotics can be developed to support a variety of these niche roles.

Public Comment and Discussion Session

The following are views expressed during the public comment and discussion session by either presenters, members of the panel, or others in attendance.

- *Roadmap level of detail.* Several participants stated that the roadmap covers a lot of ground with a very wide set of topics, which corresponds to a very high-level treatment of individual technologies. As a result, the technical challenges are also at a high level and not very focused. They recommended looking at different solutions that attempt to solve a specific problem (e.g., image/pattern recognition technologies for docking versus planetary exploration), rather than trying to solve this in general.
- *Linkage between missions and technology developments.* It was suggested that NASA was more capability-driven than goal-driven when developing TA04. As a result, it can be challenging to see how the priorities for technology development were derived, given the vast list of technologies and the poor linkage between specific technology levels and the missions that require them.
- *Autonomy versus automation.* Some attendees recommended that autonomy (cognition) be considered separately from automation (i.e., following if/then/else trees). Many of the roadmap technologies that discuss autonomy are really describing more complex automation. The NASA roadmap developers countered that it is hard to imagine a credible mission in the next 20 to 30 years that would require high

levels of autonomy, as mission controllers would prefer to wait a few hours to troubleshoot a potential issue rather than spend the money to develop on-board autonomy and avoid that delay. Entry, descent, and landing (EDL) applications may drive this technology pull because of the short timeframe available to make decisions.

- *Risk of autonomy.* During the autonomy discussion, some participants viewed autonomy as a risk to the program, because the spacecraft may take an action that puts the hardware at risk. They mentioned that NASA may be hesitant to use high levels of autonomy because of the high cost of mission and the high visibility of failure. Other participants viewed autonomy as a way to protect those assets and to speed up the sense-understand-decide-act loop by removing information transport delays and ground operators from the loop. This was a fundamental disconnect about the role of technology in future missions. Additional autonomy can be tested on existing vehicles during extended missions, where the program risk tolerance is higher.
- *Technology pull from future missions.* There was some discussion about how realistic the missions in the roadmap were. Many of the target missions may not be funded, and may not have high enough priority to become real. As a result, NASA should consider updating the roadmaps with a more realistic assessment of potential missions, so that a better rationale for market pull (versus technology push) can be identified.
- *Synergy with DOD investments.* A participant from the DOD encouraged NASA to participate in interagency technology development efforts. This could be a mechanism for NASA to communicate with their DOD counterparts and to identify high-TRL development items that NASA could productize to claim early successes for OCT.

H

TA05 Communication and Navigation

INTRODUCTION

The draft roadmap for technology area (TA) 05, Communication and Navigation, consists of six level 2 technology subareas:[1]

- 5.1 Optical Communication and Navigation
- 5.2 Radio Frequency Communication
- 5.3 Internetworking
- 5.4 Position, Navigation, and Timing
- 5.5 Integrated Technologies
- 5.6 Revolutionary Concepts

The Communication and Navigation TA supports all NASA space missions with the development of new capabilities and services that make NASA missions possible. Communication links are the lifelines to spacecraft, providing commanding, telemetry, and science data transfers as well as navigation support. Planned missions will require a slight improvement in communication data rate, as well as moderate improvements in navigation precision. However, advancement in communication and navigation technology will allow future missions to implement new and more capable science instruments, greatly enhance human missions beyond Earth orbit, and enable entirely new mission concepts. This will lead to more productivity in science and exploration missions, as well as provide high-bandwidth communications links that will enhance the public's ability to be a part of NASA programs of exploration and discovery.

The roadmap describes the communication and navigation technology developments that are necessary to meet the needs of future missions, provide enhanced capabilities, or enable new mission concepts. This includes identification of representative future missions and key capabilities and investments that will enable or enhance these missions. The roadmap focuses on several key issues for the future of communications: development of radio frequency (RF) technology while initiating a parallel path to develop optical communications capability, application of Earth's internetworking technology and processes to reduce operational costs through simplified

[1] The draft space technology roadmaps are available at http://www.nasa.gov/offices/oct/strategic_integration/technology_roadmap.html.

data handling and distribution, improvements in navigation accuracy, development of integrated communication systems, and identification of potentially revolutionary technologies.

Before prioritizing the level 3 technologies included in TA05, several technologies were renamed, deleted, or moved. The changes are explained below and illustrated in Table H.1. The complete, revised technology area breakdown structure (TABS) for all 14 TAs is shown in Appendix B.

TABLE H.1 Technology Area Breakdown Structure for TA05, Communication and Navigation

NASA Draft Roadmap (Revision 10)	Steering Committee-Recommended Changes
TA05 Communication and Navigation	Two technologies have been merged and one has been renamed.
5.1. Optical Comm. and Navigation	
5.1.1. Detector Development	
5.1.2. Large Apertures	
5.1.3. Lasers	
5.1.4. Acquisition and Tracking	
5.1.5. Atmospheric Mitigation	
5.2. Radio Frequency Communications	
5.2.1. Spectrum Efficient Technologies	
5.2.2. Power Efficient Technologies	
5.2.3. Propagation	
5.2.4. Flight and Ground Systems	
5.2.5. Earth Launch & Reentry Comm.	
5.2.6. Antennas	
5.3. Internetworking	
5.3.1. Disruptive Tolerant Networking	
5.3.2. Adaptive Network Topology	
5.3.3. Information Assurance	
5.3.4. Integrated Network Management	
5.4. Position, Navigation, and Timing	
5.4.1. Timekeeping	Merge 5.4.1 and 5.4.2:
5.4.2. Time Distribution	Rename: 5.4.1. Timekeeping and Time Distribution
5.4.3. Onboard Autonomous Navigation and Maneuver	Delete: 5.4.2. Time Distribution
5.4.4. Sensors and Vision Processing Systems	
5.4.5. Relative and Proximity Navigation	
5.4.6. Auto Precision Formation Flying	
5.4.7. Auto Approach and Landing	
5.5. Integrated Technologies	
5.5.1. Radio Systems	
5.5.2. Ultra Wideband	
5.5.3. Cognitive Networks	
5.5.4. Science from the Comm. System	
5.5.5. Hybrid Optical Comm. and Nav. Sensors	
5.5.6. RF/Optical Hybrid Technology	
5.6. Revolutionary Concepts	
5.6.1. X-Ray Navigation	
5.6.2. X-Ray Communications	
5.6.3. Neutrino-Based Navigation and Tracking	
5.6.4. Quantum Key Distribution	
5.6.5. Quantum Communications	
5.6.6. SQIF Microwave Amplifier	
5.6.7. Reconfigurable Large Apertures	Rename: 5.6.7. Reconfigurable Large Apertures Using Nanosat Constellations

NOTE: Technologies 5.4.1, Timekeeping, and 5.4.2, Time Distribution, have been merged and renamed: 5.4.1 Timekeeping and Time Distribution because the technologies are very similar and it would be most effective to develop them together. Technology 5.6.7, Reconfigurable Large Apertures, has been renamed 5.6.7 Reconfigurable Large Apertures Using Nanosat Constellations to better indicate the content of this technology as described in the TA05 roadmap.

TOP TECHNICAL CHALLENGES

The panel identified three top technical challenges for TA05. They are listed below in priority order.

1. Autonomous and Accurate Navigation: Meet the navigation needs of projected NASA missions by developing means for more autonomous and accurate absolute and relative navigation.

NASA's future missions show a diverse set of navigational challenges that cannot be supported with current methods. Precision position knowledge, trajectory determination, cooperative flight, trajectory traverse, and rendezvous with small bodies are just some of the challenges that populate these concepts. In addition, NASA spacecraft will need to do these things farther from Earth and more autonomously. Proper technology investment can solve these challenges and even suggest new mission concepts.

2. Communications Constraint Mitigation: Minimize communication data rate and range constraints that impact planning and execution of future NASA space missions.

A recent analysis of NASA's likely future mission set indicates that communications performance will need to grow by about a factor of ten every ~15 years just to keep up with projected robotic mission requirements. A second dimension of the challenge is measured simply in bits per second. History has shown that NASA missions tend to return more data with time according to an exponential "Moore's law." Missions will continue to be constrained by the legally allocated international spectral bandwidth. NASA's S-band is already overcrowded and there are encroachments at other bands.

Many of the complex things future missions will need to do are hampered by keeping Earth in the real-time decision loop. Often, a direct link to Earth may not even be available when such decisions are desired. This can be mitigated by making decisions closer to the platform—minimizing reliance on Earth operations. Advancements in communications and navigation infrastructure will allow information to be gathered locally and computation to be performed either in the spacecraft or shared with nearby nodes. Clearly this goal is coupled with the need for increased autonomy and flight computing.

3. Information Delivery: Provide integrity and assurance of information delivery across the solar system.

Future missions will include international partnerships and increased public interaction. This will imply increased vulnerability to information compromise. As mentioned in the 2012 Science and Technology Priorities Memo from the White House, NASA needs to "Support cybersecurity R&D to investigate novel means for designing and developing trustworthy cyberspace—a system of defensible subsystems that operate safely in an environment that is presumed to be compromised." As internetworking extends throughout the solar system, the communications architecture needs to operate in a safe and secure manner.

QFD MATRIX AND NUMERICAL RESULTS FOR TA05

The process used to evaluate the level 3 technologies is described in detail in Chapter 2. The results of the evaluation are shown in Figures H.1 and H.2, which show the relative ranking of each technology. The panel assessed four of the technologies as high priority. Three of these were selected based on their QFD scores, which significantly exceeded the scores of lower ranked technologies. After careful consideration, the panel also designated 5.5.1 Radio Systems as a high-priority technology.[2]

---

[2] In recognition that the QFD process could not accurately quantify all of the attributes of a given technology, after the QFD scores were compiled, the panels in some cases designated some technologies as high priority even if their scores were not comparable to the scores of other high-priority technologies. The justification for the high-priority designation of all the high-priority technologies for TA05 appears in the section titled "High Priority Level 3 Technologies," below.

	Benefit	Alignment with NASA Needs	Alignment with Non-NASA Aero Tech Needs	Alignment with Non-Aerospace National Goals	Technical Risk and Reasonableness	Sequencing and Timing	Time and Effort	QFD Score (Weighted)	Panel Priority
Multiplier	27	5	2	2	10	4	4		
	0/1/3/9	0/1/3/9	0/1/3/9	0/1/3/9	1/3/9	-9/-3/-1/1	-9/-3/-1/0		
Technology Name	Benefit	Alignment			Risk/Difficulty				
5.1.1. Detector Development	3	9	3	1	3	-3	-1	148	M
5.1.2. Large Apertures	3	9	1	0	3	-3	-3	134	M
5.1.3. Lasers	3	9	1	1	3	-3	-1	144	M
5.1.4. Acquisition and Tracking	3	9	1	0	3	-3	-1	142	M
5.1.5. Atmospheric Mitigation	3	9	1	1	3	-3	-3	136	M
5.2.1. Spectrum Efficient Technologies	1	9	3	0	3	-3	-1	92	M
5.2.2. Power Efficient Technologies	1	9	9	3	3	1	-1	126	M
5.2.3. Propagation	1	9	1	1	3	-9	-3	58	L
5.2.4. Flight and Ground Systems	1	9	3	1	3	-3	-1	94	M
5.2.5. Earth Launch and Reentry Communications	1	9	1	0	3	-9	-3	56	L
5.2.6. Antennas	3	9	3	0	3	-3	-1	146	M
5.3.1. Disruptive Tolerant Networking	3	9	3	3	3	1	-1	168	M
5.3.2. Adaptive Network Topology	3	9	3	3	9	-9	-1	188	H
5.3.3. Information Assurance	1	9	9	0	1	-9	-3	52	L
5.3.4. Integrated Network Management	3	9	3	0	3	-1	-1	154	M
5.4.1. Timekeeping and Time Distribution	3	9	9	3	9	-9	-1	200	H
5.4.3. Onboard Autonomous Navigation and Maneuvering	3	9	3	0	9	-3	-1	206	H
5.4.4. Sensors and Vision Processing Systems	3	9	3	0	3	-3	-1	146	M
5.4.5. Relative and Proximity Navigation	3	9	3	0	3	-3	-1	146	M
5.4.6. Auto Precision Formation Flying	3	3	1	0	9	-3	-1	172	M
5.4.7. Auto Approach and Landing	3	3	1	0	3	-3	-1	112	M
5.5.1. Radio Systems	3	9	3	9	3	-3	-1	164	H*
5.5.2. Ultra Wideband Communications	3	3	1	0	9	-9	-1	148	M
5.5.3. Cognitive Networks	3	3	3	3	3	-9	-3	90	M
5.5.4. Science from the Communication System	1	3	0	0	3	-3	-1	56	L
5.5.5. Hybrid Optical Communication and Navigation Sensors	1	3	1	0	3	-3	-1	58	L
5.5.6. RF/Optical Hybrid Technology	1	9	3	1	3	-9	-1	70	L
5.6.1. X-Ray Navigation	0	3	0	0	1	-9	-3	-23	L
5.6.2. X-Ray Communications	0	0	0	0	1	-9	-3	-38	L
5.6.3. Neutrino-Based Navigation and Tracking	0	0	0	0	1	-9	-9	-62	L
5.6.4. Quantum Key Distribution	0	3	1	0	1	-9	-3	-21	L
5.6.5. Quantum Communications	0	3	1	0	1	-9	-9	-45	L
5.6.6. SQIF Microwave Amplifier	1	3	3	1	1	-9	-3	12	L
5.6.7. Reconfigurable Large Apertures Using Nanosat Constellations	1	3	0	0	1	-9	-3	4	L

FIGURE H.1 Quality function deployment (QFD) summary matrix for TA05 Communication and Navigation. The justification for the high-priority designation of all high-priority technologies appears in the section "High-Priority Level 3 Technologies." H = High Priority; H* = High Priority, QFD score override; M = Medium Priority; L = Low Priority.

CHALLENGES VERSUS TECHNOLOGIES

Figure H.3 shows the relationship between the individual level 3 TA10 technologies and the top technical challenges. Note that the lowest-priority technologies as determined by QFD rankings tend not to be strongly connected to the top technical challenges. All of the high-priority technologies have a strong connection to two of the top technical challenges. This shows a good level of consistency between the evaluations and the QFD rankings.

HIGH-PRIORITY LEVEL 3 TECHNOLOGIES

Panel 2 identified four high-priority technologies in TA05. The justification for ranking each of these technologies as a high priority is discussed below.

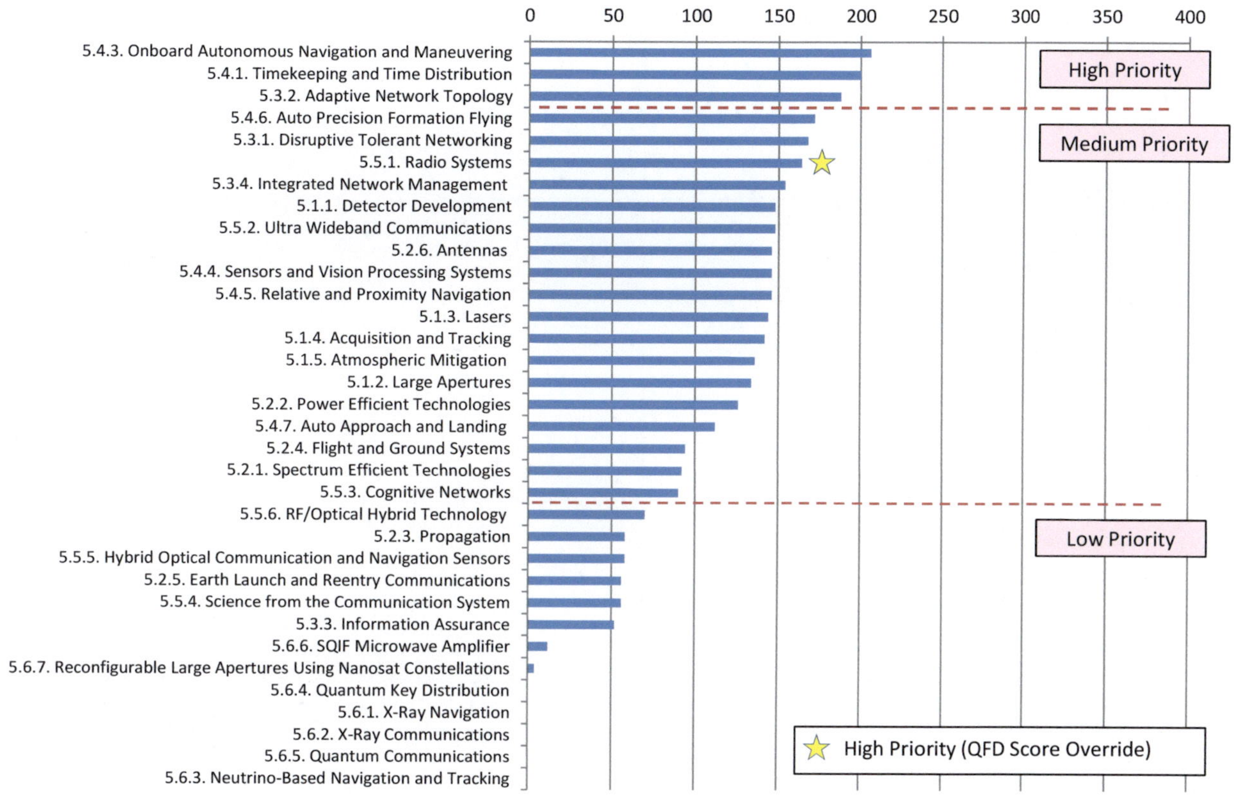

FIGURE H.2 Quality function deployment rankings for TA05 Communication and Navigation.

Technology 5.4.3, Onboard Autonomous Navigation and Maneuvering Systems

Onboard autonomous navigation and maneuvering (OANM) techniques are critical for improving the capabilities and reducing the support requirements for many future space missions. By accurately determining the position and attitude of the vehicle, the dependence on routine position fixes from Earth will be greatly reduced, freeing the communication network for other tasks. The onboard maneuver planning and execution monitoring will increase the vehicle agility, enabling new mission capabilities currently not possible given the round-trip time delay to Earth and the loss of communication during atmospheric reentry. It could also reduce costs by helping to eliminate the large work force required to support the routine operations of the spacecraft.

Given the previous flight demonstrations, e.g., DS-1 and EO-1, some basic aspects of the OAMN are at TRL 9, but the new advanced capabilities required for future missions are closer to a TRL of 3. The research and flight demonstrations needed for this technology to be accepted for future missions are well aligned with the NASA experience and capabilities and consistent with NASA's past roles.

It is possible that NASA will be able to draw upon new algorithms and capabilities resulting from the significant current interest in autonomous robotics and UAVs, but the mission and autonomy requirements in these domains differ significantly from NASA's needs. There is more overlap with Department of Defense work on autonomous underwater robotics, given the truly remote operations with limited off-board communication capabilities that often exist. However, given the length, variety, and complexity of the missions, the challenges faced by NASA are unique and it is very unlikely that they will be fully addressed by an external research organization. As such, NASA investment in this technology area is critically important.

The panel determined that the OANM systems provide a major benefit due to the technology's potential to significantly improve the capabilities/performance and reduce the operational cost of future missions. The alignment

		Top Technology Challenges		
Priority	TA 05 Technologies, Listed by Priority	1. Autonomous and Accurate Navigation: Meet the navigation needs of projected NASA missions by developing means for more autonomous and accurate absolute and relative navigation.	2. Communications Constraint Mitigation: Minimize communication data rate and range constraints that impact planning and execution of future NASA space missions.	3. Information Delivery: Provide integrity and assurance of information delivery across the solar system.
H	5.4.3. Onboard Autonomous Navigation and Maneuvering	●	●	
H	5.4.1. Timekeeping and Time Distribution	●	●	
H	5.3.2. Adaptive Network Topology	○	●	●
H	5.5.1. Radio Systems	○	●	●
M	5.4.6. Auto Precision Formation Flying	○		
M	5.3.1. Disruptive Tolerant Networking		○	○
M	5.3.4. Integrated Network Management		○	○
M	5.1.1. Detector Development		○	
M	5.5.2. Ultra Wideband Communications		○	
M	5.2.6. Antennas		○	
M	5.4.4. Sensors and Vision Processing Systems	○		
M	5.4.5. Relative and Proximity Navigation	○		
M	5.1.3. Lasers	○	○	
M	5.1.4. Acquisition and Tracking	○	○	
M	5.1.5. Atmospheric Mitigation	○	○	
M	5.1.2. Large Apertures		○	
M	5.2.2. Power Efficient Technologies		○	○
M	5.4.7. Auto Approach and Landing	○		
M	5.2.4. Flight and Ground Systems	○	○	
M	5.2.1. Spectrum Efficient Technologies		○	○
M	5.5.3. Cognitive Networks		○	○
L	5.5.6. RF/Optical Hybrid Technology		○	○
L	5.2.3. Propagation		○	
L	5.5.5. Hybrid Optical Comm. and Nav. Sensors	○	○	
L	5.2.5. Earth Launch and Reentry Commmunications		○	
L	5.5.4. Science from the Comm. System			
L	5.3.3. Information Assurance			○
L	5.6.6. SQIF Microwave Amplifier			
L	5.6.7. Reconfigurable Large Apertures using Nanosat Constellations			
L	5.6.4. Quantum Key Distribution			
L	5.6.1. X-Ray Navigation	○		
L	5.6.2. X-Ray Communications			
L	5.6.5. Quantum Communications			
L	5.6.3. Neutrino-Based Navigation and Tracking			

●	Strong Linkage: Investments by NASA in this technology would likely have a major impact in addressing this challenge.
○	Moderate Linkage: Investments by NASA in this technology would likely have a moderate impact in addressing this challenge.
[blank]	Weak/No Linkage: Investments by NASA in this technology would likely have little or no impact in addressing the challenge.

FIGURE H.3 Level of support that the technologies provide to the top technical challenges for TA05 Communication and Navigation.

to NASA's needs is high because it will impact deep space exploration with crew, robotic science missions, planetary landers and rovers. The alignment with other aerospace needs is considered to be medium, as some of the automation and planning algorithms will overlap with the DOD/commercial space missions and other robotic applications, but the alignment with national needs is considered to be low as this work will really only be relevant to spacecraft. The risk is assessed to be moderate to high, within the bounds of NASA's acceptable risk levels for technology development. Of particular importance is the observation that the autonomous navigation and maneuver planning algorithms can be added (and verified) incrementally on near-term missions to enable a solid foundation to be established for future missions. This approach could be used to appropriately scale the workforce required to address the OANM technology challenges at a moderate level of effort. Due to the potential for major mission improvements and cost savings, strong alignment with NASA needs, and reasonable risk and development effort, OANM are rated as high-priority technologies.

Technology 5.4.1, Timekeeping and Time Distribution

NASA provides communications and navigation infrastructure for its missions. Underlying this infrastructure are atomic clocks and time transfer hardware and software. New, more precise atomic clocks operating in space, as well as new and more accurate means of time distribution and synchronization of time among such atomic clocks (e.g., optical transmitters and receivers operating in the space environment), will enable the infrastructure improvements and expansion NASA requires in the coming decades.

The TRL of timekeeping and distribution systems varies over a wide range from 9 for current systems all the way down to 1 for yet-to-be-proposed atomic clocks. Ground-based work[3] at the National Institute of Standards and Technology (NIST) and other laboratories can be considered TRL 3-4. The planned ISS experiment ACES[4] is a scientific instrument, not routine PNT gear; TRL 3 is probably a good estimate for the status of ACES when considered for use in an operational scenario.

Precision timekeeping and transfer is of interest not only to NASA but also to the defense and communication industries as well as the scientific community. NASA has expertise in timekeeping and distribution in space systems via the Deep Space Network and the TDRSS. NASA collaborations with DOD, NIST, university labs, and international partners such as France and Japan will be fruitful. (Japan, France, NIST, and JPL are partners in ACES mentioned above.) NASA, with appropriate partners, could focus on miniaturizing laboratory prototypes, providing the necessary stable mechanical and thermal environment, and increasing the longevity and reliability of precision clocks and time transfer equipment. The ISS may play a role in development of timekeeping and distribution technology by providing a microgravity environment in which the technologies can be demonstrated. However, for missions beyond LEO, hardware demonstration at high radiation levels and algorithms appropriate to solar system navigation are needed.

Advances in timekeeping and distribution of several orders of magnitude were judged to provide major benefits, since increased precision of timekeeping and transfer leads to increased precision of relative and absolute position and velocity which in turn provides better starting solutions to enable autonomous rendezvous, docking, landing, and formation flying remote from Earth. In addition, precision timekeeping enables new tests of fundamental physics, time and frequency metrology, geodesy, gravimetry, and ultra-high-resolution VLBI science applications. Alignment with NASA's needs is considered high due to the substantial impact of the technologies to multiple missions in multiple mission areas including human and robotic spaceflight involving rendezvous, relative station keeping and landing missions. Similarly there is high alignment with other aerospace and non-aerospace national needs, as the benefit of precise timekeeping and synchronization of navigational and communications equipment grows with the explosive increase of data required by modern technology. The risk associated with development of precision timekeeping and distribution is judged to be moderate to high, a good fit for NASA's technology development program. Due to their major benefits, alignment with NASA and other national needs, and reasonable risk and development effort, precision timekeeping and time distribution are rated as high-priority technologies.

Technology 5.3.2, Adaptive Network Topology

Adaptive Network Topology (ANT) is the capability for a network to change its topology in response to either changes or delays in the network, or additional knowledge about the relationship between the communication paths. This technology area includes technologies to improve mission communication such as ad hoc and mesh networking, methods of channel access, and techniques to maintain the quality of signal across dynamic

[3] See http://www.nist.gov/pml/div688/logicclock_020410.cfm.

[4] See http://www.spaceflight.esa.int/users/downloads/factsheets/fs031_10_aces.pdf.

networks to assure successful exchange of information needed to accommodate increased mission complexity and achieve greater mission robustness. The goal is to improve end-to-end rather than hop-to-hop performance. This was originally pursued for commercial uses. However, optimization in terrestrial networks with this method has been mostly abandoned, replaced by adding additional tall towers and other hubs. The fall in terrestrial equipment prices has obviated the need for optimization inside of a limited network.

The TRL for these technologies are within the 2-5 range. They have been demonstrated in the laboratory, but no commercial or space implementation has been realized. While of strong academic interest, it is unlikely that any practical development will be done by commercial entities. However, government entities with extreme applications, such as underwater communications, are likely to pursue these capabilities. Investments within DOD are happening, but the constraints may be sufficiently different to be incompatible with NASA interests. While the basic research might be purchasable from academic or research interests sponsored by the National Science Foundation and others, applicable technology will not. Because of the lack of commercial need, NASA participation is necessary if this technology is to be developed to a level that can support NASA missions in the next decades.

Access to the ISS is not a requirement for development in this technology area. The applicability of this technology potentially crosscuts all missions where the end-points do not have direct line of sight communications. The distinction between disruptive tolerant networking (DTN) and ANT is debatable. The panel views DTN as a subset of ANT and this view is, in part, responsible for the high priority of ANT.

The panel rated the alignment with NASA needs very high for this technology. The benefit to NASA is derived because future multi-element missions will require advanced network topologies which will need to be adaptive in order to remain robust for their applications. A high rating in the category of technical risk and reasonableness was given since this is a significant extension of prior capability, with associated risk. However, the level of risk has been mitigated by a variety of prior, small scale research projects by a number of government agencies such as DOD, the National Science Foundation, and even the National Institute for Occupational Safety and Health in the mining industry (especially following accidents). A low score for sequencing and timing stems mainly from the fact that a mission need schedule is not well defined for this technology.

Technology 5.5.1, Radio Systems

Radio systems technology focuses on exploiting technology advances in RF communications, PNT, and space internetworking to develop advanced, integrated space and ground systems that increase performance and efficiency while reducing cost. For example, a multipurpose software defined radio might be developed that can change its function with mission phase and requirements. While this technology can benefit from individual advancements in many of the other level 3 technologies in TA05 (antennas, atmospheric mitigation, large apertures, power efficient technologies, spectrum efficient technologies, propagation, Earth launch and reentry communication systems, and information assurance), this entry focuses on the challenges associated with integration of these advancements into operational systems.

Advancements in radio systems integration focus on one of the highest priority technical challenges within TA05: Minimize communications constraints on data rate and range that impact planning and execution of future NASA space missions. Like all technologies in TA05, these advancements benefit multiple types of missions in deep space as well as near-Earth orbit.

Software defined radios (SDRs) are a prime example of the challenges of integration for radio systems. These offer frequency agility and wide frequency coverage, but require broadband circuitry, broadband antennas, and sophisticated software to drive the system. While the component technologies in each of these areas are mature, integration into flight-worthy systems is still required before widespread adoption of SDRs becomes possible. SDRs can be configured for many applications such as radar, arrays for beamforming, direction finding, signal identification, etc.

The radio astronomy community has substantial overlap with the technologies for future SDR implementations for receiving telescopes. The Allen Telescope Array (ATA) of the SETI Institute and the University of California at Berkeley is a prototype of what may become the standard in radio astronomy. In this case, Allen provided development funding to the SETI Institute through proof of principle and initial operations, but a lack of ongoing operational funding caused the array to be mothballed. The technologies implemented in the ATA can be applied to future transmitting telescopes as well, and would be a tremendous enhancement of the capabilities of the Deep Space Network. While NASA can leverage techniques and technologies from the ATA, transmitting is atypical for the radio astronomy community and will require technology investments from NASA and others.

Active antenna elements are needed for smart transmitting arrays, which could also be used for a smart receiving array. It would also be useful in future relay networks and orbiting communication networks, which would enable ground users to form beams in several different directions simultaneously. This provides the following substantial simultaneous benefits: Covering large fractions of the sky on a wide range of frequencies (wavelengths of meters to millimeters); tracking or searching for multiple targets simultaneously; and nulling-out discrete or extended sources of interference while receiving faint signals from sources with only small separations in look angle and measuring their spectra, Doppler shifts, and angular structure. With this technology, NASA could improve the data rates and efficiency of communication with deep space missions, including enabling high-rate communication with multiple missions simultaneously. It would also reduce the need for detailed scheduling of access to the antennas and receivers, as they can be operated simultaneously.

Receiving arrays have a current TRL of ~6, but transmitting arrays are closer to the TRL 2-3 range. There may be component technologies in radar systems that can be applied to transmitting arrays, but these will need to be demonstrated in an integrated fashion to achieve a higher TRL. The DOD has highly versatile, broadly applicable radar systems that can perform many of these same functions, but to our knowledge the applicability of these systems to NASA-unique needs has not been addressed.

The ISS could be used as a platform for space-based testing of new SDR integrated systems, but it is not an ideal platform due to the noisy RF environment and the number of arrays, radiators, and modules that inhibit directional line-of-sight communications. Much of the development and systems integration can be performed on the ground, especially for future Deep Space Network systems. But for space-based applications such as relay networks or formations of small satellites, the technology would get as much, if not more, benefit from initial testing on any orbiting vehicle.

NASA has the expertise to take a leading role in the development of radio systems, software defined radios, and array telescopes. While some advancements could come from DOD and the National Reconnaissance Office efforts, NASA has unique requirements for low signal levels and demanding angular resolution that will not be reflected in outside agency work. Given that the commercial cell phone industry is interested in applications for future generations of cognitive radios as well, NASA could pursue joint development projects, or at minimum remain cognizant of advances made within these industries and organizations.

The steering committee assessed the benefit of radio systems technologies to result in major mission performance improvements due to the potential to improve throughput, versatility, and reliability with lower impact on the host spacecraft in terms of size, weight, and power (SWAP). The alignment to NASA needs is high because improvements in communication systems will impact nearly every NASA spacecraft, including near-Earth, deep space, and human exploration missions. The alignment with other aerospace needs is moderate due to its potential impact on DOD and commercial spacecraft. The alignment with other national needs is high, as it could impact any terrestrial industry that uses radios, WiFi, cable, internet, and other communications technologies. The risk is assessed to be low to moderate, as previously developed prototypes have already demonstrated key technologies at the component or subsystem level. The remaining challenge is to integrate these with NASA communications and complete space qualification of the integrated systems. The panel overrode the QFD score for this technology to designate it as a high-priority technology because the QFD scores did not capture the value of this technology in terms of the system level advances that would occur with major improvements in radio systems.

MEDIUM- AND LOW-PRIORITY TECHNOLOGIES

TA05 includes 30 level 3 technologies that ranked low or medium priority.

In this roadmap, two technologies (5.4.6 Autonomous Precision Formation Flying and 5.5.2 Ultra Wideband) were assessed to be medium priority because of limited alignment within multiple NASA mission areas. While several potential missions (e.g., fractionated satellites or missions with multiple spacecraft, burst transmissions between nearby spacecraft) were identified, the technologies were not broadly applicable to multiple missions in multiple NASA mission areas.

Seven other medium-priority technologies were judged to provide major improvements in mission performance, but were assessed to have a lower risk level than is usually appropriate for NASA OCT technology investments. These technologies included 5.3.1 Disruptive Tolerant Networking, 5.3.4 Integrated Network Management, 5.2.6 Antennas, 5.4.4 Sensors and Vision Processing Systems, 5.4.5 Relative and Proximity Navigation, 5.1.3 Lasers, and 5.1.5 Atmospheric Mitigation. For most of these technologies, the development risk is low because the underlying technology is well understood, and a gradual development program focused on advances in these areas could be formulated, tested, and applied incrementally.

Five additional medium-priority technologies were also judged to provide major improvements in mission performance with broad applicability to multiple NASA missions, but development cost and cost risk were a concern. These technologies included 5.1.1 Detector Development, 5.1.4 Acquisition and Tracking, 5.1.2 Large Apertures, 5.4.7 Auto Approach and Landing, and 5.5.3 Cognitive Networks. Despite the potential for significant benefits, these technologies were assessed to have a high probability of encountering pitfalls which could complicate the effort and cause additional development problems, possibly leading to significant cost growth and schedule delays.

The remaining three medium-priority technologies and six low-priority technologies were judged to have only minor benefits to NASA missions within the next 20 to 30 years. These technologies included 5.2.2 Power Efficient Technologies, 5.2.4 Flight and Ground Systems, 5.2.1 Spectrum Efficient Technologies, 5.5.6 RF/Optical Hybrids, 5.2.3 RF Propagation, 5.2.5 Earth Launch and Reentry Communication, 5.5.5 Hybrid Optical Communication and Navigation Sensors, 5.5.4 Science from the Communication System, and 5.3.3 Information Assurance. Each of these technologies falls into at least one of two categories: (1) technologies that are fairly mature with limited room for additional improvements (e.g., 5.2.3 RF Propagation) or (2) technologies where even substantial improvements in component or subsystem technology will not easily translate into order-of-magnitude improvements in mission performance or mission risk (e.g., 5.5.3 Information Assurance).

The remaining seven low-priority technologies were judged to be so immature or impractical that no benefit is projected in the 20- to 30-year timeframe of interest. These technologies were all part of the Revolutionary Concepts section of the roadmap, and include 5.6.6 SQIF Microwave Amplifiers, 5.6.7 Reconfigurable Large Apertures Using Nanosat Constellations, 5.6.4 Quantum Key Distribution, 5.6.1 X-Ray Navigation, 5.6.5 Quantum Communications, 5.6.2 X-Ray Communications, and 5.6.3 Neutrino-Based Navigation and Tracking.

The panel noted technologies where investment from other organizations outside NASA would overshadow any potential NASA investments. When this is the case, NASA's limited funds could be better spent on areas where DOD, international space agencies, or commercial industry are not making substantial financial contributions. This impacts technologies such as 5.3.1 Disruption Tolerant Networking, 5.3.4 Integrated Network Management, and 5.1.5 Atmospheric Mitigation. NASA's key work in these areas could focus on adapting advances made by other organizations to the specific requirements of NASA missions.

OTHER GENERAL COMMENTS ON THE ROADMAP

All NASA missions require communication and navigation to some degree, so the priorities developed in this section are mostly independent of the mission mix. For some missions, advances in communication throughput could enable them to carry more advanced, high-data rate instruments with significant science benefit (e.g., hyperspectral imagers). Similarly, improvements in navigation accuracy, when coupled with coordinated improvements in guidance, autonomy, sensors, and EDL technologies, will enable high precision pinpoint landing on planetary surfaces. But in most cases, the prioritization of communication and navigation technologies is not impacted by specific missions in the mission model.

PUBLIC WORKSHOP SUMMARY

The workshop for the TA05 Communication and Navigation technology area was conducted by the Robotics, Communications, and Navigation Panel on March 29, 2011, at the Keck Center of the National Academies,

Washington, D.C. The discussion was led by panel chair Stephen Gorevan. He started the day by giving a general overview of the roadmaps and the NRC's task to evaluate them. He also provided some direction for what topics the invited speakers should cover in their presentations. After this introduction, the day started with an overview of the NASA roadmap by the NASA authors, followed by several sessions addressing the key areas of the roadmap. For each of these sessions, experts from industry, academia, and/or government were invited to provide a 30 to 45 minute presentation/discussion of their comments on the NASA roadmap. At the end of the day, there was approximately 1 hour for open discussion by the workshop attendees, followed by a concluding discussion by the panel chair summarizing the key points observed during the day's discussion.

Roadmap Overview by NASA

The presentation by the NASA roadmap development team identified challenges faced by future interplanetary missions, and described how improved communications and navigation can mitigate these challenges. From a technology development portfolio standpoint, the key challenge is ensuring that advancements in communication and navigation are out ahead of new missions, so that the technology is sufficiently advanced that it can be incorporated in future missions without adversely impacting the critical path.

The existing roadmap was developed using a previous decadal survey, but a new one was released a few weeks before the workshop. NASA recognized the need to update the roadmap to reflect the new decadal survey, but they did not feel this would substantially change the technologies in the roadmap. Even without pull from specific future missions, NASA indicated that they would pursue the same technology list to ensure that current technology would stay ahead of potential communication and navigation demand. The discussions mostly focused on the challenges for deep-space (interplanetary) communication and navigation, though some near-Earth-space solutions for optical communication repeaters were described.

NASA outlined the issues for each of the level 2 WBS subareas: optical communication and navigation; RF communication; internetworking; position, navigation, and timing (PNT); and revolutionary concepts. Within each subarea, the presentation described the current state of the art, recent advancements, desired future developments, and mission-level impact. Challenges identified include laser pointing and stabilization (optical communication subarea), communication through challenging RF environments such as rocket plumes and reentry ionization (RF communication), utilizing arrays of antennas to increase the effective aperture (RF communication), onboard computing's mass/power/volume constraints as they relate to onboard routing (internetworking), and autonomy for navigation (PNT).

NASA mentioned that the integration of optical communications and disruption-tolerant networking (DTN) offers combined benefits that are stronger than each one alone. The DTN program is structured around near-term improvements in single-hop DTN, which will demonstrate benefits to automation rather than networking. Rather than scripting data transfers with manual downlink plans, DTN would allow for this to happen automatically onboard the vehicles. This provides a motivation for DTN even without a vision for a deep-space multi-node network of relays.

After NASA's presentation, a brief discussion period allowed the panel and interested members of the general public to ask questions of the NASA presenters. There were questions about the composition of future missions in the latest roadmaps, internetworking simulators, deep-space infrastructure for internetworking, networking interoperability/compatibility with international partners, technologies and architectures for space-to-ground optical communications, technology hurdles for pinpoint landing, and prioritization/linking of technologies within the roadmap.

Briefing 1: Satellite Networking

Eylem Ekici (Ohio State University) provided a brief background of his work in wireless communication networks, wireless network analysis and modeling, and protocol development. His comments on the TA05 roadmap focused primarily on the internetworking section. He stated that the roadmap captured the most important existing projects and test cases, but also emphasized the room for growth in this subarea. He believes there is significant potential for research and development in space-based networking, which can take advantage of terrestrial improvements in autonomous network operations for ad hoc mesh networks.

Ekici identified disruption tolerant networking as the most critical challenge and a game-changing capability within the roadmap. However, he feels that existing work on DTN is insufficient to address needs of actual space missions due to poorly partitioned objectives, assets, functions, and interoperability. In addition, he believes that work on adaptive network topology should be a part of a forward-looking vision for networking, including work on requirements analysis, general network architectures, and protocol and solution development. While he felt that information assurance was important for medium-term applications, existing solutions (or new solutions developed by other agencies) are likely to address about 90 percent of NASA's challenges at very low cost to the agency.

In the question and answer session following his presentation, there were questions about the suitability of space-qualified, radiation-hardened electronics for networking applications. Existing commercial and government hardware typically have issues with radiation including single event upsets and latchups. Because NASA deep space missions operate in a much higher radiation environment than missions in LEO and GEO, there may be hardware-based challenges to implementing space networking solutions. Ekici indicated that there are likely to be hardware/processing issues that will have to be addressed, and recommended that this work occur in parallel with the networking and protocol development.

Briefing 2: Optical Communications

Gary Swenson (University of Illinois at Urbana Champaign) followed Ekici's presentation with a discussion on observations generated by himself and a group of professors at UIUC with backgrounds in fiber lasers, quantum electronics, and optical remote sensing. Based on the background of the involved professors, this assessment focused on TA 5.1 Optical Communication and Navigation. They identified laser transmitters as a key challenge for optical communication, including desired improvements in weight, reliability, and efficiency. Secondary challenges in this area are laser beam expanders, the uniformity of the beam in single mode, and methods for mitigating atmospheric disturbance of the beam. For high bandwidths, sensor efficiencies, dark current (cooling), and amplifiers are all areas recommended for improvement.

As a technology development effort, Swenson recommended that NASA address methods to extend the lifetime of optical communication components, including reducing susceptibility to damage by energetic particles and reducing fiber degradation/darkening over time. However, Swenson also stated that NASA's highest near-term priority should be performing an in-space demonstration of optical communications systems, rather than component-level improvements in apertures or atmospheric turbulence mitigation.

Swenson also identified optical communications as low risk, but during the discussion period he clarified that this was low risk for near-term demonstration objectives. For long-term objectives with very large receivers, NASA needs to better understand the sensitivity and efficiency issues before a realistic assessment of the development risk can be made.

There was additional discussion about the need for a local sync for generating path lengths between optical arrays on the ground. NASA indicated that they had some discussion on this subject while generating the roadmap, but the technical details were not addressed in the roadmap document itself. This was recognized as an area that will require additional simulation before an appropriate assessment of risk can be made.

Briefing 3: Guidance and Control

Mimi Aung (JPL) presented an assessment of TA05 from a guidance, navigation, and control perspective. She identified four classes of next-generation missions that will drive improvements in guidance and control: (1) precision landing for large bodies (reducing the landing error ellipse, autonomous safety); (2) landing on small bodies and primitive bodies; (3) formation flying (including swarms and distributed clusters); and (4) autonomous rendezvous and docking. Together, these missions will drive the need for onboard autonomous target-relative GN&C, multiple-spacecraft GN&C, and onboard target-relative navigation sensors. In all of these, target-relative navigation was identified as a key technology challenge. Aung indicated that improvements in GN&C will stress current onboard computing elements, and recommended that NASA invest in advanced computing (including multi-core processing and field-programmable gate arrays, or FPGAs) to keep pace with the demand for additional processing.

Aung stated that improvements in this area will require both subsystem and system demonstrations, including development and utilization of ground-based integrated testbeds and on-orbit demonstrations. Currently, each mission has to develop its own testbed; future efforts could focus on developing an open-architecture adaptable hardware-in-the-loop integrated testbed which could be re-used for multiple applications. Aung also mentioned that JPL feels that the integration of individual technology pieces into a high-functioning system will be a challenge on par with individual technology developments. JPL believes that an integrated test of precision EDL should be a higher priority than advancement of specific pieces. (This echoed the importance of system-level testing from Swenson's previous presentation.) Several other workshop participants echoed this response, emphasizing the possible advantages of moderate-performance technologies that work well when integrated over the absolute best component technology.

However, other comments from the public described a desire to focus the roadmap on technologies that OCT could bring to TRL 6, rather than integrated capabilities that were more program-specific (e.g., radiation-hardened FPGAs rather than terrain-relative navigation (TRN)). TRN as a discipline is more important for NASA to invest in, but each mission will still require substantial customization for their applications and operational concepts, which limits the effectiveness of investments in general TRN capabilities. Similarly, other public comments recommended that general-purpose testbeds could be developed by the programs/missions, rather than an independent technology program. The panel pointed out that science missions are focused much more on science than on technology, and are not generally interested in spending limited program resources on technology maturation. If the testbed or general-purpose research is not funded by a technology organization, the science program is unlikely to do it.

Aung recommended adding a new section to the roadmap dedicated to autonomous integrated GN&C. This led to significant discussion on this approach, and highlighted the need to address the topic of GN&C in an integrated fashion. Currently, navigation is carried under TA05, but guidance and control is split between TA04 (Robotics, Tele-Robotics, and Autonomous Systems) and TA09 (Entry, Descent, and Landing).

Briefing 4: X-Ray Navigation

Darryll Pines (University of Maryland at College Park) described the X-ray NAVigatation and Autonomous Position (XNAV) program that he ran while he was at DARPA from 2003 to 2006. This presentation focused on only one technology: 5.6.1 X-Ray Navigation. Pines described the motivation for using pulsars as navigation beacons, and described potential accuracies of 24 to 36 meters in Earth orbit for long-duration observations, and accuracies less than a kilometer at other locations in the solar system. Newly researched navigation algorithms can provide accurate along-track solutions using the measured Doppler delay in the X-ray pulses. According to Pines, XNAV offers benefits over GPS due to the higher theoretical accuracy from short wavelengths (with appropriate clock updates from the ground), utility throughout the solar system, inherent radiation hardening, and robustness to radiation damage, jamming, and contamination. Because of these benefits, there have been several follow-on programs including the X-ray Timing program at DARPA and several smaller NASA programs. One drawback of XNAV is that past systems required instruments on the order of a few square meters in cross-section and approximately 25 to 50 kg in weight. Recent developments, however, have improved detector signal-to-noise sensitivity and reduced the size, weight, and volume of prototype navigation instruments. Nevertheless, greater position and attitude accuracy will require improvements in X-ray photon detection and optics to produce instruments small enough for deep space navigation.

Briefing 5: Deep-Space Navigation

Lincoln Wood (JPL) provided a briefing with comments on specific sections of the roadmap that impact accuracy and performance of deep space navigation. Wood assessed the roadmap as conservative in estimates of deep space navigation accuracy for single antenna applications (e.g., line-of-sight Doppler and ranging) as well as multi-antenna applications (very long baseline array). In addition to new technologies related to frequency and timing, Wood stated that advancements in SWAP for existing capabilities/components should be technologies in themselves, and should be reflected in the roadmap alongside new capabilities. He stated that any game-changing capability would likely result from evolutionary advances with existing technologies, rather than revolutionary technologies.

Wood's assessment of the top technical challenges (within the frequency and timing section of the roadmap) include improvements in high-performance clocks (stability, SWAP, and reliability), oscillators, and space qualified lasers. These imply high-priority technology advancements in (1) improving frequency and timing reference sources by advancing the TRL, making SWAP advancements, and improving reliability, and (2) making reliability and SWAP improvements (rather than improving performance) for optical metrology hardware. Mercury ion atomic clocks were identified as a component technology near a tipping point, and neutrino navigation was identified as unfeasible on any kind of realistic development timescale.

During the question and discussion phase, Wood recommended that NASA place navigation beacons around places of interest and locations that will be visited at higher frequencies, such as the Moon, Mars, Titan, and other locations.

Briefing 6: Mission Design and Navigation

Alberto Cangahuala (JPL) provided inputs on the roadmap's treatment of navigation for lunar and interplanetary applications. In this area, Cangahuala identified top technical challenges as: (1) high-fidelity modeling to mitigate the impact of long round-trip light times on numerical precision, (2) reducing onboard resources (mass, power, and delta-V) required for guidance and navigation, and (3) boot-strapping required to accrue necessary detailed environmental characterization information for target bodies. He also identified several trends for modern navigation systems, including the challenges of route planning for low-thrust systems, low-energy transfers, proximity operations, satellite and planetary tours, and operations in unstable and/or unknown dynamical environments. As a result of these trends, modern missions require a stronger integration of flight path and attitude control (i.e., coupling of control knowledge and thrust), highlighting the need for more onboard autonomy (both fully integrated systems and navigation "apps" that take advantage of existing flight system capabilities).

Cangahuala identified mission/trajectory design as a potential technology gap in the roadmap. This is a cross-cutting capability with parts in modeling and simulation and navigation. Navigation and mission/trajectory design are connected by a common need for consistent modeling for planetary and spacecraft dynamics, and mission/trajectory design can minimize navigation uncertainties as part of a design process. Within this field, high-priority challenges include improving the speed, agility, and robustness of trajectory optimizers and investigating new trajectory mechanisms (e.g., invariant manifolds, cycler trajectories).

Public Comment and Discussion Session

The following are views expressed during the public comment and discussion session by either presenters, members of the panel, or others in attendance.

- *Synergy with DOD investments.* Some technologies in the roadmap may be synergistic with elements in the DOD. This includes a near-term experiment on atmospheric dynamics, knowledge about space-based IP-networking (though information assurance is a challenge), specific TWTA and SSPA component technologies, and minimization of SWAP for existing technologies. While some of these were tied to the now-cancelled TSAT program, other efforts in these areas are ongoing. NASA and DOD may benefit from working together on these technologies, or at least remaining cognizant of developments made by the other organization. NASA was encouraged to participate in the Space Industrial Council's Critical Technologies Working Group, which addresses inter-agency technology development efforts. This could be a mechanism for NASA to communicate with their DOD counterparts and identify high-TRL development items that NASA could productize to claim early successes for OCT.

- *International involvement.* Based on a question about international cooperation, a representative from NASA OCT clarified that the near-term focus of the NRC is prioritization of the technologies, regardless of who matures them. At some point in the future, OCT will address the potential overlap between NASA's priorities and other organizations, but this will not be part of the NRC's charter.
- *Systems engineering framework.* A workshop participant stated that the roadmaps illustrate component technologies, but lack a systems engineering framework to identify synergies between technologies (especially between different roadmaps). OCT responded that they will do this kind of strategic integration over the next 12 months, and that the NRC is not being asked to do this.
- *Technology pull from future missions.* There was some discussion about how realistic the missions in the roadmap were. Many of the target missions may not be funded, and may not have high enough priority to become real. Participants discussed the value of using a more realistic assessment of potential missions, so that a better rationale for market pull (versus technology push) can be identified.
- *Industrial base.* For navigation components (e.g., fiber-optic gyros, atomic clocks), there are potential industrial base/market size issues that may drive NASA toward a particular solution. There are only a few vendors in these areas, and some are not doing well and could strongly benefit from NASA developmental funding.

I

TA06 Human Health, Life Support, and Habitation Systems

INTRODUCTION

The draft roadmap for technology area (TA) 06, Human Health, Life Support, and Habitation Systems, consists of five level 2 technology subareas:[1]

- 6.1 Environmental Control and Life Support Systems and Habitation Systems
- 6.2 Extravehicular Activity Systems
- 6.3 Human Health and Performance[2]
- 6.4 Environmental Monitoring, Safety, and Emergency Response
- 6.5 Radiation

The draft NASA roadmap for TA06 includes technologies necessary for supporting human health and survival during space exploration missions. These missions can be short suborbital missions, extended microgravity missions, or missions to various destinations. These missions experience extreme environments with reduced gravity (less than 1 *g*); high levels of several types of radiation and ultraviolet light (space weather); vacuum or significantly reduced atmospheric pressures; micrometeoroids, and/or orbital debris. While many TA06 technology solutions will have broad application to designs used during transit to a number of destinations, assuming that transit is always at a microgravity level, destination environments could drive different functional requirements for surface missions. Designs that are independent of destination are worthwhile goals, but this can lead to design requirements with no feasible or cost-effective solutions. Also, human exploration missions to destinations beyond the Moon will not have early return or abort options, so testing and certifying systems in flight-like environments and developing certified models will be critical to mission success and safety.

The draft TA06 roadmap is divided into 20 level 3 technologies, which are subdivided into and 78 level 4 items. As with some of the other TAs, the level 3 "technologies" in TA06 typically have a broad scope that encompasses a variety of systems, subsystems, and components with multiple potential design solutions.

[1] The draft space technology roadmaps are available at http://www.nasa.gov/offices/oct/strategic_integration/technology_roadmap.html.

[2] During the execution of this study, the NRC completed its report *Recapturing a Future for Space Exploration: Life and Physical Sciences Research for a New Era* (April 2011), which represents a more in-depth review of subject matter covered in TA06.3, Human Health and Performance.

TABLE I.1 Technology Area Breakdown Structure for TA06, Human Health, Life Support, and Habitation Systems

NASA Draft Roadmap (Revision 10)	Steering Committee-Recommended Changes
TA06 Human Health, Life Support, and Habitation Systems	One technology has been renamed.
6.1. Environmental Control and Life Support Systems and Habitation Systems	
6.1.1. Air Revitalization	
6.1.2. Water Recovery and Management	
6.1.3. Waste Management	
6.1.4. Habitation	
6.2. Extravehicular Activity Systems	
6.2.1. Pressure Garment	
6.2.2. Portable Life Support System	
6.2.3. Power, Avionics and Software	
6.3. Human Health and Performance	
6.3.1. Medical Diagnosis/Prognosis	
6.3.2. Long-Duration Health	
6.3.3. Behavioral Health and Performance	
6.3.4. Human Factors and Performance	
6.4. Environmental Monitoring, Safety and Emergency Response	
6.4.1. Sensors: Air, Water, Microbial, etc.	
6.4.2. Fire: Detection, Suppression	
6.4.3. Protective Clothing/Breathing	
6.4.4. Remediation	
6.5. Radiation	
6.5.1. Risk Assessment Modeling	
6.5.2. Radiation Mitigation	
6.5.3. Protection Systems	
6.5.4. Space Weather Prediction	Rename: 6.5.4. Radiation Prediction
6.5.5. Monitoring Technology	

Prior to prioritizing the level 3 technologies included in TA06, one of them was renamed. The change is explained below and illustrated in Table I.1. The complete, revised technology area breakdown structure (TABS) for all 14 TAs is shown in Appendix B.

It was the consensus of the panel that technology topic 6.5.4., Space Weather, be removed from this roadmap and possibly identified as a separate interagency roadmap (for example, including NASA, NOAA, NSF, and DOD) outside the scope of Panel 4 and NASA. Level 3, 6.5.4, was then restructured and renamed "Human Radiation Prediction." As described in the roadmap, this technology is focused on monitoring, modeling, and predicting ionizing radiation from solar particle events (SPEs) and galactic cosmic radiation (GCR). This radiation is a subset of space weather, which includes many other phenomena. The new name better describes the limited scope of this technology as applied to this roadmap.

TOP TECHNICAL CHALLENGES

The panel identified five top technical challenges for TA06, listed below in priority order.

1. Space Radiation Effects on Humans: Improve the understanding of space radiation effects on humans and develop radiation protection technologies to enable long-duration human missions.

Missions beyond low Earth orbit (LEO) present an expanded set of human health hazards. Lifetime radiation exposure is already a limiting flight assignment factor for career astronauts on the International Space Station

(ISS). Still, human health radiation models for predicting health risks are currently hampered by large uncertainties based on the lack of appropriate in situ data. At the present time, these models predict that crewed missions beyond LEO would be limited to 3 months or less because of adverse health impacts, either during the mission or during a crewmember's lifetime. Without the collection of in situ biological data to support the development of appropriate models, as well as the development of new sensors, advanced dosimetry instruments and techniques, solar event prediction models, and radiation mitigating designs, extended human missions to the Moon, Mars, or near-Earth asteroids (NEAs) may be beyond acceptable risk limits for both human health and mission success. An integrated approach is needed to develop systems and materials to monitor radiation in near-real time and protect crewmembers. In order to implement these technologies, existing radiation protection technologies must be upgraded and new technologies deployed as needed so that the radiation environment is well characterized and solar events can be forecast from at least Earth to Mars. Game changers that will help address this technical challenge include decreased transit times through new propulsion systems to lower exposure; new materials for EVA suits, spacecraft, rovers, and habitats; and new ISRU capabilities to build protective habitats in situ.

2. Environmental Control and Life Support Closed Loop Systems: Develop reliable, closed-loop environmental control and life support systems (ECLSS) to enable long-duration human missions beyond low Earth orbit.

ECLSS for missions beyond Earth orbit (for spacesuits, spacecraft, and surface habitats) are critical for safety and mission success. It was a loss of an oxygen tank and subsequently a compromise of a portion of the ECLSS loop (CO_2 removal) that nearly cost the Apollo 13 crew their lives. In missions without early return capability or remote safety depots, the ECLSS system must be as close to 100 percent reliable as possible and/or easily repairable with little or no resupply. Because air and liquid systems are sensitive to gravity level, extended testing of systems in reduced gravity may be necessary before they are integrated into exploration spacecraft. Current ISS experience with both U.S. and Russian ECLSS systems shows significant failure rates that would be unacceptable for an extended human exploration mission. In many cases, ISS ECLSS equipment has been launched and implemented without microgravity testing. Even with ISS testing, data on the performance of ECLSS systems in the reduced gravity of the Moon (~1/6 *g*) and Mars (~3/8 *g*) is not and will not be available without suitable reduced/variable-gravity test facilities. This will be a major impediment to maturing ECLSS technologies. New propulsion capabilities that reduce mission duration would reduce exposure to failures.

3. Long-Duration Health Effects: Minimize long-duration crew health effects.

The accumulated international experience with long-duration missions indicates that physical and behavioral health effects and adverse events will occur on long-term exploration missions. In some cases, health effects could be life threatening in the absence of effective diagnosis and treatment. Some of these health-related effects and events can be predicted and planned for, but it is highly likely that others cannot. In such a situation, autonomous, flexible, and adaptive technologies and systems will help promote long-duration health and effectively restore it when accident or illness occurs. Areas of interest include adverse effects of reduced gravity (such as bone loss, muscular and cardiovascular deconditioning, and neurovestibular disorders), in-flight surgery capability in reduced gravity, autonomous medical decision support and procedures management, and in-flight medical diagnosis enabled by a new generation of solid state, non-invasive, wireless biomedical sensors and "laboratory on a chip" technologies.

4. Fire Safety: Assure fire safety (detection and suppression) in human-rated vehicles and habitats in reduced gravity.

Current fire safety technologies for 1 g and microgravity environments are well understood and have an excellent history for longevity as will be needed for future human exploration missions beyond LEO. However, the space shuttle experience included two cases where smoldering electrical fires were detected by crew members working in close proximity to the problem and not by electronic sensors. Also, Russia's Mir space station experienced a fire in its oxygen-generating candle system that proved difficult to extinguish and required several days to achieve full recovery. Microgravity fire suppression systems currently use water or CO_2 as the extinguishing agent. Dumping large amounts of CO_2 into a small cabin environment is hazardous to the crew and puts significant strain on the ECLSS to remove the excess CO_2. Using water as the extinguishing does not pose a crew hazard but does have a significant impact on mission mass. Research and testing are needed to understand why current sensors failed to detect a smoldering electrical fire, develop more efficient and less hazardous fire suppression systems, and develop remediation capabilities that do not impair ECLSS components and/or processes.

5. EVA Surface Mobility: Improve human mobility during extravehicular activity (EVA) in reduced gravity environments (0 *g* up to 3/8 *g*) in order to assure mission success and safety.

Two closely related versions of space suits were used during the Apollo lunar missions; relatively little supported research has taken place on suits for environments other than microgravity in the intervening four decades. In the interim, experience with EVA during the Space Shuttle Program focused primarily on modifications for maintainability and on-orbit resizing; the only major advances in pressure garment mobility during that time was in the development of more dexterous pressure gloves. Differences in Apollo and future planetary suits will include the effects of long-term exposure to microgravity en route, prior to reduced gravity EVA operations for surface durations many times that of the 3 days in J-class Apollo missions. Apollo suits were further restricted by the mandate that they also be suitable for launch and entry use, but the Space Shuttle Program demonstrated the utility of separating the launch and entry function from EVA operations. Critical issues for research in this area include the effects of reduced gravity levels—at least lunar (1/6 *g*) and martian (3/8 *g*)—on gait, posture, and suited biomechanics; the use of advanced materials and techniques for extending life, enabling ease of maintenance, and reducing the effect of surface dust on bearings, seals, and closure mechanisms. Operationally there are substantial benefits from thorough integration of rovers, pressurized habitats, and robotic assist vehicles in extended surface operations: these need to be further researched to provide a sufficient database to make quantitative trade-offs for specific mission objectives. Finally, innovative technologies providing sensory, data management, and actuation assistance to the suit wearer must be developed and assessed for potential augmentation of future human EVA system architectures.

QFD MATRIX AND NUMERICAL RESULTS FOR TA06

Figure I.1 summarizes the consensus scores of the 20 level 3 technologies in NASA's draft roadmap for TA06. The panel evaluated each of these to level 4, based on the description of level 4 items in the roadmap because this was the only level to which TRLs were assigned. The scores shown below for each technology reflect the highest priority level 4 items within each level 3 technology. For the high-priority level 3 technologies (see Figure I.2), key level 4 items are discussed in the section below that covers high-priority technologies.

Figure I.2 plots the overall rankings for each level 3 technology. The panel assessed 14 of the technologies as high priority. Twelve of these were selected based on their QFD scores, which significantly exceeded the scores of lower ranked technologies. After careful consideration, the panel also designated two additional technologies as high-priority technologies.[3] The quanitity of "high-priority" topics can be misleading because the 14 high-priority level 3 technologies actually break down into five high-priority theme areas: Radiation, ECLSS/Habitation, Human Health/Performance, EVA Systems, and Envrionmental Monitoring/Safety (EMS).

Nine of the 14 high-priority technologies are in the radiation or life support (ECLSS/Habitation) theme areas, which are the most critical for crew survival beyond Earth orbit.

[3] In recognition that the QFD process could not accurately quantify all of the attributes of a given technology, after the QFD scores were compiled, the study panels in some cases designated technologies as high priority even if their scores were not comparable to the scores of the other high-priority technologies. The justification for the high-priority designation of all the high-priority technologies appears in the section "High Priority Level 3 Technologies," below.

Technology Name	Benefit	Alignment with NASA Needs	Alignment with Non-NASA Aero Tech Needs	Alignment with Non-Aerospace National Goals	Technical Risk and Reasonableness	Sequencing and Timing	Time and Effort	QFD Score (Weighted)	Panel Priority
Multiplier	27	5	2	2	10	4	4		
	0/1/3/9	0/1/3/9	0/1/3/9	0/1/3/9	1/3/9	-9/-3/-1/1	-9/-3/-1/0		
	Benefit	Alignment			Risk/Difficulty				
6.1.1. (ECLSS) Air Revitalization	9	3	3	1	3	1	0	300	H
6.1.2. (ECLSS) Water Recovery and Management	9	3	3	3	9	1	-3	352	H
6.1.3. (ECLSS) Waste Management	9	3	3	1	9	1	-1	356	H
6.1.4. Habitation	9	3	3	9	9	1	-3	364	H
6.2.1. (EVA) Pressure Garment	9	3	1	3	9	1	-3	348	H
6.2.2. (EVA) Portable Life Support System	3	9	9	3	3	1	-1	180	H*
6.2.3. (EVA) Power, Avionics, and Software	3	9	9	9	9	-3	-9	204	M
6.3.1. Medical Diagnosis / Prognosis	1	9	3	1	3	1	-1	110	L
6.3.2. Long-Duration (Human) Health	9	9	1	0	9	-3	-3	356	H
6.3.3. Behavioral Health and Performance	3	9	1	1	9	-3	-3	196	M
6.3.4. Human Factors and Performance	3	9	9	9	3	1	-3	184	M
6.4.1. Sensors: Air, Water, Microbial, etc.	3	3	3	3	9	1	-3	190	M
6.4.2. Fire: Detection, Suppression	9	9	9	9	1	1	0	338	H
6.4.3. Protective Clothing / Breathing	1	3	0	0	1	1	0	56	L
6.4.4. Remediation	3	3	0	0	3	1	-3	118	H*
6.5.1. (Radiation) Risk Assessment Modeling	9	9	3	3	9	-3	-3	366	H
6.5.2. Radiation Mitigation	9	9	3	3	9	-9	-3	342	H
6.5.3. (Radiation) Protection Systems	9	9	3	3	9	-3	-1	374	H
6.5.4. Radiation Prediction	9	9	9	9	3	1	-3	346	H
6.5.5. (Radiation) Monitoring Technology	9	9	9	9	9	-3	-3	390	H

FIGURE I.1 Quality function deployment (QFD) summary matrix for TA06 Human Health, Life Support, and Habitation Systems. The justification for the high-priority designation of all high-priority technologies appears in the section "High-Priority Level 3 Technologies." H = High Priority; H* = High Priority, QFD score override; M = Medium Priority; L = Low Priority.

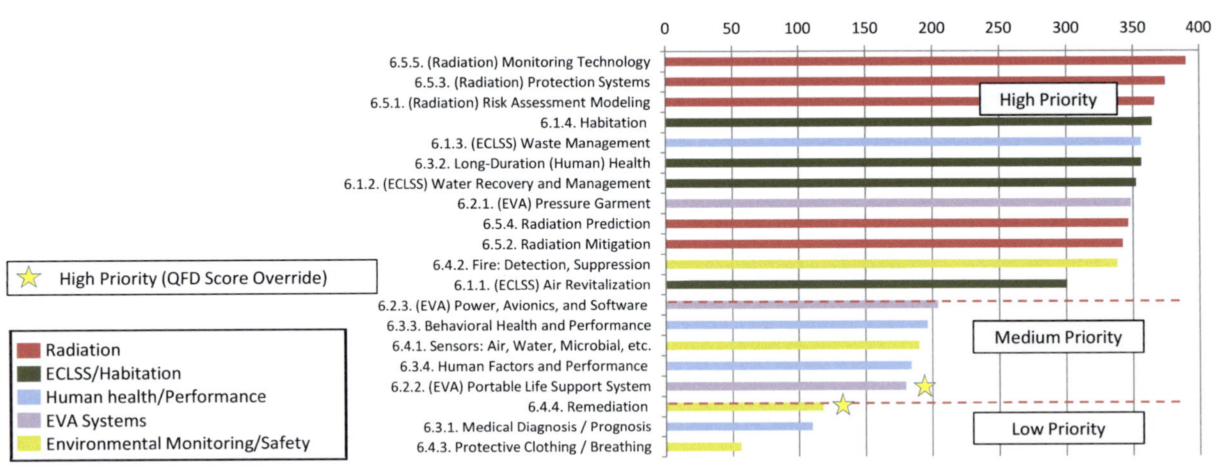

FIGURE I.2 Quality function deployment rankings for TA06 Human Health, Life Support and Habitation Systems.

CHALLENGES VERSUS TECHNOLOGIES

Figure I.3 shows the relationship between the individual level 3 TA04 technologies and the top technical challenges. In general, the mapping validates the major focus areas.

HIGH-PRIORITY LEVEL 3 TECHNOLOGIES

Panel 4 identified 14 high-priority technologies in TA06, grouped into five high-priority theme areas: Radiation, ECLSS/Habitation (life support), Human Health/Performance, EVA Systems, and Environmental Monitoring/Safety (fire safety). The justification for ranking each of these technologies as a high priority is discussed below.

Radiation

Space radiation poses a grave risk to human health for long-duration space missions (NRC, 2008). Thus, the high-priority technologies for TA06 include five level 3 technologies related to radiation, as follows:

- 6.5.5. Radiation Monitoring Technology: Measuring the exposure
- 6.5.3. Radiation Protection Systems: Reducing the exposure (shielding)
- 6.5.1. Radiation Risk Assessment Modeling: Understanding the effects of radiation
- 6.5.4. Radiation Prediction: Forecasting radiation exposure
- 6.5.2. Radiation Mitigation: Reducing the impacts of exposure (countermeasures)

Technology 6.5.5, Radiation Monitoring Technology

The ability to monitor the local radiation environment at and even within the crew members on long-duration space missions will be critical to ensure human health and mission success. This technology specifically addresses the need to measure and report on the ionizing particle environment (including neutrons) wherever humans may travel beyond Earth orbit (including the surface of the Moon and Mars, during unshielded EVAs and inside shielded vehicles and habitats). Measuring the local radiation environment, including the secondary radiation generated in the shielding, is necessary to ensure that astronauts keep their total exposure "as low as reasonably achievable" (a key tenet of radiation safety practices). Passive radiation monitors have been employed by NASA throughout its history of human spaceflight. However, established technologies are not sensitive to the full range of radiation that will be encountered beyond Earth orbit, nor do they give details about the types of particles contributing to the indicated dose. This is important because the most recent assessment of known radiation exposure for astronauts would limit them to approximately 90 days of exposure during missions outside of Earth orbit. This would preclude Mars and NEA missions without some revolutionary advance in in-space propulsion capabilities.

Advances are needed for small, low-power dosimeters with active readout that can sense and distinguish among a broad range of radiation, especially neutron dose and dose rate. In addition, NASA may be introducing a new approach to estimate the biological effectiveness of radiation which may call for a new family of dosimeters (see discussion below on technology 6.5.1, Radiation Risk Assessment. The current paradigm is based on the notion that biological impact is largely due to the distribution of energy deposited as a particle moves through the body. Thus, different particles that deposit energy at the same rate would have similar impacts. A new paradigm gaining favor proposes that both the energy and charge distribution of the particles impacting the body affect human health. Current dosimeters do not provide this type of information. Another area of technology development relates to biodosimetry: non-invasive means of measuring the dose and dose rate inside the human body. One example is to analyze blood samples, perhaps without even drawing blood.

Traditional dosimeters currently in use are at TRL 9. Improved versions that are lower power and provide better active readout exist in the TRL range of 4 to 6, and next generation dosimeters and biodosimeters to meet the requirements of long-duration spaceflight are just now emerging with TRLs 1 to 3.

While NASA needs dosimeters sensitive to a unique radiation environment, improvements in active dosimeters may have applications beyond NASA. Examples include radiological medicine, miners exposed to radioactive

Priority	TA06 Technologies, Listed by Priority	Top Technical Challenges: 1. Space Radiation Effects on Humans: Improve the understanding of space radiation effects on humans and develop radiation protection technologies to enable long-duration human missions.	2. Environmental Control and Life Support Closed Loop Systems: Develop reliable, closed-loop environmental control and life support systems (ECLSS) to enable long-duration human missions beyond low Earth orbit.	3. Long-Duration Health Effects: Minimize long duration crew health effects.	4. Fire Safety: Assure fire safety (detection and suppression) in human-rated vehicles and habitats in reduced gravity.	5. EVA Surface Mobility: Improve human mobility during extravehicular activity in reduced gravity environments in order to assure mission success and safety (0 *g* up to 1 *g*).
H	6.5.5. (Radiation) Monitoring Technology	●		○		
H	6.5.3. (Radiation) Protection Systems	●		○		
H	6.5.1. (Radiation) Risk Assessment Modeling	●		○		
H	6.1.4. Habitation		●			
H	6.1.3. (ECLSS) Waste Management		●			
H	6.3.2. Long-Duration (Human) Health	○	○	●		
H	6.1.2. (ECLSS) Water Recovery and Management		●			
H	6.2.1. (EVA) Pressure Garment					●
H	6.5.4. Radiation Prediction	●				
H	6.5.2. Radiation Mitigation	●		○		
H	6.4.2. Fire: Detection, Suppression		○		●	
H	6.1.1. (ECLSS) Air Revitalization		●		○	
H	6.2.2. (EVA) Portable Life Support System		●			●
H	6.4.4. Remediation		○		●	
M	6.2.3. (EVA) Power, Avionics, and Software					○
M	6.3.3. Behavioral Health and Performance			○		
M	6.4.1. Sensors: Air, Water, Microbial, etc.		○		○	
M	6.3.4. Human Factors and Performance			○		○
L	6.3.1. Medical Diagnosis / Prognosis					
L	6.4.3. Protective Clothing / Breathing					

●	Strong Linkage: Investments by NASA in this technology would likely have a major impact in addressing this challenge.
○	Moderate Linkage: Investments by NASA in this technology would likely have a moderate impact in addressing this challenge.
[blank]	Weak/No Linkage: Investments by NASA in this technology would likely have little or no impact in addressing the challenge.

FIGURE I.3 Level of support that the technologies provide to the top technical challenges for TA06 Human Health, Life Support, and Habitation Systems.

minerals, monitoring nuclear power plants and surrounding areas (both during routine operation and during emergency contingencies), commercial airlines, and first responders dealing with radiological emergencies.

The ISS has been and will continue to be a significant testbed for improved radiation monitors. Testing on the ISS enables cross calibration with dosimeters that have been used for decades in a realistic and well-characterized work environment. However, the radiation environment in deep space is different than in LEO, where Earth's magnetosphere provides some protection. To better characterize the deep space radiation environment, inert and biological-based sensors should be sent to higher orbits, lunar-Earth Lagrange points, or the lunar surface.

Technology 6.5.3, Radiation Protection Systems

Radiation protection systems include materials and other approaches to limit astronauts' radiation exposure. This technology complements 6.5.2, Radiation Mitigation, which addresses physiological countermeasures to alleviate the impact of radiation exposure, and it may benefit from advanced materials developed by other roadmaps. Although discrete technology options were not provided in the draft TA06 roadmap for radiation protection systems, and even though the earliest start date for this technology suggested in the roadmap was 2014, the panel endorses near-term investments in this technology.

Shielding is a critical design criterion for many elements of human exploration, including deep space transport vehicles, surface habitats, surface rovers, EVA suits, and other transportation elements. Shielding alone is unlikely to eliminate radiation exposure from GCR (including secondary exposure), but a well-shielded vehicle or habitat could substantially reduce the exposure from SPEs.

It may also be possible be to emulate the electromagnetic field in Earth's magnetosphere. This has been suggested as an attribute to the advanced Variable Specific Impulse Magnetoplasma Rocket (VASIMR), which will be tested on the ISS within the next 3 years. The essential challenge is to reduce radiation exposure while meeting overall mission allowances for mass, cost, and other design considerations. A comprehensive radiation mitigation strategy will also address the need for adequate SPE warning and detection (technologies 6.5.4 and 6.5.5) so that astronauts can take shelter in a timely fashion.

Advanced radiation protection technology is at low TRL (~1-3). Low-atomic-number (low z) elements (such as hydrogen) provide better shielding from particle radiation than high z elements (such as heavy metals). Thus, water and polyethylene, which contains a lot of hydrogen atoms, are effective shielding materials. Recent research has been exploring multifunctional composites that would have structural and shielding properties suitable for incorporation directly into spacecraft structures. Advanced composites of interest include hydrogen-loaded carbon or boron nitrate nanotubes for multilayer coatings or components. These approaches are at low TRL.

Developments required for progress in this technology include the following:

- Identify advanced shielding materials and approaches.
- Advance the TRL of promising materials.
- Integrate shielding approaches into spacecraft and habitat design and operations concepts.
- Explore innovative shielding techniques, including active shielding concepts (magnetic shields).

NASA would benefit from improved radiation shielding concepts that reduce the radiation risk to astronauts, potentially at lower mass, and thus at lower mission cost. This effort could provide a new materials for incorporation across a wide range of human exploration mission architectures. Improved integration of shielding concepts early in the design process will enhance the mission design process.

The ISS has been and will continue to be a significant testbed for improved shielding material. Concepts can be deployed and tested in realistic conditions, including, where necessary, outside the station for durability testing in the space environment. However, because of the shielding effects of Earth's magnetosphere, certification testing for exploration missions beyond LEO may require testing in higher orbits, at lunar-Earth Lagrange points, or on the lunar surface.

While shielding against GCR and SPEs is a uniquely NASA requirement, advanced radiation shielding materials and approaches could have terrestrial applications, for example, for shielding in nuclear power plants, high-altitude aircraft, radiation medicine, and in radioactive contingency response against terrorist threats.

Technology 6.5.1, Radiation Risk Assessment Modeling

The major contributor to estimated mission dose is from highly penetrating GCR. Additional exposure is possible from periodic, intense episodes of solar activity known as SPEs. While reasonable shielding can significantly limit SPE exposure, shielding is largely ineffective at reducing the GCR risk.

There are several layers of risk limits included in NASA's permissible exposure limits. The major mission constraint is to limit the astronauts "risk of exposure induced death" (REID) from cancer to no more than 3 percent at the 95 percent confidence level. This risk limit is dominated by a significant uncertainty in quantifying the radiation impacts. It is estimated that the uncertainties in the cancer REID is about a factor of 3.5. There are no quantifiable limits to radiation impacts on the central nervous system, the cardiovascular system, and the immune system. However, it is believed that the cancer risk to these organs can be limited by maintaining whole body radiation dose within established limits, which are higher for professional astronauts than for the general public. The commercial spaceflight industry will need to carefully evaluate acceptable passenger exposure levels, provide appropriate protection, and seek federal approval to increase radiation dose limits for "space tourists, " if necessary.

Doubling the nominal shielding reduces the GCR-induced dose by only 15 to 20 percent, and reducing the cancer uncertainty by fifty percent could require increasing the shielding by up to a factor of five. Reducing biological and other 50 about radiation health risks is needed to quantify the value of alternative shielding (see 6.5.3, Radiation Protection Systems) and the efficacy of possible countermeasures (see 6.5.2, Radiation Mitigation). Attaining this reduction will require continuation or expansion of a substantial research program that includes data collection in space; ground-based fundamental research, data analysis, and technology development; and in-space validation of new models in micro- and reduced-gravity environments to explore the synergistic effects of various *g*-levels. The ISS can contribute to the validation of risk models, but this would require the development of appropriate systems, such as the bioreactor and other facilities. Final validation may require lunar surface facilities which can be periodically accessed.

No other agencies or organizations have the responsibility to understand the impacts of the unique space radiation environment on human health, and NASA research in this area could make substantial contributions to understanding fundamental aspects of radiation's role in carcinogenesis. In addition, the techniques developed for the risk assessment methodology could have broad applicability to terrestrial occupational and environmental health.

Technology 6.5.4, Human Radiation Prediction

The ability to forecast the radiation environment will be critical to ensure the safety of astronauts and mission success. This technology specifically addresses the need to forecast SPEs, which are periods of intense ionizing radiation associated with solar storms. A related technology area, 6.5.5, Radiation Monitoring Technology (see above), addresses the requirement to measure the local radiation environments.

There is no capability to forecast the onset of SPEs, and only limited ability to forecast the evolution of an SPE once it is underway. Today, the strategy to manage the risks associated with SPEs can be categorized as "cope and avoid," meaning that conservative flight rules are developed that enable astronauts to quickly take shelter in a shielded segment of a spacecraft or habitat after observing the onset of an SPE. This has several drawbacks: it limits the time astronauts can prepare for an event, it restricts exploration timelines to stay within a narrow response window, and it leads to over-reaction to small events or false alarms. All of this overly constrains mission operations. In addition, it leaves astronauts susceptible to an unusually severe event that surprises a crew, with potentially severe health risks.

Developments required for progress in this technology include the following:

- Improve understanding of the physics of SPEs.
- Develop better applied models for forecasting the timing and impacts of SPEs.
- Develop advanced dosimetry to more accurately measure the natural radiation environment in locations that are or will be frequented by astronauts (especially the very highest energy particles, hundreds of MeV). (See technology 6.5.5. Radiation Monitoring Technology.)

- Develop lighter instruments and satellites to enable the deployment of a cost-effective radiation monitoring architecture.
- Develop more efficient communication systems to relay data from disparate monitors around the solar system.

The implementation of an improved SPE forecast system (including physics-based forecasts of 8- to 24-hour "all clear periods") would improve mission effectiveness and enable more cost effective mitigation strategies by giving astronauts more time to respond, by reducing the time they spend under shelter, and by reducing or eliminating false alarms.

Monitoring the radiation environment near, but external to, the ISS would contribute to models that attempt to understand the dynamic response of Earth's magnetosphere to the complex series of events that occur during major SPEs. In particular, it helps identify the extent to which the intensity of high-energy particles extends to mid to low latitudes, increasing the radiation exposure to spacecraft in low Earth orbit with medium to low inclination.

No other agency has the need or responsibility to forecast the intensity or nature of SPEs except in the neighborhood of Earth. However, if NASA succeeds in implementing an effective strategy to forecast SPEs throughout the inner solar system (to support missions to Mars or NEAs) then it will benefit forecasts of events that affect Earth.

This technology will have a significant benefit, as the impact of SPEs goes far beyond astronaut health: the high-energy particles in a SPE can damage or degrade instruments and other equipment on satellites of all types, including communications, navigation, and remote sensing satellites. This technology will also support the emerging commercial spaceflight industry and aviation, particularly for flights at high latitude. Polar routes have increased substantially in the past 10 years. During SPEs, it may be prudent for aircraft on polar routes to fly at reduced, less economical altitudes or diverted to alternate routes or destinations to avoid the effects of SPE interference with polar communication networks. Accurate forecasts would improve flight planning and could avoid mid-flight corrections.

Technology 6.5.2, Radiation Mitigation

Radiation Mitigation addresses countermeasures to alleviate the impact of radiation exposure. This technology area complements 6.5.3, Radiation Protection Systems, which includes shielding and other approaches to reduce astronauts' radiation exposure. It is generally considered that shielding alone will not eliminate GCR exposure; especially on spacecraft where mass is at a premium. (ISRU technologies may be able to provide substantially better shielding on surface habitats using in situ materials.) In any case, exploring biological and/or pharmacological countermeasures may be able to mitigate the effects of continuous, long-term radiation exposure. In addition, in spite of best intentions and careful planning, an astronaut could still be exposed to a significant dose of radiation during an SPE. In such an event, medical measures to limit the severity of acute radiation effects could be invaluable. Unfortunately, at present there are few pharmacological countermeasures to acute doses of radiation. Also, these countermeasures were developed for use after nuclear emergencies, and it is assumed that they would be provided in conjunction with extensive medical care. The side effects of these countermeasures may limit their usefulness in space. Also, there are no effective countermeasures to chronic GCR exposure.

It is not likely that countermeasures designed to mitigate one radiation endpoint (say, carcinogenesis) will be effective against another (degeneration of the cardiovascular system). Diets rich in antioxidants can reduce some radiation effects, but the full efficacy of such as diet has not been quantified for various types and levels of radiation exposure. In fact, it may be difficult or impossible to quantify accurately the value of countermeasures before the uncertainty in radiation levels and the health impacts of that radiation has been reduced (see technology 6.5.1, Radiation Risk Assessment Modeling). Countermeasures are still at low TRLs (1 to 3).

Developments required for progress in this technology include the following:

- Identify potential countermeasures to specific radiation-induced adverse health effects.
- Conduct appropriate clinical studies to prove efficacy.
- Explore synergistic effects of multiple countermeasures.
- Increase the TRL of effective countermeasures.

NASA would benefit from improved radiation countermeasure concepts by reducing the radiation risk to astronauts, by enabling longer missions, and provide protection from accidental exposure to acute doses of radiation.

The ISS has been and will continue to be a significant test bed for countermeasure studies. Concepts can be deployed and tested under realistic conditions, especially to confirm the lack of side effects or unintended consequences.

While countermeasures against GCR and SPEs are a uniquely NASA requirement, some radiation mitigation techniques could likely be used in broader terrestrial applications, such as response to nuclear power plants emergencies, high-altitude aircraft, radiation medicine, and in response to terrorist threats.

ECLSS/Habitation

The ECLSS/Habitation theme area has four high-priority technologies:

- 6.1.4. Habitation: Includes food production and processing as well as hygiene.
- 6.1.3. Waste Management: Includes liquid and solid waste.
- 6.1.2. Water Recovery and Management: Provides recycling for long missions.
- 6.1.1. Air Revitalization: Removes CO_2, particulates, and contaminants, and provides thermal control.

As shown in Figures I.1 and I.2, these four technologies received approximately the same QFD scores.

Technology 6.1.4, Habitation

The habitation technology is focused on functions that closely interface with life support systems. The level 4 items in this technology are as follows: food production (6.1.4.1), food preparation/processing (6.1.4.2), crew hygiene (6.1.4.3), metabolic waste collection and stabilization (6.1.4.4), clothing/laundry (6.1.4.5), and re-use/recycling of logistics trash (6.1.4.6). This technology is predicated on and provides necessary functions for human spaceflight. It provides food, sanitation, comfort, and protection for space-faring crew members.

While a portion of this technology, such as crew hygiene and waste collection and stabilization, are at TRL 9, other activities such as food production, food preparation, and trash recycling are TRL 6, at most. The clothing and laundry area is a mix, as clothing is at a TRL 9, while laundry is at a TRL 4.

This technology would directly support human spaceflight for the complete range of missions lasting from several hours to months, such as long-duration stays on the ISS, or long-duration exploration missions to Mars or an NEA. The ISS is the ultimate test platform for this technology. It would allow habitation technology to ultimately be offered for use in other applications such as commercial human spaceflight. NASA involvement in the development of habitation technology is critical; without NASA involvement there will be no new habitation development, and it is essential to NASA for long-duration human spaceflight missions. There is no reason for NASA to partner with other federal agencies to develop this technology.

The use of the ISS would prove valuable in maturing the habitation technologies, increasing their reliability, and providing a means to demonstrate their functional performance. A good example would be the development of a plastic waste melt compactor, in which plastic trash and other trash material is melted and compacted to form a tile that can be used for radiation protection. As part of this technology, the effectiveness such tiles for radiation shielding could be tested Food production technology could also benefit from using the ISS as a test bed for scientific plant growth and for testing food production systems.

Habitation technology is a high priority because additional work is needed to advance from current LEO missions to long-duration missions beyond LEO. Food production would provide significant mass savings for such missions, augment life support systems, and provide psychological crew benefits. Food processing would allow a safe means for a crew to consume the food that is produced, reducing the likelihood of food borne illnesses and increasing crew health. The recycling of logistics trash converts otherwise disposed material into useful products to increase crew safety and/or capabilities.

Technology 6.1.3, ECLSS Waste Management

Waste management technology safeguards crew health, increases safety and performance, recovers resources, and protects planetary surfaces. This technology includes disposal, storage, and resource recovery from trash and crew waste. Key areas of concern include volume reduction, stabilization, odor control, and recovery of water, oxygen and other gases, and minerals.

Current ISS technology for trash disposal is very basic, and involves crew effort and duct tape. Feces disposal, stabilization, and odor control systems on the ISS have considerable spaceflight heritage and are at TRL 9. Still, there is room for continuous improvement. Automated trash compaction and resource recovery of spaceflight waste are in the early stages of development.

NASA has been developing and funding development of automated trash compaction and water recovery. NASA has the expertise, capabilities, and facilities to further develop these capabilities and to advance waste stabilization and recovery of additional resources. NASA spaceflight may be the only application for the compaction and resource recovery capabilities that must be developed, but other government and industrial organizations may be interested in waste stabilization.

The ISS is the ideal test bed for advanced waste management. EXPRESS Racks could be used for early technology development test beds, and successful technology could then undergo long-term testing on the ISS before they are adopted for a long-duration human exploration mission.

Resource recovery from waste is vital to closing the loop for long-duration human spaceflight. Until dumping large amounts of trash overboard is generally accepted, stabilization and volume reduction of trash is necessary to maintain safe and comfortable living conditions for crews on long-duration missions. As long as long-duration spaceflight is a goal, effective waste management must also be a goal. Trash stabilization, volume reduction, and water recovery are already showing promise in their early development, but more effort is needed to mature these technologies. Recovery of additional resources, such as O_2, CO_2, N_2, and minerals from trash and human waste may require hardware with significant mass, power, and volume requirements, and a considerable amount of development may be needed until there is an equivalent system mass advantage to recovering these resources from waste, even for long-duration missions.

Technology 6.1.2, ECLSS Water Recovery and Management

This technology provides a safe and reliable supply of potable water to meet crew consumption and operational needs. Water recovery from wastewater is essential for long-duration transit missions due to the tremendous launch mass of water that would be required for an entire transit mission without recycling, and the impracticality of resupply from Earth. Both short-duration and long-duration missions require some degree of wastewater stabilization to protect equipment and facilitate potable water disinfection for storage.

Primary water recovery from urine, hygiene water, and humidity condensate is currently performed on the ISS and is at TRL 9. This technology needs additional development to improve reliability for long-duration missions where resupply from Earth is not possible. Primary water processers recover approximately 90 percent of usable water from a wastewater stream and expel the remaining 10 percent as waste brine. Technologies to recover water from wastewater brines are in the early stages of development, and are challenged by resource limitations, odor requirements, and the urine pretreatment chemicals that make brines acidic, corrosive, and sticky.

NASA has experience with primary wastewater processors and has been funding development of technologies to recover water from wastewater brines and alternatives to the current urine pretreatment methods that use sulfuric acid and an oxidizing agent. NASA has the expertise, capabilities, and facilities to continue development of these technologies. The demands and limitations for water recovery for long-duration missions are unlike those of any other industrial or government need. It is likely that NASA will pursue these developments independent of any other organization for the foreseeable future.

The ISS is the ideal test bed for this technology, as operation in microgravity is one of the primary challenges of any liquid processing technology. EXPRESS Racks could be used for early technology development test beds, and new technology could be used on the ISS long term before committing to transit missions.

Long-duration transit missions will not be possible without recovering water from wastewater. The importance of water recovery increases as the mission's length increases. Even the difference between 98 percent recovery and 99 percent recovery has a tremendous impact on launch mass for a mission with a duration of several years with no resupply. Achieving 99 percent recovery would be difficult, but it is achievable with a coordinated effort among wastewater treatment, primary water processor, and brine processor development. An integrated approach is essential because each step in the water recovery process impacts downstream steps.

Technology 6.1.1, Air Revitalization

Air Revitalization (6.1.1) includes carbon dioxide removal, carbon dioxide reduction, oxygen supply, gaseous trace contaminant removal, particulate removal, temperature control, humidity removal, and ventilation. Several systems that provide one or more of these functions currently operate on the ISS; they are at TRL 9. The major shortcoming with the current state of the art of air revitalization technology is in carbon dioxide reduction. The ability to recover oxygen from waste carbon dioxide will be very important to reduce mass requirements for long-duration human exploration missions. To date, various technologies are being demonstrated in the laboratory, but none have yet been packaged or tested for flight. Other key objectives for this technology include reducing overall system mass and power consumption and decreasing acoustic emissions. A far-reaching goal is to use plants as a way to provide significant air revitalization functions. This has been demonstrated to varying degrees of success in the laboratory, but not in space. To do so would require developing a large amount of support equipment.

NASA has experience with every aspect of this technology and has been funding research in areas with low maturity for some time. NASA has the expertise, capabilities, and facilities to continue development of these technologies. The demands and limitations for revitalizing air for long-duration missions are unlike those of any other industrial or government need. It is likely that NASA will pursue these developments independent of any other organization for the foreseeable future.

The ISS is the ideal test bed for this technology, as operation in microgravity is one of the primary challenges of life support equipment. EXPRESS Racks could be used for early technology development test beds, and new technology could be used on the ISS long term before committing to transit missions. Attaching a larger structure, such as an inflatable module, to the ISS as a technology test bed would enable the use of the ISS on a larger scale.

Long-duration missions will not be possible without robust and comprehensive air revitalization capabilities, especially carbon dioxide reduction or some element of a hybrid physicochemical and/or biological air revitalization architecture for long-term sustainability. The difference between 50 percent oxygen recovery and 75 percent to 100 percent oxygen recovery has a tremendous impact on launch mass for a mission that must last several years with no resupply. The required development effort would be difficult, but achievable with a coordinated effort between the NASA centers and advanced air revitalization system providers, including physicochemical and biological.

Human Health/Performance

The human health/performance theme area has one high-priority technology:

- 6.3.2. Long-Duration Health

Technology 6.3.2, Long-Duration Health

As stated in the roadmap, the focus of technology 6.3.2, Long-Duration Health, is to create "validated technologies for medical practice to address the effects of the space environment on human systems." The accumulated international experience with long-duration missions to date (in LEO) reveals and predicts a simple, compelling truth about future exploration-class crewed missions: physical and behavioral health effects and adverse events will occur. Some of these health-related effects and events can be predicted and planned for, but it is highly likely that others cannot. Thus, autonomous, flexible, and adaptive systems to promote long-duration health, and effectively restore it when accident or illness occurs, are a high priority. This determination is consistent with the findings of a series of prior Academy studies, including Safe Passage: Astronaut Care for Exploration Missions (2001), Review of NASA's Longitudinal Study of Astronaut Health (2004), A Risk Reduction Strategy for Human Exploration of Space: A Review of NASA's Bioastronautics Roadmap (2006), Managing Space Radiation Risk in the New Era of Space Exploration (2008) and Review of NASA's Human Research Program Evidence Books (2008).

The panel identified artificial gravity evaluation/implementation as a game-changing capability that would greatly mitigate many adverse health effects that would otherwise occur during long-duration habitation in transit (or Earth orbit). These adverse effects include bone loss, muscular and cardiovascular deconditioning, and neurovestibular disorders. However, the prospect of a rotating a space station or exploration vehicle to produce 1 *g* or a significant fraction thereof seems unattainable in the time frame envisioned in the technology roadmap. Thus, the ability to generate artificial gravity using a facility such as a large centrifuge on the ISS would be a high priority if it could be achieved. Such a facility would enable research and testing on small mammals and other biological and spaceflight systems in trying to understand the effects of reduced gravity on humans, other biological systems, and spaceflight systems.

Maintaining human health on long-duration human exploration missions would be greatly facilitated by advances in this technology. The panel would place the highest priority on developing in-flight surgery capability in microgravity environments (6.3.2.4), autonomous medical records, informatics, and procedures management (6.3.2.5), and in-flight medical diagnosis (6.3.2.6). These capabilities are essential and require solutions that are highly tailored to the space environment. Trauma is the most highly prevalent medical issue in long-duration flight, and the ability to perform life-saving surgery after major trauma and other unpredictable life threatening conditions (e.g., appendicitis) will be very important for exploration class missions to improve crew survivability. Compact, low-mass imaging technologies (6.3.2.3) will support precise anatomic diagnosis and could potentially spur development of similar health technologies for healthcare on Earth. Similarly, a new generation of solid state, non-invasive, wireless biomedical sensors (6.3.2.2) to monitor physiologic processes could be catalytic for all forms of in-flight monitoring, disease prediction, diagnosis, and treatment monitoring. It might also lead to dramatic improvements in home monitoring of chronic health conditions worldwide. A highly reliable low mass "clinical laboratory on a chip" (6.3.2.3) could dramatically expand diagnostic and care possibilities. Countermeasures (6.3.2.11, 6.3.2.12), including technology advances for exercise-based approaches to mitigating physiologic deconditioning, as well as pharmacologic countermeasures will be very important for maintaining crew health during long-duration missions. The Advanced Test Platform (6.3.2.1) will be required to develop and assess various approaches to countermeasures. For each of the high-priority technologies noted above, the ISS will be a critical technology evaluation platform.

EVA Systems

The EVA Systems theme area has two high-priority technologies:

- 6.2.1. Pressure Garment: Provides an advanced EVA suit.
- 6.2.2. Portable Life Support System: Contains many of the same components of the ECLSS for a spacecraft.

Technology 6.2.1, Pressure Garment

Pressure garments constitute the anthropomorphic articulated spacecraft in which each EVA crew member works and survives. The ideal pressure garment is easy to don and doff, highly articulated and readily adjustable to the kinematics of the wearer's body, and minimizes additional forces and torques which the wearer must overcome to accomplish all tasks.

Space shuttle EMUs and Apollo A7L and A7L-B suits represent TRL 9 technologies. While they provide adequate functionality at pressures of 4.3 and 3.5 psi (respectively), substantial fractions of crew physiological workload (estimated variously between 40 percent and 75 percent) go into moving the pressurized suit joints.

Advanced suits are functional in laboratory and field situations (TRL 4-5), with near-term plans for neutral buoyancy and vacuum chamber testing given sufficient development funding.

While there are some applications of pressure garment technology outside of NASA, such as HAZMAT suits and biowar garments for the military, it is clear that no progress will be made in this field without significant and ongoing research support from NASA. NASA has retained a basic research and development capability in pressure garments throughout the shuttle program at the Johnson Space Center and at a small number of industrial suppliers; there was in the past a vibrant community of EVA researchers in academia and smaller companies, but this community has been largely starved to death over the past decade. A significant and sustained program of research and development will be required to create the next generation of extravehicular suit, whether for microgravity missions such as near-Earth objects, or for reduced gravity operations on the Moon or Mars.

The ISS represents the obvious location for the first flight tests of new pressure garment technologies, with benefits to enhanced crew performance for ISS maintenance tasks. At the same time, the real test of future pressure suits will be in their ability to support prolonged legged locomotion in the reduced-gravity environment of the Moon and Mars, and ISS testing will not validate this critical functionality.

EVA pressure garments are pivotal to all aspects of human spaceflight. They are required for protection against cabin depressurization for launch and entry, and for extravehicular activities in any class of human missions. While Apollo-era suits were a compromise between these two categories of use, future programs will almost certainly require (as did the Space Shuttle Program) specialized garments for the launch and entry and the extravehicular roles. While pressure suits have been included on every single human mission since Alan Shepard's flight, the current operational technology represents incremental changes to the shuttle extravehicular mobility unit (EMU), which was developed more than 30 years ago. Thus, significant potential exists for substantial increases in performance and operational capabilities as compared to the current state of the art.

Modeling and analysis have advanced considerably since the development of the EMU; modern capabilities in biomechanical modeling and fundamental understanding of the behavior of pressurized fabric structures will provide a rigorous foundation for future pressure garment development. In parallel, current and upcoming materials, including "smart" materials with integrated sensors, controls, and even actuators will provide enhanced suit functionality. However, the primary focus of next-generation pressure garments must be on enhanced mobility. While specific estimates vary based on testing protocols and measurement techniques, there is widespread agreement that the space suit wearer requires a substantial fraction of their physiological workload for simply moving the pressure garments. Fatigue is a defining issue for procedures development and EVA planning, particularly in the hands and wrists for heavily dexterous activities. Advanced approaches which can improve the state-of-the-art in pressure garment mobility closer to nude-body performance will substantially reduce EVA crew workload, and subsequently expand the range of potential EVA applications in future space missions.

Note that the technologies for a launch and entry pressure garment are essentially mature, unless there is an operational requirement to use that suit for nominal or contingency EVA. In that case, the suit would need to be designed largely as an EVA suit capable of use for launch and entry (as in Apollo), rather than as a launch and entry suit which does EVA on the side.

Technology 6.2.2, EVA Portable Life Support Systems

The panel overrode the QFD scores and designated this technology a high priority because two level 4 topics were felt to be of critical importance and, despite low scores (since they are not critical to basic functionality, and all personal life support system (PLSS) functions are limited in non-spaceflight applications), both thermal control and CO_2 capture were assigned high priority for special attention in future research funding decisions.

The current thermal control system (TRL 9) is based on water sublimation, which increases consumables usage and is not feasible in atmospheres, even the tenuous atmosphere of Mars. Heat sink approaches are typically massive, and required radiator area makes that approach infeasible for pressure suits. CO_2 capture in PLSS applications is currently performed by expendable LiOH canisters or rechargeable metal oxide canisters, which require significant mass, volume, and power for recharging ovens to drive out the captured CO_2.

Advanced portable life support systems are applicable to firefighters, hazmat suits, bio-warfare gear, and underwater breathing systems. The particular focus for NASA technology development (in terms of thermal control without sublimation and extremely high-reliability systems where cost is relatively unimportant) is unique to the NASA mission.

Initial flight testing of advanced PLSS technologies is well performed at the ISS, and could represent significant increase to ISS EVA capabilities. Challenges for future PLSS systems, such as thermal control in the martian atmosphere or extended functionality in the dusty environment of surface exploration, are not well represented by ISS testing.

Increasing the capacity, reliability, and maintainability of a PLSS while extending duration and reducing on-back weight for the user are important, but difficult goals.

The greatest challenge in this technology area is in environmental effects on the PLSS systems. These include the effects of low but discernible atmospheric pressure on Mars, long-term effects of dust on all solar systems bodies from asteroids to the Moon and Mars, and charging effects in geostationary orbit. The PLSS must be designed as an integrated system with necessary support equipment for replenishing the unit between sorties, and the overall system optimized including the impact of the support systems. This effect is exacerbated if PLSS components are used for life support in small pressurized rovers or space exploration vehicles, as with (for example) the requirement for METOX recharging spikes at the end of each extended mission, rather than the more steady demand if functionality is limited to local EVA support.

Environmental Monitoring/Safety

The Environmental Monitoring/Safety theme area has two high-priority technologies:

- 6.4.2. Fire Detection and Suppression
- 6.4.4. Remediation: Restores air quality following a fire or other contaminating event

Technology 6.4.2, Fire Detection and Suppression

Level 3 technology 6.4.2 Fire Detection and Suppression is concerned with ensuring crew health and safety by reducing the likelihood of a fire and, if one occurs, minimizing risk to crew, mission, and/or systems. Areas of research include fire prevention, fire detection, fire suppression, and a proposed free-flying fire test bed with reduced gravity, lower total pressure, and higher oxygen partial pressure environments capabilities. Fire suppression and in-space fire test bed concept are the two areas that drove the high scores for Fire: Detection and Suppression

Fire prevention technology maturation, primarily materials flammability testing, is required for reduced-gravity environments and cabin total and oxygen partial pressures expected in the next generation human space systems. Understanding material flammability and combustion products in spacecraft operational environments is critical to fire prevention. A new round of materials testing can be avoided if microgravity cabin environments continue to operate at sea level conditions. However, there is still a need to understand the impact of reduced gravity environments on flammability and combustion products.

Fire detection systems have proven remarkably reliable in the space shuttle and the ISS programs, experiencing only one failure to date. Again, new validation is required for reduced gravity and new atmospheric pressure and oxygen partial pressure environments. Fire and/or smoke on a station or vehicle is not a hypothetical event: the Apollo One crew was lost on the pad in 1967, and one serious event on the Russian space station, Mir, in 1997, nearly resulted in loss of both crew and vehicle.

Current fire prevention approaches and technologies are at TRL 9 for 1-*g* and microgravity environments in sea level equivalent atmospheres. However, destinations to reduced gravity environments in spaceships with lower total pressure and higher oxygen partial pressure relative to sea level conditions offer new challenges. A new fire suppression approach using fine mist water spray has been demonstrated on ground fires that offers significant advantages to human space systems relative to current water spray and CO_2 solutions (NRC, 2011, pp. 276, 277, 327). Since fine water mist sprays have not been tested in space environments it must be considered to be at TRL

4. Testing in reduced gravity and pressure environments expected for human exploration missions beyond LEO is needed to mature this technology.

Ground applications have demonstrated the ability to extinguish fires using approximately one third the amount of water as traditional approaches. Applying fine mist water spray fire suppression techniques to human space systems will eliminate adding large CO_2 quantities into the atmosphere and significantly reduce fire suppression system mass relative to current water or CO_2 systems. This technology is critical to NASA, foreign national space agencies, and commercial space programs as a means to reduce deep-space mission mass and improve fire suppression capabilities. NASA is well positioned to mature this technology for space applications and make it available for any and all space system providers.

The ISS can play a major role in safely maturing fire detection and suppression studies in the space environment by employing existing expendable visiting cargo vehicles. Expendable cargo vehicles can be outfitted with detection and suppression systems plus recording and transmission capabilities. Once the visiting vehicle has left the ISS, the internal pressure and oxygen partial pressure can be adjusted and a fire ignited and observed for a short time then extinguished using the new fine mist water spray. This provides a safe method to test sensors and suppression methods in a relevant environment prior to incorporating them into new human space systems.

Benefits include reduced fire suppression system mass and eliminating the hazard of adding significant CO_2 to the human space system atmosphere. The risk of maturing this technology is relatively low as it has already been demonstrated in ground tests and can be tested in space away from humans after an expendable cargo vehicle is on its deorbit trajectory. Ultra-high-pressure fire suppression is game-changing because it can reduce fire suppression water mass by 10 to 33 percent as well as time to extinguish and it will be impossible to "abandon ship" on long-duration deep space exploration missions. As an example, Mir was almost abandoned due to the difficulty in extinguishing an oxygen generation candle fire and it took days to recover from its effects. In addition, there were two space shuttle incidents with smoldering electrical fires that were not detected by fire sensors but were noticed by crew members.

The space station offers a chance for testing in a relevant environment for NEO missions and transit periods between Earth and the Moon or Mars. Testing at Moon and Mars gravity levels on reduced-gravity aircraft flights would also be valuable. This will help understand the impact gravity has on sensors that rely on airflow due to buoyancy. Results can also be used to validate computer simulations.

Technology 6.4.4, Remediation

The panel elevated this technology to high-priority status based on Mir, the ISS, and space shuttle experiences with fire and post-fire remediation. Mir was almost abandoned due to an oxygen generation candle fire and several days were required to recover from its effects. Alcohol in surface wipes used on the ISS to clean surfaces had a deleterious effect on the ECLSS components. On the space shuttle there were two smoldering electrical fires detected by crew members through their sense of smell that electronic sensors failed to detect.

The issues behind these failures need to be thoroughly understood and corrected before long-duration missions are conducted where vehicle abandonment is not an option, systems must operate throughout the mission, and situational awareness is critical to survival, not just mission success.

The ISS will be a critical component of technology development and certification.

MEDIUM- AND LOW-PRIORITY TECHNOLOGIES

TA06 includes eight level 3 technologies that ranked low or medium priority. Of course, many of these technologies will be integral to a final successful design of future operational systems. However, investments in some technologies will produce greater benefits that investments in others. In fact, the division of TA06 technologies was driven by the panel assessment of their expected benefit: all of the high-priority technologies were assigned a maximum score of 9 for benefit; all of the medium-priority technologies were assigned a benefit score of 3, and all of the low-priority technologies were assigned a benefit score of 1. Specific factors that contributed to this scoring are detailed below for some of the medium and low technologies.

Technology 6.2.3, EVA Power, Avionics, and Software, includes important areas of development, but they are likely to be paced by non-EVA, and primarily non-NASA, domain requirements.

Technology 6.3.3, Human Health and Performance Behavioral Health and Performance, was identified as a medium priority based on the knowledge that behavioral health issues will inevitably occur during exploration class missions based on prior experience, but the roadmap is unclear on the nature of the technologies envisioned for some elements of this technology.

Technology 6.3.1, Human Health and Performance, includes a strong emphasis in the roadmap on prediction of future health events by comprehensive screening, and newer processes such as DNA variant screening could indeed incrementally improve criteria for crew selection, but such predictions are unlikely to eliminate the need for planning for in-flight care of those same health conditions. Also, much of the knowledge to be gained in this area will accrue from large cohort studies and from harvesting of clinical data from electronic medical records systems, combined with high-throughput laboratory measurements of the human genome (DNA) and other biologically important classes of molecules, but this will require many more study participants than NASA can reasonably afford, so advances in this technology will also be driven by research conducted outside of NASA.

Technology 6.4.1, Sensors: Air, Water, Microbial, etc., received lower scores due to the relatively small mass, power, and volume impacts that advances in this technology would make on human space systems and adequacy of current systems. (A possible exception may be related to recent data on water-borne molds or mildews emerging on the ISS, similar to reported and observed growths on the Russian Mir space station.)

Technology 6.4.3, Protective Clothing/Emergency Breathing (including long-duration clothing, protective coveralls, oxygen masks, etc.), is a low priority because of the adequacy of current and alternative methods for achieving goals in this technology. For example, clothing mass can be minimized by incorporating a clothes washer and dryer rather than requiring longer-wear outfits, and improvements in emergency breathing capability can be made by increasing the number of portable and umbilical breathers throughout the habitable volume and increasing their operable life or providing more spares.

DEVELOPMENT AND SCHEDULE CHANGES FOR TECHNOLOGIES COVERED BY THE TA06 ROADMAP

Appendix J reports on the draft roadmap for TA07, Human Exploration Destination Systems. The section in Appendix J that corresponds to this section addresses four topics:

- Requirements linked to destinations,
- Microgravity testing on the International Space Station,
- Microgravity testing on the Moon, and
- Asteroid missions.

The discussion of those topics in Appendix J also pertains to the roadmap for TA06 Human Health, Life Support, and Habitation Systems.

OTHER GENERAL COMMENTS ON THE ROADMAP: HUMAN FACTORS

There was one evident gap within the TA06 roadmap, which involved Human Factors technologies. The panel encourages NASA to continue research in microgravity and reduced gravity human factors (and related technologies), and to maintain and update NASA Standard 3001 in order to ensure, among other things, that variable gravity environments are captured in spacecraft design, EVA suit design, and habitat design. Ideally, past human factors "lessons learned" since the beginning of the human spaceflight program would be added to the electronic database and shared with the commercial community (as recommended to the panel at the public session by the commercial spaceflight community). The panel had no evidence that the commercial community was generating any of this measurement technology or human factors information independently, and access to the current and future data (with appropriate export controls) was stated during the panel's public sessions as critical to their success.

In addition, if future mission requirements place an emphasis on ensuring that exploration missions can involve as large a segment of the population as possible, laser scanning technology and relevant (military and civilian) human anthropometric databases could be used to define the required anthropometric measurements which will be utilized for human systems designs (spacecraft, EVA, rovers, surface habitats). This could include advanced technologies which would allow custom sized EVA suits to be cost effectively manufactured directly from laser scan electronic databases. Although, this is a matter of national policy, it seems prudent that spacecraft and EVA suits be developed and designed so that future crew anthropometric requirements are no less than those which were established for the space shuttle in the 1970s. The panel was presented with data which showed that original astronaut selection size standards have been narrowing since the late 1980s due to budget pressures, (e.g., reducing EVA suit sizes) and, more recently, to the acceptance of Russian Soyuz vehicle for crew transportation and the Russian Orlan EVA suit standards, which are sized primarily for a narrow range of Russian males. Current crew selection requirements for the ISS would now preclude many of the former space shuttle era astronauts from being considered (both men and women). This will also have a significant impact for the commercial community as they move into a LEO presence with humans.

PUBLIC WORKSHOP SUMMARY

The Human Systems Panel covered the Human Health, Life Support, and Habitation Systems technology area on April 26, 2011. The discussion was led by Panel Chair Bonnie J. Dunbar. Dunbar started the day by giving a general overview of the NRC's task to evaluate the roadmaps along with some direction for what topics the invited speakers should cover in their presentations. After this introduction, several sessions were held addressing the key areas of each roadmap or representatives of key areas of interest. For each of these sessions, experts from industry, academia, and/or government were invited to provide a brief presentation/discussion of their comments on the NASA roadmap. At the end of each session, there was a short open discussion by the workshop attendees focusing on the recent session. At the end of the day, there was a concluding discussion by the panel chair summarizing the key points observed during the day's discussion.

Session 1: NASA Human Exploration Framework Team (HEFT) Status

The first session of the day was aimed at providing an overview of NASA's latest studies regarding the future direction of human spaceflight and exploration.

Christopher Culbert (NASA-JSC) started the session by providing background on the work HEFT had completed and key findings. One significant finding was that HEFT could not find an architecture that could close technically and financially within an appropriate timeframe that was politically sustainable. Other key findings suggested that no single architecture could achieve all objectives, satisfying all stakeholders is not feasible, and the politically proposed 15-year analysis horizon is too short. Culbert also provided a brief introduction to the Human Spaceflight Architecture Team (HAT) and indicated that both the HEFT and HAT identified technologies align well with the NASA roadmaps.

Scott Vangen (NASA-HQ) provided additional insight into the findings of the HEFT efforts. Based on the technologies evaluated by HEFT, he indicated that extended duration missions on the lunar surface and all missions beyond the Moon require substantially more technology investments than LEO, Cis-lunar, and short-duration lunar surface missions. Additionally, while a majority of necessary technologies can be matured in 3 to 8 years, some key technologies for Mars missions require longer lead time. He then cited radiation protection/shielding as an example of a long-lead technology investment.

Session 2: Roadmap Overview by NASA

The presentation by the NASA roadmap development team described the five level 2 subareas within TA06, including specific examples of the level 3 technologies within each subarea. The specific examples of each level 3 technology provided a description of the particular solution along with a discussion of critical test facilities, technology readiness levels, and mission applicability. The briefing included a chronologically sorted list of 17 top technical challenges, though the briefing did not describe a direct one-to-one mapping between the challenges and the level 3 technologies. At the conclusion of the briefing, the panel questioned whether or not a comprehensive survey of available technical solutions had been completed. The roadmap team responded that the technology examples used in the roadmaps were based on what was known or readily available to the NASA team.

Session 3: Environmental Control and Life Support Systems (ECLSS) and Habitation Systems

Jordan Metcalf (NASA-JSC) started the session with an overview of ECLSS from an operational perspective. In reviewing the TA06 ECLSS Roadmap, he identified the key technology drivers to be high-reliability processes and integrated systems, increased self-sufficiency, and minimized logistics supply. Additionally, he identified regenerative ECLSS as game-changing for long-duration human spaceflight and suggested that further development of the ISS regenerative ECLSS shows the greatest promise for a point of departure for exploration beyond LEO.

Daniel Barta (NASA-JSC) provided an assessment of the current state of the art of ECLSS and habitation technologies conducted by the NASA Exploration Technology Development Program (ETDP). This assessment included TRLs for specific items within each level 3 technology, issues with current state-of-the-art systems, and new technology requirements for various destinations/applications. Although some technologies may have been used in flight, Barta noted that the TRL of these technologies has to be degraded if they are to be operated in a different environment, such as a planetary surface.

In the discussion session, the panel raised the issue of reduced-gravity considerations in technology development and whether or not ETDP was considering these issues in their assessment. Barta indicated that although reduced-gravity considerations were important for some technologies, *g*-neutral solutions were optimal. Additionally, there was some discussion on the use of technologies that can perform multiple functions and whether or not from a systems engineering perspective this was good or bad. It was generally agreed upon by both the speakers and the panel that maximizing commonality was highly desirable for human exploration spaceflight.

Session 4: Human Health and Performance

Jeffrey Davis (NASA-JSC), Director of Space Life Sciences, started the session with an overview of Space Life Sciences and their portfolio of work. His group had developed an evidence-based risk management system through which 65 human system risks had been identified. In reviewing the TA06 roadmap, he identified numerous key opportunities to synergize the technology developments of the OCT roadmaps with the investment portfolio of the NASA Human Research Program.

Jeffrey Sutton (National Space Biomedical Research Institute) provided a brief overview of his institute's activities. Regarding the TA06 Human Health and Performance Roadmap, he cited imaging technology as a high priority across multiple organizations. Additionally, Sutton identified the top technical challenges to be individual susceptibility to radiation, access to the ISS, and barriers to international cooperation.

In the discussion session, the conversation centered on radiation protection and environment characterization, topics closely tied to human health. Sutton indicated that although the radiation environment outside LEO, specifically between Earth and Mars, is adequately well understood, important factors must be taken into account such as the exposure levels that are acceptable and the amount of shielding needed. Furthermore, there is a complex trade space between human health/safety and the added cost/mass of radiation shielding.

Session 5: Environmental Monitoring, Safety and Emergency Response

Nigel Packham (NASA-JSC) started the session with his assessment of this section of the TA06 roadmap. While the roadmap adequately addresses the challenges associated with environmental monitoring and control, fire prevention, detection, and suppression, Packham noted that it fails to suggest partnering with other government agencies that have similar challenges in terms of closed systems and long-duration sorties. Additionally, he suggested that the TA06 roadmap identified target dates for technology demonstration on the ISS that require station operations until 2030, 10 years longer that what is currently planned. Lastly, when asked about challenges, Packham recognized the largest challenge with remediation was trading venting and cleanup versus compartmentalization and habitable volume.

Ralph Cacace (Honeywell Defense and Space) provided a brief overview of current state-of-the-art smoke detectors used on the ISS. He indicated a growing focus on miniaturization of sensors. Beyond size, other desired sensor characteristics included high accuracy, low power, and what was termed as smart sensors. These smart sensors combine multiple simple measurements with applied physics to produce more complex measurements.

Session 6: Extra Vehicular Activity (EVA)

Jim Buchli (Oceaneering) started the session with a programmatic perspective of the roadmap. He identified three key considerations for the EVA Roadmap: a clear definition of the mission and requirements, a critical mass of technical skills and experience, and the deployment of technologies that are flexible, producible, and supportable. Additionally, Buchli identified the top EVA technical challenges to be gloves, mobility, modularity of the portable life support system (PLSS), and miniaturization of electronics.

Brian Johnson (NASA-JSC) began with an overview of NASA's recent EVA technology development efforts by ETDP. He also noted that the HEFT studies reaffirmed the development of space suits as one of the top "destination system" elements to be addressed. He then presented a set of strategic objectives for EVA technology advancement: increased safety and reliability, lower system mass, autonomous operations, expanded anthropometric limits, and lower cost. Lastly, Johnson provided a list of gaps in the current NASA EVA portfolio which included battery specific energy, radiation protection, alternative heat rejection, suit materials/dust, and advanced PLSS packaging and materials.

In the discussion session, the panel posed a question regarding the "academic pipeline" to sustain the development of EVA systems in the future. Buchli responded saying that the number of universities focusing on suit development has definitely been reducing. The panel then questioned the latest advancements in dust mitigation for space suits. Johnson referenced the studies and progress of the former Constellation Program indicating that significant work had been done in this area prior to being cancelled.

Session 7: Radiation

Martha Clowdsley (NASA-LaRC) started the session with a description of radiation protection as an integrated approach consisting of active shielding, forecasting, detection, bio/medical measures, and structure/materials/configuration optimization. She then went on to provide a series of recommendations. For radiation shielding, Clowdsley recommended continued basic materials research and a broad effort to raise the TRLs of existing shielding materials. For exposure analysis tools, she recommended multiple ways to improve both space radiation transport calculations and vehicle/habitat analyses to improve NASA's radiation modeling capabilities for beyond LEO long-duration exploration.

Edward Semones (NASA-JSC) provided an overview of radiation monitoring technologies. Regarding the difference between LEO and exploration missions, he informed the panel that the radiation dosage rates were significantly higher (by a factor of 2 to 3) for exploration missions as well as being much longer, emphasizing that exploration missions would likely challenge the established human radiation risk limits set by NASA. Additionally, Semones identified the key challenges associated with radiation monitoring technologies to be improved battery technology for personal dosimeters, fail-safe data storage and transmission, in situ active warning and monitoring, and data for forecasting models (particularly for forecasting SPEs).

In the discussion session, the panel asked what international assets were available for space weather monitoring. Semones responded saying that there were none outside of those the United States had already collaborated on: ACE, GOES, SOHO, and STEREO space weather monitoring satellites. The panel also posed the question of which technology showed the most promise, biological countermeasures or radiation shielding. Clowdsley responded by saying that the initial focus should be to reduce the biological uncertainty associated with the effects of the space radiation environment.

Session 8: Industry Panel

Paul Zamprelli (Orbital Technologies Corporation) started the session with a summary of the technology developments being conducted by ORBITEC. Pertaining to the technologies covered by TA06, Zamprelli discussed the company's hybrid ECLSS resource recovery development. This system is being designed to demonstrate 90 percent oxygen and 98 percent water resource recovery closure compared to the 60 percent closure demonstrated by the ISS ECLSS. Currently, according to Zamprelli, the ORBITEC hybrid ECLSS has demonstrated approximately 84 percent closure at the time of the workshop.

Barry Finger (Paragon Space Development Corporation) provided an overview of the current state of the art of ECLSS on the ISS. He then identified the need for a simple, reliable, and maintainable ECLSS for long-duration human spaceflight beyond LEO. Finger then went on to identify the ISS as a necessary test bed for ECLSS technologies. Regarding the TA06 Roadmaps, he believes that ECLSS advancements are true game-changing technologies for deep space human missions.

Greg Gentry (Boeing) provided a series of lessons learned from developing and maintaining the ECLSS on both the space shuttle and the ISS. These lessons learned were extensive and ranged from general "philosophical" lessons learned to specific component-level lessons learned. His overall lesson learned was that when changes are made in operational systems, be ready to deal with "unintended consequences."

Edward Hodgson (Hamilton Sundstrand) provided his assessment of the TA06 EVA roadmap. He identified several possible gaps in the roadmap such as "on-back" mass and volume reduction to support Mars surface missions. Additionally, Hodgson suggested that the roadmap assumes comparable EVA sortie durations to history; however, radiation environments could significantly change the EVA architecture due to lack of protection/shielding leading to insignificant EVA operations.

REFERENCES

NRC (National Research Council). 2008. *Managing Space Radiation Risk in the New Era of Space Exploration.* The National Academies Press, Washington, D.C.

NRC. 2011. *Recapturing a Future for Space Exploration: Life and Physical Sciences Research for a New Era.* The National Academies Press, Washington, D.C.

J

TA07 Human Exploration Destination Systems

INTRODUCTION

The draft roadmap for technology area (TA) 07, Human Exploration Destination Systems, addresses six level 2 technology subareas:[1]

- 7.1 In-Situ Resource Utilization (ISRU),
- 7.2 Sustainability and Supportability,
- 7.3 Advanced Human Mobility Systems,
- 7.4 Advanced Habitat Systems,
- 7.5 Missions Operations and Safety, and
- 7.6 Crosscutting Technologies.

The TA07 roadmap includes technologies necessary for supporting human operations and scientific research during space exploration missions, both in transit and on the destination surfaces. The missions identified in this roadmap will experience extreme environments. These environments include reduced gravity (less than 1 g); high levels of several types of radiation as well as ultraviolet light (space weather); vacuum or significantly reduced atmospheric pressures; dusty surfaces (Moon and Mars); and micrometeoroids and/or orbital debris. Many of the elements of the roadmap could have early application to the International Space Station (ISS) and commercial human spaceflight operations, and the ISS is an important test platform for some of the technologies to be deployed beyond low Earth orbit (LEO).

The technology area breakdown structure (TABS) for TA07 includes 19 level 3 technologies, which are subdivided into 70 level 4 items. The envisioned schedule for this roadmap extends to 2035 with a human exploration mission to the surface of Mars.

The requirements for design and development of exploration destination technologies and their integrating systems will be driven by the varying environments of the selected destination, such as reduced gravity levels; the characteristics and chemistry of surface dust; atmospheric pressure and composition; etc. There are many commonalities among the various destinations which may result in common technology solutions, but there are also important differences that will drive the development of divergent technologies. As a result, human surface

[1] The draft space technology roadmaps are available at http://www.nasa.gov/offices/oct/strategic_integration/technology_roadmap.html.

exploration will require prior robotic mapping missions as well as in situ data collection such as was performed with the Moon before human lunar surface missions were conducted during the Apollo Program (e.g., Lunar Orbiter 1, Surveyor 1, 3, 5, 6, and 7). Successful ISRU technologies will depend upon the return of in situ samples to Earth for chemical analyses and subsequent development of processing equipment. For example, the approximately 722 pounds of lunar rocks and regolith returned to Earth during the Apollo missions has provided the basis for research involving the processing of regolith into structural and radiation protection materials. Similar materials research is required to develop the processes for water and oxygen extraction.

Prior to prioritizing the level 3 technologies included in TA07, several technologies were renamed, deleted, or added. The changes are briefly explained below and illustrated in Table J.1. The complete, revised TABS for all 14 TAs is shown in Appendix B.

Technology 7.2.1, Logistics Systems, has been renamed Autonomous Logistics Management to more fully include all elements of inventory and stowage control and to encourage development of technologies needed for autonomous capabilities that would ideally begin at the inception of a project and include all supporting vendors and suppliers.

TABLE J.1 Technology Area Breakdown Structure for TA07, Human Exploration Destination Systems

NASA Draft Roadmap (Revision 10)	Steering Committee-Recommended Changes
TA07 Human Exploration Destination Systems	A number of technologies have been added, deleted, and renamed.
7.1. In Situ Resource Utilization	
7.1.1. Destination Reconnaissance, Prospecting, & Mapping	
7.1.2. Resource Acquisition	
7.1.3. Consumables Production	Rename: 7.1.3 ISRU Products/Production
7.1.4. Manufacturing & Infrastructure Emplacement	
7.2. Sustainability & Supportability	
7.2.1. Logistics Systems	Rename: 7.2.1 Autonomous Logistics Management
7.2.2. Maintenance Systems	
7.2.3. Repair Systems	
	Add: 7.2.4 Food Production, Processing and Preservation (formerly a level 4 item under 7.2.1)
7.3. Advanced Human Mobility Systems	
7.3.1. Extravehicular Activity (EVA) Mobility	
7.3.2. Surface Mobility	
7.3.3. Off-Surface Mobility	
7.4. Advanced Habitat Systems	
7.4.1. Integrated Habitat Systems	
7.4.2. Habitat Evolution	
	Add: 7.4.3 Smart Habitats (formerly a level 4 item under 7.4.1)
7.5. Mission Operations & Safety	
7.5.1. Crew Training	
7.5.2. Environmental Protection	Delete: 7.5.2 Environmental Protection
7.5.3. Remote Mission Operations	Delete: 7.5.3 Remote Mission Operations
7.5.4. Planetary Safety	Delete: 7.5.4 Planetary Safety
	Add: 7.5.5 Integrated Flight Operations Systems
	Add: 7.5.6 Integrated Risk Assessment Tools
7.6. Crosscutting Systems	
7.6.1. Modeling, Simulations & Destination Characterization	Delete: 7.6.1 Modeling, Simulations, and Destination Characterization
7.6.2. Construction & Assembly	
7.6.3. Dust Prevention & Mitigation	

In the draft roadmap, Food Production, Processing and Preservation is just one element of technology 7.2.1, Logistics Systems. Given the importance and complexity of this topic, Food Production, Processing, and Preservation has been established as a new level 3 technology (7.2.4).

In the draft roadmap, technology 7.4.1, Integrated Habitat Systems, includes several elements, including Smart Habitats. The technologies associated with smart habitats are ubiquitous across all human space vehicles, and so Smart Habitats has been established as a new level 3 technology (7.4.3).

Technology 7.5.2, Environmental Protection, has been deleted because all elements of this technology are being treated in other roadmaps (e.g., radiation protection and thermal protection) or they are adequately handled by currently available technologies and design processes (e.g., electromagnetic interference and UV protection).

Technology 7.5.3, Remote Mission Operations, has been deleted because relevant technologies identified in this topic are more appropriately included in the roadmap for TA11 Modeling, Simulation, and Information Technology & Processing. However, the provision of training for and providing real time support for human missions has been added in 7.5.5, Integrated Flight Operations Systems.

Technology 7.5.4, Planetary Safety, has been deleted from this roadmap, but is captured within the Robotics roadmap. The content of this technology category, as described in the draft TA07 roadmap, focused on planetary protection involving robotic missions—that is, ensuring that robotic missions do not contaminate planetary destinations with biological agents from Earth (forward), and ensuring that robotic sample return missions do not contaminate Earth with alien biological agents (backward)). Similarly, it was observed that NASA planetary protection policies are limited to robotic missions. Until those policies are updated to provide guidance on human exploration, in compliance with recent COSPAR planetary protection policies, it would be premature to invest in new technologies relevant to planetary safety in TA07. With respect to Mars, relevant statements from the COSPAR Planetary Protection Policy of October 2002, as amended to March 2011, appear below:

> Crewmembers exploring Mars, or their support systems, will inevitably be exposed to martian materials. In accordance with these principles, specific implementation guidelines for human missions to Mars include:
>
> - Human missions will carry microbial populations that will vary in both kind and quantity, and it will not be practicable to specify all aspects of an allowable microbial population or potential contaminants at launch. Once any baseline conditions for launch are established and met, continued monitoring and evaluation of microbes carried by human missions will be required to address both forward and backward contamination concerns.
> - A quarantine capability for both the entire crew and for individual crewmembers shall be provided during and after the mission, in case potential contact with a martian life-form occurs.
> - A comprehensive planetary protection protocol for human missions should be developed that encompasses both forward and backward contamination concerns, and addresses the combined human and robotic aspects of the mission, including subsurface exploration, sample handling, and the return of the samples and crew to Earth.
> - Neither robotic systems nor human activities should contaminate "Special Regions" on Mars, as defined by this COSPAR policy.
> - Any uncharacterized martian site should be evaluated by robotic precursors prior to crew access. Information may be obtained by either precursor robotic missions or a robotic component on a human mission.
> - Any pristine samples or sampling components from any uncharacterized sites or Special Regions on Mars should be treated according to current planetary protection category V, restricted Earth return, with the proper handling and testing protocols.
> - An onboard crewmember should be given primary responsibility for the implementation of planetary protection provisions affecting the crew during the mission.
> - Planetary protection requirements for initial human missions should be based on a conservative approach consistent with a lack of knowledge of martian environments and possible life, as well as the performance of human support systems in those environments. Planetary protection requirements for later missions should not be relaxed without scientific review, justification, and consensus.

Technology 7.5.5, Integrated Flight Operations Systems, has been added to support the development of capabilities to provide real-time support for spaceflight operations between a crewed vehicle and a mission control center with reduced ground-based staffing coupled with communications latency and/or extended loss of signal periods. The focus of this technology would be operational data management and related technologies to improve integrated vehicle-ground decision making to ensure mission success and safety of flight for missions beyond LEO. This technology represents an intersection between flight software development; Earth-based command and control, models, and crew training; and simulation, as applied to crewed vehicles and ground control systems.

Technology 7.5.6, Integrated Risk Assessment Tools, has been added to support development of new software tools for assessing integrated safety risks for varying exploration scenarios or Design Reference Missions (DRMs). These tools would improve the ability to assess the risk of various exploration and vehicle development strategies (for example, in terms of destinations, habitat deployment strategies, and the role of ISRU). Although the scoring did not result in a high-priority score, the panel recognized the need to complete this assessment prior to determining the sequence of human exploration destinations.

Technology 7.6.1, Modeling, Simulations & Destination Characterization, has been deleted because relevant technologies are more appropriately included in the roadmap for TA11, Modeling, Simulation, Information Technology & Processing. Destination characterization could be performed by robotic missions in order to safely inform the technologies and design of human missions.

TOP TECHNICAL CHALLENGES

The panel identified six top technical challenges for TA07. They are listed below in priority order.

1. ISRU Demonstration: Develop and demonstrate reliable and cost beneficial ISRU technologies for likely destinations (e.g., the Moon and Mars) to reduce the costs of and to enhance and/or enable productive long-duration human or robotic missions into the solar system.

ISRU capabilities directly impact the deployment and success of some future exploration missions. In planetary human space mission planning, the high cost of both up-mass and resupply has been a major hurdle. ISRU has the potential to greatly reduce these costs. ISRU also can greatly increase the human safety margin and likelihood of mission success and extend mission lifetimes for robotic missions. Key technology challenges are the in situ characterization of the raw resources, demonstration of resource recovery and beneficiation, establishment of the optimum processes under the right *g*-environment (using, for example, reduced-gravity aircraft, the ISS centrifuge, a free-flying tethered artificial-gravity satellite, or the lunar surface), and production of the strategic products necessary to support future explorations missions. The priority order of use areas is propellant, life support, and habitat construction. System analysis for a given transport scenario is required to validate the benefit of an ISRU capability to a particular mission. This system analysis needs to be conducted for every ISRU technology being considered for development. Positive benefits can result in smaller spacecraft, increased payload, lower cost, extension of mission life, and increased safety for human crews. Future human planetary surface exploration missions will require large launch masses and, as a result, high launch costs. ISRU technology development would seek to significantly reduce the launch masses and costs of these missions by producing the return-trip propellants (fuels and oxygen) at the exploration site. Additional benefits of developing ISRU technologies include the provision of life support resources (oxygen), volatiles for growing food (nitrogen, carbon, hydrogen, and oxygen), production of metals (aluminum, iron, and titanium), bricks, and other materials (concrete, ceramics, and glasses) for habitats, radiation protection, structures, other surface infrastructures, and other products. In order for ISRU to proceed, samples from prospective exploration sites must first be returned to Earth, simulants must be created, and testing of ISRU processes and technologies must be conducted in relevant environments.

2. Dust: Characterize and minimize the impact that dust in destination environments will have on extravehicular activity (EVA), rover, and habitat systems.

Dust is a critical environmental hazard for hardware tribology, surface solar power systems, instruments, and human habitat atmospheric systems. During the Apollo missions, dust was a problem for both EVA suit systems (clogging cooling sublimator) and human health (lung ingestion). Dust samples from the Apollo landing sites have been well characterized, but more information is needed about the composition and particle size of unexplored areas of the Moon and Mars. This information is needed to develop dust-mitigating technologies for EVA (self-shedding suit fabrics), design requirements for rover treads, and simulants for ISRU. Researchers have defined needs for Earth-based test chambers and ISS testing as appropriate.

3. Supportability: Invest in autonomous logistics management (ALM), maintenance, and repair strategies in order to reduce mission costs and improve probabilities of mission success.

Improving supportability (ALM, repair systems, and maintenance systems) for long-duration missions requires a "launch to end of mission" concept of operations that incorporates highly reliable, maintainable, and repairable systems with fully integrated ALM. Reuse and recycling also will be required to reduce the logistics burden of resupply (if resupply is factored in the design reference missions (DRMs) at all). Ideally, supportability systems should be integrated into the design of the systems themselves at the outset to ensure that vehicle systems can be easily maintained with a minimum of crew. Without resupply, with limited up-mass capabilities, and limited crew time for supportability tasks, requirements for future missions to distant destinations will surely require a very high level of reliability (greater than the ISS).

4. Food Production, Preservation, and Processing: Develop a food subsystem, as part of a closed-loop life support system, to provide fresh food and oxygen and to remove atmospheric CO_2 during long-duration missions.

Food systems for long-duration missions are required in order to reduce the costs of up-mass and resupply, habitat volume, and consumables storage requirements at exploration sites. The production of fresh food would also address concerns that preserved foods may lose nutritional value during long missions. NASA and the Russian Space Agency have invested in both closed loop 1 *g* and microgravity food growth. Little work has proceeded to the point of processing fresh food in reduced-gravity environments. Human spaceflight to distant destinations requires that the nutritional needs of the crew be met for long periods of time. Enabling the production of food onboard and at destinations could greatly increase the probability of maintaining crew health throughout the mission.

5. Habitats: Develop space and surface habitats that protect the crew, implement self-monitoring capabilities, and minimize crew maintenance time.

Future human missions to distant destinations will almost certainly involve mission durations equal to or beyond those attempted on the Mir and the ISS, and mass will be much more highly constrained. While much is known in microgravity biomechanics, practically nothing is known about humans living, working, and being productive for long periods of time in reduced gravity environments such as the Moon and Mars. There is no data on neutral body postures, unsuited gaits, or work station configurations in reduced gravity, or even in such mundane design details as how high the ceiling should be for lunar or martian habitats. Future habitats will need to provide radiation shielding, accommodate long-term exposure to dust from surface environments, and provide a highly reliable habitable volume for months or perhaps years (in the event of an emergency). Future habitat designs will also need accommodate serious medical and surgical intervention, provision for world-class research equipment, and yet provide a comfortable and sustainable living environment.

6. Surface Mobility (Rovers and EVA): Develop advanced rovers and EVA systems for large-scale surface exploration.

The later Apollo missions clearly demonstrated the functionality of integrating rovers with human surface exploration. In the case of much longer missions to the Moon and ultimately Mars, enhanced surface mobility at all levels, whether on foot, in unpressurized or pressurized roving vehicles, or using innovative solutions such as ballistic "hoppers," will improve the science return of exploration missions. Current robotic missions to Mars provides us with the experience that there is little overlap between surface regions of greatest scientific interest (craters, hills, etc.) and areas suitable for safe landing (flat and expansive). A comprehensive program of geological exploration needs access to high slopes, loose and unstable surfaces, and the subsurface via drilling or excavation. Technology issues such as wheel-soil interactions, optimum mobility platform design, and high-reliability mechanisms with high tolerance for dust and exposure to extreme environments must be addressed to develop the systems that can achieve these goals. These technology challenges also exist for robotic telepresence systems (e.g., rover mounted cameras and robotic arms) which could be used for extending human exploration from surface habitats while constraining total EVA exposure.

QFD MATRIX AND NUMERICAL RESULTS FOR TA07

Figure J.1 summarizes the consensus scores of the 19 level 3 technologies in NASA's draft Human Exploration Destination Systems. The panel evaluated each of these technologies based on the description of its content (at level 4) in the roadmap and in light of changes that the panel would make to the roadmap, as detailed above and in the discussion of individual high-priority technologies below. The scores shown below for each technology reflect the highest priority assigned to one or more level 4 topics within each level 3 technology. For the high-priority level 3 technologies (see Figure J.2), these key level 4 items are discussed in the section below on high-priority technologies.

	Benefit	Alignment with NASA Needs	Alignment with Non-NASA Aero Tech Needs	Alignment with Non-Aerospace National Goals	Technical Risk and Reasonableness	Sequencing and Timing	Time and Effort	QFD Score (Weighted)	Panel Priority
Multiplier	**27**	**5**	**2**	**2**	**10**	**4**	**4**		
	0/1/3/9	**0/1/3/9**	**0/1/3/9**	**0/1/3/9**	**1/3/9**	**-9/-3/-1/1**	**-9/-3/-1/0**		
Technology Name	**Benefit**		**Alignment**			**Risk/Difficulty**			
7.1.1. (ISRU) Destination Reconnaissance, Prospecting, and Mapping	3	9	3	1	9	1	-1	224	M
7.1.2. (ISRU) Resource Acquisition	9	9	1	0	9	1	-3	372	H
7.1.3. ISRU Products/Production	9	9	3	3	9	1	-1	390	H
7.1.4. (ISRU) Manufacturing and Infrastructure Emplacement	9	9	3	0	9	1	-3	376	H
7.2.1. Autonomous Logistics Management	9	9	3	3	9	1	-1	390	H
7.2.2. Maintenance Systems	3	9	9	9	9	-3	-3	228	H*
7.2.3. Repair Systems	3	9	9	9	1	1	-9	140	L
7.2.4. Food Production, Processing and Preservation	9	9	3	9	3	1	-1	342	H
7.3.1. EVA Mobility	3	9	0	1	9	1	0	222	M
7.3.2. Surface Mobility	9	9	1	1	9	-3	-3	358	H
7.3.3. Off-Surface Mobility	3	3	0	0	9	-1	-3	170	L
7.4.1. Integrated Habitat Systems	3	9	3	9	3	-9	-1	140	L
7.4.2. Habitat Evolution	9	9	1	0	9	-1	-9	340	H
7.4.3. Smart Habitats	9	3	1	9	3	-3	-3	284	H
7.5.1. Crew Training	1	9	9	1	3	1	-1	122	L
7.5.5. Integrated Flight Operations Systems	3	9	3	3	3	1	-1	168	L
7.5.6. Integrated Risk Assessment Tools	3	9	9	9	3	1	-1	192	M
7.6.2. Construction and Assembly	9	9	3	3	9	1	-1	390	H
7.6.3. Dust Prevention and Mitigation	9	9	3	1	9	1	-1	386	H

FIGURE J.1 Quality function deployment (QFD) summary matrix for TA07 Human Exploration Destination Systems. The justification for the high-priority designation of all high-priority technologies appears in the section "High-Priority Level 3 Technologies." H = High Priority; H* = High Priority, QFD score override; M = Medium Priority; L = Low Priority.

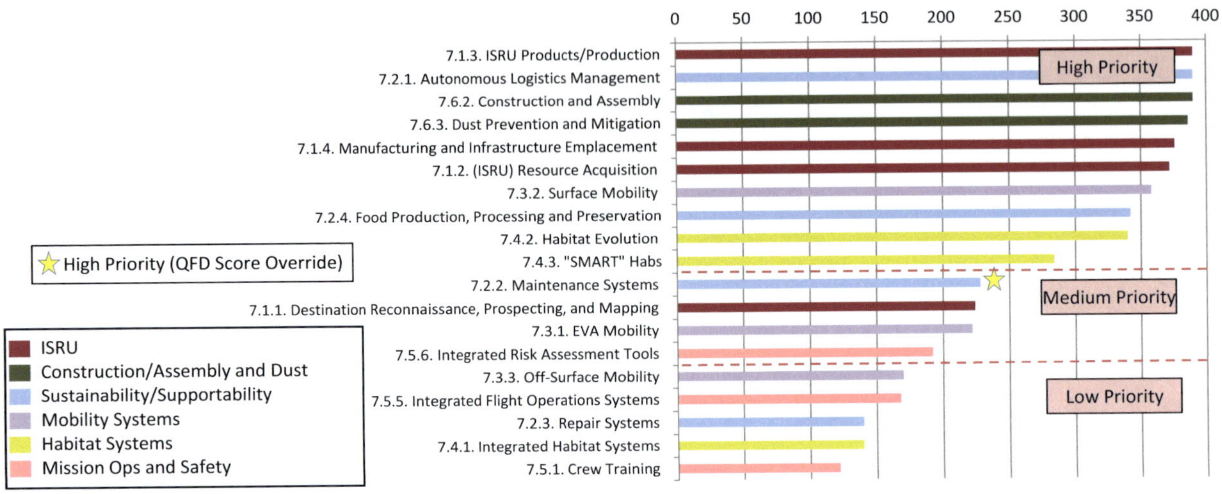

FIGURE J.2 Quality function deployment rankings for TA07 Human Exploration Destination Systems.

Figure J.2 graphically shows the relative ranking of each of the technologies. The panel assessed 11 of the technologies as high priority. Ten of these were selected based on their QFD scores, which significantly exceeded the scores of lower ranked technologies. After careful consideration, the panel also designated one additional technology as a high-priority technology.[2] Note that the 11 "high-priority" technologies fall within the following groups: ISRU (3), Crosscutting Systems (2), Sustainability and Supportability (3), Advanced Human Mobility (1), Advanced Habitat Systems (2).

CHALLENGES VERSUS TECHNOLOGIES

The top technical challenges for TA07, as defined by the panel, were mapped against the panel's high-priority technologies in this area (see Figure J.3). In general, there was good correlation, validating that investments in the high-priority technologies have the potential to make substantial progress in meeting the challenges.

HIGH-PRIORITY LEVEL 3 TECHNOLOGIES

Panel 4 identified 11 high-priority technologies in TA07. These technologies may be grouped into five high-priority theme areas: ISRU (3), Crosscutting Systems (2), Sustainability and Supportability (3), Advanced Human Mobility (1), and Advanced Habitat Systems (2). The justification for ranking each of these technologies as a high priority is discussed below.

In Situ Resource Utilization

The concept of using destination resources to enable a robust, sustainable, human and robotic exploration program is not new. It was an integral component of all prior successful exploration on Earth's surface, and this philosophy continued into early lunar exploration development. NASA-sponsored work at Aerojet concerning the possibility of extracting oxygen from the regolith of the Moon occurred several years before Apollo 11 landed at Tranquility Base in 1969. In a paper written in 1961, scientists speculated that the cold traps at the lunar poles could

[2] In recognition that the QFD process could not accurately quantify all of the attributes of a given technology, after the QFD scores were compiled, the panels in some cases designated some technologies as high priority even if their scores were not comparable to the scores of other high-priority technologies. The justification for the high-priority designation of all the high-priority technologies for TA07 appears in section on High Priority Level 3 Technologies, below.

Priority	TA07 Technologies, Listed by Priority	Top Technical Challenges: 1. ISRU Demonstration: Develop and demonstrate reliable and cost beneficial ISRU technologies for likely destinations (e.g., the Moon and Mars) to reduce the costs of and to enhance and/or enable productive long-duration human or robotic missions into the solar system.	2. Dust: Characterize and minimize the impact that dust in destination environments will have on extravehicular activity (EVA), rover, and habitat systems.	3. Supportability: Invest in autonomous logistics management (ALM), maintenance, and repair strategies in order to reduce mission costs and improve probabilities of mission success.	4. Food Production, Preservation, and Processing: Develop a food subsystem, as part of a closed-loop life support system, to provide fresh food and oxygen and to remove atmospheric CO_2 during long-duration missions.	5. Habitats: Develop space and surface habitats that protect the crew, implement self-monitoring capabilities, and minimize crew maintenance time.	6. Surface Mobility (Rovers and EVA): Develop advanced rovers, and EVA systems for large-scale surface exploration.
H	7.1.3. ISRU Products/Production	●			○		
H	7.2.1. Autonomous Logistics Management			●			
H	7.6.2. Construction and Assembly	○	○			●	
H	7.6.3. Dust Prevention and Mitigation		●			○	
H	7.1.4. (ISRU) Manufacturing and Infrastructure Emplacement	●	○			○	
H	7.1.2. (ISRU) Resource Acquisition	●	●				
H	7.3.2. Surface Mobility		●				●
H	7.2.4. Food Production, Processing, and Preservation				●		
H	7.4.2. Habitat Evolution		○			●	
H	7.4.3. Smart Habitats		○			●	
H	7.2.2. Maintenance Systems			●			
M	7.1.1. (ISRU) Destination Reconnaissance, Prospecting, and Mapping	○	○				○
M	7.3.1. EVA Mobility		○				○
M	7.5.6. Integrated Risk Assessment Tools						
L	7.3.3. Off-Surface Mobility						
L	7.5.5. Integrated Flight Operations Systems						
L	7.2.3. Repair Systems						
L	7.4.1. Integrated Habitat Systems						
L	7.5.1. Crew Training						

●	Strong Linkage: Investments by NASA in this technology would likely have a major impact in addressing this challenge.
○	Moderate Linkage: Investments by NASA in this technology would likely have a moderate impact in addressing this challenge.
[blank]	Weak/No Linkage: Investments by NASA in this technology would likely have little or no impact in addressing the challenge.

FIGURE J.3 Level of support that the technologies provide to the top technical challenges for TA07 Human Exploration Destination Systems.

harbor water ice. This was determined conclusively in 2009 when LCROSS impacted the Moon in October of that year. Some of the processes have their challenges, but many of the basic chemical reactions trace their heritage back hundreds of years. It is clear that the technologies are within reach, and, some involving the Moon, have already garnered commercial interest. For example, over the past 5 years, NASA's ISRU program, in partnership with small business, has built and demonstrated lunar oxygen production in a lunar analog environment. Early demonstration of lunar oxygen production capability could impact decisions regarding destinations and the manner in which missions are conducted. The importance of ISRU has been identified and confirmed by NASA-supported research and by a series of NRC studies over the past decades. The high-priority ISRU technologies are as follows:

- 7.1.3 ISRU Products/Production: Specific products to be produced from raw materials available.
- 7.1.4 Manufacturing/Infrastructure: Ability to make physical structures and complex products.
- 7.1.2 Resource Acquisition: Collection and acquisition of the raw materials needed to manufacture products.

Technology 7.1.3, ISRU Products/Production

ISRU Technology, Consumables Production (7.1.3), has been renamed ISRU Products/Production because ISRU products are expected to consist of more than just consumables (for example, radiation protection materials made from lunar or martian regolith). Future human planetary surface exploration missions will require large launch masses and concurrently high launch costs. ISRU technology development has the potential to significantly reduce the launch masses and costs of these missions by producing key products on the destination surface, such as the return-trip propellants (H_2, HC, etc.); oxygen; water; metals (Si, Fe, Al, etc.); concrete materials; glass, ceramics, windows; fiber (from basalt), fabrics, textiles; other volatile materials (CO_2, NH_3, CH_4, N_2, He, etc.); plastics; and other hydrocarbon materials.

Mission analyses indicate that the use of lunar- and martian-derived propellants could reduce launch masses by over 60 percent. Additional benefits of developing the ISRU technologies include solar energy conversion, volatiles for growing food, and production of metals, bricks and materials for building habitats, structures, and general planetary surface infrastructures.

The panel's assessment of this technology assumes that its content, as described in the draft roadmap, is modified somewhat. In particular, research in this technology would be enhanced by replacing the current level 4 items in Solids and Liquids Processing (7.1.3.1 and 7.1.3.2) with level 4 items that are focused on specific high-priority products to be made via ISRU, including oxygen; water; fuels (H_2, HC, etc.); metals (Si, Fe, Al, etc.); concrete materials; glass, ceramics, windows; fiber (from basalt), fabrics, textiles; volatiles other than water (CO_2, NH_3, CH_4, N_2, He, etc.); plastics; and other hydrocarbon materials. As noted above, the processes for these products are generally different for the Moon and Mars due to differences in both surface chemistry and gravity.

Oxygen is likely the top priority for recovery and use on the Moon and Mars. Water is known to exist at the poles, but its accessibility is unknown. There are several potential methods for producing oxygen, including the use of carbon or methane reduction (carbothermal process), hydrogen reduction (lower yield), molten oxide electrolysis and several other methods. Water recovery is also vital for future exploration programs. Oxygen, hydrogen, and water can be used in propellants (fuels and oxidizers) and life support systems.

Key volatile (non-water) recovery from the regolith and cold traps (e.g., CH_4, CO, CO_2, NH_3, Ar, N_2, etc.) would be used by life support systems and other applications. Fuel production by processing water (H_2O) and the Mars atmosphere (CO_2) could enable production of H_2, CH_4, and other hydrocarbon fuels production on the Mars. Metals produced from the regolith (e.g., Si, Fe) could be used for solar cell production for power plants. Iron can be used for wire in a low-oxygen environment and as a structural material. Aluminum can be used for a wide variety of products.

Other building materials will be important for future low-cost infrastructure on planetary surfaces. New fabric technology using lunar and Mars materials could be of great benefit for future space settlement structures.

Oxygen production has a TRL in the range of 4 to 5. Advancing this TRL for lunar applications would require development of reduced-gravity excavation technologies, mineral beneficiation technology, and oxygen extraction technology. The TRL of other elements of this technology are generally lower. Lunar and Mars concrete technologies are more viable than previously considered now that significant water appears to be available on the Moon and Mars. Efforts to develop production of water, fuels, metal feedstock, and building materials for in situ manufacturing could be integrated to accelerate their TRL and to give NASA the tools that it needs to inform exploration roadmaps beyond LEO.

This technology is well-aligned with NASA's expertise, capabilities, and facilities, and it is a game changer for exploration. NASA, along with its industry and academic partners, can make a significant difference. Currently, it is not likely that commercial organizations or other federal agencies will invest in ISRU. Once NASA commits to a destination with a specified timeline, more investment by industry will likely take place. Other nations may invest is selected elements of ISRU technology (such as oxygen production and water recovery on the Moon).

The ISS can provide some environmental factors (e.g., most notably reduced gravity and vacuum) that would contribute to development of some ISRU technology and processes. An assessment of process effects as a function of gravity is necessary to mature these technologies, and the ability of the ISS to support ISRU technology development would be greatly enhanced if it were equipped with a substantial variable-gravity centrifuge facility.

This technology is considered game-changing because it would significantly reduce the cost of and enhance the productivity of long-duration human or robotic missions. The production of oxygen, water, fuel, metals, and building/construction materials would be particularly beneficial, and these capabilities would be in strong alignment with NASA's human exploration program needs. Development of system components and autonomous plant operations also ranks high in benefits and alignment. Technical risks are considered a good fit for NASA: moderate-to-high risk. The greatest benefit would likely result from an initial focus on the production of polar water on the Moon. If the lunar focus bears fruit, it could change the architecture to explore Mars and, at the same time, potentially enable new commercial markets. A large quantity of water produced in situ and collected in a lunar depot could fundamentally change space exploration. If transferred to an orbital fuel depot, this water could be used to refuel orbit transfer stages, enabling the injection of large payloads to the Moon or Mars. The water could eventually form the basis of a cis-lunar economy, providing the impetus for commercial entities to supply NASA with ISRU products for ongoing space exploration missions.

Technology 7.1.4, Manufacturing/Infrastructure

The panel's assessment of ISRU manufacturing/infrastructure technology assumes that its content, as described in the draft roadmap, is expanded somewhat. According to the roadmap, this technology would include in situ infrastructure (7.1.4.1), in situ manufacturing (7.1.4.2), and in situ derived structures (7.1.4.3). Research in this technology would be enhanced by expanding its scope to include regolith deep excavation for infrastructure (7.1.4.4), spare parts manufacturing (7.1.4.5), and regolith stabilization (7.1.4.6).

Excavating rock for emplacement of infrastructure (7.1.4.1) could likely involve mechanical excavators like rotary cutter heads or explosives. Fixed cutter heads currently used on Earth are optimized to excavate one particular type of rock. A variable-geometry cutter head can adapt its configuration to efficiently operate over a range of consolidated regolith and rocks to minimize energy requirements or maximize excavation rate. This capability is estimated to be at TRL 4. To move forward, a small, lightweight cutter head should be tested in simulated rock deposits expected on the Moon and Mars. Eventually, adequate operating lifetime and reliability will need to be demonstrated under simulated operating conditions at targeted destinations. The use of explosives could also be used as a low-mass and low-energy approach to excavate rock, but there are safety concerns with flying rock fragments. This is a common practice on Earth, but it has not been demonstrated for space applications (TRL 2).

The initial application of in situ manufacturing capabilities (7.1.4.2) is typically the manufacture of photovoltaic cells from lunar regolith. Vacuum deposition of silicon and iron feedstock material derived from lunar regolith is the key aspect of the process (TRL 3).

In situ derived structures (7.1.4.3), such as building materials created from regolith, could involve several methods, including the production of concrete, sintered blocks, processed glass or ceramics, and melted regolith. The ability to produce large quantities of water on the Moon and Mars would make the concrete option more feasible. Techniques to minimize the amount of water required have been advanced to TRL 4. Manufacturing sintered regolith blocks in an oven has been demonstrated using electric resistance heating elements where the temperature of the regolith is uniform and controlled (TRL 4). Processed regolith from the carbothermal reduction process cast into glass or ceramic components is another option for making building materials (TRL 3). Processed regolith from the molten oxide electrolysis process can also be cast into glass or ceramic components. Molten oxide electrolysis can produce oxygen and alloys of Fe-Si-Al-Ti in the processed regolith that can be further refined into metallic products. The resulting processed regolith can also be cast into glass or ceramic components (TRL 3). Regolith melted with concentrated solar energy, microwave energy, or electrical resistance heating can be cast into glass or ceramic components. Molten regolith cast into glass or ceramic components is much stronger than sintered regolith. Melting the regolith requires more thermal energy than sintering, but sintering of regolith requires even heating and precise temperature control. Melting and casting of regolith does not require even heating or precise temperature control due to the higher thermal conductivity of molten regolith (TRL 4).

Deep excavation of dry, consolidated regolith for surface construction (Regolith Deep Excavation for Infrastructure, 7.1.4.4) can be achieved by vibration-assisted excavation, a variable-geometry cutter head, backhoe, etc. Vibration of cutting surfaces causes the regolith particles to fluidize, significantly reducing the forces required for excavation (TRL 4). The cutter head can adapted to the properties of the consolidated regolith to minimize the energy requirement or maximize the production rate (TRL 4). Backhoes are widely used on Earth to excavate holes, but they would require extensive modifications to survive and operate on the surface of the Moon or Mars (TRL 3).

Advanced techniques for spare parts manufacturing (7.1.4.5) on demand from ISRU materials include selective laser sintering, electron beam melting, fused deposition modeling, and 3-D printing. Metals from the regolith can be extracted as feedstock material by the carbothermal reduction or molten oxide electrolysis processes. The metallic iron produced by the carbothermal reduction process is immiscible in the molten regolith, so it separates itself into iron globules. Iron separation from the processed molten regolith has been demonstrated in crucible processing container in a laboratory environment. Silicon in the processed regolith would require additional processing to separate (TRL 3). The molten oxide electrolysis approach creates an alloy of Fe-Si-Al-Ti in the processed regolith that could be refined into metallic products (TRL 2-to-3). On Mars, the atmospheric CO_2 can be used to grow excess plants to make components (TRL 4). The manufacturing of textiles, paper, insulation blankets, and structural beams from inedible biomass could be adapted from pulp molding technology currently used on Earth. This technology could be used to convert inedible biomass from plants grown for food production into useful components. These components can be continuously recycled to make new components using the same technology. Key challenges are to (1) optimize the required processing equipment to operate in the space or planetary environment by minimizing the mass, volume, and energy requirements and minimizing any Earth-supplied consumables required; (2) evaluate the various fireproofing techniques available; (3) investigate the incorporation of multiple waste materials (e.g., organics, plastics, etc.) into the final products; and (4) demonstrate the processing equipment in a relevant operating environment.

Regolith stabilization (7.1.4.6) to prepare roads, landing pads, berms, etc., to reduce dust and prevent damage from blast ejecta presents many potential areas of investigation:

- Sintering and/or melting of the regolith surface using concentrated solar energy (TRL 4). Direct use of solar energy is more efficient than conversion of solar energy to electricity to microwave energy or thermal energy. Sintering of regolith with concentrated solar energy can only be done in thin layers due to the low thermal conductivity of the regolith.
- Sintering and/or melting of the regolith surface using microwave energy (TRL 4). Microwave energy can sinter or melt a thicker layer of regolith than concentrated solar energy. However, there are inefficiencies in converting solar energy to electricity to microwave energy to thermal energy.
- Sintering and/or melting of the regolith surface using electric resistance heating (TRL 4). Sintering of regolith with electric resistance heating elements can only be done in thin layers due to the low thermal conductivity of the regolith. The efficiency of converting electricity to thermal energy is higher than with microwave heating.

- Adding compounds to the regolith to significantly increase its cohesion and create a consolidated surface (TRL 3). A low-energy process that does not require a thermal energy source is sought. However, this approach probably requires Earth-supplied additives such as polymers.
- Placing flexible textile or rigid covers on the regolith surfaces (TRL 3). This may be a low-energy process if the cover material is shipped from Earth. In situ production of the cover material from regolith may be energy intensive. In situ production of the cover material from inedible biomass would be less energy intensive.

This technology is well-aligned with NASA needs. It is not likely that others will invest in this technology.

A study evaluating which processes relevant to this technology could be developed using the ISS would be very beneficial.

This technology is rated high because of the potential for reducing launch costs through reduction of up mass volume and mass. Alignment to NASA programs is also very high, and success in this technology could enable new mission capabilities.

Technology 7.1.2, Resource Acquisition

This technology pertains to collecting and acquiring the raw materials to be used and/or processed into the appropriate product or use. The panel's assessment of ISRU manufacturing/infrastructure technology assumes that its content, as described in the draft roadmap, is expanded somewhat. According to the roadmap, this technology would include regolith and rock acquisition (7.1.2.1), atmospheric acquisition (7.1.2.2), and material scavenging and resource pre-processing (7.1.2.3). Research in this technology would be enhanced by expanding its scope to include cold-trap technologies (7.1.2.4), shallow excavation of dry regolith (7.1.2.5), and excavation of icy regolith (7.1.2.6).

Regolith and rock acquisition and mining machines include cutting tools and drills, scoops, lifting and rotating gears, seals, bearings, actuators, dust filters, electric motors, containers and storage handling equipment, crushers and grinders, and beneficiation devices. For atmosphere acquisition of CO_2, Ar, etc. on Mars, advanced compressors, sorption tools, or freezers are required.

Development of recycling and pre-processing technologies for the ISS, lunar habitats, and Mars habitats would minimize or eliminate the waste storage problem currently experienced by the ISS.

Recovery of water and other gases from the cold trap regions of the Moon will require new technologies because of the extreme environmental conditions expected. In the permanently shadowed areas at the lunar poles, typical ambient temperatures may be less than 70 K. New technologies will be needed to enable mechanical, electrical, and fluid systems to function in temperatures that are much colder than those experienced by conventional machinery designed for terrestrial operations. Subsystems and components of concern include bearings and lubricants, electrical motors, electrical components, computers, microprocessors, chips, contacts, rotating joints, fluid lines, fluid valves, and fluid regulators. Planetary surface hoppers that can quickly fly in and out of the cold trap regions may mitigate problems associated with these cold environments.

Shallow excavation of dry regolith will be a major part of surface and infrastructure preparation on planetary surfaces. Pneumatic regolith excavation reduces the amount of moving components exposed to the abrasive regolith. However, a source of gas is required. In contrast, the only direct consumable required is energy for mechanical excavation. However, the cutting surfaces and moving components will require periodic replacement due to wear from the regolith. Cutter heads are widely used on Earth, but they would require extensive modifications to survive and operate on the Moon or Mars.

Excavation of icy regolith (frozen water with the regolith) will likely be a challenge in the recovery of water on the Moon and Mars. Cutter heads which can adapt to changes in icy regolith deposits will minimize energy requirements and maximize production rate.

Regolith mining devices are generally at TRL 3 to 5. Previous low-level funding resources have been applied to some mining devices over the years, but much more detailed technology development is required to achieve higher TRLs. Technology for Mars atmosphere collection is currently at TRL 6.

This technology is well-aligned with NASA needs. It is not likely that others will invest in this technology, although some advances may have terrestrial applications. For example, a variable geometry cutter head configuration designed for use on the Moon may useful on Earth in certain situations. Advances made by this technology in acquisition devices, mining machines, volatile extraction devices, heat recovery systems, and resource pre-processing/beneficiation may also improve some processes and practices on Earth.

The ability of the ISS to support development of this technology would be greatly enhanced if it were equipped with a substantial variable-gravity centrifuge facility.

This technology is critical to ISRU as it relates to the human exploration program. Without resource acquisition the cost and other benefits of ISRU would not be available. In some cases, this technology may have terrestrial applications. For example, a variable geometry cutter head configuration designed for use on the Moon, might be applicable to Earth in certain situations. In addition, success with this technology could potentially lead to commercial mining of Helium-3 from the lunar regolith as an energy source.

Crosscutting Technologies

Crosscutting technologies contains two high-priority level 3 technologies:

- 7.6.3 Dust Prevention and Mitigation
- 7.6.2 Construction and Assembly

Both of these technologies are dependent on the properties and environments of the destinations.

Technology 7.6.3, Dust Prevention and Mitigation

Dust prevention and mitigation addresses a potential human health and systems performance risk for missions to the Moon and Mars. In addition, recently returned data on asteroids, which may also consist of a mixture of dust and rubble, indicate that dust may also pose a risk to asteroid missions as well. The development of technologies that mitigate the deleterious effects of dust will require knowledge of the chemistry and particle size distribution of the dust. Surface rovers will require detailed knowledge of surface chemistry and morphology in order to design traction devices and to understand tribological effects. Interviews with Apollo crewmembers have shown that the lunar dust was not effectively dealt with during the Apollo program. Dust infiltrated every part of the lunar module (even underneath the fingernails of the crew). Given the relatively short duration of the lunar EVAs and the lunar stay, the risks to crew health and mission success were tolerable, but not without risk. In one incident, lunar dust covered the EVA suit sublimator, the crewmember overheated, and the surface EVA was nearly terminated. The threat posed by dust will increase for missions that entail longer stays and/or more EVAs, or which involve dust properties that humans have not yet encountered (e.g., on Mars).

The draft roadmap for TA07 does not make any substantial progress on crew-related dust prevention until approximately 2029 for crewed Mars missions, after a mission to a near-Earth asteroid (NEA). Dust mitigation for a crewed NEA is insufficiently defined and could introduce poorly characterized dust into the spacecraft following an asteroid EVA. The roadmap notes the technology needs for both prevention and mitigation. Mitigation is defined as "remove or tolerate excessive dust build-up." In the interest of ensuring crew safety and health, a firmer universal requirement to prevent dust intrusion into habitable areas may be advisable. In addition, the draft roadmap for TA06, which includes the design and development of EVA systems and components, does address dust prevention or mitigation, even though lunar or martian dust may be a design driver for suit components such as shoulder, waist, and glove bearings and interfaces. Further, the adverse impacts of dust could drive human-rover interface design and functionality to maximize crew safety and health as well as science return.

The ability to effectively simulate the effect of martian soil on a suited EVA crewman is not readily available. Understanding and reproducing the nature of the soil, including its magnetic and microscopic properties, is essential to effectively preclude dust intrusion. Section 7.6 of TA07 identifies the need for "simulant beds inside vacuum chambers" as a needed technology, but it is unclear whether this includes the capability to demonstrate suited crewmember operations in the dusty environment, at Mars atmosphere (let alone at Mars gravity-level), all of which can impact EVA mission success. Robotic precursor missions will be important to advance the TRL of dust prevention and mitigation technology tailored to meet the needs of specific landing sites.

Habitat designs will need to include technologies for the prevention of dust intrusion and for sequestering dust that does intrude into the habitable volume. The behavior of generic dust has been researched in a microgravity environment on prior space shuttle and Spacelab flights, and this data may be useful in determining properties if dust invades future exploration vehicles prior to their return. The ability of the ISS to support development of this technology would be greatly enhanced if it were equipped with a substantial variable-gravity centrifuge facility.

Dust prevention and mitigation could be a game changer for planetary surface operations and crew health by significantly reducing mission risk, increasing crew safety, and increasing the potential for mission success.

Technology 7.6.2, Construction and Assembly

Other than large module berthing and utilities installation, which has been performed routinely in the construction of the ISS, most of the functionality of the Construction and Assembly technology are readily available on Earth, but they have not been adapted to spaceflight. This technology includes assembling structures in space that are too large and/or too massive for a single launch vehicle. This technology would allow future missions to move beyond deployable structures (e.g., the James Webb Space Telescope) or modular assembly (e.g., the ISS) to erectable structures, including possibly significant use of structural components obtained and fabricated in situ. This technology is also relevant to reduced gravity environments and the use of ISRU technologies for large-scale movement and positioning of regolith to cover components for environmental protection, to improve landing sites to enhance safety and mitigate dangers from secondary ejecta, etc.

Earth-based construction and assembly techniques typically assume the presence of Earth-normal gravity and atmosphere, as well as plentiful water. These techniques need to be revised to accommodate the harsh environment of extraterrestrial destinations and the mass, power, and consumables limitations of spaceflight. Given these unique demands, NASA will need to lead the development of this technology, while drawing on applicable expertise in industry, academia, and other government agencies. The vast majority of technology transfer in this area will be from Earth-based processes and technologies into NASA.

The ISS is both an existing example and an excellent test bed for microgravity assembly techniques. However, it is of limited or no utility for testing surface construction techniques such as regolith moving and collection.

Construction and assembly technology is essential for many future space missions, both manned and unmanned. Large-scale science facilities in both the optical and radio spectrum are pushing the limits of current capabilities with deployable structures and may need to rely on on-orbit assembly to advance in size and capabilities. Surface construction capabilities will be required to establish permanent facilities on the Moon or Mars. Current understanding of the effects of galactic cosmic rays reveal few feasible approaches to significant reduction of deleterious effects on biological organism other than the use of mass shielding, which would be much more economical using construction with in situ materials. Sintering of landing pads would reduce the threat posed by secondary ejecta. In-space assembly of advanced systems, whether astronomical telescopes or Mars transport vehicles, requires productivity levels and reliability currently obtainable only in ground-based facilities. Hardware developed for this technology will need to accommodate the harsh, dusty environments of extraterrestrial destinations. Most construction hardware on Earth is made robust and accurate by the use of massive body components, which will not be feasible for space systems. Active controls and highly redundant actuators may be needed to reduce system and component mass while maintaining system stiffness and strength.

Sustainability and Supportability

Missions beyond LEO will be limited in the amount of logistical support available, including spare parts and systems. Requirements for spacecraft reliability and maintainability will likely increase as mission duration and transit time to the destination increases. The ISS must be continually maintained and resupplied. This has been facilitated by its close proximity to Earth; it can be reached within 48 hours after launch, and in an emergency, the crew can evacuate and return to Earth in less than 2 hours. However, the life support systems on the ISS must be continually resupplied and maintained, and while some atmospheric and water systems are reaching a closed loop status, the food system is not. Also, the ISS is continually refueled to enable control of attitude and altitude. ISRU capabilities will meet some of these requirements for missions to the Moon or Mars, but does not mitigate the risk during the transit mission. The high-priority sustainability and supportability technologies are as follows:

- 7.2.1. Autonomous Logistics Management
- 7.2.4. Food Production/Processing/Preservation
- 7.2.2. Maintenance Systems

Technology 7.2.1, Autonomous Logistics Management

For the purpose of this study, autonomous logistics management (ALM) includes the integrated tracking of location, availability, and status of mission hardware and software to facilitate decision making with respect to consumables usage, spares availability, and the overall health and capability of human exploration mission vehicles and habitats and their subsystems. An ALM system might incorporate radio frequency identification (RFID), electro-optical scanning devices, built-in-test systems, and automatic health monitoring. The system would automatically update the location of hardware items as they were moved around the vehicle or habitat, track life cycle times and condition of equipment, and inform the mission team of resupply needs based on that information as well as mission recycle and reuse strategies. Although integrated ALM systems have been demonstrated in various ground systems, none have been used in NASA programs. For space applications, ALM technology is at approximately TRL 4.

Experience onboard the ISS has shown the significant benefit that ALM technology could provide, given the mass and volume of logistics delivered to the ISS on an annual basis. The ISS would be an ideal test bed for this technology and provide a basis for defining requirements for future transit and planetary habitats.

Exploration missions will have small crews and limited ground communications, necessitating numerous and reliable autonomous systems. ALM technologies would significantly reduce requirements for crew time and for ground-based flight controller time. By having all of this information readily available in a single repository and automatically updated would allow the crew and mission team to quickly assess the effects of system or subsystem failures on the logistics train and allow reprioritization of resupply and/or rationing of critical spares and consumables. The potentially long duration of future missions coupled with long response times for resupply makes it imperative that the mission team know the failure tolerance of the integrated system with respect to available resources and the resupply chain. An ALS system could also factor in the ability of ISRU systems to generate supplies locally.

Technology 7.2.4, Food Production/Processing/Preservation

In the draft roadmap for TA07, this technology is a level 4 item (7.2.1.2) under 7.2.1 Logistics Systems. The scope of this technology includes light sources, water delivery methods, harvesting methods, and preservation methods. Ideally, technologies for the processing of food waste should be coordinated with development of ISRU technologies because food waste is a potential source of organic material. The management of food resources is critical to provide the crew with a healthy diet and reduce the launch mass (and cost) of exploration missions, including resupply missions. The nutritional value of foods on the ISS appears to degrade with time, and some fresh food may be required for a long-duration mission. Food production can also consume CO_2 and generate oxygen for the crew. However, the effects of microgravity on plant growth have been documented, but the effects of reduced gravity at 1/6 *g* or 3/8 *g* are largely unknown.

NASA has extensive experience in terrestrial production, processing, and preservation of food for use in LEO. In addition, plants have been grown and harvested in space-analog 1-*g* closed loop laboratories. Some plants have been grown in microgravity aboard the space shuttle, the Russian space station Mir, and the ISS, but food has not been processed in space, and the technology for harvesting, packaging, and preserving food in reduced gravity is at a very low TRL. Plant growth at 1/6 *g* and 3/8 *g* has not yet been demonstrated.

Development of this technology for extended exploration and reduced gravity is of interest to NASA alone.

The ISS is an excellent location for microgravity testing of this technology in the microgravity environment that would be experienced during a transit to Mars, for example. The ability of the ISS to support development of this technology for surface applications would be greatly enhanced if the ISS were equipped with a substantial variable-gravity centrifuge facility. In any case, development of this technology may require more dedicated equipment racks on the ISS to allow for the processing and preservation of food products, as well as oxygen capture.

The primary benefits of this technology include the ability to sustain crew health for extended periods through augmentation of the oxygen supply, recycling waste for ISRU applications, and reducing resupply and launch mass requirements and associated costs. This technology may have applications for remote stations, such as in Antarctica, and improved food preservation technology might also have military applications (for combat rations that are stored for long periods of time).

Technology 7.2.2, Maintenance Systems

This technology includes four level 4 items: 7.2.1.1 Intelligent/Smart Systems; 7.2.2.2 Non-Destructive Evaluation and Analysis (NDE); 7.2.2.3 Robotic Systems for Maintenance, and 7.2.2.4 Contamination Control and Clean-up. Maintainability and reliability are key requirements for any system or subsystem within exploration vehicles and habitats. The inability to return faulty equipment to Earth for repair during a mission, coupled with potentially long resupply times, enhances the value of equipment designs that facilitate servicing by the crew—or eliminate the need for crew servicing. Each element of this technology would achieve those goals.

Robotic systems have matured to a high TRL with the deployment of Robonaut, but the other elements of this technology have generally not been tested in space or in the reduced-gravity environments of the Moon or Mars (or capabilities demonstrated to date are problematic), and so most of this technology remains at low TRLs (less than 4).

Some elements of this technology may be of interest to non-NASA industries and could lead to teaming with companies that require similar technology development for the maintenance of complex systems.

The ISS would be an ideal test bed for some intelligent maintenance systems technologies and for contamination and control. It currently is a test bed for robotic EVA support.

The panel overrode the QFD score for this technology to designate it as a high-priority technology because the QFD scores did not capture the value of this technology in terms of value to habitat and vehicle development, especially since designs should incorporate these capabilities early in the process.

Advanced Human Mobility Systems

Surface Mobility, 7.3.2, is the only level 3 technology in the Advanced Human Mobility Systems theme area that was determined to be high priority. As detailed below, EVA mobility (7.3.2) is important, but it should probably be integrated with EVA suit design in the roadmap on TA06, Human Health, Life Support, and Habitation Systems. Crew-operated vehicles for use with asteroids will also require new technology development, especially as it relates to station keeping on a rotating body, as most asteroids are expected to be in varying rates of rotation about one or more axes.

Technology 7.3.2, Surface Mobility

As described in the draft roadmap for TA07, Surface mobility technology (7.3.2) is composed of four topics: 7.3.2.1 Rovers and Pressurized Rover, 7.3.2.2 Hoppers, 7.3.2.3 Advanced Surface Transport (such as rail guns, anchoring, zip lines, and maglev pneumatic tubes), and 7.3.2.4 Berthing and Anchoring for application to NEAs. Surface mobility technologies are a high priority for missions to the Moon and Mars because they enable several mission critical capabilities. For example, they allow scientific research over a large area from a single landing site and they make dispersed landing areas acceptable.

Investment in unpressurized rovers and pressurized rovers (life support, thermal control, power supply, communications, traction devices, etc.) should be considered the highest priority element of this technology. Experience with the Apollo program's lunar rover in the early 1970s was relatively successful, and it generated lessons learned applicable to future lunar and Mars vehicles carrying humans. Experience with the Mars surface has been gained by robotic rovers, but at very low velocities and with limited range. Additionally, prior rovers had no pressurized compartments and were not designed with human radiation protection in mind.

Technology for advanced surface mobility systems for humans is approximately at TRL 4. Berthing and anchoring technologies for NEA applications is at or below TRL 3. However, anchoring may be problematic if the asteroid surface is composed primarily of dust and rubble, as was recently determined for Phobos. Berthing with an asteroid having a gravity level of approximately 0.001 *g* is probably better described as station keeping.

NASA is the only entity which has requirements for these technologies. NASA has produced lunar and Mars rovers, and development of this technology is well aligned with agency goals.

Because surface mobility is primarily associated with the reduced-gravity environment on the surface of the Moon or Mars, the utility of the ISS as a test bed is limited. However, if human piloted vehicles are envisioned for NEA proximity operations where there is no significant gravitational attraction, then the ISS would provide an early and relatively safe test bed for related technologies.

Enhanced surface mobility would provide access to large portions of the surface of the Moon or Mars from a single landing site, it would greatly simplify the logistics plan and reduce the resources needed for exploration (in support of scientific objectives and establishing ISRU capabilities), reducing the need for multiple habitats and backup life support systems, and reducing the risk associated with requiring additional launches to place assets where they are needed. Also, if multiple surface facilities are established for whatever reason, enhanced surface mobility would facilitate the movement of personnel and cargo for routine and emergency operations.

Advanced Habitat Systems

The high-priority technologies in the advanced habitat systems theme area are as follows:

- 7.4.2 Habitat Evolution
- 7.4.3 Smart Habitats

The panel rated these technologies as high priority because of their importance to enabling the exploration vision with respect to space-based habitats as well as surface habitats. Both of these areas are also closely aligned with investments in two other high-priority theme areas: Sustainability and Supportability and Crosscutting Technologies (see above). Therefore the technology investments in each of these theme areas should be closely coordinated so that they can be effectively integrated.

Technology 7.4.2, Habitat Evolution

Habitat evolution consists of four level 4 items: deployable habitat destination structures; interplanetary space habitats; artificial gravity; and advanced, integrated habitat shells. Potential applications described in the roadmap include habitats in high Earth orbit (HEO) that would be constructed in LEO and gradually moved to a higher orbit. Eventually, if moved closer to the lunar surface, they could incorporate the use of substantial in situ resources (such as lunar regolith) for shielding to reduce cosmic radiation exposure. Similarly, if feasible at some point in the future, a rotating artificial gravity habitat could allow research into the long-term effects of reduced gravity levels on biological organisms. A rotating spacecraft that provides Earth-normal gravity levels in crew quarters is the one method currently known to avoid musculoskeletal deconditioning in long-term spaceflight, but it is not currently feasible due to the required rotation radius, variable gravity along the radius, and induced coriolis effects on the human vestibular system. However, technology demonstrators could be evaluated and pursued once experiments determine the minimum gravity-level necessary to prevent adverse health effects.

The various elements of this technology are generally at a very low TRL (1 to 2). Advances are needed in materials, environmental control and life support systems (ECLSS), radiation protection, and smart systems.

This technology is unique to NASA, but some elements could conceivably have applications to the commercial sector and the broader research community in the future.

The ISS, as currently configured, is not well suited to initial research and early flight testing of new habitats. The addition of docking adapters could improve the utility of the ISS for in situ research and technology testing. The ISS could also become a staging platform for the release of habitats which would cycle between the lunar Lagrange points and temporary habitats at the ISS, which could be used for quarantining crews returning from missions to Mars or other offworld destinations.

Habitat evolution is of critical importance as a solution to problems in long-duration human spaceflight. Advanced conceptual habitat systems would advance the state of the art, provide a higher level of safety and reliability, and mitigate the long-term effects of microgravity and/or radiation exposure to crew on prolonged transits to and from remote destinations.

Technology 7.4.3, Smart Habitats

This topic area involves the development of advanced avionics, knowledge-based systems, and potential robotic servicing capabilities to create long-term habitats with significantly reduced demands on human occupants for diagnosis, maintenance, and repair. In the draft roadmap for TA07, Smart Habitats was a level 4 item in technology 7.4.1, Integrated Habitat Systems. The panel established Smart Habitats as a distinct level 3 technology because of the importance of this technology to future space exploration.

This technology is currently at TRL 3 to 5, primarily in "concept homes" for Earth-based applications. Some limited efforts are underway to incorporate smart habitat capabilities into experimental habitats for field simulations of NASA-relevant missions. NASA is the logical choice to pursue smart habitats for space applications.

The ISS could be used as a test-bed for development and initial operational assessment of smart habitat technologies.

Studies of crew time during Skylab showed that the three crews typically worked one 8-hour day performing science experiments and a second 8 hours each day to maintain and service habitat systems. When the ISS had three-person crews, about 2.5 crew were required to maintain station systems, with only one crewmember available half-time for science activities. In addition, a large ground staff is required around the clock for ISS system monitoring, stowage tracking, and crew operations planning. Smart habitats would minimize these requirements, and they will be increasingly valuable for habitats on remote missions far from Earth.

This technology will advance habitat systems that augment the crew by providing many of the functions currently performed by mission control or the crew itself. Systems parameters will be internally monitored, and adaptive expert systems will detect off-nominal conditions and provide immediate diagnosis onboard. Sensors will detect the use and stowage locations of the thousands of components onboard the habitat, and will aid the crew in locating specific components needed for tasks. Onboard planning and optimization programs will provide simple graphical user interfaces to allow the crew to have real-time input on workload planning, easing crew concerns about a lack of control of their efforts. In the ultimate form of the "smart" habitat, robotic systems could perform routine maintenance during crew sleep cycles, or even on a "not-interference" basis with the nominal crew activities.

MEDIUM- AND LOW-PRIORITY TECHNOLOGIES

Human Exploration Destination Systems includes eight level 3 technology topics that ranked low or medium priority. The technology areas in this grouping included one from ISRU, two from Advanced Human Mobility Systems, three from Mission Operations and Safety, one from Sustainability and Supportability, and one from Advanced Habitat Systems. Two technology areas were added to ensure completeness of the TABS for Mission Operations and Safety. Integrated Flight Operations Systems has been added to the roadmap to develop capabilities for real-time software/data management tools to support spaceflight operations between a crewed vehicle and a mission control center beyond LEO where the communications environment is characterized by reduced ground-based staffing, significant communications latency, and/or extended loss of signal (LOS) periods. The focus of this technology would be to support integrated vehicle-ground decision making to ensure mission success and safety of flight. This technology represents an intersection between vehicle flight software development; Earth-based command and control software; and advanced training simulators for both the crewed vehicles and ground-based mission control centers. Integrated Risk Assessment Tools was added to support development of new software tools for assessing integrated safety risks for varying exploration design reference missions (DRMs) beyond LEO. It extends beyond the current PRA (probabilistic reliability assessment) tools by improving the ability to assess the relative risks for varying destinations, to include ECLSS, reliability, human health risks, software vulnerabilities, and use of ISRU. This was assessed as medium priority by the steering committee.

Five of the eight were assessed as low priority and placed in this category either because investments could be deferred for at least 5 years until other related technologies mature or destinations are selected, or because there were marginal benefits within the next 20 years. Vehicle and habitat repair materials and adhesives will depend upon the vehicles and habitats selected. Off Surface Mobility is highly dependent upon knowledge of the environments for defined destinations. Atmospherically buoyant transports and martian atmospheric fliers are marginally beneficial in the near term and will require supporting feasibility studies. Integrated Habitat Systems largely addressed new materials which also should be incorporated into the materials roadmaps. (Smart Habitats was moved out of this category and rated "high.") Crew Training and Integrated Flight Operations were rated low because they are considered to be evolutionary rather than revolutionary research technology areas, but the steering committee also recognized that early investments in these areas could help to inform decisions in future spacecraft and habitat design.

Two remaining highly diverse technology areas were rated medium priority: EVA mobility is considered extremely important to provide the maximum flexibility to the crew to perform planned and unplanned mission critical tasks outside of the primary vehicle, but also is dependent upon vehicle and destination selection. Mobility should be designed in coordination with EVA suits. Supporting technologies include power-assisted exoskeletons, EVA transition systems and mobility aids, tools, and telerobotic support. Destination Reconnaissance, Prospecting and Mapping is important to the ISRU and science effort, but is expected to be accomplished through unmanned spacecraft before crews are utilized via rovers. Sample Collection and Characterization (e.g., Dust) had the highest rating among the level 4 categories.

DEVELOPMENT OF TECHNOLOGIES, DESTINATIONS, AND NOTIONAL SCHEDULES PRESENTED IN THE TA07 ROADMAP

This section addresses the maturity of the technologies (their TRLs) in the draft roadmaps for TA07, Human Exploration Destination Systems, in the context of the strategy for integrating them within the framework of the exploration mission sequence and identified destinations and schedules. These comments are also generally applicable to the draft roadmap for TA06, Human Health, Life Support, and Habitation Systems.

The draft roadmap for TA07 contains few detailed development schedules and milestones at level 3, but it does include broad goals and exploration mission destinations and missions in a one-page chart. These are summarized in Figure J.4. This mission profile does not include mission to the Moon, but it does include a HEO habitat in 2020, an NEA/near-Earth object (NEO) mission in 2025, a mission to Phobos (the moon nearest Mars) or Mars orbit in 2030, and a mission to Mars in 2035.

The panel mapped each of destinations in the roadmap to known destination environments as shown in Figure J.5. For comparison, this figure also includes additional destinations: Earth, Lagrange points, and the Moon. Many environmental variables are yet to be defined for some of these destinations, particularly with respect to Mars, Phobos, and NEAs/NEOs. However, these are the key variables that drive technology development and subsequent system design. Dust is not included in Figure J.5, but it is a factor for missions to the Moon, Mars, and NEA visits.

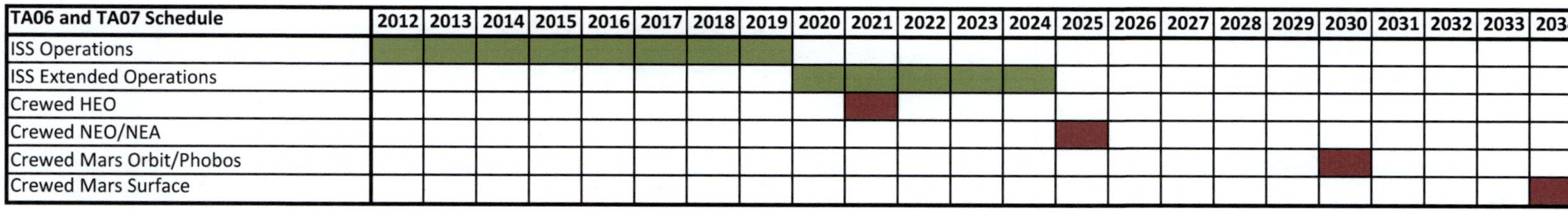

FIGURE J.4 Human exploration mission schedule defined in TA07 roadmap.

Generic Reference Missions	Mission Characteristics								Mission Experience
Identifier	Destination	Radiation	Atmosphere	Gravity	Surface Composition	Total Mission Duration	Transit Duration (micro g)	Abort Return to Earth	
Terrestrial	Earth	L	O_2/N_2 14.7 psi	1 g	Geological soil/H_2O	N/A	N/A	N/A	
Human LEO		M	Vacuum	~0	N/A	90-270 days	Continuous	90 minutes	Apollo, Skylab, ISS, Shuttle
Human HEO		H	Vacuum	~0	N/A	?	?	?	None
Human Lagrange Points		H	Vacuum	~0	N/A	?	?	?	None
Human Lunar	Moon	H	Vacuum	1/6 g	Lunar regolith/H_2O	12 days	3.5 days	3.5 days	Apollo (6 landings)
Human Mars	Mars	H	CO_2 0.17 psi	3/8 g	Martian reolith/H_2O	520 days	180 days		Robotic
Human Asteroid / Small Body	Asteroid / Small Bodies	H	Vacuum	~0	Unknown	> 80 days	180 days		Robotic

FIGURE J.5 Human exploration design environments.

HEFT DRM 4	DRM 4	Other Crew Destination			
Technology Area	Near-Earth Objects	EM-L1 / Lunar Orbit	Mars Orbit	Lunar Surface (Long Dur.)	Mars Surface
Technologies for Human Health & Habitation					
Life Support and Habitation	~	[~]	●	●	●
Exploration Medical Capability	~	[~]	●	●	●
Space Radiation Protection	~	~	●	~	●
Human Health and Countermeasures	~	[~]	●	●	●
Behavioral Health and Performance	~	[~]	~	●	●
Space Human Factors & Habitability	~	[~]	~	●	●
EVA & Robotics Technologies					
EVA Technology	~	~	~	●	●
Human Exploration Telerobotics	~	[~]	~	~	●
Human Robotic Systems	~	[~]	~	●	●
Surface Mobility	< ⊘ >	< ⊘ >	< ⊘ >	⊘	⊘

	Symbol		
Technology Development Complete	~	(green)	Technology Required for this Destination
Additional Technology Development Required	●	[ ]	Technology is Applicable to this Destination
Technology Not Developed	⊘	< >	Not Applicable

FIGURE J.6 Consolidated chart from the 2010 NASA HEFT report *Technology Progress Towards Other Destinations.*

Atmospheric pressure at the destinations in Figure J.5 range from 760 mm on Earth with an oxygen/nitrogen gas mixture to vacuum on the surface of the Moon. Gravity levels range from 1 *g* on Earth to 3/8 *g* on Mars to 1/6 *g* on the Moon, to 0.000001 *g* on the ISS. Radiation protection is best provided on Earth: physiological effects from radiation exposure is slightly more risky in LEO and significantly more risky once the protection of Earth's electromagnetic field is left behind on the way to all other destinations. Adverse health effects from both microgravity and radiation increase with mission duration and distance from Earth. Radiation protection and the reliability of life support systems are pacing technologies with respect to distance from Earth and mission duration.

The panel reviewed the NASA Human Exploration Framework Team (HEFT) Phase 1 Closeout report (2010), which maps technology development status against possible future exploration destinations (Figure J.6).

Requirements Linked to Destinations

Developing single technologies to satisfy the needs of multiple missions without regard to the different environmental characteristics at those destinations is problematic. From an engineering and technology development perspective, destination characteristics will drive engineering design requirements, science objectives, development schedules, and test facility requirements. The history of NASA space missions confirms that differences in gravitational levels, atmospheric composition, surface compositions, and radiation environments frequently calls for different design solutions. For example, the parametric relationship between 1 *g* and microgravity (or zero gravity) is not known for fluid physics, combustion, and human health (bioastronautics), or for those technologies related to crew health, life support systems, spacesuit cooling, etc. Assumptions that these processes are linear between 0 and 1 *g* are probably incorrect for most phenomena. In fact, recent research at NASA Glenn in combustion with drop towers indicates that combustion processes vary in a nonlinear fashion at different gravity levels. Current knowledge of liquid behavior at microgravity is still incomplete as is continually demonstrated by the failures in the ISS Life Support Systems. ISRU processes, which could be developed on the Moon for extraction of oxygen and water resources may not work on Mars because of differences in gravity, regolith chemistry, and heat transfer characteristics. Even so, data acquired from reduced-gravity operations on the Moon could be instrumental to development of oxygen and water extraction processes on Mars, by enhancing diffusion and heat transfer models that are currently validated only at 1 g and microgravity. Similar data is required for understanding human bone loss in microgravity: it is suspected that there is a threshold for g, but it is not known.

In the Apollo program, NASA made incremental steps in technology development, addressing each risk and testing at the highest TRLs in flight like environments before deploying humans. Lunar missions were preceded by multiple LEO missions in which life support systems were tested, bio-medical tests were conducted, and the effects of microgravity were evaluated on gas and liquid based systems. A similar strategy would benefit the current technology development program as it prepares for an unprecedented mission to Mars.

From a budget perspective, it would seem appropriate to organize technology development so that projects intended to support the same mission will move forward in a coordinated fashion. If destinations are not identified and scheduled early, the goals and schedules of various technology development efforts could be out of sync with each other and with larger agency goals. If funding is sufficient to support a broad technology effort, then research focused on multiple destinations could be conducted simultaneously.

Microgravity Testing on the Space Station

The ISS and its hypothetical LEO successor(s), like its predecessors Skylab and the space shuttle/Spacelab, will be required for microgravity research and technology development as long as the United States continues to explore beyond LEO. In addition to its general research mission for terrestrial applications by NASA, industry, and academia, the ISS also supports exploration research and technology development. During transit between Earth and destinations such as the Moon, Mars, and NEAs, astronauts will experience microgravity for periods ranging from 3 days to 9 months. Space stations in LEO provide the only near-Earth, long-duration microgravity environment for human bioastronautics research, technology development, and systems certification.

Space stations in LEO may also serve as future transfer points for astronauts returning to Earth from the Moon, Mars, or other destinations, especially if quarantine is required. The ability of a space station in LEO to support development of technology for surface missions to the Moon or Mars would be greatly enhanced if it were equipped with a substantial variable-gravity centrifuge facility. If rotating space vehicles are to be developed, some of the fundamental research into the behavior of systems and biology in reduced gravity could also be conducted on station-based centrifuges. Fundamentally, as long as exploration continues beyond LEO, LEO research stations will probably be required for advancing research and development, and perhaps for providing operational support. The ISS and its successors provide the optimum high-fidelity environment for microgravity research which is also close enough to Earth to provide astronauts with the ability to quickly return to Earth in case of an emergency.

Microgravity Testing on the Moon

The Moon is not included as a destination in the draft roadmap for TA07, although lunar missions are included in the 2010 NASA HEFT report. Also, the Moon is the closest destination to Earth that enables long-term research and testing in a reduced-gravity environment (significantly less than the 1 g of Earth and significantly more than the microgravity of the ISS). Other possible destinations, such as Phobos or a NEA, also experience essentially zero gravity), and so do not lend themselves to reduced gravity testing of equipment for a Mars surface mission. Developing analog surface operations on the Moon in a vacuum, high-radiation environment with reduced (1/6) gravity provides the best technology research and development environment for many technologies in TA06 and TA07, including those related to bioastronautics, ISRU, EVA, human mobility, habitat system design, sustainability and supportability, and radiation protection. Testing and operational experience gained from an extended mission on the surface of the Moon could significantly reduce risk for a Mars mission in which there is little to no opportunity for abort over the course of a 2-year mission. This incremental strategy is not without precedent. Before U.S. astronauts first landed on the Moon on July 20, 1969, many earlier flights were conducted to develop and test hardware and software, to collect physiological data, and to design environmental control and life support systems.

Asteroid Missions

Attempting to develop technology to support a mission to an as-yet-unidentified asteroid in ~2025 as NASA's next major milestone in human exploration is a difficult challenge, especially because the schedule does not appear to support the development of required technologies. Several technologies must be developed and tested in the next 14 years for a human mission that may last 6 months or longer in a high-radiation environment at great distances from Earth. Subsystems would need to reach TRL 6 to support PDR by 2019 (6 years before flight).

An estimated 6-month NEA mission would significantly stretch the reliability of the currently available life support systems and radiation protection systems. By 2019, the candidate life support systems should have been tested on the ISS in a microgravity environment for at least 2 years to allow for both modifications and lifetime data. This implies deployment in about 2017. The radiation data required to design the protection systems will had to have been collected at orbits higher than LEO and then integrated into a test bed system prior to 2017. The TA06 and TA07 roadmaps do not clearly state when these data collection missions would occur or how the effects of the expected radiation environment on biological systems would be determined.

The NEA mission is not yet focused on a specific destination. Rather, it is focused on a hypothetical, notional asteroid which has not yet been discovered, in a hypothetical orbit co-incident with Earth's orbit about the Sun. Thus, the location, composition, rotational rate, and other characteristics of the destination for an NEA mission are largely unknown. However, its gravitational level is almost certain to be so low that an NEA mission will not enable testing of systems in gravity level that is significantly different than the microgravity level found on the ISS.

A human mission to Phobos would be as challenging as a mission to Mars in many respects, particularly with regard to life support and radiation protection. Phobos may be covered with a thick layer of fine-grained regolith, and the gravity on Phobos is about 0.001 *g*. These conditions could preclude human surface operations, and it requires characterization of the surface dust of Phobos before it is possibly introduced into a spacecraft. The draft TA07 roadmap does not identify advanced human mobility for station keeping, but this may be required.

Pages 25 and 26 of the 2010 NASA HEFT report note that the following technologies are "technology development complete" for near-Earth objects (NEO): life support and habitation, exploration medical capability, space radiation protection, human health countermeasures, behavioral health and performance, space human factors and habitability, EVA technology, human exploration telerobotics, and human robotic systems. Assuming that a deep space NEA mission would last 6 months or more without abort capability or resupply, several technologies in the list above appear to conflict with the technology status reported in the draft TA06 and TA07 roadmaps. Additional consideration regarding forward and backward contamination also needs to be addressed.

PUBLIC WORKSHOP SUMMARY

The Human Systems Panel covered the Human Exploration Destination Systems technology area on April 27, 2011. The discussion was led by Panel Chair Bonnie Dunbar. Dunbar started the day by giving a recap of Day 1 of the workshop and providing additional direction for what topics the invited speakers should cover in their presentations. After this introduction by the panel chair, subsequent sessions focused on either the key areas of each roadmap or the speakers' key areas of interest. For each of these sessions, experts from industry, academia, and/or government were invited to provide a brief presentation of their comments on the draft NASA roadmap. At the end of each session, there was a short open discussion by the workshop attendees focusing on the recent session. At the end of the day, there was a concluding discussion by the panel chair summarizing the key points observed during the day's discussion.

Session 1: Roadmap Overview by NASA

A presentation by the NASA roadmap development team provided an overview of the destinations considered in developing the draft roadmap for TA07. The briefing included a chronologically sorted list of 13 top technical challenges, though the briefing did not describe a direct one-to-one mapping between the challenges and the level 3 technologies in this roadmap. The presentation also included a discussion of what facilities would be required to develop the various technologies, which prompted a question from the panel requesting a distinction between the facilities that do and do not currently exist. The roadmap team responded that the efforts of the Constellation Program to catalog existing facilities had been useful to investigate that question, but that this effort was on-going.

Session 2: In Situ Resource Utilization (ISRU)

The ISRU session started with a written statement from Bill Larson (NASA-KSC). The statement stressed the importance of pursuing ISRU technologies for human exploration beyond LEO and noted that the ISRU technology development program has historically taken a "back seat" to launch vehicle and spacecraft technology development. Additionally, Larson stressed the need to go to the Moon first to demonstrate and mature ISRU technologies, particularly oxygen production from lunar polar ice.

Leslie Gertsch (Missouri University of Science and Technology), represented by Diane Lenne (NASA Glenn Research Center), provided a review of technologies for detection, characterization, and acquisition of minerals, rocks, and soil materials. She stated that the environmental constraints of various destinations differ from conventional practices (on Earth) and that some technologies that work efficiency on Earth may fail in other gravity environments. Gertsch then provided her assessment of top technology choices in the areas of prospecting, acquisition, and beneficiation.

In the discussion session, the panel questioned how the differences in gravitational environments would impact ISRU technologies. One of the session speakers responded that the differences between microgravity, 1/6 *g*, 1/3 *g*, and 1 *g* create substantial differences in many chemical processes. When asked if technology demonstrations on the Moon would be beneficial to development of ISRU technologies suitable for use at other destinations, one of the speakers responded that although many of the processes will still differ, studying the chemical processes at 1/6 *g* on the Moon will at least add another data point between current information obtained in environments with microgravity (on the ISS and other platforms in LEO) and 1 *g* (on Earth).

Session 3: Sustainability and Supportability

Laura Duvall (NASA, Johnson Space Center) began the Sustainability and Supportability session by providing an overview of the ISS crew post-flight debrief process. She reported that human-related data collected during post-flight debriefs is stored in the Flight Crew Integration ISS Life Sciences Crew Comment Database, and the data are analyzed to generate summary reports covering topics ranging from habitable volume to exercise. These reports are then used for a number of applications such as crew training, hardware and software design, and requirements development.

Margaret Gibb (NASA, Johnson Space Center) provided a brief overview of the ISS inventory and stowage officer responsibilities, challenges, and lessons learned. The key takeaway relative to the draft roadmap for TA07 was that logistics management and cargo storage for a long-duration mission beyond LEO, where resupply is not possible, would be a significant challenge.

Kyle Brewer (NASA, Johnson Space Center) provided an overview of the ISS operations support officer responsibilities, challenges, and lessons learned. The key lessons learned were that in many cases miniaturization of system elements leads to increased complexity and difficulty in performing maintenance, commonality is the best strategy to prevent a large number of spares and tools, and a robust maintenance and diagnostic toolkit will be critical for exploration missions.

Session 4: Advanced Human Mobility Systems

Rob Ambrose (NASA, Johnson Space Center), represented by Brian Wilcox, started the Advanced Human Mobility Systems session with a briefing highlighting various examples of human mobility technologies, both historical and those currently in development. These included a zero-gravity jet pack, a pressurized vehicle for zero-gravity mobility, and a pressurized vehicle for surface mobility. Wilcox identified the top technical challenges associated with human mobility to be mobility element sustainability, dust control, and human interaction with autonomous robotics.

David Wetergreen (Carnegie Mellon University) provided a commentary via telephone on the crosscutting nature of robotic mobility throughout the draft roadmaps and stressed that an emphasis was needed on robotic mobility in support of human exploration. He considered this to be a technology gap. He suggested that the top challenges were wheel design and terrain modeling, mechanisms design, and communication between humans and robotic mobility systems.

In the discussion session that followed these presentations, the panel asked about the latest status of suitports in architecture trades. One of the speakers responded that determining whether suitports or airlocks are better suited to a particular vehicle depends heavily on the environmental conditions of the vehicle.

Session 5: Advanced Habitat Systems

Marc Cohen (ex-NASA) started the Advanced Habitat Systems session with a briefing on the need for evidence-based performance requirements for advanced habitat systems. He also identified five showstoppers for deep space exploration by humans: the effects of reduced gravity, radiation, dust, the need for regenerative and bioregenerative life support, and planetary protection (to prevent both forward and backward contamination).

Larry Bell (University of Houston) provided a commentary on design aspects necessary for deep space habitats. He emphasized the need for commonality, simplicity, and autonomy, and noted that the level of acceptable risk is linked to mass requirements and, hence, system design. Regarding risk, Bell stated that NASA and the public would need to accept a higher level of crew risk to enable long-duration exploration missions. He also suggested that the top challenges to be overcome are simplifying complex systems, increasing commonality, adapting high-TRL technologies for exploration purposes, and improving space suit durability.

Session 6: Mission Operations and Safety

Paul Hill (NASA-JSC), Director of Mission Operations, started the Mission Operations and Safety session with a commentary on the current status of mission operations at Johnson Space Center. Relative to the draft roadmap for TA07, he noted that autonomy of mission operations is not driven by the concept of operations but rather by the autonomy built into the spacecraft and that the associated certification requirements drive the design and costs. Additionally, Hill suggested the top challenges to be overcome are the communications lag time on deep space missions, autonomous recovery software to address these lag times, and virtual reality and on-board simulation and training.

Nigel Packham (NASA Johnson Space Center) provided his assessment of the mission operations and safety section of the draft roadmap for TA07, indicating that it correctly identifies crew autonomy to be a vital aspect of the current plan-train-fly approach to crew training for missions beyond LEO. He also notes, however, that the draft roadmap fails to identify quantitative models that could account for the risks posed to crews beyond LEO.

Session 7: Crosscutting Systems: Dust Mitigation

Mark Hyatt (NASA-GRC) started the Dust Mitigation session with a brief overview of the Dust Management Project and a review of the dust issues encountered during Apollo missions. He identified the major operational challenges associated with dust as surface obscuration during descent, coating and contamination of surfaces, reduced crew efficiency, and human exposure to dust (and potential health effects). Lastly, Hyatt provided a description and TRL status for three promising dust mitigation technologies: electrodynamic dust shields, lotus coatings, and space plasma alleviation of regolith concentrations in lunar environments by discharge (also known as SPARCLED).

In the discussion session, the panel asked if there were significant differences between lunar and martian dust. Hyatt answered in the affirmative, although some dust mitigation technologies are likely to work at both destinations. The panel then directed a question to the TA07 Roadmap team asking if simulant production was included in the roadmap. The TA07 team answered no, as they were specifically directed not to include simulants in the roadmap.

Session 8: Industry Panel

Brad Cothran (Boeing) started the industry session with an overview of ISS lessons learned as they pertain to long-duration spaceflight. Based on his experience with the ISS, he identified the top challenges for deep space missions to be logistics stowage, ECLSS reliability and maintainability, waste disposal, and real-time vehicle health management.

William Pratt (Lockheed, Plymouth Rock/Red Rocks) provided an overview of the Plymouth Rock and Red Rocks mission concepts to explore near-Earth asteroids and the martian moon Deimos, respectively. He then provided a summary of the enabling technologies for those missions which included microgravity exposure mitigation, radiation exposure mitigation, high reliability ECLSS, dust mitigation, and advanced EVA and mobility.

Laurence Price (Lockheed, Orion Project) provided a presentation summarizing the latest technology advances incorporated in the Orion crew module. Relative to the draft roadmap for TA07, Orion uses a closed loop life support system capable of supporting a 21 day mission with a crew of four. Additionally, Price informed the panel that the Orion crew module includes advanced thermal control, waste management, and fire detection and suppression technologies. When asked how Orion would handle EVAs, he responded that because Orion had no dedicated airlock, the entire cabin would depressurize and serve as the airlock.

Ken Bowersox (SpaceX) provided a brief commentary on the status of the development of the SpaceX Dragon crew module. According to Bowersox, the first crewed flight of the Dragon is expected within the next 3 years. Regarding technologies being used on Dragon, Bowersox reported that SpaceX was planning to use currently available ECLSS technologies.

K

TA08 Science Instruments, Observatories, and Sensor Systems

INTRODUCTION

The draft roadmap for technology area (TA) 08, Science Instruments, Observatories, and Sensor Systems, consists of three level 2 technology subareas:[1]

- 8.1 Remote Sensing Instruments/Sensors
- 8.2 Observatories
- 8.3 In Situ Instruments/Sensors

The TA08 roadmap addresses technologies that are primarily of interest for missions sponsored by NASA's Science Mission Directorate. They are directly relevant to space research in Earth science, heliophysics, planetary science, and astrophysics; and many areas also have potential applications for the National Oceanic and Atmospheric Administration (NOAA), the Department of Defense (DOD), and commercial remote sensing missions. NASA's science program technology development priorities are generally driven by science goals and future mission priorities recommended in NRC decadal survey strategy reports, and the panel considered those priorities in evaluating the roadmap's level 3 technologies (NRC, 2011, 2010, 2007, 2003).

Before prioritizing the level 3 technologies included in TA08, several technologies were added, renamed, deleted, or moved. The changes are explained below and illustrated in Table K.1. The complete, revised technology area breakdown structure (TABS) for all 14 TAs is shown in Appendix B.

8.1.3 Optical Components was merged with 8.2.1. Mirror Systems and renamed: 8.1.3 Optical Systems because the technologies are very similar and it would be most effective to develop these technologies together.

8.1.7 Space Atomic Interferometry has been added to fill a gap in the roadmap. Atomic interference of laser-cooled atoms has enabled fundamental physics laboratory experiments (at technology readiness level (TRL) 4), including gravitational measurements with greatly improved precision. Advances in this technology could lead to extremely sensitive space detectors for acceleration and, thus, gravity waves.

8.2.4 High Contrast Imaging and Spectroscopy Technologies has been added to fill a gap. Development of advanced approaches to high-dynamic-range imaging would be a game-changing technology to support exoplanet imaging, which is a priority initiative in the Astro2010 decadal survey for astronomy and astrophysics (NRC,

[1] The draft space technology roadmaps are available at http://www.nasa.gov/offices/oct/strategic_integration/technology_roadmap.html.

TABLE K.1 Technology Area Breakdown Structure for TA08, Science Instruments, Observatories, and Sensor Systems

NASA Draft Roadmap (Revision 10)	Steering Committee-Recommended Changes
TA08 Science Instruments, Observatories and Sensor Systems	Several technologies have been added or merged.
8.1. Remote Sensing Instruments / Sensors	
8.1.1. Detectors and Focal Planes	
8.1.2. Electronics	
8.1.3. Optical Components	Rename: 8.1.3. Optical Systems (now includes substance of 8.2.1)
8.1.4. Microwave/Radio	
8.1.5. Lasers	
8.1.6. Cryogenic/Thermal	
	Add: 8.1.7 Space Atomic Interferometry
8.2. Observatories	
8.2.1. Mirror Systems	Delete: 8.2.1. Mirror Systems (merged into 8.1.3)
8.2.2. Structures and Antennas	
8.2.3. Distributed Aperture	
	Add: 8.2.4 High Contrast Imaging and Spectroscopy Technologies Add: 8.2.5 Wireless Spacecraft Technologies
8.3. In Situ Instruments/Sensors	
8.3.1. Particles: Charged and Neutral	
8.3.2. Fields and Waves	Merge 8.3.2 into a renamed 8.3.1, Particles, Fields, and Waves: Charged and Neutral Particles, Magnetic and Electric Fields
8.3.3. In Situ (Instruments and Sensors)	Delete 8.3.2. Fields and Waves (merged into 8.3.1) Add: 8.3.4. Surface Biology and Chemistry Sensors: Sensors to Detect and Analyze Biotic and Prebiotic Substances

2010). This technology would provide unprecedented sensitivity, field of view, and spectroscopy of exoplanetary systems, with many subsidiary applications such as solar physics and the study of faint structures around bright objects (such as jets, halos, and winds).

8.2.5 Wireless Spacecraft Technologies has been added to fill a gap in the roadmap. The use of wireless systems in spacecraft avionics and instrumentation will usher in a new and game-changing methodology in the way spacecraft and space missions will be designed and implemented. Wireless avionics could provide numerous improvements over hard-wired architectures, such as inherent cross-strapping, an architecture that is extensible and reliable; reduction in cable mass; and a significant reduction in the cost and time of system integration and test.[2]

Two technologies in the roadmaps (8.3.1 Particles: Charged and Neutral and 8.3.2 Fields and Waves) seem to have so much overlap that they have been combined to form one entry. The title of the new technology is 8.3.1 Particles, Fields, and Waves: Charged and Neutral Particles, Magnetic and Electric Fields.

TOP TECHNICAL CHALLENGES

The panel identified the following list of six top technical challenges that help provide an organizing framework for setting priorities. Two pertain to crosscutting technologies, and the other four relate to specific important scientific goals. They are listed below in priority order.

1. Rapid Time Scale Development: Enable the exploration of innovative scientific ideas on short time scales by investing in a range of technologies that have been taken to sufficiently high TRLs and that cover a broad class of applications so that they can be utilized on small (e.g., Explorer and Discovery-class) missions.

[2] Avionics (including wireless avionics) is a crosscutting gap that is not addressed in the draft roadmaps (see Tables 4-1 and 4-2).

Innovative ideas need to be tested and evaluated on a rapid time scale, so that the best of them can be brought to maturity. To accomplish this goal, there needs to be inexpensive and routine access to space for technology demonstration. Continuing cooperative programs for instrument development within university engineering and science departments also can be a key asset. This type of program needs to promote development of appropriate management tools and of engineering parts kits that utilize standard interfaces, which can make instruments significantly easier to integrate and test.

2. Low-Cost, High-Performance Telescopes: Enhance and expand searches for the first stars, galaxies, and black holes, and advance understanding of the fundamental physics of the universe by developing a new generation of lower-cost, higher-performance astronomical telescopes.

Cosmologically important astronomical objects are very distant, producing faint signals at Earth. Measurement requires much larger effective telescope collecting areas and more efficient detector systems, spanning the wavelength range from far infrared into the gamma-ray region. This goal requires new, ultra-stable, normal and grazing incidence mirrors with low mass-to-collecting area ratios. A challenge will be to maintain or extend the angular and spectral resolution properties for such mirror systems, which must be coupled to advanced, large-format, low-noise focal plane arrays. Advanced detector systems will require sub-Kelvin coolers and high-sensitivity camera systems.

3. High-Contrast Imaging and Spectroscopy: Enable discovery of habitable planets, facilitate advances in solar physics, and enable the study of faint structures around bright objects by developing high-contrast imaging and spectroscopic technologies to provide unprecedented sensitivity, field of view, and spectroscopy of faint objects.

Among the highest-priority and highest-visibility goals of the space science program is the search for habitable planets and life upon them; only technologies that are fully developed and demonstrated to a high level will facilitate the large, expensive missions needed to achieve this goal. Such technology, once implemented, will set the stage for detailed study of planetary systems, their formation, nature, evolution, and death. The new capabilities will also be of fundamental value for a wide variety of high-contrast targets such as active galactic nuclei and their relativistic jets and subtle but scientifically important features on the sun.

4. Sample Returns and In Situ Analysis. Determine if synthesis of organic matter may exist today, whether there is evidence that life ever emerged, and whether there are habitats with the necessary conditions to sustain life on other planetary bodies, by developing improved sensors for planetary sample returns and in situ analysis.

The needed technologies include integrated and miniaturized sensor suites, sub-surface sample gathering and handling, unconsolidated-material handling in microgravity, temperature control of frozen samples, portable geochronology, and instrument operations and sample handling in extreme environments. In order to enable missions to surfaces of Venus and outer planet satellites, geological, geophysical, and geochemical sensors and instrumentation that survive in extreme environments will be necessary.

5. Wireless Systems. Enhance effectiveness of spacecraft design, testing, and operations, and reduce spacecraft schedule risk and mass, by incorporating wireless systems technology into spacecraft avionics and instrumentation.

To make wireless systems ready for application in spacecraft, current ground-based network technologies will need to be adapted and improved to accommodate very high data rates, provide high throughput and low latency wireless protocols, support a myriad of avionics interfaces, and be immune to interference.

6. Synthetic Aperture Radar. Enable the active measurement from space of planetary surfaces and of solid-Earth and cryosphere surface deformation and monitoring of natural hazards by developing an affordable, lightweight, deployable synthetic aperture radar antenna.

Synthetic aperture radar can provide unique information having both scientific and beneficial applications value regarding earthquakes, volcanoes, landslides, ground subsidence, floods, glacier surges, and ice sheet/shelf collapse. In addition, synthetic aperture radar can enable measurements of planetary surfaces, such as geologic features on the cloud-shrouded surfaces of Venus or Titan. Major advances can come either via a large single structure or apertures distributed across two or more spacecraft. The technology also will depend on advances in high-performance computing in space.

QFD MATRIX AND NUMERICAL RESULTS FOR TA08

Figures K.1 and K.2 show the relative ranking of each technology. The panel assessed seven of the technologies as high priority. Four of these were selected based on their QFD scores, which significantly exceeded the scores of lower ranked technologies. After careful consideration, the panel also designated three additional technologies as a high-priority technology.[3]

The TA08 technologies are displayed in Figure K.2 in order of priority.

	Benefit	Alignment with NASA Needs	Alignment with Non-NASA Aero Tech Needs	Alignment with Non-Aerospace National Goals	Technical Risk and Reasonableness	Sequencing and Timing	Time and Effort	QFD Score (Weighted)	Panel Priority
Multiplier	27	5	2	2	10	4	4		
	0/1/3/9	0/1/3/9	0/1/3/9	0/1/3/9	1/3/9	-9/-3/-1/1	-9/-3/-1/0		
Technology Name	Benefit	Alignment			Risk/Difficulty				
8.1.1. Detectors and Focal Planes	9	9	1	1	9	-1	-1	374	H
8.1.2. (Instrument and Sensor) Electronics	3	9	3	9	3	1	-1	180	H*
8.1.3. Optical Systems	9	9	3	0	9	1	-3	376	H
8.1.4. Microwave / Radio (Sensors)	3	9	9	9	9	-1	-1	244	M
8.1.5. (Instrument and Sensor) Lasers	3	9	1	9	9	-3	-1	220	H*
8.1.6. (Instrument and Sensor) Cryogenic / Thermal	3	9	1	3	9	-1	-1	216	M
8.1.7. Space Atomic Interferometry	3	3	1	1	9	1	-1	190	M
8.2.2. (Observatories) Structures and Antennas	3	9	3	1	9	-3	-1	208	M
8.2.3. (Observatories) Distributed Aperture	9	9	3	0	1	1	-3	296	M
8.2.4. High Contrast Imaging and Spectroscopy Technologies	9	9	1	3	9	1	-1	386	H
8.2.5. Wireless Spacecraft Technology	3	9	9	0	9	1	-1	234	H*
8.3.1. Particles, Fields, and Waves (Sensors)	3	9	1	1	3	1	-1	160	M
8.3.3. In-Situ (Instruments and Sensors)	9	3	3	9	9	1	-1	372	H
8.3.4. Surface Biology and Chemistry Sensors	9	3	9	3	3	1	-1	312	M

FIGURE K.1 Quality function deployment (QFD) summary matrix for TA08 Science Instruments, Observatories, and Sensor Systems. The justification for the high-priority designation of all high-priority technologies appears in the section "High-Priority Level 3 Technologies." H = High Priority; H* = High Priority, QFD score override; M = Medium Priority.

[3] In recognition that the QFD process could not accurately quantify all of the attributes of a given technology, after the QFD scores were compiled, the panels in some cases designated some technologies as high priority even if their scores were not comparable to the scores of other high-priority technologies. The justification for the high-priority designation of all the high-priority technologies for TA08 appears in section "High-Priority Level 3 Technologies."

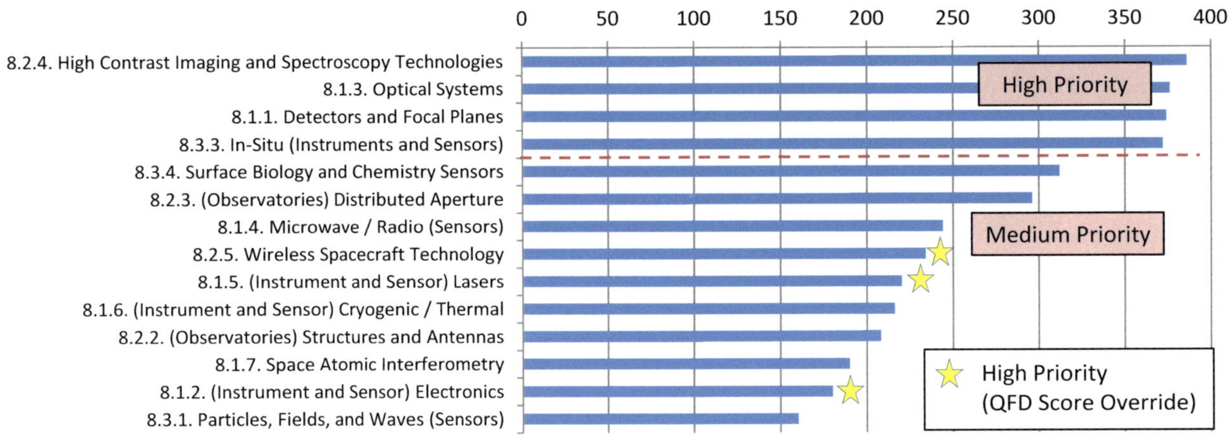

FIGURE K.2 Quality function deployment rankings for TA08 Science Instruments, Observatories, and Sensor Systems.

CHALLENGES VERSUS TECHNOLOGIES

Figure K.3 provides an overview of the linkages between the level 3 technologies and the panel's list of top technical challenges for space science instruments, observatories, and sensor systems.

HIGH-PRIORITY LEVEL 3 TECHNOLOGIES

Panel 3 identified seven high-priority technologies in TA08. The justification for ranking each of these technologies as a high priority is discussed below.

Technology 8.2.4, High-Contrast Imaging and Spectroscopic Technologies

Development of these technologies would enhance high-dynamic-range imaging and support the 2010 astronomy and astrophysics decadal survey priority initiative for exoplanet imaging. There is a strong linkage between this technology and making progress on the top technical challenge to enable discovery of habitable planets, facilitate advances in solar physics, and enable the study of faint structures around bright objects. Technical approaches identified in the decadal survey include star shades (external occulters), interferometry, and coronagraphy (NRC, 2010). The technical challenges (currently at TRL 3 to 4) are well understood; progress would come from merging of several crosscutting technologies.

This technology is well-aligned with NASA's expertise, capabilities, and facilities and with its major industrial partners. The primary focus is on direct imaging and spectroscopy of exoplanets in the habitable zone, but the technology is relevant to any application requiring high dynamic range measurements including potential national defense applications. Use of the ISS is not required for this.

This new technology is game-changing because it would provide substantially increased sensitivity, field of view, and spectroscopy of exoplanetary systems, with many subsidiary applications such as solar physics and the study of faint structures around bright objects (jets, halos, winds, etc.). This area received the panel's highest score due to its high scientific value, relevance to multiple NASA science mission areas, and high ratings for risk and reasonableness.

Technology 8.1.3, Optical Systems

Two optical systems technologies are of particular interest: active wavefront control and grazing-incidence optical systems. There is a strong linkage between these technologies and making progress on the top technical challenge regarding development of a new generation of lower-cost astronomical telescopes.

Priority	TA 08 Technologies, Listed by Priority	Top Technology Challenges					
		1. Rapid Time Scale Development: Enable the exploration of innovative scientific ideas on short time scales by investing in a range of technologies that have been taken to sufficiently high TRLs and that cover a broad class of applications so that they can be utilized on small (e.g. Explorer and Discovery-class) missions.	2. Low-Cost, High-Performance Telescopes: Enhance and expand searches for the first stars, galaxies, and black holes, and advance understanding of the fundamental physics of the universe by developing a new generation of lower-cost, higher-performance astronomical telescopes.	3. High-Contrast Imaging and Spectroscopy: Enable discovery of habitable planets, facilitate advances in solar physics, and enable the study of faint structures around bright objects by developing high-contrast imaging and spectroscopic technologies to provide unprecedented sensitivity, field of view, and spectroscopy of faint objects.	4. Sample Returns and In Situ Analysis: Determine if synthesis of organic matter may exist today, whether there is evidence that life ever emerged, and whether there are habitats with the necessary conditions to sustain life on other planetary bodies, by developing improved sensors for planetary sample returns and in-situ analysis.	5. Wireless Systems: Enhance effectiveness of spacecraft design, testing, and operations, and reduce spacecraft schedule risk and mass, by incorporating wireless systems technology into spacecraft avionics and instrumentation.	6. Synthetic Aperture Radar: Enable the active measurement from space of solid-Earth and cryosphere surface deformation and monitoring of natural hazards by developing an affordable, lightweight, deployable synthetic aperture radar antenna.
H	8.2.4. High Contrast Imaging and Spectroscopy Technologies	○	●	●			
H	8.1.3. Optical Systems	○	●	○			
H	8.1.1. Detectors and Focal Planes	○	●	○		○	
H	8.3.3. In-Situ (Instruments and Sensors)	○			●		
H	8.2.5. Wireless Spacecraft Technology	○	○	○		●	
H	8.1.5. (Instrument and Sensor) Lasers	○					
H	8.1.2. (Instrument and Sensor) Electronics	●	○	○	○	○	○
M	8.3.4. Surface Biology and Chemistry Sensors	○			●		
M	8.2.3. (Observatories) Distributed Aperture	○				○	●
M	8.1.4. Microwave / Radio (Sensors)	○					
M	8.1.6. (Instrument and Sensor) Cryogenic / Thermal	○					
M	8.2.2. (Observatories) Structures and Antennas	○	○			○	●
M	8.1.7. Space Atomic Interferometry	○	○				
M	8.3.1. Particles, Fields, and Waves (Sensors)	○					

●	Strong Linkage: Investments by NASA in this technology would likely have a major impact in addressing this challenge.
○	Moderate Linkage: Investments by NASA in this technology would likely have a moderate impact in addressing this challenge.
[blank]	Weak/No Linkage: Investments by NASA in this technology would likely have little or no impact in addressing the challenge.

FIGURE K.3 Level of support that the technologies provide to the top technical challenges for TA08 Science Instruments, Observatories, and Sensor Systems.

Active wavefront control enables the modification of mirror figure and alignment in response to external disturbances. It allows automated on-orbit alignment of optical systems and the use of lightweight mirrors and telescopes. Current state of the art for alignment of individual mirror segments (e.g., James Webb Space Telescope) is TRL 6. Technology for active adjustment of individual mirror segments is TRL 4 to 5. Although ground-based telescopes routinely use wavefront compensation to correct for atmospherically induced disturbances, this approach cannot be applied readily to a space observatory. Because lightweight, actively controlled telescope systems will be challenging to test fully in a 1-*g* environment, low-cost access to space, possibly including use of the ISS, will open key opportunities for maturing the TRL of the technology.

Active wavefront control aligns closely with NASA's need to develop the next generation of large-aperture astronomical telescopes, lightweight laser communication systems, and high-performance orbiting observatories for planetary missions. NASA has built capabilities and expertise in this technology by a long history of space-borne electro-optical sensors. There also may be interest and expertise in the DOD for this technology. The challenge in developing reliable techniques will be in demonstrating them in microgravity environments.

As evaluated by the 2010 astronomy and astrophysics decadal survey, the TRL ranges from 2 to 3 for grazing incidence systems and from 2 to 5 for normal-incidence technology. These technologies are well-aligned with NASA's expertise, capabilities, and facilities and with its major industrial partners. Access to the ISS is not required for work on grazing-incidence mirror systems. Access to the ISS may be helpful as a test bed for development of active wavefront control.

Further development in grazing-incidence optical systems to improve spatial resolution by at least a factor of 10, without increasing mass per unit area, is critical for future x-ray astronomy missions. This will involve improvements in production systems for piezo adjustment of thin slumped glass and in mounting and testing the sets of optics. Applications are for x-ray and far ultraviolet (UV) (<500 Angstrom) astronomy, and may be extended into the soft gamma/hard x-ray region (to ~100 keV). Two-dimensional adjustment capability may also benefit the synchrotron community.

These are game-changing technologies that would enable direct imaging of stars and detailed imaging of energetic objects such as active galactic nuclei. Adjustable optics based on thin, slumped glass, is a game-changing technology. Normal incidence mirrors with diameters of four meters and beyond that could operate to wavelengths as low as 300 angstroms also would be a game-changing technology.

Technology 8.1.1, Detectors and Focal Planes

Development of sub-Kelvin coolers and high-sensitivity detectors (covering three different spectral energy bands: far-infrared (IR), far and extreme UV, and few-keV x-rays) are very-high-priority efforts for future space astronomy missions. There is a strong linkage between these technologies and making progress on the top technical challenge regarding development of a new generation of lower-cost astronomical telescopes. These devices have reached TRL 4-5 in the laboratory, but further work is needed to make the technologies space-qualified. The necessary expertise is well aligned with NASA's in-house, university, and industry-based capabilities. The DOD also supports work in these areas. Access to the ISS is not required.

This technology is game-changing for the following reasons: The availability of capable sub-K refrigerators could enable long-duration space missions that would be important for many NASA science disciplines including astronomy and planetary studies. Development of these refrigeration technologies could also enable entire new categories of devices that could have enormous commercial and social impact, such as superconducting and quantum computing and superconducting electronics. The proposed technology development would permit 10 times greater sensitivity than current IR satellite observatories for wavelengths above 250 microns and much higher pixel count detectors. Improvement of nearly an order of magnitude in sensitivity and pixel count for FUV and EUV wavelengths would enable new missions, for example to study star formation in galaxies with unprecedented sensitivity and resolution. The proposed improvement in X-ray detectors could significantly decrease the cost and/or increase the capability of next-generation x-ray observatories.

Technology 8.3.3, In Situ Instruments and Sensors

There is a strong linkage between these technologies and making progress on the top technical challenge to determine if synthesis of organic matter may exist today, whether there is evidence that life ever emerged, and whether there are habitats with the necessary conditions to sustain life on other planetary bodies. Geological, geophysical, and geochemical sensors and instrumentation need to be designed that will survive in extreme environments, such as high atmospheric pressure, high or low temperature, and adverse chemistry. Examples of instrument types include cameras and imagers, spectrometers, radiation sensors, seismometers, magnetometers, and entry-descent-landing instruments. The potential benefits of this technology are very high, and in view of the current TRL of 4, the risk of full development to flight is low.

NASA's role in the development of this technology is crucial for advancing it and tailoring it for specific missions that will investigate the composition of solar-system bodies and their atmospheres and conduct searches for life. Access to the space station is not required.

This new technology is game-changing because it would enable missions to the surface and atmosphere of Venus and the surface and sub-surface of outer planet satellites such as the Jovian and Saturnian moons.

Technology 8.2.5, Wireless Spacecraft Technology

Wireless spacecraft technology was added to the observatories subarea of the original NASA roadmap because the use of wireless systems in spacecraft avionics and instrumentation can usher in a new, game-changing methodology in the way spacecraft and space missions will be designed and implemented. Wireless avionics can provide reliable subsystem-to-subsystem communications that facilitate a new fault-tolerant, inherently cross-strapped, extensible, and reliable architectural approach.

To make wireless systems ready for application in spacecraft, current ground-based network technologies will need to be adapted and improved to:

- Accommodate very high data rates as well as low data rates within and between spacecraft subsystems.
- Provide high-throughput and low-latency wireless protocols.
- Support a myriad of avionics interfaces (serial/parallel interfaces, RS-422, 1553, etc.).
- Be immune to interference, including multi-path self-interference.

Ultimately, these systems will require flight testing and demonstration.

The current TRL is estimated to be 3. Several R&D groups in NASA centers, industry, and academia, are investigating the wireless approach. Simple sensors and devices have been flown on shuttles and the ISS. While access to the ISS is not required, testing of new wireless sensors and systems on the ISS would greatly benefit their development.

The panel overrode the QFD score for this technology to designate it as a high-priority technology because it directly relates to meeting the top technical challenge to enhance effectiveness of spacecraft design, testing, and operations, and reduce spacecraft schedule risk and mass, by incorporating wireless systems architecture into spacecraft avionics and instrumentation.

Wireless avionics will provide reliable subsystem-to-subsystem communications with the following improvements over hard-wired architectures:

- Reduces cable mass,
- Reduces integration and test schedule along with maintenance, upgrade, and turnaround time,
- Allows multiple simultaneous communications between subsystems, improving response,
- Significantly reduces the cost and time of system integration and test,
- Provides inherent electrical isolation between subsystems, and
- Provides redundancy against cable and connector failures: increased spacecraft avionics reliability.

Technology 8.1.5, Lasers

Lasers are fundamental components of topographic lidars, atmospheric composition probes (e.g., for CO_2 concentration), and Doppler wind instruments. Key enabling technologies for space are increased laser efficiency and long life. Increased output power, resulting from increased efficiency would enable missions such as multi-beam topographic lidars and high-power, multi-frequency lasers for 3D wind and aerosol and ozone mapping and profiling. A challenge for space is avoiding contamination, which can result in long-term damage where intensity is high—a serious issue for pulsed lasers with high peak power. As noted in the Earth science and applications decadal survey, a "hybrid (combination of two DWL systems, coherent and non-coherent, operating in different wavelength ranges that have distinctly different but complementary measurement advantages and disadvantages). Hybrid Doppler wind lidar (HDWL) in LEO could have a transforming effect on global tropospheric-wind analyses" (NRC, 2007, p. 138).

The panel overrode the QFD score for this technology to designate it as a high-priority technology because the QFD scores did not capture its applications value—a NASA laser demonstration mission to obtain Earth's 3-D wind field would define Earth's atmospheric momentum field with unprecedented quality and add to numerical weather prediction capabilities, especially for severe storms.

The laser industry is a multi-billion dollar enterprise characterized by strong innovation over a broad technical area. It is clear that NASA investment, only if focused by the above specific mission needs, can make a significant contribution to the industry as a whole. Laser developments would also have high potential payoff for future chemical-weather and air-pollution applications. The panel concurs with the decadal survey statement that recommended "an aggressive program to design, build, aircraft-test, and ultimately conduct space-based flight tests of a prototype HDWL" (NRC, 2007, p. 138). NASA would be well served by evaluating and encouraging emerging laser technologies as needed to support the ongoing needs of space missions identified in decadal survey reports and by focusing on approaches for qualifying laser systems for space. In doing so, NASA technology development can respond to the needs of the science community, serve society, and bring later-tier decadal missions that depend on this technology maturation closer to hand.

Technology 8.1.2, Electronics

The state of the art in readout circuitry supports detector sizes currently in flight use (~4 k × 4 k pixels). The design of future readout integrated circuitry (ROIC) to support larger detector sizes will require appropriate design, layout, simulation tools, and fabrication, making use of state-of-the-art ASIC technology. Note that among the attributes of speed, power noise, and capacity (as in number of pixels, channels, etc. read) not all are needed for every detector application. This technology is directly aligned with NASA's expertise, capabilities, facilities, and role in development, and is directly in concert with the thrust toward large detectors and arrays. Access to the ISS is not required to support development in this technology.

The panel overrode the QFD score for this technology to designate it as a high-priority technology because the QFD scores did not capture the value of this technology in terms of its broad applicability to many categories of NASA missions, commercial imaging, NOAA missions, and the national security space communities. There is a strong linkage between these technologies and making progress on the top technical challenge regarding development and maturation of technologies for small missions in short time scales, and the technologies will be valuable in facilitating progress in all of the top technical challenges. Future instruments using large arrays of many types, such as charge-coupled devices (CCDs), photon detectors, or spectrometers, will require new ROIC to fully realize the performance gains from these arrays. High-density, high-speed, low-noise, and low-power ROIC will enable instruments of reduced power and mass, reduced parts count, and simpler designs, with commensurate increased reliability. Instruments intended for extreme environments (high radiation or low temperature) will also benefit from radiation hardening or low temperature capability.

MEDIUM-PRIORITY TECHNOLOGIES

TA08 includes 10 level 3 technologies that were ranked at medium priority. As a group, these technologies generally had lower benefit and/or alignment scores than the high-priority technologies. As noted above, three of these technologies were elevated to high priority, and so they have been discussed above.

A new level 3 technology for space atomic interferometry (8.1.7) was added because atomic interference of laser-cooled atoms has enabled fundamental physics laboratory experiments (at TRL 4) including gravitational measurements of greatly improved precision, and this technology could potentially lead to extremely sensitive space detectors of acceleration and thus of gravity waves.

A new level 3 technology for surface biology and chemistry sensors (8.3.2) was added because low-mass, low-power, and small-volume envelope technologies for ultra-high-resolution mass spectrometry and automated microchip electrophoresis will allow highly sensitive analyses of organic compounds everywhere in the solar system, including Mars, Europa, Titan, and small bodies. This technology received the highest medium-priority score because of its high scientific value and potential alignment with users outside NASA.

Two medium-priority technologies—8.3.2 Surface Biology and Chemistry Sensors and 8.2.3 Distributed Aperture—were considered to be potentially game changing, but their scores fell below the panel's high-priority cutoff. This illustrates the fact that defining a set of realistic and affordable priorities required the panel to make some difficult choices, and it does not reflect a view that these two technologies are unimportant. The other medium-priority technologies all tended to lack identified users or missions or to have only niche roles. Particles, fields, and waves sensor technologies were assigned the lowest medium-priority ranking because, while they are aligned with NASA heliophysics mission areas, they have limited linkage to other aerospace or non-aerospace applications and because they need only incremental improvements.

PUBLIC WORKSHOP SUMMARY

The Instruments and Computing Panel (Panel 3) for the NASA Technology Roadmaps study held a workshop on Science Instruments, Observatories, and Sensor Systems (NASA Technology Roadmap TA08) on March 29, 2011, at the National Academies Beckman Center in Irvine, California. The workshop was attended by members of Panel 3, one or more members of the Steering Committee for the NASA Technology Roadmaps study, invited workshop participants, study staff, and members of the public who attended the open sessions. The workshop began with a short introduction by Jim Burch, the panel chair, in which he emphasized the need for inputs from the scientific community. What followed was a series of six hour-long panel discussions, then a 45-minute session for public comment and general discussion, and finally a short summary and wrap-up by the panel chair. Each panel discussion was moderated by a Panel 3 member. Experts from industry, academia, and/or government were invited to present.

Panel Discussion 1: Technology Tests and Demonstrations via Suborbital and Low-Cost Orbital Flight

The first session focused on the crosscutting issue of using low-cost suborbital and orbital flights to advance in-space technologies. The session was moderated by Webster Cash.

The first presentation was given by Christopher Martin (California Institute of Technology), chair of the Astrophysics Sounding Rocket Assessment Team (ASRAT). He emphasized the importance of sounding rockets as a platform to develop technology for future missions and provided a summary of their historical success in doing so. He also noted that they have been used to train the next generation of space experimentalists and technologists by providing an end-to-end experience in carrying out a space science mission. ASRAT recommends that NASA initiate an Orbital Sounding Rocket (OSR) program to launch science payloads of up to 1,000 lb. into low Earth orbit with mission durations from 1 to 100 days, at a rate of at least one launch per year. They believe that an OSR program would allow for additional tests and science that could not be done using current short suborbital flights. An OSR program would fill the gap between suborbital flights that cost a few million dollars and orbital missions that start at around $100 million. ASRAT believes that OSR would be a game-changing platform for technology innovation, workforce development, and science for NASA and other government agencies.

Alan Stern (Southwest Research Institute) discussed the use of emerging commercial reusable suborbital vehicles to fly experiments. These systems may provide three to four minutes of microgravity (10 times the microgravity time of zero-gravity aircraft and a 100 times cleaner microgravity environment) at 140 km altitude, enabling additional applications that are not possible on other platforms. The human suborbital tourism market may lead to routine flights and low costs. It could allow flying off-the-shelf laboratory equipment with researchers along (since these vehicles would be human-rated for tourism) at a tenth the cost of sounding rockets. Many of the commercial companies have proposed payload bays that would be exposed to the flight environment. Stern said that it will likely be a challenge to NASA to accept these new systems, but they will eventually change the way NASA can develop new systems.

Ray Cruddace (Naval Research Laboratory) discussed routine low-cost access to space (RLCAS). The objectives are to make science observations quickly in order to test new scientific ideas and predictions, to subject new instruments to flight testing, and to expand the envelope for scientific research and technology development beyond the 5 to 10 minutes provided by suborbital sounding rockets. RLCAS is envisioned as developing from the current sounding rocket program into an OSR. The essence of an OSR program is to accept a certain level of risk, maintain an experienced engineering team, use off-the-shelf components, conduct thorough environmental testing, and use a proven launch vehicle. The most promising current candidate launch vehicle is the Falcon 1e at a cost of $10.5 million in fiscal year 2009. The envisioned characteristics of an OSR mission in general include a mission duration of 1 to 3 months, a launch frequency of at least once per year, and a mission cost target of $30 million. The potential user community includes NASA, DOD, and universities.

Panel Discussion 2: Observatories

The next session focused on next generation space-based observatories as a whole. The session was moderated by John Hackwell.

Jim Anderson (Harvard) focused his presentation on observatories needed to study climate change. He discussed the impact on national priorities—especially the link between global energy demand and climate. For the trillion dollar industry associated with global energy demand, the time to adjust is decades. For climate, the timescale is years, and there are questions of feedback and irreversibility. He suggested that climate feedback loops such as those influencing arctic sea ice should drive technology decisions. He also discussed the ability to perform some of these observations using both robotic aerial platforms and small spacecraft.

Tony Hull (L-3 Integrated Optical Systems) discussed programmatic issues that have plagued NASA technology development efforts. His experience with the Terrestrial Planet Finder project has shown him the importance of stable and continuous technology development efforts with thorough plans. He stated that there needs to be early selection and stabilization of a consensus minimal science requirement, definition of an associated realistic budget, and a convergent process for stabilization of a baseline technical approach with redundant capabilities. He expressed concerns about the health of the U.S. industrial base and the repercussions of dwindling development budgets. He also noted the potential impact of International Traffic in Arms Regulations (ITAR) in effect restricting of multi-national organizations from partnering with NASA. The net result is that offshore space technologies are surpassing U.S. technological capabilities. In reference to specific technology priorities, he suggested that wavefront control via active mirrors should have an increasing role as systems get larger and less rigid, but actuated mechanisms are an anathema for managers of many space observatory programs. He believes that adaptive mirrors that use wavefront sensing and control is a game-changing technology for large aperture systems. He sees optical substrate fabrication (to reduce mirror mass) and optical finishing techniques as near a tipping point that might result in a breakthrough for mass and performance.

Panel Discussion 3: Photon Detectors (IR, Visible, UV)

The next session focused on photon detectors for the visible and near visible regions of the electrodynamic spectrum and was moderated by Joel Primack.

Terry Lomheim (The Aerospace Corporation) gave a presentation that focused on focal plane arrays. He began his talk by pointing out the differences between commercial terrestrial needs and in-space applications while explaining why in-space arrays are so much more expensive. In-space array requirements lead to fewer arrays per silicon wafer and less wafers per design. He believes that this leads to a key challenge of keeping the fabrication/manufacturing supply chain going. He also discussed his views of the technology challenges by emphasizing the need for even larger complementary metal oxide semiconductor (CMOS) arrays and their corresponding readout integrated circuits (ROICs). He believes that all interested organizations (including NASA and other government organizations) would have to have major coordinated development efforts to push the technology forward.

Christopher Martin (California Institute of Technology) discussed the potential impacts of UV/optical photon-counting detector technology developments. He said that current state-of-the-art detectors have a quantum efficiency of around 10 percent, which is far below the theoretical limit, and development has hit a wall within the cost scope. He believes that photon-counting detectors will vastly improve spectral, spatial, and temporal resolution, higher pixel count, and provide high dynamic range. These advances are required for several future science missions and could provide the potential for large cost savings of others. One example is Cosmic Web Baryon Mapping, which looked at three different technologies to reach a given signal-to-noise-ratio and photon counting required the smallest telescope diameter and the least money. He noted one specific promising technology of back-illuminated, delta-doped, AR-coated electron multiplying CCDs that are reaching up to 50 percent quantum efficiency with a TRL of 4-5.

Oswald Siegmund (University of California, Berkeley) gave the final presentation which focused on micro-channel plate (MCP) photon counting detectors. He said that MCPs have been involved in many NASA and ESA ultraviolet space-based missions and continue to be proposed and selected for future missions. Siegmund described some of the current MCP detector developments at Berkeley including increasing the size, improvements in photocathode quantum efficiency (QE) to >50 percent, and improving the operability and cost. He noted that the supporting electronics need to be improved to achieve the full performance benefits of these detector advances. He noted one particular technology, Borosilicate Glass MCPs, with large areas could be a game changer, but this technology is still at an early stage.

Panel Discussion 4: Photon Detectors (X-Ray, Gamma-Ray)

The next session focused on photon detectors for the higher energy region of the electrodynamic spectrum and was moderated by Alan Title.

Steve Murray (Johns Hopkins University) gave the first presentation on x-ray detectors needed for future missions. He said that current envisioned missions have identified very specific technical needs. For instance, the International X-ray Observatory wants 1,024 elements, while the current state-of-the-art is tens of elements. He believes that current CCD technology is pushing up against limits of improving the number of pixels and speed. Some of the approaches he mentioned to meet the needs of next generation observatories include a mosaic of many smaller arrays and ASICs for low power processing. He said that CMOS technology is newer and a potentially game-changing technology where it may be possible to event trigger at the pixel. However, current x-ray CMOS production is very unreliable.

Kent Irwin (National Institute of Standards and Technology) gave the second presentation. He started by saying that from a technology perspective, low-temperature detectors (LTD) used in x-ray applications have enormous leverage/coupling to those used at longer wavelengths and their investments should be considered together. He believes that single pixel LTDs are almost as good as they need to be and can be in terms of energy resolution (both long wavelength and x-ray). He thinks that most of the technology advancements will come with increasing array size with the current size of LTD arrays doubling about every 20 months. The state of the art for sub-millimeter LTD is 10,000 pixels and 256 for x-ray. He said that the supporting electronics is the area with the greatest potential for cost improvements. As arrays increase in size there is a need for improved techniques to modulate and encode the detector signals.

Panel Discussion 5: Earth and Planetary Remote Sensing Observation

The session covering remote sensing of Earth and other planetary bodies within the solar system was moderated by David Kusnierkiewicz.

Chris Webster (Jet Propulsion Laboratory) and Keith Raney (Applied Physics Laboratory) jointly gave the presentation, which was a broad survey of Earth and planetary remote sensing instruments and sensors. The presentation began with a summary of the challenges inherent in remote planetary missions including developing for mission-unique environments. They showed examples of typical payload suites of Earth and planetary science spacecraft to emphasize the diversity of the instruments carried. They also cataloged the current and proposed NASA Earth and Planetary science missions.

The presentation gave a summary review of the NASA roadmap with focus on the Remote Sensing Instruments/Sensors technology area. They noted potential gaps such as the possibility of including high-bandwidth downlink under Electronics (8.1.2). They also noted that innovative architectures (as opposed to specific technology elements) might be missing using the Lunar Reconnaissance Orbiter mission as an example of a paradigm shift that occurred under 8.1.4 Microwave and Radio Transmitters and Receivers. They also suggested more discussion of the impacts of enabling technologies from other roadmaps using examples of improved communications and reduced launch costs. They believe that NASA should strive to identify technology push factors for every one of the 11 pathways (in 6 themes) in the TA08 roadmap. Finally, they suggested that Tier 1 missions not now in NASA's funding plan should also be considered in regard to new technologies.

The presenters then gave their views on the direction that NASA should take with technology development. They emphasized the comment in the roadmap that a healthy technology R&D program requires three elements: competition, funding, and peer review. They noted that the recent Planetary Decadal Survey suggested that 6 to 8 percent of the total NASA Planetary Science Division budget should be dedicated to technology development and that resource allocation should be carefully protected. They noted that affordability is a fundamental factor given the recent descopes suggested by the Planetary Decadal Survey and NASA's Earth Science budget proposal.

Panel Discussion 6: In Situ Surface Physical, Chemical, and Biological Sensors

The final session covered interactive measurements of non-Earth bodies within the solar system and was moderated by Daniel Winterhalter (JPL).

Jeffrey Bada (University of California, San Diego) was the first presenter and he focused on the challenge of detecting organic and biological matter. He noted that on Earth, a pre-biotic phase led to a pre-RNA world, which led to an RNA world. The RNA world would have been considered an early stage of life on Earth. That led to a DNA/protein world that is the basis for all terrestrial biology. All of this was assumed to take place in liquid water which is considered the most abundant solvent in the universe. Bada suggested that the search for weird life based on some other biochemistry or solvent should not be included because we do not know what to look for. Instead he suggested that future missions follow the nitrogen. He noted that in carbonaceous meteorites as well as in prebiotic simulation synthesis experiments, there are at least 50 to 70 different amino acids, while in biology as we know it there are only 20 different amino acids in proteins. He suggested that a simple total amine detector the size of a box of stick matches would be a good initial instrument. He believes that a Mars organic analyzer the size of a shoebox is the next step as it can address the issue of homochirality, which is a unique characteristic of amino acids on Earth and presumably life elsewhere as well.

Michael Hecht (Jet Propulsion Laboratory) gave the second presentation, which began with a quick review of the in situ instrument roadmap. He noted some sensor challenges that were not in the roadmap include in situ geochronology and ultra-high-resolution mass spectroscopy (resolve isobars). He felt there were also system challenges not covered in the roadmap including the need to avoid alteration in Mars sample return curation (in situ); extreme environments (Venus, Titan); kW and mW power sources; non-solar, non-nuclear power sources (e.g., wind, thermal, chemical); and full spacecraft sterilization for planetary protection and contamination control. He identified non-biological high-priority sensor technology areas as liquid phase analysis (wet chemistry, lab-on-a-chip, and ice/water analysis); mass spectroscopy (isobar-resolving with >100 K resolving power, laser ablation mass spectroscopy, and geochronology); and chemical microscopy (scanning electron microscope/energy dispersive X-ray microanalysis, small spot scanning x-ray fluorescence, spectroscopic imaging, and chromophormicroscopy). He also identified game-changing technologies near the tipping point, including the ability to do things related to sample return (e.g., in situ geochronology, advanced life detection, and micro-analysis). He also identified a broadened access to deep space (flying instruments as discussed earlier in the day) as game-changing.

Public Comment Session and General Discussion

The day concluded with a public comment session moderated by Robert Hanisch. Most of the questions and comments focused on general issues of technology development. Topics included the challenges of maintaining a qualified workforce capable of advancing the state of the art and the difficulties of maintaining the knowledge and capabilities that have already been developed.

REFERENCES

NRC (National Research Council). 2003. *The Sun to the Earth—and Beyond: A Decadal Research Strategy in Solar and Space Physics*. The National Academies Press, Washington, D.C.

NRC. 2007. *Earth Science and Applications from Space: National Imperatives for the Next Decade and Beyond*. The National Academies Press, Washington, D.C.

NRC. 2010. *New Worlds, New Horizons in Astronomy and Astrophysics*. The National Academies Press, Washington, D.C.

NRC. 2011. *Vision and Voyages for Planetary Science in the Decade 2013-2022*. The National Academies Press, Washington, D.C.

L

TA09 Entry, Descent, and Landing Systems

INTRODUCTION

The draft roadmap for technology area (TA) 09, Entry, Descent, and Landing Systems, consists of four level 2 subareas:[1]

- 9.1 Aeroassist and Entry
- 9.2 Descent
- 9.3 Landing
- 9.4 Vehicle Systems Technology

Entry, descent, and landing (EDL) systems is a critical technology that enables many of NASA's landmark missions, including Earth reentry, manned Moon-landings, and robotic landings on Mars. EDL technologies support all of the systems and demonstration thereof necessary to perform any or all of the three mission phases defined by entry, descent, and landing. NASA's draft roadmap for TA09 defines entry as the phase from arrival through hypersonic flight, with descent being defined as hypersonic flight to the terminal phase of landing, and landing being from terminal descent to the final touchdown. EDL technologies can support all three of these mission phases or just one or two of them. For example, aerocapture or aerobraking technologies support only the entry phase.

Entry, Descent, and Landing are the three main level 2 technology subareas; the fourth, Vehicle Systems Technology, encompasses technologies that cover multiple phases of EDL.

Before prioritizing the level 3 technologies included in TA09, several technologies were renamed, deleted, or moved. The changes are explained below and illustrated in Table L.1. The complete, revised technology area breakdown structure (TABS) for all 14 TAs is shown in Appendix B.

Technology 9.1.5, Instrumentation and Health Monitoring, is applicable to descent and landing as well as entry, and so it has been moved to technology subarea 9.4, Vehicle Systems Technology, which encompasses technologies that cover multiple phases of EDL, and has been redesignated 9.4.6.

Modeling and Simulation appears as separate line items in Entry (9.1.6), Descent (9.2.5), and Landing (9.3.6). However, there is so much overlap among these three areas, and the factors that determine priority vary so little from one to another, that they have been combined into a new level 3 technology (9.4.5 EDL Modeling and Simulation) in technology subarea 9.4, Vehicle Systems Technology.

---

[1] The draft space technology roadmaps are available at http://www.nasa.gov/offices/oct/strategic_integration/technology_roadmap.html.

TABLE L.1 Technology Area Breakdown Structure for TA09, Entry, Descent, and Landing Systems

NASA Draft Roadmap (Revision 10)			Steering Committee-Recommended Changes
TA09 Entry, Descent, and Landing Systems			Several items have been merged and/or relocated or deleted.
9.1.	Aeroassist and Atmospheric Entry		
	9.1.1.	Rigid Thermal Protection Systems	
	9.1.2.	Flexible Thermal Protection Systems	
	9.1.3.	Rigid Hypersonic Decelerators	
	9.1.4.	Deployable Hypersonic Decelerators	
	9.1.5.	Instrumentation and Health Monitoring	Move: 9.1.5 into added 9.4.6, Instrumentation and Health Monitoring
	9.1.6.	Entry Modeling and Simulation	Merge 9.1.6 with 9.2.5 and 9.3.6 and move to added 9.4.5, EDL Modeling and Simulation
9.2.	Descent		
	9.2.1.	Attached Deployable Decelerators	
	9.2.2.	Trailing Deployable Decelerators	
	9.2.3.	Supersonic Retropropulsion	
	9.2.4.	GN&C Sensors	Merge 9.2.4 with 9.3.4 and move to added 9.4.7, GN&C Sensors and Systems
	9.2.5.	Descent Modeling and Simulation	Merge 9.2.5 with 9.1.6 and 9.3.6 and move to added 9.4.5, EDL Modeling and Simulation
9.3.	Landing		
	9.3.1.	Touchdown Systems	
	9.3.2.	Egress and Deployment Systems	
	9.3.3.	Propulsion Systems	
	9.3.4.	Large Body GN&C	Merge 9.3.4 with 9.2.4 and move to added 9.4.7.
	9.3.5.	Small Body Systems	
	9.3.6.	Landing Modeling and Simulation	Merge 9.3.6 with 9.1.6 and 9.2.5 and move to added 9.4.5.
9.4.	Vehicle Systems Technology		
	9.4.1.	Architecture Analyses	Delete: 9.4.1. Architecture Analyses
	9.4.2.	Separation Systems	
	9.4.3.	System Integration and Analyses	Note: In some places in the roadmap, 9.4.3 "Systems Integration and Analyses" is titled "Vehicle Technology." "Systems Integration and Analyses" more accurately describes the content of this technology.
	9.4.4.	Atmosphere and Surface Characterization	
			Add: 9.4.5. EDL Modeling and Simulation Add: 9.4.6. Instrumentation and Health Monitoring Add: 9.4.7. GN&C Sensors and Systems

GN&C sensors (9.2.4) are applicable to entry and landing as well as descent. In addition, there are entry and descent aspects to large-body GN&C (9.3.4). Therefore, these items have been combined into a new level 3 technology (9.4.7, GN&C Sensors and Systems) in technology subarea 9.4, Vehicle Systems Technology.

Technology 9.4.1 Architecture Analyses appears in the TABS and in two summary figures in the TA09 roadmap, but it does not appear in the roadmap table of contents or in the text in the main body of the roadmap. It has been deleted.

Technology 9.4.3 in the TABS is titled Systems Integration and Analyses. In some places in the roadmap it is titled Vehicle Technology. Systems Integration and Analyses more accurately describes the content of this technology.

TOP TECHNICAL CHALLENGES

EDL is commonly a challenging aspect of NASA missions; EDL problems have been associated with some significant failures as well as many near misses.

NASA's draft EDL roadmap may be too narrow because it is focused on the development of human class, large payload delivery to Mars as the primary emphasis, even though such a mission may be three decades away. While this mission beneficially stresses and challenges EDL technology development, it would be prudent to consider the benefit of advanced EDL technologies to other possible applications to ensure that the technology under development is not tied too closely to a specific mission or destination. EDL technologies that enable the broadest spectrum of future missions by accommodating the widest range of variations in destination and timing would be of particular value. This is reflected in the broad set of six top technical challenges and the discussion of generic reference missions that follows. The top technical challenges defined by the panel are listed below in priority order. The first four challenges would make EDL systems more technically capable, the fifth challenge would make them safer and more reliable, and the sixth challenge would make them more affordable.

1. Mass to Surface: Develop the ability to deliver more payload to the destination.

NASA's future missions will require ever greater mass delivery capability in order to place scientifically significant instrument packages on distant bodies of interest, to facilitate sample returns from bodies of interest, and to enable human exploration of Mars. For a given launch system and trajectory design, the maximum mass that can be delivered to an entry interface is fixed. Hence, increasing the mass delivered to the surface (or other destination, such as a planetary orbit or a mobile flight platform) will require reductions in spacecraft structural mass; more efficient, lighter thermal protection systems; more efficient, lighter propulsion systems; and/or lighter, more efficient deceleration systems. In a sense, increasing mass delivery to a planet surface is "the name of the game" for EDL technology because it may enable missions that are presently impossible (such as a human Mars landing) and/or provide enhancements such as more sophisticated science investigations and sample return capability for currently planned missions.

2. Surface Access: Increase the ability to land at a variety of planetary locales and at a variety of times.

Ideally, any exploration mission would have the ability to land at a variety of locales, including those at higher latitudes or elevations that may be difficult to access, at whatever time best satisfies other mission requirements and goals. Access to specific sites can be achieved by landing at one or more specific locations or by transiting (e.g., via a rover) from a single designated landing location to other locations of interest. However, it is not currently feasible to transit long distances and through extremely rugged terrain on Mars. In addition, improving the robustness of entry systems to better withstand a variety of environmental conditions (atmospheric winds, solar incident angle, etc.) could aid in reaching more varied landing sites. Alternatively, uncertainties in the entry environment could be better dealt with if the entry vehicle first went into orbit. Increased surface access could be achieved by tailoring the mission entry (i.e., the ability to control the inclination of entry and/or cross range capability during entry). Systems that have higher lift-to-drag ratios are an area for potential investigation in improving surface access on exploration destinations, such as Mars, that have a significant atmosphere.

3. Precision Landing: Increase the ability to land space vehicles more precisely.

A precision landing capability allows a vehicle to land closer to a specific, predetermined position in order to assure that the vehicle lands safely (without damage to itself or other personnel that may already be on the surface), or in order to meet other operational or science objectives. The level of precision (e.g., 1000 m, 100 m, etc.) that is achievable at touchdown is a function of the design of the guidance, navigation, and control (GN&C) system, the control authority of the vehicle, and the entry environment. Precision landings require accurate GN&C performance throughout the entire descent and landing phases. This requires accurate control of vehicle position, velocity, attitude, and other vehicle states (Paschall et al., 2008).

4. Surface Hazard Detection and Avoidance: Increase the robustness of landing systems to surface hazards.

The surface hazards associated with exploration destinations remain uncertain to some degree until the site has been visited. Relying on passive systems alone to characterize a landing site can be problematic, as was evident during the Apollo Program, where each of the six landing missions faced potentially mission-ending hazards at the landing sites. Hazardous rocks, craters, and slopes were perilously close to each of the successfully landed missions and brought to light the incredible challenge each mission faced (Brady and Paschall, 2010). Active hazard detection methods can quickly optimize safe sites and reduce fuel costs while directly characterizing a landing surface in real time, but technology development is needed to improve key capabilities in this area (Brady et al., 2009).

5. Safety and Mission Assurance: Increase the safety, robustness, and reliability of EDL.

Loss-of-mission events during EDL for NASA and the international community have been unacceptably high for Earth-entry and especially planetary entry missions. For example, the failure rate for U.S. missions to Mars over the past 20 years is 27 percent (i.e., 3 of 11). U.S. lander missions to Mars have a failure rate of 20 percent (i.e., 1 of 5) (NASA, 2011). High-profile U.S. failures include the Mars Polar Lander and the Mars Climate Orbiter. Other nations have also experienced failure during EDL (e.g., Beagle 2 and numerous Soviet missions), especially in the earlier years of planetary exploration. These events are costly setbacks for high-profile robotic missions, which are the result of many years of effort in design, development, flight, and operations resources. For crewed missions, EDL failures can result in tragedy, such as the Columbia accident.

Safety and mission assurance are necessary constraints for mission and vehicle design. Some level of risk is unavoidable with planetary exploration missions. This challenge seeks to improve safety and mission assurance while achieving important mission objectives in an affordable manner.

6. Affordability: Improve the affordability of EDL systems.

EDL missions with large payloads are expensive. Improving EDL system affordability would allow more missions to be flown within fixed and predictable budgets, and it would enable new missions previously deemed unaffordable. In fact, the issue of affordability led the Planetary Decadal Survey to question whether a Mars sample return mission belongs in its roadmap of future planetary missions (NRC, 2011). The affordability of EDL systems can be improved either by (1) improving EDL capabilities so that it is less expensive to get a payload of some particular mass to the surface or (2) improving payload technologies so that the same mission objectives can be achieved with a smaller payload mass delivered to the surface. Technology development in TA09 is focused on the first approach; the second approach will occur naturally as a result of advances in other TAs.

QFD MATRIX AND NUMERICAL RESULTS FOR TA09

Figure L.1 shows the QFD scores for each technology in TA09. Figure L.2 shows the relative ranking of each technology, grouped into high, medium, and low priorities. The process by which the QFD scores were generated is described in Chapter 2. The panel assessed eight of the technologies as high priority. Four of these were selected based on their QFD scores, which significantly exceeded the scores of lower-ranked technologies. After careful consideration, the panel also designated four additional technologies as high-priority technologies.[2]

The four technologies selected based on their QFD scores were: 9.4.7 GN&C Sensors and Systems, 9.1.1 Rigid Thermal Protection Systems, 9.1.2 Flexible Thermal Protection Systems, and 9.1.4 Deployable Hypersonic Decelerators. These technologies received that ranking based on their significant overall benefit, their ability to meet NASA needs (generally driven by supporting a wide range of missions), and their risk and reasonableness. 9.4.7 GN&C Sensors and Systems was also noted to meet non-NASA aerospace needs. The four additional technologies did not score as high on the QFD scale, but investments in these technologies will support a wide range of expected future EDL missions.

[2] In recognition that the QFD process could not accurately quantify all of the attributes of a given technology, after the QFD scores were compiled, the panels in some cases designated some technologies as high priority even if their scores were not comparable to the scores of other high-priority technologies. The justification for the high-priority designation of all the high-priority technologies for TA09 appears in the section "High-Priority Level 3 Technologies."

	Benefit	Alignment with NASA Needs	Alignment with Non-NASA Aero Tech Needs	Alignment with Non-Aerospace National Goals	Technical Risk and Reasonableness	Sequencing and Timing	Time and Effort	QFD Score (Weighted)	Panel Priority
Multiplier	27	5	2	2	10	4	4		
	0/1/3/9	0/1/3/9	0/1/3/9	0/1/3/9	1/3/9	-9/-3/-1/1	-9/-3/-1/0		
Technology Name	**Benefit**	**Alignment**			**Risk/Difficulty**				
9.1.1. Rigid Thermal Protection Systems	9	9	3	1	9	1	-3	378	H
9.1.2. Flexible Thermal Protection Systems	9	9	3	1	9	-1	-3	370	H
9.1.3. Rigid Hypersonic Decelerators	3	9	1	0	3	-1	-3	142	M
9.1.4. Deployable Hypersonic Decelerators	9	9	1	0	9	-3	-3	356	H
9.2.1. Attached Deployable Decelerators	3	3	1	0	9	-1	-1	180	M
9.2.2. Trailing Deployable Decelerators	3	9	1	0	9	-1	-1	210	M
9.2.3. Supersonic Retropropulsion	1	3	1	0	3	-1	-3	58	L
9.3.1. Touchdown Systems	3	9	1	1	1	1	-1	140	M
9.3.2. Egress and Deployment Systems	1	3	0	0	1	1	-1	52	L
9.3.3. (EDL) Propulsion Systems (Interaction)	3	3	1	0	3	-1	-1	120	M
9.3.5. (EDL) Small Body Systems (No Gravity)	1	3	1	0	9	-1	-1	126	M
9.4.2. (EDL) Separation Systems	1	9	3	0	1	1	-1	88	L
9.4.3. (EDL) System Integration and Analyses	3	9	3	1	9	-1	-1	216	H*
9.4.4. Atmosphere and Surface Characterization	3	9	3	3	9	-1	-1	220	H*
9.4.5. EDL Modeling and Simulation	3	9	3	1	9	1	-1	224	H*
9.4.6. (EDL) Instrumentation and Health Monitoring	3	9	3	0	9	1	-1	222	H*
9.4.7. GN&C Sensors and Systems (EDL)	9	9	9	3	9	1	-1	402	H

FIGURE L.1 Quality function deployment (QFD) summary matrix for TA09 Entry, Descent, and Landing Systems. The justification for the high-priority designation of all the high-priority technologies appears in the section "High-Priority Level 3 Technologies." H = High Priority; H* = High Priority, QFD score override; M = Medium Priority; L = Low Priority.

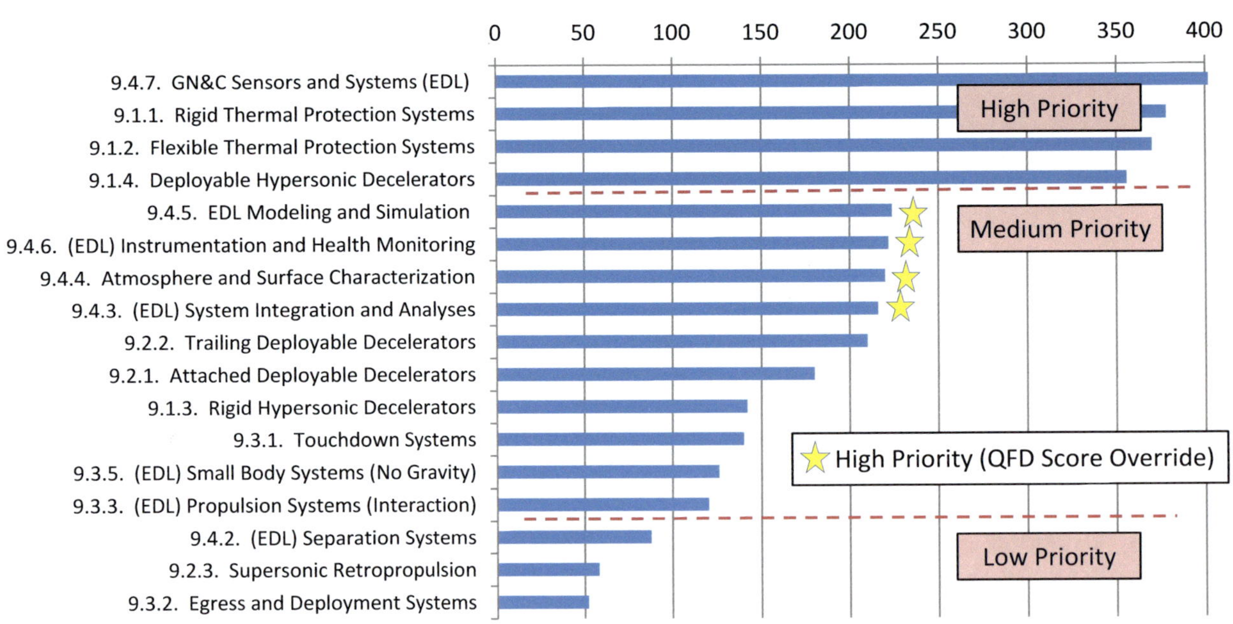

FIGURE L.2 Quality function deployment rankings for TA09 Entry, Descent, and Landing Systems.

CHALLENGES VERSUS TECHNOLOGIES

Figure L.3 shows the relationship between the level 3 technologies in TA09 and the top technical challenges. The high-priority technologies have a strong relationship to many of the technical challenges. The top-rated level 3 technology, 9.4.7 GN&C Sensors and Systems, would likely have a major impact on four of the six technical challenges and a moderate impact on the remaining two technical challenges. The next two highest ranked technologies (9.1.1 Rigid Thermal Protection Systems and 9.1.2 Flexible Thermal Protection Systems) would likely have an impact on five of the six technical challenges. For both of these technologies, improvements in reusable TPS could specifically help meet the affordability challenge, especially for human return from low Earth orbit. The fourth ranked technology, 9.1.4 Deployable Hypersonic Decelerators, contributes to four of the six technical challenges. The four additional high-priority technologies also play an important role in meeting the technical challenges. In fact, three of them could contribute toward meeting all six technical challenges. These technologies can in particular help to improve knowledge of the EDL system, thereby reducing required margins.

GENERIC REFERENCE MISSIONS FOR TA09

Development of EDL technologies with a broad focus will help prevent potentially important missions from being eliminated for consideration because they are perceived as unachievable due to their EDL requirements. Additionally, technology that can enable only a few missions may have less payoff than a building block approach to technology development that supports a series of progressively more challenging missions. Furthermore, because of resource constraints, EDL technology investments must be time-phased. EDL technologies that enable the broadest spectrum of future missions by accommodating the widest range of variations in destination and timing will tend to be the most highly valued. This is reflected in the broad set of technology challenges above and in the discussion of generic reference missions (GRMs) below and in Figure L.4, which is provided as a replacement to Table 1 in NASA's draft roadmap; Figure L.4 is more comprehensive.

The GRMs in Figure L.4 capture the broad spectrum of EDL missions that have flown or could be flown in the foreseeable future. For example, there is not a human mission to Pluto GRM because this type of mission is too far out on the horizon to be reasonably achieved. The 35 GRMs in Figure L.4 are distinguished from each other in terms of destination, local environmental discriminators (presence of atmosphere, gravity, and extreme environments), and mission-defined discriminators (entry or landing, hard or soft landing, and class of lander, such as human, cargo, or robotic).

Given the GRM definitions in Figure L.4, Figure L.5 shows the linkage between the GRMs and the EDL technologies evaluated by the panel. Figure L.5 confirms that the high-priority technologies (and many of the medium-priority technologies) are applicable to a wide range of GRMs.

HIGH-PRIORITY LEVEL 3 TECHNOLOGIES

Panel 6 identified eight high-priority technologies in TA09. The justification for ranking each of these technologies as a high priority is discussed below.

EDL technologies in general do not benefit from access to the International Space Station (ISS). However, advanced EDL technology could lead to new vehicles with improved capabilities for returning crew or payloads from the ISS, and return flights from the ISS could provide flight testing opportunities for EDL technologies.

Technology 9.4.7, Guidance, Navigation, and Control (GN&C) Sensors and Systems

A primary objective of many EDL missions is to safely land a vehicle in new destinations such that human or robotic exploration can be achieved. The ability to accurately hit entry corridors, to control the vehicle during entry and descent, to navigate the vehicle during all phases of EDL, and to safely and precisely land a vehicle in hazardous terrain are examples of a high-performing EDL GN&C system.

Priority	TA 09 Technologies, Listed by Priority	Top Technology Challenges: 1. Mass to Surface: Develop the ability to deliver more payload to the destination.	2. Surface Access: Increase the ability to land at a variety of planetary locales and at a variety of times.	3. Precision Landing: Increase the ability to land space vehicles more precisely.	4. Surface Hazard Detection and Avoidance: Increase the robustness of landing systems to surface hazards.	5. Safety and Mission Assurance: Increase the safety, robustness, and reliability of EDL.	6. Affordability: Improve the affordability of EDL systems.
H	9.4.7. GN&C Sensors and Systems (EDL)	○	●	●	●	●	○
H	9.1.1. Rigid Thermal Protection Systems	●	●	○		○	●
H	9.1.2. Flexible Thermal Protection Systems	●	●	○		○	●
H	9.1.4. Deployable Hypersonic Decelerators	●	●	○			○
H	9.4.5. EDL Modeling and Simulation	○	○	●	○	●	●
H	9.4.6. (EDL) Instrumentation and Health Monitoring	○	○	○		●	○
H	9.4.4. Atmosphere and Surface Characterization	○	●	●	●	○	○
H	9.4.3. (EDL) System Integration and Analyses	○	○	○	○	●	●
M	9.2.2. Trailing Deployable Decelerators	●	●	○	○		○
M	9.2.1. Attached Deployable Decelerators	●	●	○			○
M	9.1.3. Rigid Hypersonic Decelerators	○	○	○			○
M	9.3.1. Touchdown Systems	○			○	○	○
M	9.3.5. (EDL) Small Body Systems				○		
M	9.3.3. (EDL) Propulsion Systems				○	○	
L	9.4.2. (EDL) Separation Systems						
L	9.2.3. Supersonic Retropropulsion	○					
L	9.3.2. Egress and Deployment Systems						

●	Strong Linkage: Investments by NASA in this technology would likely have a major impact in addressing this challenge.
○	Moderate Linkage: Investments by NASA in this technology would likely have a moderate impact in addressing this challenge.
[blank]	Weak/No Linkage: Investments by NASA in this technology would likely have little or no impact in addressing the challenge.

FIGURE L.3 Level of support that the technologies provide to the top technical challenges for TA09 Entry, Descent, and Landing Systems.

	EDL MISSION LIST	MISSION CHARACTERISTIC							EXAMPLE MISSIONS
GRM ID#	Identifier	Destination	Human, Cargo, or Robotic?	Entry or Landing?	Hard or Soft Landing?	Atmosphere?	Gravity?	Extreme Environment?	
1	Human Earth Low L/D Return	Earth	H	L	S	Y	Y	N	Apollo, Orion, Commercial Crew
2	Human Earth High L/D Return			L	S	Y	Y	N	Shuttle, Xcor
3	Human Earth Retro Return			L	S	Y	Y	N	SpaceX Dragon
4	Earth Cargo Low L/D Return		C	L	S	Y	Y	N	SpaceX Dragon
5	Earth Cargo High L/D Return			L	S	Y	Y	N	Dreamchaser
6	Earth Cargo Retro Return			L	S	Y	Y	N	
7	Robotic Earth Capsule Return		R	L	S	Y	Y	N	Hayabusa, Stardust, Genesis
8	Robotic Earth High L/D Return			L	S	Y	Y	N	X-37
9	Robotic Earth Retro Return			L	S	Y	Y	N	Masten, Armadillo
10	Human Mars	Mars	H	L	S	Y	Y	N	
11	Cargo Mars		C	L	S	Y	Y	N	
12	Robotic Mars		R	L	S	Y	Y	N	Viking, Phoenix, Pathfinder, MER
13	Penetrator Mars			L	H	Y	Y	N	DS2
14	Robotic Entry Mars			E	-	Y	Y	N	Mars Airplane, MGS, Odyssey
15	Human Lunar	Lunar	H	L	S	N	Y	N	Apollo, Altair
16	Cargo Lunar		C	L	S	N	Y	N	Altair
17	Robotic Lunar		R	L	S	N	Y	N	Lunar Sample Return
18	Penetrator Lunar			L	H	N	Y	N	Lunar-A
19	Human Asteroid / Small Body	Asteroid / Small Bodies	H	L	S	N	N	N	
20	Robotic Asteroid / Small Body		R	L	S	N	N	N	NEAR
21	Penetrator Asteroid / Small Body			L	H	N	N	N	Hayabusa
22	Robotic Comet Sample Return	Comet	R	L	S	N	N	N	Stardust
23	Robotic Comet Lander			L	S	N	N	N	
24	Penetrator Comet			L	H	N	N	N	Deep Impact
25	Robotic Venus/Titan Entry	Venus / Titan	R	E	-	Y	Y	Y	Huygens, Titan Balloon
26	Robotic Venus/Titan Lander			L	H	Y	Y	Y	Pioneer Venus, Venera 7
27	Robotic Venus/Titan Penetrator			L	H	Y	Y	Y	
28	Robotic Icy Moon Lander	Icy Moon	R	L	S	N	Y	Y	
29	Penetrator Icy Moon			L	H	N	Y	Y	
30	Robotic Mercury Lander	Mercury	R	L	S	N	Y	Y	BepiColombo
31	Mercury Penetrator			L	H	N	Y	Y	
32	Robotic Saturn / Jupiter Entry	Giant Planet	R	E	-	Y	Y	Y	Galileo
33	Robotic Uranus/Neptune Entry	Outer Planet	R	E	-	Y	Y	Y	
34	Robotic Uranus/Neptune Lander			L	S	Y	Y	Y	
35	Penetrator Uranus/Neptune			L	H	Y	Y	Y	

FIGURE L.4 Generic reference missions for TA09 Entry, Descent, and Landing Systems.

	EDL MISSION LIST	EDL TECHNOLOGY MAPPING																
		High Priority								Medium						Low		
GRM ID#	Identifier	GNC Systems and Sensors	Deployable Hypersonic Decelators	Rigid Thermal Protection Systems	Flexible Thermal Protection Systems	EDL Modeling and Simulation	Atmosphere and Surface Characterization	Instrumentation and Health Monitoring	System Integration and Analysis	Trailing Deployable Decelerators	Attached Deployable Decelerators	Rigid Hypersonic Decelerators	Touchdown Systems	Small Body Systems	Propulsion Systems	Separation Systems	Supersonic Retropropulsion	Egress & Deployment Systems
1	Human Earth Low L/D Return	Y	Y	Y	Y	Y	Y	Y	Y	Y	Y	Y	Y	-	-	Y	-	-
2	Human Earth High L/D Return	Y	Y	Y	Y	Y	Y	Y	Y	Y	Y	Y	Y	-	-	Y	-	-
3	Human Earth Retro Return	Y	Y	Y	Y	Y	Y	Y	Y	Y	Y	Y	Y	-	Y	Y	Y	-
4	Earth Cargo Low L/D Return	Y	Y	Y	Y	Y	Y	Y	Y	Y	Y	Y	Y	-	-	Y	-	-
5	Earth Cargo High L/D Return	Y	Y	Y	Y	Y	Y	Y	Y	Y	Y	Y	Y	-	-	Y	-	-
6	Earth Cargo Retro Return	Y	Y	Y	Y	Y	Y	Y	Y	Y	Y	Y	Y	-	Y	Y	Y	-
7	Robotic Earth Capsule Return	Y	Y	Y	Y	Y	Y	Y	Y	Y	Y	Y	Y	-	-	Y	-	-
8	Robotic Earth High L/D Return	Y	Y	Y	Y	Y	Y	Y	Y	Y	Y	Y	Y	-	-	Y	-	-
9	Robotic Earth Retro Return	Y	Y	Y	Y	Y	Y	Y	Y	Y	Y	Y	Y	-	Y	Y	Y	-
10	Human Mars	Y	Y	Y	Y	Y	Y	Y	Y	Y	Y	Y	Y	-	Y	Y	Y	Y
11	Cargo Mars	Y	Y	Y	Y	Y	Y	Y	Y	Y	Y	Y	Y	-	Y	Y	Y	Y
12	Robotic Mars	Y	Y	Y	Y	Y	Y	Y	Y	Y	Y	Y	Y	-	Y	Y	-	Y
13	Penetrator Mars	Y	Y	Y	Y	Y	Y	Y	Y	Y	Y	Y	-	-	-	Y	-	-
14	Robotic Entry Mars	Y	Y	Y	Y	Y	Y	Y	Y	Y	Y	Y	-	-	-	Y	-	-
15	Human Lunar	Y	-	-	-	Y	Y	Y	Y	-	-	-	Y	-	Y	Y	-	Y
16	Cargo Lunar	Y	-	-	-	Y	Y	Y	Y	-	-	-	Y	-	Y	Y	-	Y
17	Robotic Lunar	Y	-	-	-	Y	Y	Y	Y	-	-	-	Y	-	Y	Y	-	Y
18	Penetrator Lunar	Y	-	-	-	Y	Y	Y	Y	-	-	-	Y	-	-	Y	-	-
19	Human Asteroid / Small Body	Y	-	-	-	Y	Y	Y	Y	-	-	-	Y	Y	Y	Y	-	Y
20	Robotic Asteroid / Small Body	Y	-	-	-	Y	Y	Y	Y	-	-	-	Y	Y	Y	Y	-	Y
21	Penetrator Asteroid / Small Body	Y	-	-	-	Y	Y	Y	Y	-	-	-	-	Y	-	Y	-	-
22	Robotic Comet Sample Return	Y	-	-	-	Y	Y	Y	Y	-	-	-	Y	Y	Y	Y	-	Y
23	Robotic Comet Lander	Y	-	-	-	Y	Y	Y	Y	-	-	-	Y	Y	Y	Y	-	Y
24	Penetrator Comet	Y	-	-	-	Y	Y	Y	Y	-	-	-	-	Y	-	Y	-	-
25	Robotic Venus/Titan Entry	Y	Y	Y	Y	Y	Y	Y	Y	Y	Y	Y	-	-	-	Y	-	-
26	Robotic Venus/Titan Lander	Y	Y	Y	Y	Y	Y	Y	Y	Y	Y	Y	Y	-	Y	Y	-	Y
27	Robotic Venus/Titan Penetrator	Y	Y	Y	Y	Y	Y	Y	Y	Y	Y	Y	-	-	-	Y	-	-
28	Robotic Icy Moon Lander	Y	-	-	-	Y	Y	Y	Y	-	-	-	Y	-	Y	Y	-	Y
29	Penetrator Icy Moon	Y	-	-	-	Y	Y	Y	Y	-	-	-	-	-	-	Y	-	-
30	Robotic Mercury Lander	Y	-	-	-	Y	Y	Y	Y	-	-	-	Y	-	Y	Y	-	Y
31	Mercury Penetrator	Y	-	-	-	Y	Y	Y	Y	-	-	-	-	-	-	Y	-	-
32	Robotic Saturn / Jupiter Entry	Y	Y	Y	Y	Y	Y	Y	Y	Y	Y	Y	-	-	-	Y	-	-
33	Robotic Uranus/Neptune Entry	Y	Y	Y	Y	Y	Y	Y	Y	Y	Y	Y	Y	-	Y	Y	Y	Y
34	Robotic Uranus/Neptune Lander	Y	Y	Y	Y	Y	Y	Y	Y	Y	Y	Y	Y	-	Y	Y	Y	Y
35	Penetrator Uranus/Neptune	Y	Y	Y	Y	Y	Y	Y	Y	Y	Y	Y	-	-	-	Y	Y	-

FIGURE L.5 Mapping of generic reference missions to level 3 technologies for TA09 Entry, Descent, and Landing Systems.

Fundamentally, an EDL mission is supported by a design architecture to achieve its goals. The ability of the GN&C system to achieve its mission objectives is a function of GN&C sensor performance, vehicle actuator ability, and the designer's ability to craft them sensibly together onboard a capable, real time, computing platform.

The technology readiness level (TRL) widely varied among GN&C sensors and systems. Tried and true GN&C sensors such as inertial measurement units (IMUs) and star cameras typically are at a high TRL, but they could benefit from reduced size, weight, and power, while also increasing performance and noise immunity. EDL capable velocimeters and altimeters generally have a lower to mid-level TRL and would benefit from increased accuracies, range, and update rates while also decreasing size, weight, and power. Finally, GN&C sensor and system advancement in the following currently low-TRL items would significantly improve or even enable future EDL missions.

- Terrain relative navigation systems, sensors, and algorithms
- Precision landing systems, sensors, and algorithm design
- Hazard relative navigation systems, sensors, and algorithm design
- Hazard detection sensors and systems
- Adaptive control systems
- Autonomous GN&C sequencing and mission managers
- Inertial swarm sensing methods and instrumentation
- Enhanced fault tolerance

As shown in Figure L.5, GN&C Sensors and Systems are common to *all* of the foreseen EDL generic reference missions. They align extremely well with NASA's expertise, capabilities, and facilities. Given their broad applicability, other non-NASA agencies (European Space Agency, Japanese Aerospace Exploration Agency) and military organizations (National Reconnaissance Office, Missile Defense Agency, U.S. Navy, U.S. Air Force, etc.) will and have improved the state of the art of some GN&C sensors, but additional work is needed to improve operational systems. NASA could lead by example and invest in an aggressive, planned, and sustained NASA technology development effort to advance GN&C sensors and systems given such broad applicability to a multitude of their missions. It will take a sustained and coordinated effort, possibly shared among organizations, to raise the TRL of necessary GN&C sensors and systems, particularly across the "valley of death" levels of TRL 5-6 where validation in the relevant environment is required.

This technology is game-changing because it significantly enhances the ability to increase mass to the surface, the ability to land anywhere, and the ability to land at any time. This technology was evaluated as a high-priority technology since significant technology advancement is likely to provide transformational capabilities that would enable important new missions that are not currently feasible during the next 20 years, and because it is broadly applicable across the entire aerospace community and in multiple NASA mission areas. Furthermore, the technical risk associated with development of this technology is moderate to high, which is a good fit to NASA's level of risk tolerance for technology development, and the likely cost to NASA and the timeframe to complete technology development is not expected to substantially exceed that of past efforts to develop comparable technologies.

Technology 9.1.1, Rigid Thermal Protection Systems

Thermal protection systems (TPS) are used to protect the payload of the entry vehicle (both human and robotic) from the high-temperature and high-shear flow environment experienced during the hypersonic entry phase. Rigid TPS materials are typically separated into two major classes, reusable and non-reusable, and some missions will use a combination of the two. An example of reusable rigid TPS is the thermal protection tiles used on the space shuttle. Though reusable, these materials do have a finite lifetime due to the thermal and mechanical environments experienced during re-entry. For high-energy entries, like an Earth return beyond low Earth orbit (LEO) or entry into the atmosphere of another planet or moon, non-reusable or ablative thermal protections systems have been historically used since the currently available reusable systems cannot handle such high heat loads. Several materials (e.g., AVCOAT, PICA, and SLA-561V) have been flight qualified and are at TRL 7 to 9. However, the process for making AVCOAT had to be redeveloped for the Orion vehicle and its TRL is likely less (i.e., 5 to 7). Materials such as carbon phenolic, which are also widely used for military applications, are also in the TRL 7 to 9 range. However, carbon phenolic in particular presents availability issues because of a lack of U.S. suppliers of rayon.

Most NASA flight experience has been with rigid thermal protection systems, where the TPS is installed onto a rigid aeroshell structure. These systems can handle both high velocities and high heat fluxes. Rigid TPS, however, can account for a large percentage of the entry vehicle mass. Vehicle designs use conservative estimates of heating to account for uncertainties, which contributes to the large mass of a TPS system. Recent research has been focused on the development of lower density ablators. For higher speed entries into the outer planets or their moons that have atmospheres, new materials will need to be developed that can also handle extreme environments that include both high convective and radiative components.

While it is clear that flight quality heritage materials exist (or need to be re-constituted) for return-from-LEO applications, next-generation EDL systems for return from LEO would benefit from additional research and development to improve reliability and maintainability and to reduce cost. However, the draft roadmap's description of rigid TPS focuses on ablative, single-use TPS with application to planetary entry. As noted during the TA09 EDL workshop, the value of this technology would be enhanced if its scope were increased to include return-from-LEO applications (Grantz, 2011; Picetti, 2011). For new ablative and reusable materials, the TRL is between 1 and 3.

Commercial applications are primarily focused on lower-energy LEO return, while rigid TPS for high-energy entries primarily have application to NASA or military missions. Therefore, there should be opportunities for NASA to partner with other organizations in the development of this technology. Also, since many other nations are entering the civilian space arena, there are also opportunities to partner with them. An example is the ESA EXPERT (European Experimental Reentry Testbed) mission (Thoemel et al., 2009; Muylaert, 2011). Additionally, there could be spin-off opportunities for other high-temperature applications requiring thermal protection (i.e., rocket motor nozzles or nuclear reactors). Because the thermal protection technology is unique to these applications and requires both small-scale ground testing and large-scale flight testing, it is unlikely that industry would take the lead in either developing or qualifying new materials or systems. Therefore, NASA (and perhaps the Department of Defense) is likely the best option for maintaining facilities such as arc-jets and investing in state-of-the-art computational tools (and the human capital to develop and apply these tools).

This technology is game-changing because advances in this area would enable new missions in extreme thermal environment or reduced mass to increase vehicle payload and performance, far beyond what has been previously achieved. Because of the applicability of this technology to the military, it can also have high impact on non-NASA aerospace needs. Moderate to high levels of risk are involved to further this technology.

Technology 9.1.2, Flexible Thermal Protection Systems

Like rigid TPS, flexible TPS can be reusable or ablative (or some combination thereof). Because of their flexible nature, these TPS systems could be packaged into tighter volumes, applied to irregular surfaces, and deployed when necessary. In addition to thermal protection, these systems can also be expected to carry significant aerodynamics loads (primarily for deceleration). Because of their flexibility, it might be possible to tailor the shape of the TPS to improve both the aerodynamic performance during the hypersonic entry phase (to provide lifting and cross range capability). It may also be possible to use these flexible materials to control local boundary layer state (i.e., laminar versus turbulent) and heating loads. While flexible TPS has been used on the leeside of the space shuttle (i.e., advanced flexible reusable surface insulation blankets), they have not been demonstrated for high-energy entries or where significant aero-thermal-structural interactions might occur. The TRL for advanced flexible TPS materials is 1 to 2.

Like their rigid counterparts, flexible TPS will have use in the commercial, civilian, and military space market. The ability to morph the shape of the TPS could have application in long-range strike applications where precise aerodynamic control is essential. Similar to the rigid TPS, it is expected that NASA will be the prime agency involved in maturing this technology with industry eventually being able to commercialize the capabilities developed. Advanced flexible TPS will require significant research as the primary TPS for atmospheric entry. Significant challenges exist with most any deployable decelerator concept, including the associated thermal protection in terms of handling, manufacturing, packing, and deployment following interplanetary transit or LEO storage. Like rigid TPS, furthering flexible TPS technology will require significant advances in computational capabilities (for example, to model fluid-structure interactions in high heating environments). Additionally, being able to do both quantitative and reliable ground and flight testing of these systems will post significant challenges.

This technology is game-changing because advances in flexible TPS may be able to reduce TPS size and weight below that which can be achieved with rigid TPS. The military aerospace community could also benefit from this technology. Due to the lack of experience in this regime, the risk is moderate-high, but likely higher than that of the rigid TPS, particularly in the category of reusable flexible TPS in high heat flux/high dynamic pressure regions, a subject that was not addressed in the discussion of this technology in the draft NASA roadmap for TA09. Further developments in both materials and manufacturing of these systems are required before flexible TPS can be considered as a candidate technology for future use. This will increase the time necessary to prove the technology over its rigid counterparts.

Technology 9.1.4, Deployable Hypersonic Decelerators

Current entry systems employ traditional rigid decelerator architectures to provide thermal protection and deceleration. The shape and size of rigid devices define their aerodynamic performance. Launch vehicle dimensions define the maximum size and hence the maximum drag area in traditional decelerators. Deployable decelerators, which could use flexible or rigid components, would provide the ability to utilize much larger drag areas and novel vehicle shapes relative to traditional rigid decelerators, because they are not limited to the size of the spacecraft.

Deployable decelerators could enhance the drag area of the spacecraft during the early phase of EDL, thereby reducing the altitude required and increasing the time available to establish the landing configuration. These technologies could be used for the safe landing of larger objects from suborbital terrestrial trajectories, as well as enabling heavier payloads to successfully arrive at Mars, where low-density atmospheres are particularly challenging. Deployable decelerators can also be deployed in space vacuum to enhance aerocapture development which has significant mass fraction benefits for science missions.

Rigid decelerators can be deployed with servos, thrusters, stored gas, and/or other mechanical devices that are relatively simple in design, construction, and implementation, but which come with a relatively high price in terms of mass budget. Mechanical deployment systems that do not rely on inflatables do however require the integration of flexible TPS and the associated complexities.

Inflatable decelerators are very appealing in terms of mass fractions but require some form of internal pressurization to inflate them. Inflation systems may use stored gas, pyrotechnic gas generators, or a hybrid approach. Ballutes use ram-air (atmospheric) inflation gas to establish their shape. As a result, they do not require auxiliary gassing systems or their associated mass, but neither can they be deployed in the vacuum of space. This feature complicates the use of ballutes as aerocapture or aerobraking devices.

Many of the technologies employed as inflation systems have relatively high TRLs, but the TRL of inflations systems in combination with a decelerator structure is much lower. Overall, deployable hypersonic decelerator technology is in the range of TRL 2 to 4.

This technology is game-changing because it provides the ability to utilize much larger drag areas and novel vehicle shapes relative to traditional rigid decelerators. These features can enhance thermal protection and deceleration during entry and thus provide one approach for enabling new missions (e.g., landing large masses on Mars). The technology is well aligned with NASA's needs and can be employed over a broad range of exploration missions. Commercial launch service providers may also take advantage of these capabilities. Significant cost reductions can be realized by enabling booster and cargo delivery vehicle recovery for re-use. However, it is unlikely that commercial providers will make the required investment to advance this technology past TRL 6 and therefore NASA should lead development. It is certainly a difficult undertaking, but well within the grasp of the agency.

Technology 9.4.5, EDL Modeling and Simulation

EDL Modeling and Simulation (M&S) technology would provide the ability to conduct computational predictions for robust and efficient design in all phases of EDL missions. This technology includes computational fluid dynamics analysis, finite element modeling, fluid-structural interaction analysis, aerothermodynamics modeling (including ablative surface and thermal radiation physics, coupled stability, and trajectory analysis), multi-disciplinary analysis tools, and other high-fidelity analysis required for EDL missions.

Because of the limited range of test conditions available in experimental tests, and because of the high cost of such testing, M&S tools are highly valued in every phase of design and analysis of EDL systems. In addition to development of physical models, numerical methodologies, and software tools to conduct M&S, this technology also includes development and application of experimental validation including flight tests. Only if high-fidelity models are also well-validated can they be useful in reducing margins, thereby increasing mission capability without a loss in safety. Legacy flight data is often not sufficient to validate the codes, particularly because uncertainties in the measurements were often not well understood and the boundary conditions required for M&S were not well characterized. The entire process is further complicated by the wide range of operational conditions experienced by EDL systems.

NASA conducts primarily mission-related modeling and simulation work in the Aeronautics Research Mission Directorate and in ESMD. This technology would build on the well-known foundational research carried out by NASA's Aeronautics Research Mission Directorate to advance the state of the art in critical areas for missions planned over the next 20 years.

Current M&S technology is at TRL 4 to 7; in that it can already conduct accurate, steady aerodynamic analysis of rigid configurations. EDL missions, however, require accurate analysis in areas for which current predictive capability is insufficient, and is generally at TRL 3. For example, because the ability to reliably predict TPS recession rate at every location in a radiative environment does not exist, heat shields tend to be over-designed, thereby increasing weight and reducing mission capability.

This technology is well aligned with NASA's expertise, capabilities, and facilities. NASA has taken a lead role in the development of thermo-chemical nonequilibrium modeling for aerospace applications but would benefit by working closely with Department of Defense and Department of Energy laboratories and industry partners who conduct research in this technology for weapon systems in continuum and rarefied high-temperature flow environments. NASA has significant in-house expertise in physical modeling and software development to conduct and help guide research conducted by academic, federal, and industry partners in this technology. NASA's investment in high-performance-computing facilities also makes it well suited to conduct this work, and it currently possesses unique ground and flight test capabilities to conduct experimental validation required for EDL Modeling and Simulation. Continued investments in ground test facilities, such as large scale wind tunnels, arc-jet facilities, and supersonic and hypersonic wind tunnels, will ensure that the means to validate codes are available when required. NASA is uniquely motivated to pursue this technology and major investments from industry are not expected in the absence of NASA involvement.

The panel overrode the QFD score for this technology to designate it as a high-priority technology because the QFD scores did not fully capture the value of this technology in terms of how widely applicable it is to EDL missions and to the successful development and implementation of other high-priority technologies in TA09, particularly the TPS and decelerator technologies. Furthermore, EDL modeling and simulation supports all six of the top technology challenges. It also is characterized by the appropriate level of risk and difficulty in physical modeling, numerical technique development and experimental validation, yet it builds upon a long-standing core competency which NASA possesses in these subjects. The development plan is clear and there is a likelihood of joint funding by federal agencies and the industrial space industry.

Technology 9.4.6, Instrumentation and Health Monitoring

NASA draft roadmap for TA09 notes that, "Entry instrumentation for both engineering data and vehicle health monitoring provides a critical link between predicted and observed performance of entry vehicle systems." This is particularly true for entry thermal protection systems because complete simulation of the entry environment is impossible in ground-based test facilities. Hence, while ground-based test facilities are indispensable in developing thermal protection systems, the complete, rigorous validation of TPS design algorithms can only be achieved through comparison of predictions with flight data. Also, health monitoring instrumentation can provide system performance data as well as evidence that vehicle systems are operating properly prior to entry. The value of investments in this technology would be greatly enhanced if NASA heeded the call in the draft roadmap to "develop a NASA policy for required EDL instrumentation and data acquisition in order to advance and build confidence in models that are essential to EDL system qualification."

As pointed out in the draft roadmap, major technical challenges in entry instrumentation include high-temperature systems capable of direct heat flux measurements in situ, measurements (temperature and strain) in flexible TPS, advanced optical and other non-intrusive measurement techniques, and shock layer radiation measurements in ablative TPS. Challenges in health monitoring include development of low-data, low-power networks, elimination of false positives, and the ability to initiate and monitor repair of detected damage.

The TRL of this technology could be said to be 9 since such instrumentation has been successfully flown on previous missions (e.g., Apollo, space shuttle). However, current instrumentation systems are heavy and, in some cases, unacceptably intrusive. What are needed are new lighter, smaller, less intrusive, and more accurate approaches. Such systems are currently at TRL 2 or 3.

The panel overrode the QFD score for this technology to designate it as a high-priority technology. The QFD scores did not fully capture the value of this technology in terms of how widely applicable it is to all EDL missions, as well as the contribution it would make to improving the safety and reliability of EDL missions. This technology is well aligned with NASA's expertise and capabilities and requires NASA involvement for its successful development. The data obtained and the resulting improved heat shield design algorithms will be of great interest to NASA, DOD, DOE, university researchers, and the commercial space transportation community.

The scope of this technology overlaps the scope of some of the technologies in the autonomy subarea in TA04 Robotics, TeleRobotics, and Autonomous Systems.

Technology 9.4.4, Atmosphere and Surface Characterization

The goal of this technology is to provide a description of the atmosphere and surface of a planet in sufficient detail to facilitate the planning and execution of planetary missions. In the case of planetary atmospheres, a predictive model is required that will define the spatial and temporal atmospheric characteristics on global, zonal, and local scales, including annual, seasonal, and daily variations. Such models exist for the Moon, Mars, and Venus, but they do not provide the needed level of detail. The current state of the art is represented by the Mars Global Reference Atmospheric Model (Justus et al., 2005; Justh et al., 2011). This model allows predictions of the needed atmospheric profiles and dynamics but its accuracy is limited by an inadequate data base. For other planets, the models that exist provide only gross descriptions with very little detail. Surface models are extremely important for rover missions and are critical for missions involving human landings where the grain size and abrasive character of the regolith can cause serious damage to bearings, drive motors, and space suits. Atmosphere models are of critical importance for entry missions that involve aeromaneuvering for increased landing accuracy and aerocapture to increase landed mass. The inaccuracies of the present atmospheric models result in conservative designs and large mass margins that degrade performance and have contributed to the lack of acceptance of these techniques by project managers.

Future investments in this technology could include distributed weather measurements (short and long duration) on Mars; the development of a standard, low-impact measurement package for all Mars landed missions to provide future Mars landers with surface pressure and upward-looking wind measurements; the development of orbiter instruments for wind and atmospheric property characterization at altitudes relevant to aerobraking, aerocapture, and aeromaneuvering; and the development of higher fidelity atmospheric models based on these data. Another major contribution would be the development of automated data fusion for visible stereo imagery, multispectral imagery, and altimetry and its conversion to onboard topography and albedo surface maps suitable for use by a terrain tracker. Other high-priority work areas include the development of "scout" probes that would measure atmospheric properties ahead of an entry vehicle and the development of vehicle-based sensors for making real-time measurements of local and far-field atmospheric properties.

The TRLs of the current models range from 6 (for the surface characterization of the Moon and the atmosphere of Mars) to 2 or 3 for the more distant planets. The associated programs, described above, are currently at relatively low TRLs from 2 to 4. The panel overrode the QFD score for this technology to designate it as a high-priority technology because the QFD scores did not capture the value of this technology in terms of its importance in achieving all of the EDL top technical challenges, and having a strong linkage to three of the top six challenges.

There is a critical need for this technology in designing and carrying out most future planetary missions. This technology is well aligned with NASA's expertise and capabilities and requires NASA involvement for its successful development. These models will be of interest to the scientific community and basic science investigations can provide important inputs in their development. Both basic science investigations and the development of predictive engineering models should be carried out in parallel to maximize the benefits.

Technology 9.4.3, System Integration and Analyses

EDL systems are a highly coupled and interdependent set of capabilities consisting of software and hardware components as well as multiple disciplines. The nature of this problem lends itself to technologies that develop improved methods of performing systems integration and analysis such as multidisciplinary design optimization. Optimizing an EDL system involves various disciplines (e.g., thermal, fluid dynamics, and trajectories), multiple flight phases (entry, descent, and landing), overall system reliability and cost, and a host of tools that address all of the above. This technology is closely coupled with 9.4.5. Modeling and Simulation.

System performance is enhanced, life cycle costs are reduced, and development times are shortened when the interplay between system requirements, system concepts, and the potential benefit of new technologies is explored as early as possible in the development of new systems, especially when overly restrictive or arbitrary requirements are filtered out (Mavris and DeLaurentis, 2000).

Existing systems integration and analyses technologies have not been widely applied to EDL mission design, certainly not to the degree that will likely be necessary to design an EDL system for a human mission to Mars. The whole EDL technology portfolio would benefit from the expanded use of these technologies across a larger mission set since they can help to understand the benefits that other technologies can bring to a given mission or a whole set of missions once the systems integration and analysis techniques are validated. Like 9.4.5. Modeling and Simulation, systems integration and analysis would benefit immensely from flight engineering data to validate and improve systems integration and analysis tools. However, as the fidelity of systems integration and analysis techniques improves, the need for further (expensive) testing can be reduced.

This understanding of and the need for this technology largely resides within NASA and, to some extent, in DOD, but its nature lends itself to projects that universities could perform.

While systems integration and analyses is not expected to be game-changing technology, the panel overrode the QFD score for this technology to designate it as a high-priority technology because it supports the complete mission set and all six of the EDL top technology challenges.

MEDIUM- AND LOW-PRIORITY TECHNOLOGIES

One group of medium- and low-priority technologies in TA09 had a lower overall benefit than the high-priority technologies as well as development challenges. Technologies 9.2.2 Trailing Deployable Decelerators, 9.2.1 Attached Deployable Decelerators, 9.1.3 Rigid Hypersonic Decelerators, and 9.2.3 Supersonic Retropropulsion all fell into this category. Trailing and Attached Deployable Decelerators were the highest ranked of the medium- and low-priority technologies. Both of these technologies were judged to have limited benefit, however, particularly because they apply to the descent phase and were not deemed to be game-changing. Rigid Hypersonic Decelerators were the next highest ranked technology. While additional improvements can be made in this technology, benefits are generally limited by the maximum payload size of current launch vehicles. Supersonic Retropropulsion was also judged to have limited benefit, and it was technically the most challenging. While often described as a game-changing technology, the panel deemed that its applicability was limited mainly to landing large payloads on the surface of Mars, but in that application disadvantages associated with transporting extra propellant on a mission to Mars would likely outweigh any other possible mission improvements.

Another group of medium- and low-priority technologies are less technically critical and typically include engineering developments that normally are addressed in the development of individual missions. Technologies 9.3.1 Touchdown Systems, 9.3.3 Propulsion Systems, 9.4.2 Separation Systems and 9.3.2 Egress and Deployment Systems all fall into this group.

Technology 9.3.5 Small Body Systems was judged to have a low benefit because of its very limited mission applicability. This technology may actually fit better in the roadmap for TA04, Robotics, TeleRobotics, and Autonomous Systems, because it is essentially a rendezvous and docking problem.

DEVELOPMENT AND SCHEDULE CHANGES FOR THE TECHNOLOGIES COVERED BY EACH ROADMAP

The ebb and flow of EDL missions make it difficult for the EDL community to maintain core capabilities and knowledge. EDL technology development requires continuous effort and sustained funding over a number of years in order to be successful and to generate industry participation (Peterson, 2011; Grantz, 2011; Rohrschnedier, 2011).

After the Apollo program, Viking and other planetary probes capitalized on the ablative heat shield technologies developed during Apollo. However, in more recent years, the focus has been more on the reusable TPS used on the space shuttle for return from low Earth orbit, and momentum was lost in the ablative material development and supply chain. Today, given the end of the Space Shuttle Program (for human spaceflight) and the long gap between development of the Mars Science Laboratory and the Mars 2018 mission (for robotic exploration), the process of advancing reusable TPS could itself lose momentum in the coming years. For example, key materials suppliers are terminating production and high-temperature coating developments for ceramic tiles are under-funded (Grantz, 2011).

Ideally, EDL research and technology development by NASA would build on past work to meet future requirements. Since EDL is not a high demand opportunity for industry, it is important that NASA maintain these capabilities. A successful technology program would preserve test capabilities and advance key technologies at a steady pace that does not depend solely on flight mission approvals. By ensuring knowledge capture, NASA will not have to relearn lessons from the past. Struggles with Avcoat are a good example of loss of knowledge, experience, and lessons learned.

PUBLIC WORKSHOP SUMMARY

The workshop held by the Entry, Descent, and Landing Panel for the NASA Technology Roadmaps study took place on March 23-24, 2011, at the Beckman Center in Irvine, California. The discussion was led by panel chair Todd Mosher. Mosher started the day by giving a general overview of the roadmaps and the NRC's task to evaluate them. He also provided some direction for what topics the invited speakers should cover in their presentations. This introduction was followed by an overview of the NASA draft EDL technology roadmap, presented by the roadmap authors. Each panel session began with a brief introduction by the panel moderator, followed by a presentation from each of the invited panel members. Time was then left for open discussion among all workshop participants regarding the topics addressed by the panelists. At the end of each day, there was additional time for general discussion among all of the workshop participants.

Roadmap Overview by NASA

The workshop began with a presentation on the NASA draft roadmap for TA09 by Mark Adler. Other roadmap authors were also in attendance. Adler addressed the general EDL challenges: "not burning up, slowing down in time, hitting the target, and surviving the impact." Related to the challenges, Adler presented the benefits of EDL technologies, particularly focusing on enabling and enhancing capabilities: increased mass to destination, increased planet surface access, increased delivery precision to the surface, expanded EDL timeline to accomplish critical events during entry, increased robustness of landing system to surface hazards, enhanced safety and probability of mission success, human safety during return from beyond low Earth orbit (LEO), and sample return reliability and planetary protection. In developing the roadmap, Adler commented that the authors tried to take a snapshot of future missions in order to choose the technologies and map them to those missions. Only technologies that were deemed feasible in the 2010-2030 timeframe were considered, although they were influenced by longer term missions (e.g., human Mars surface mission) due to their very long lead times. The remainder of the presentation presented an overview of each of the level 2 technology areas in the roadmap and the technologies found in each: Entry, Descent, Landing, and Vehicle Systems. Finally, one slide was presented on crosscutting EDL technologies, with the comment that there are very few applications outside of NASA for many of the technologies in the roadmap. A few examples were some TPS overlap with DOD systems, launch abort for commercial crew, and a potential emerging market for commercial applications of EDL technologies for suborbital and orbital vehicle recovery.

Adler's presentation was followed by a question and answer session. One question was asked about the affordability of testing, to which the roadmap authors answered that for many EDL technologies, testing must be done at the target (e.g., Mars), because it is very difficult to test in a relevant environment at Earth. Because of the cost of doing this, most Mars missions have built on the technology demonstrated during Viking and Mars Pathfinder—very little new technology development has occurred since. Another workshop participant asked about the status of NASA's arcjet facilities, in light of the cancellation of the Constellation program (arcjet facilities became an ongoing topic of discussion throughout the 2-day workshop). Other questions focused on specific technologies, particularly lift-to-drag for aeroshells, initiator technologies, and materials.

Panel A: Non-NASA Government Agencies

James Keeney (AFRL) began this session with a presentation that focused on how AFRL in particular (and private industry in general) could benefit from the EDL technologies proposed in the roadmap and how AFRL could play a role in the development of those technologies. He commented that most of the agreements and sharing of expertise, technologies, and assets occur at the principal investigator level through personal associations, but that it would be beneficial to formalize these associations so that they can occur more at the corporate level. In terms of EDL, AFRL would be interested in teaming up with NASA to bring some of their experiments back to the surface from LEO (space weather research was given as a prime example). Commercial entities focus primarily on LEO, and Keeney was concerned as to how NASA is going to bleed their technologies into COTS efforts to help keep costs down. A key deficiency in the roadmap, as seen by Keeney, is the lack of interdependencies between other agencies, national labs, and international partnerships. Keeney also pointed out that AFRL has a lot of facilities that NASA could potentially leverage, such as wind tunnels, and he does not see NASA looking beyond their current facilities in the draft roadmap.

Audience questions focused on specifics of the experiments (200 kg class), AFRL sensors (lidar and radiofrequency-based units), and expertise in building terrain maps (something the DOD excels at and could translate to planetary applications for NASA). Another audience member asked if there is any cross-correlation with NASA in terms of modeling activities. Keeney commented that while the roadmap specifically points out that there is a problem with testing (which he agrees with), there are no solutions proposed to fix this. Most of AFRL's testing is Earth-based, while expansion to other planets' atmospheres falls more within NASA's purview.

Carl Peterson (Sandia, retired) discussed the perceptions that have charted the course for NASA EDL thus far. Peterson asserted that the expense of the "test what you fly, fly what you test" approach has restricted NASA from testing and using new technologies instead of relying on Viking-era EDL technology and methodology (e.g., entry vehicle shape, parachutes). Furthermore, Peterson continued, engineering technology often does not get enough investment, as technologies are developed specific to each particular mission. If NASA wants future missions that expand current capabilities (to enable new destinations and sample return, for example), past EDL technologies are not good enough. Peterson commented on the possibility of teaming up with the Air Force (or DOD in general), since there are overlapping technology needs. Peterson suggested that continuity of technology development funding is critical, and bureaucratic oversight and reporting requirements should be kept to a minimum. Also, the need to re-evaluate and make changes to the R&D goals is periodically necessary. Peterson also stressed that EDL technology development is a strategic objective, which requires continuous effort and should not be revised by every new administration. A NASA participant in the workshop agreed that the constantly shifting priorities and budgets within NASA have made this a challenge in the past, and that the goal of the roadmaps is to provide that continuity over the long term. In terms of sequencing and schedule, Peterson believes that the near-term schedule in the draft roadmap is overly optimistic. Finally, Peterson believes that the predominantly test-based approach is no longer going to be affordable. Instead, NASA will need to capitalize on advances in computing and in modeling and simulation, and full-scale flight tests should be limited in number and used as a qualification tool (not a design tool).

Several key discussion points came up in the Panel A question and answer session. First was a discussion over collaboration with other agencies. Suggested collaborations included national labs for supercomputing, and Missile Defense Agency for reentry modeling. A key point was that contractual agreements should be signed to jointly maintain facilities. With regards to facilities, there was also significant discussion on the need to emphasize facilities issues in the roadmap. Second was a discussion over modeling and simulation and its relationship to testing. Several speakers commented that models must be validated with tests—you cannot believe your computational models until you have some verification. Every opportunity should be used to gather data, including instrumenting actual flights. The role of testing has changed though—instead of being used for design, it now needs to be tied into validating prediction methods. Several audience members also commented on the need for integrated system demonstrations to avoid the TRL valley of death (aerocapture is a popular example), and the roadmap should include milestones for dedicated technology demonstration missions. Third was a discussion over the focus of the roadmap—i.e., should it focus on Mars, or was it already too focused on Mars? A NASA staff member commented that there should be a core capability of investment that is not tied to a single program office, since programs come and go. However, in a constrained budget environment, it is difficult to have parallel paths and technologies must be justified by their use in missions. There was concern over the roadmap focusing too much on a human Mars landing. Instead, many believed that the roadmap should focus on more near-term goals, since technology may become obsolete by the time NASA is ready to do a human Mars mission. Finally, in terms of industry, long-term objectives and funding continuity are also needed—it is the only way to get industry to invest some of their own money into technology development.

Panel B: Industry I

Arthur Grantz, Boeing, focused his presentation on entry from LEO, which he saw as a gap/weakness in the draft roadmap. Just as TPS materials and technologies were "lost" after Apollo, he fears that the same will happen now that the space shuttle has been retired. Return from LEO will continue to be important in the coming years, and should be made a priority. He believes that reusable TPS materials, manufacturing processes, and maintainability are at a critical tipping point. He believes that the roadmap, however, focuses too much on new ablative TPS materials. He also discussed areas of overlap between high-speed Earth return (e.g., from Moon or Mars) and LEO return. Additionally, Grantz believes that instrumentation should be required on EDL missions to improve/validate computation models. Infrastructure also needs to be maintained, particularly arcjet facilities. The roadmap should also include more Earth atmospheric flight testing of LEO and higher entry velocity systems.

Al Herzl, Lockheed Martin, discussed the EDL technologies that he believes are needed: TPS materials, high-temperature insulation and structures, aeroshell systems, mechanical separation and deployment systems, parachutes, propulsion systems, landing and hazard avoidance sensors and algorithms, and landing gear. He believes that missions must be identified to pull more mature technologies. Along the same lines, he believes that every mission has the responsibility to further technology development. He also sees seeking out commercial markets as beneficial, and commented that academia and industry both want to work on technology development, so NASA should try and collaborate. Finally, be believes that test programs are the key to confidence—you need test data to build the analysis and then to validate your models.

Steve Jolly, Lockheed Martin, discussed what areas of the roadmap he agreed with and what areas need modification. Overall, he agreed with the key recommended areas of technology development with the exception of the small body technologies (should potentially go in a separate roadmap). He also believes that the draft roadmap does not focus enough on integrated system approaches, which he believes is critical for EDL systems that tend to be highly integrated and highly coupled. He also believes that the focus is too much on planetary missions, and not on Earth return from LEO, HEO, Cis-Lunar or NEO destinations, which he commented "hasn't been solved." He agreed with the recommended immediate actions listed in the roadmap, as it is important to pick out the low-hanging fruit. He listed was he believes to be the top EDL technical challenges: qualification of TPS (and the need for arcjet facilities), qualification of decelerators (parachutes), re-contact threats, hazard avoidance, horizontal velocity and touchdown, and ground surface interaction. Finally, he also listed what he sees as the game-changing technologies: decoupling terminal descent propulsion from touchdown gear, GNC sensors, steerable decelerator technology, critically damped airbags for touchdown, aerocapture, terminal descent retro propulsion, and supersonic deployment of FBC-like EDL structures to avoid re-contact.

Don Picetti, Boeing, believed that the roadmap is comprehensive in scope, and a good balance between near-term and far-term investments. He thought that having parallel paths and quantitative targets for technology development were strengths of the roadmap. Picetti was also concerned with reusable Earth entry systems, which could help improve operations by commercial providers. He also emphasized the importance of high-fidelity modeling and integrated system simulations. In particular, he discussed the need to coordinate with other NASA technology programs (e.g., Aeronautics) to acquire data, the need to upgrade and maintain ground test facilities, and the need to instrument future NASA missions. He saw the top technical challenges as follows: deployable and inflatable decelerators (game-changer), supersonic retropropulsion (game-changer), and precision landing. The high-priority areas that he identified were: flight testing/ground testing and facilities, rigid aeroshells, deployable and inflatable decelerator systems, supersonic retropropulsion, and integrated high-fidelity M&S. He believes that near-term investments should be guided by potential for high impact on future missions. For LEO return, this would include robust TPS and health monitoring. For solar system exploration, this includes rigid/deployable aeroshells and TPS, supersonic retropropulsion, adaptive GN&C, integrated system M&S, and ground facilities.

Open Discussion

The first day's discussion featured several key themes:

1. There was a lot of discussion during the first day of the workshop with regard to industry's role in the development/implementation/sharing of EDL technology development. One area of concern was how NASA would transition these technologies into industry. Some workshop participants did not see a big push for commercial applications. Others saw applications to venture tourism, and want NASA to provide a push for the commercial space industry. Furthermore, there was discussion over what the best role for NASA is, what the best role for industry is, and how the two can effectively work together. Many speakers said that for planetary exploration, NASA should be in the lead with industry participating (since it's a science endeavor and there is no current business case). However, in terms of LEO missions (particularly improved performance, lower cost, higher reliability), NASA should spin these technologies off to industry. Additionally, some comments were made that part of NASA's role should be to maintain core technology and facilities.
2. Another common theme was the issue of flight testing. A question was posed as to how game-changing technologies are going to be tested, in particular with respect to SRP. There was subsequent discussion about how much testing could be done in Earth's atmosphere or in ground facilities, or if a technology demonstration mission at Mars would be required.
3. Related to testing, there was a lot of discussion with regard to facilities, in particular arcjet facilities. There is significant concern over the potential shutdown of these facilities, in that they are needed for testing and qualifying TPS materials. There was some commentary about needing mission pull to keep these facilities up and running, and that testing for technology development and/or model validation can be used to fill the downtime between missions.

Panel C: Industry II

Neil Milburn, Armadillo Aerospace, focused on the role that Armadillo Aerospace (and other small, start-up companies) could play in the EDL technology development. Armadillo Aerospace's strength is in rapid prototyping and flight testing. They also have a strong simulation capability that they have a lot of faith in because it is backed with flight testing. Milburn discussed the test-bed capabilities and vehicles of Armadillo Aerospace. Some areas that they could do testing in are soft landing, plume mitigation and impingement, and suborbital testing of EDL technologies like ballutes and parachutes. In particular, they are working on the development of a reusable sounding rocket that could launch 10 to 20 kg to almost 500 km. Essentially, Armadillo could provide a very inexpensive test platform at the subscale level for EDL technology development.

Colin Ake, Masten Aerospace, also discussed the capabilities of Masten Aerospace and how they could contribute to EDL technology development. As with Armadillo Aerospace, he believes that small start-up companies can play a role in helping to test and demonstrate EDL technologies. Ake provided an overview of some of the Masten vehicles, which like Armadillo, also participated in the lunar lander challenge and the NASA Cruiser program. Masten's experience is mostly focused on descent and landing, and they've been used before as a vertical touch-down test platform. Ake's roadmap recommendations focused on the landing aspect of EDL, and include precision landing, validating plume impingement computational fluid dynamics, propulsion (not adequately covered in the roadmap), and integrated vehicle health monitoring. Ake emphasized the need to validate models with test data, and to do as much Earth-based testing as possible. Ake also emphasized facilities and suggested using industry resources where possible. He suggested that the roadmap should include milestone for technology demonstrations. Finally, he addressed the issue of whether there is a business case for small industry—he believes that small companies can make a business case out of being a testbed, technology developer and technology demonstrator.

Reuben Rohrschneider, Ball Aerospace, discussed the benefits of creating the technology roadmaps with regard to industry participation. If industry can see that there is a long-term plan and long-term funding, it is more likely that they will participate in the technology development. However, one criticism he had of the roadmap was that it was too focused on Mars. He believes that NASA needs to perform architecture studies (similar to what's been done for Mars) for other destinations—this will help to define what the EDL technology requirements are for other missions. Rohrschneider then discussed the technologies that he believes to be important: safe and precise landing (particularly terrain relative navigation and hazard detection), deployable aerodynamic decelerators (these have broad applicability beyond Mars, despite the roadmap focusing the details for these at Mars), and material testing and development. In terms of materials, Rohrschneider commented that while materials are being developed in other industries, they often do not test or provide data for the conditions required for EDL. Therefore, he believes test facilities for materials (beyond just rigid, ablative TPS) are also critical. Rohrschneider also emphasized the need to reduce margins—this can be achieved through better knowledge of the environment, improving modeling capabilities, and reducing testing uncertainties. He also commented on the need for instrumenting flight missions to collect data.

Al Witkowski, Pioneer Aerospace, focused his discussion on decelerators, highlighting two current large parachute development programs. He commented on the fact that NASA is still using essentially the same parachute technology that was used on Viking (for Mars) and Apollo (for Earth). The primary user of these specialized decelerator systems is NASA and, historically, funding associated with a given mission must be used to reduce risk, thereby precluding the time and cost of a new decelerator development program. However, heritage decelerator technology in its current state cannot be used to land heavy payloads on Mars. Witkowski sees the top technology challenges as follows: lack of materials data for accurate modeling, lack of validation data for modeling of flexible systems (parachutes, inflatables, etc.), and affordable ways to do full-scale testing. Overall, Witkowski believes the draft roadmap is good, but needs consistent and persistent long-term funding for it to be successful. In addition to technology development, foundational basic research of flexible materials is also needed, which is where universities can play a role. Finally, he believes that decelerator measurement test capabilities are needed (e.g., stress, strain, shape, etc.) to develop models and to provide model validation. This can start at small scales, but full-scale testing will eventually be needed.

Following the four panelists presentations, there was a long question and answer session, with a focus on materials, in particular with the application toward flexible TPS (although one audience member commented that the problem is bigger than just flexible TPS). In general, NASA and the United States are not developing new materials, but instead are borrowing materials from other industries, which are often manufactured overseas. With regard to modeling and simulation, there was a general sentiment that flexible materials (e.g., parachutes and inflatables) cannot be currently modeled. While tests are needed to help build and validate models, there is also technology development that is required with regard to testing techniques (the example of measuring strain in a flexible material was given as one particular challenge). There were also several comments with regard to the need for NASA to develop a common materials database—in particular, with materials properties and regimes that are needed for EDL.

With regard to materials, a question was asked about why inflatable decelerators have not yet reached flight readiness despite years of funding—it is a technical or funding issue? One panelist answered that there is a technical hurdle to flexible TPS, particularly with regard to materials. Also, most technology development up to now has been done as part of flight missions, and no mission is willing to accept the risk of a low-TRL item like inflatables (aerocapture was another example given in this category). Since most EDL technologies are single-point failures, it is almost impossible to fly an unproven technology on a science mission. A technology demonstration mission would be required, where science is not the crux of the mission.

One audience member asked about instrumenting high-altitude flight tests to collect data and suggested that NASA should provide seed money for this instrumentation. Several comments were made re-iterating the need to instrument test flights. While some things can be tested in a laboratory setting, flight tests will still be needed.

There was further discussion about the role of NASA versus industry. Several comments were made that NASA should not be competing with industry, but should be sharing knowledge and encouraging the growth of small companies in particular. However, technology transfer is difficult, without actually moving the people with the expertise. Another comment was made with regard to facilities—NASA has to maintain certain facilities because industry cannot afford to do it.

There was also some discussion on the importance of reducing margins. What mostly impacts margins are the assumptions going into the modeling. Generally, margin is being added on top of the baseline design, which already has margin hidden in it. Rohrschneider commented that sensitivity analyses to the assumptions are rarely done.

Finally, the focus of Mars on the roadmap was discussed. The roadmap authors commented that when the roadmap was being written, Mars missions were continuing and directed, while other targets were competed, so there was no guarantee in which targets would be visited. However, there was also an attempt to focus on Earth return in the draft roadmap. Another roadmap author commented that the hardest EDL problem is landing humans on Mars, so it warrants attention. There was also discussion about stepping stone technologies that can be done in the near term, which will eventually contribute to a human Mars mission (decelerators, SRP, inflatable reentry vehicle experiment). The general sentiment in the audience was that the Earth-return segment needed to be strengthened, in particular with regard to supporting commercial missions.

Panel D: Academic Organizations

Robert Bishop, Marquette University, focused his discussion on the GN&C portion of the roadmap. He focused on three areas:

1. Aeroshells—need more lift; guidance should be integral in the design of aeroshells, not an afterthought
2. EDL = GN&C—need smart sensors and need to think about their role in navigation (and vice versa)
3. Education—roadmap did not address education very well, but need to keep students excited and engaged to develop the next generation of engineers

Bishop then addressed what the current state of EDL is: at entry (hypersonic), vehicles have a lot of lift and controllability, but don't have good knowledge of state; once the parachute is deployed, you have a good knowledge of your state, but no longer have sufficient controllability. Therefore, he believes the focus should be on more sensors and location knowledge at high altitudes. He believes that a tipping-point technology is robust modular GN&C algorithms, where these can move from mission to mission without software holding up the development. He believes a game-changing technology to be aeroshells with high lift. Echoing many other presenters, Bishop also emphasized the need to instrument EDL missions. Overall, Bishop believes that the draft roadmap needs a strong GN&C focus.

Jean Muylaert, von Karman Institute for Fluid Dynamics, began his presentation with general comments with regard to the draft roadmap. First, he believes that the link between industry, academia, and NASA is important. Second, he believes in the need to return to a vigorous ground and flight test program (but how?). Third, he emphasized the importance of physical model validation with testing (and the need for upgrading ground-based test facilities), risk analysis, and qualification at the integrated EDL level. Muylaert then discussed the in-flight experimentation strategy carried out in Europe. He believes that in-flight research test-beds should be emphasized more—in Europe, this is done on cheap launches, suborbital flights, etc., in order to bridge the gap between ground-based tests and flight data. EXPERT is one example of an in-flight test bed that he discussed in his presentation.

Tayfun Tezduyar, Rice University, focus was on parachutes and, in particular, on fluid-structure interaction modeling. This is one of the most difficult problems to test, Tezduyar explained: because parachutes are so light, many of the classical fluid-structure interaction techniques do not work, although much progress has been made over the past several years. Tezduyar discussed a set of methods developed at Rice University that have produced good results thus far. He also emphasized the need for flight test data in order to benchmark computational modeling.

In the question and answer session that followed the Panel D presentations, Muylaert was asked about ESA's EXPERT program and how NASA could fund similar technology-dedicated missions. Muylaert discussed how ESA has a technology directorate and a program directorate—in getting EXPERT funded, they included the programs directorate in the discussion, which he claims helped tremendously. Essentially, it created a new vision/strategy with regard to in-flight research.

The panelists were then asked about their view on game-changing technologies. Tezduyar answered that there needs to be a bigger role for computational modeling in the overall process (particularly fluid-structure interaction). Bishop believes that developing GN&C technology to do a pinpoint landing at Mars would be game-changing. However, without a technology demonstration mission, vehicles with more lift will not be utilized.

Finally, there was discussion with regard to education, and how to link products from graduate students to NASA.

REFERENCES

Brady, T., and Paschall, S. 2010. The challenge of safe lunar landing. IEEE Aerospace Conference, Big Sky, Montana, March 6-13, 2010. Available at http://ieeexplore.ieee.org/.

Brady, T., Paschall, S., Zimpfer, D., Epp, C., and Robertson, E. 2009. Hazard detection methods for lunar landing. IEEE Aerospace Conference, Big Sky, Montana, March 7-14, 2009. Available at http://ieeexplore.ieee.org/.

Brady, T., Paschall, S., and Crain, T. 2010. GN&C development for future lunar landing missions. AIAA Guidance, Navigation, and Control Conference, Toronto, Ontario, Canada, August 2-5, 2010. American Institute of Aeronautics and Astronautics, Reston, Va.

Grantz, A.C., Experimental Systems Group, Boeing Company. 2011. "TA-09 Entry, Descent, and Landing Panel Discussion," presentation at the National Research Council NASA Technology Roadmaps Panel 6 Workshop, Irvine, Calif., March 23, 2011. National Research Council, Washington, D.C.

Justh, H.L., Justus, C.G., and Ramey, H.S. 2011. Improving the precision of MARS-GRAM. Fourth International Workshop on the Mars Atmosphere: Modelling and Observation, February 8-11, 2011, Paris, France. Available at http://www-mars.lmd.jussieu.fr/paris2011/abstracts/justh_paris2011.pdf.

Justus, C.G., Duvall, A., and Keller, V.W. 2005. Mars aerocapture and validation of MARSGRAM with TES data. 53rd Joint Army-Navy-NASA-Air Force (JANNAF) Propulsion Meeting, December 5-8, 2005. Available at http://ntrs.nasa.gov/archive/nasa/casi.ntrs.nasa.gov/20060005595_2006003663.pdf.

Mavris, D.N., and DeLaurentis, D.A. 2000. Methodology for examining the simultaneous impact of requirements, vehicle characteristics, and technologies on military aircraft design. 22nd Congress of the International Council on the Aeronautical Sciences (ICAS), Harrogate, England, August 27-31, 2000. Paper ICAS-2000-1.4.5. Available at http://www.icas-proceedings.net.

Muylaert, J., von Karman Institute for Fluid Dynamics, Belgium. 2011. "EDL Workshop: Review of the NASA Technology Roadmap TA09," presentation at the National Research Council NASA Technology Roadmaps Panel 6 Workshop, Irvine, Calif., March 24, 2011. National Research Council, Washington, D.C.

NASA (National Aeronautics and Space Administration). 2011. Programs and Missions: Missions, Mars Exploration Program. NASA. Available at http://mars.jpl.nasa.gov/programmissions/missions/.

NRC (National Research Council). 2011. *Vision and Voyages for Planetary Science in the Decade 2013-2022.* The National Academies Press, Washington, D.C.

Paschall, S., Brady, T., and Cohanim, B. 2008. A self contained method for safe and precise lunar landing. IEEE Aerospace Conference, Big Sky, Montana, March 1-8, 2008. Available at http://ieeexplore.ieee.org/.

Peterson, C., Sandia National Laboratories (retired). 2011. "Thoughts on NASA's Future Investments in EDL Technology," presentation at the National Research Council NASA Technology Roadmaps Panel 6 Workshop, Irvine, Calif., March 23, 2011. National Research Council, Washington, D.C.

Picetti, D., Boeing Company. 2011. "NASA OCT Technology Roadmap Study," presentation at the National Research Council NASA Technology Roadmaps Panel 6 Workshop, Irvine, Calif., on March 23, 2011. National Research Council, Washington, D.C.

Rohrschnedier, R., Ball Aerospace Corporation. 2011. "Panel 6: Entry, Descent, and Landing," presentation at the National Research Council NASA Technology Roadmaps Panel 6 Workshop, Irvine, Calif., March 24, 2011. National Research Council, Washington, D.C.

Thoemel, J., Muylaert, J-M., Ratti, F., and Gavira, J. 2009. In-flight testing of critical technologies and experimentation of aerothermodynamics phenomena. AIAA Paper 2009-7232, 16th AIAA/DLR/DGLR International Space Planes and Hypersonic Systems and Technologies Conference, Bremen, Germany, October 19-22, 2009. American Institute of Aeronautics and Astronautics, Reston, Va.

M

TA10 Nanotechnology

INTRODUCTION

The draft roadmap for technology area (TA) 10, Nanotechnology, addresses four level 2 technology subareas:[1]

- 10.1 Engineered Materials and Structures
- 10.2 Energy Generation and Storage
- 10.3 Propulsion
- 10.4 Sensors, Electronics, and Devices

Nanotechnology manipulates matter and forces at the atomic and molecular levels. The accepted structure size for nanotechnology is between 1 and 100 nanometers in a minimum of one dimension, and includes materials or devices that possess at least one dimension within that size range. Quantum mechanical forces become important at this scale, which means that the properties of nano-sized materials or devices can be substantially different than the properties of the same material at the macro scale. Nanoscale materials or incorporation of them into a matrix have the promise of substantially improving the thermal, electrical, optical, and mechanical properties of a component. Nanotechnology can provide great enhancement in properties, and opens new possibilities due to the novel phenomena that occur only at the nanoscale. Materials engineered at the nano-scale will shift the paradigm in space exploration, sensors, propulsion, and overall system design.

Before prioritizing the level 3 technologies included in TA10, several technologies were renamed or moved. The changes are illustrated in Table M.1. The complete, revised technology area breakdown structure (TABS) for all 14 TAs is shown in Appendix B.

TOP TECHNICAL CHALLENGES

The panel identified five top technical challenges in nanotechnology, listed below in priority order.

1. Nano-Enhanced Materials. Reduce spacecraft and launch vehicle mass through the development of lightweight and/or multifunctional materials and structures enhanced by nanotechnologies.

[1] The draft space technology roadmaps are available at http://www.nasa.gov/offices/oct/strategic_integration/technology_roadmap.html.

TABLE M.1 Technology Area Breakdown Structure for TA10, Nanotechnology

NASA Draft Roadmap (Revision 10)	Steering Committee-Recommended Changes
TA10 Nanotechnology	The steering committee made no changes to the structure of this roadmap, although NASA's draft roadmap renamed or moved seven technologies in the TABS.
10.1. Engineered Materials and Structures	
10.1.1. Lightweight Structures	Rename: 10.1.1. Lightweight Materials and Structures
10.1.2. Damage Tolerant Systems	
10.1.3. Coatings	
10.1.4. Adhesives	
10.1.5. Thermal Protection and Control	
10.2. Energy Generation and Storage	
10.2.1. Energy Storage	Move 10.2.1 to 10.2.2 Energy Storage
10.2.2. Energy Generation	Move 10.2.2 to 10.2.1 Energy Generation
10.2.3. Energy Distribution	
10.3. Propulsion	
10.3.1. Propellants	Rename: 10.3.1. Nanopropellants
10.3.2. Propulsion Components	Rename: 10.3.2 Propulsion Systems
10.3.3. In-Space Propulsion	
10.4. Sensors, Electronics and Devices	
10.4.1. Sensors and Actuators	
10.4.2. Nanoelectronics	Rename: 10.4.2 Electronics
10.4.3. Miniature Instruments	Rename: 10.4.3 Miniature Instrumentation

Development of advanced materials using nanotechnology can improve performance in the following areas: electrical energy generation and storage, propulsion, sensors, instrumentation, signal and power transmission, thermal protection, and active structures sensing, healing, and shape control. Nano-enhanced composites have the capability to enhance mission performance by increasing the strength and stiffness of materials and reducing structural weight. Weight reduction with added material functionality (such as increased strength and stiffness) using carbon nanotube technology has already been demonstrated in numerous materials, and nano-enhanced materials are finding their way into commercial products. Nano-enhanced advanced composites could reduce structural weight in launch vehicles, cryotanks, propulsion systems, and spacecraft, thus increasing the payload mass fraction. Nano-enhanced multifunctional materials and structures may exhibit unique failure modes and thus will require new design analysis tools. Multi-scale models that are valid over scales ranging from nano to macro are needed to understand nano-enhanced composite materials failure mechanisms and interfaces in order to design with them. Multi-physics models are needed to address fabrication processes, operation in extreme environments, and designing with active materials. Additional challenges to the broad use and incorporation of nano-engineered materials into useful products are the limited availability of certain raw nanomaterials and their variable quality. New production methodologies are required, not only to manufacture the raw nanomaterials, but also to controllably incorporate them into other materials. The particular end application may require specific dispersion and ordering of the nanoparticles.

2. Increased Power. Increase power for future space missions by developing higher efficiency, lower mass, and smaller energy systems using nanotechnologies.

Energy generation and energy storage will remain a top technical challenge for all future space-related missions. Batteries and power generation account for a significant amount of weight in any launch vehicle. Efficient methods to generate and recover energy, reduce overall power requirements, and reduce weight will benefit future NASA missions. For long-duration space missions, improved energy generation and storage will play a significant role in the mission success. Nanotechnology can improve performance for energy generation, energy storage, and energy distribution. Nano-enhanced electrode materials used in batteries where the surface area is significantly increased will allow for faster charge/discharge rates, higher power densities, new battery and fuel cell materials, and increased safety. Nano-engineered devices can also improve existing storage and energy generation technologies making previously inefficient technologies competitive. Large numbers of sensors monitoring system and structural health will result in larger energy needs not only to transmit the information, but also process it. Nanotechnology will enable sensors to be self-powered and allow for distributed sensing in a networked fashion. Nanomaterials are being aggressively studied to improve methods for solar energy harvesting, thermal scavenging, and harvesting energy from the structures themselves. Newer technologies such as nano-structured metamaterials and photonic or phononic crystals with spectral compression will improve collection efficiencies and provide new capabilities.

3. Propulsion Systems. Improve launch and in-space propulsion systems by using nanotechnologies.

Advances in nanotechnology will enable new propellants and improved propulsion technologies. Nanotechnology may impact propellant technology by providing higher combustion efficiency and enabling alternative fuel materials that are less hazardous and require less cooling. More energetic propellants will reduce fuel mass in solid motors, and provide tailorable ignition and reaction rates. Higher-temperature and lower-erosion structural materials based on nanomaterials could reduce the weight of engine nozzles and propulsion structures. In-space electric propulsion (EP) systems, which couple high efficiency and large specific impulse within a relatively small package, will also benefit from performance improvements in nanoparticle propellants and nano-fabricated emission thrusters

4. Sensors and Instrumentation. Develop sensors and instrumentation with unique capabilities and better performance using nanotechnologies.

The success of NASA space missions relies heavily on a variety of sensing methods and sensor technologies for numerous environments in addition to scientific data collection. Structural monitoring of the space vehicle and internal systems self-monitoring, in addition to astronaut health monitoring, will be required as vehicle complexity and mission durations increase. The ability of systems or structures to alert operators and spacecraft systems to changing conditions allows for a proactive approach to maintain capability. Nano-sensor technology allows the incorporation of sensors in structures and systems that are smaller, more energy efficient, and more sensitive, allowing for more complete and accurate health assessments. Nanotechnology also permits targeted sensor applications that improve functional efficiency. Future space missions will also require development of smaller, more efficient excitation sources (photon or electron) for scientific work. Nanotechnology also allows miniaturization of instruments with enhanced performance.

5. Thermal Management. Improve performance of thermal management systems by using nanotechnology.

Thermal management is a key technology area that enables and impacts all of NASA's missions. Proper thermal management can reduce overall system cost and weight with direct benefit to reducing overall launch vehicle weight. Thermal control is often required at the system level as well as at the subsystem and component level. Often these requirements are at odds; for instance, the samples loaded into an instrument need to be heated, but the detector needs to be cooled. Nanotechnology can be used to tailor the thermal conductivity of materials, making them more efficient conductors or insulators. The use of nanomaterials as fillers in TPS ablators may enhance both char formation and ablator cohesion. This will reduce spalling and total erosion of the TPS materials. Such nanomaterial inclusions can result in a weight savings from using less TPS material.

QFD MATRIX AND NUMERICAL RESULTS FOR TA10

A quality function deployment (QFD) matrix was employed to assist in the ranking of the nanotechnology level 3 technologies and to capture comments and concerns of the panel regarding the technology evaluation

	Benefit	Alignment with NASA Needs	Alignment with Non-NASA Aero Tech Needs	Alignment with Non-Aerospace National Goals	Technical Risk and Reasonableness	Sequencing and Timing	Time and Effort	QFD Score (Weighted)	Panel Priority
Multiplier	27	5	2	2	10	4	4		
	0/1/3/9	0/1/3/9	0/1/3/9	0/1/3/9	1/3/9	-9/-3/-1/1	-9/-3/-1/0		
Technology Name	Benefit	Alignment			Risk/Difficulty				
10.1.1. (Nano) Lightweight Materials and Structures	9	9	9	9	3	-1	-3	338	H
10.1.2. (Nano) Damage Tolerant Systems	3	9	3	3	3	-3	-3	144	M
10.1.3. (Nano) Coatings	1	9	3	1	3	-3	-1	94	L
10.1.4. (Nano) Adhesives	1	9	3	3	3	-3	-1	98	L
10.1.5. (Nano) Thermal Protection and Control	3	9	3	3	3	-3	-1	152	M
10.2.1. (Nano) Energy Generation	9	9	9	9	3	-3	-3	330	H
10.2.2. (Nano) Energy Storage	1	9	3	3	3	-3	-3	90	L
10.2.3. (Nano) Energy Distribution	1	3	3	3	3	-3	-3	60	L
10.3.1. Nanopropellants	9	9	3	3	3	-1	-1	322	H
10.3.2. (Nano) Propulsion Systems	3	9	3	3	9	-3	-3	204	M
10.3.3. (Nano) In-Space Propulsion	3	9	3	3	3	-1	-1	160	M
10.4.1. (Nano) Sensors and Actuators	3	9	9	9	3	1	-1	192	H*
10.4.2. (Nano) Electronics	1	3	3	9	3	-3	-3	72	L
10.4.3. Miniature Instrumentation	3	9	3	3	3	-1	-1	160	M

FIGURE M.1 Quality function deployment (QFD) summary matrix for TA10 Nanotechnology. The justification for the high-priority designation of all high-priority technologies appears in the section "High-Priority Level 3 Technologies." H = High Priority; H* = High Priority, QFD score override; M = Medium Priority; L = Low Priority.

areas. The QFD matrix is shown in Figure M.1. The weighted scores for all level 3 technologies evaluated with the QFD approach are listed in Figure M.2. As shown in Figure M.2, the weighted scores of three technologies significantly exceeded those of the other technologies, and thus these three were selected as high priority. One technology, 10.4.1 Sensors and Actuators, fell into the medium-priority rating based purely on its QFD score, but the panel chose to designate it as high priority as well; its selection is discussed below in the 10.4.1 Sensors and Actuators individual technology evaluation area.

CHALLENGES VERSUS TECHNOLOGIES

Figure M.3 shows the relationship between the 14 individual level 3 TA10 technologies and the top technical challenges.

Note that the lowest-priority technologies as determined by the QFD rankings tend not to be strongly connected to the top technical challenges. (These are identified by an "L" in the left-most column, and are linked to the top challenges mainly by open circles.) All of the high-priority technologies and many of the medium-priority ones have a strong connection to at least one of the top technical challenges. This correlation shows a good level of consistency in the evaluations by the panel.

Furthermore, many of the TA10 roadmap technologies are connected to each other in support of a common top technical challenge or a crosscutting roadmap technology. For instance, many of the roadmap technologies support challenges related to propulsion, sensors, and instrumentation, and thermal management.

HIGH-PRIORITY LEVEL 3 TECHNOLOGIES

Panel 5 identified four high-priority technologies in TA10. The justification for ranking each of these technologies as a high priority is discussed below.

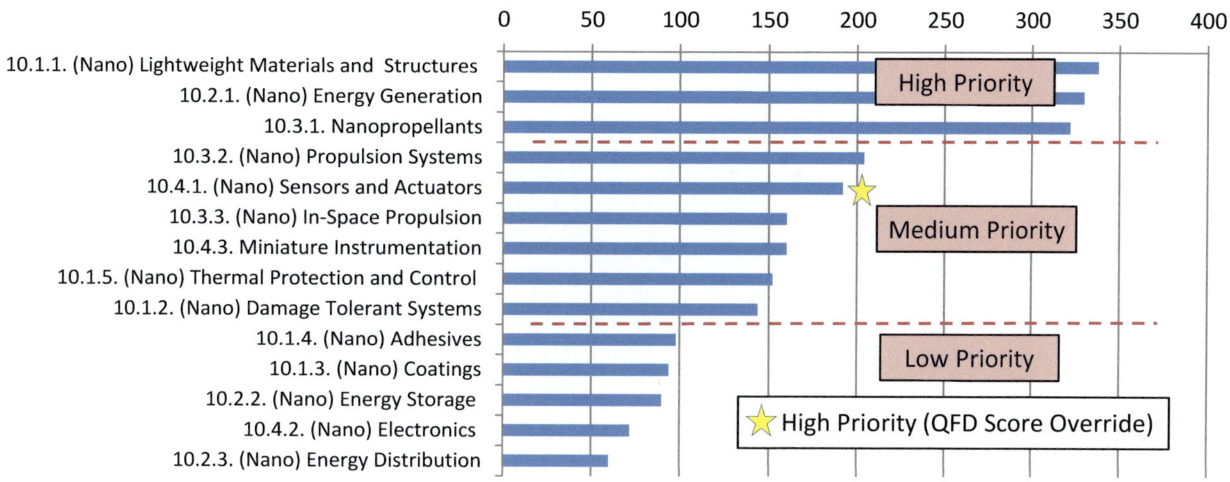

FIGURE M.2 Quality function deployment rankings for TA10 Nanotechnology.

Technology 10.1.1, Lightweight Materials and Structures

Nano-sized materials have the promise of substantially improving the thermal, electrical, and/or mechanical properties of components and structures while reducing weight. The ability of nanotechnology to impact performance on numerous levels allows for the development of multi-functional, lightweight materials and structures that will revolutionize aerospace system design and capability. A clear need for future NASA missions, manned or unmanned, is new advanced nano-enhanced composite materials. The impact of structures based on these materials is broad. Weight reductions can be readily realized in many areas, while improving the strength, damage tolerance, and properties such as thermal, mechanical, and electrical. For future missions, weight reductions and material performance increases that go beyond current capability that uses only carbon fibers are being sought for use in cryotanks and launch vehicles, for radiation protection, and for damage tolerance.

The technology readiness level (TRL) for lightweight materials and structures is difficult to assess as the state of the art is rapidly changing. Any TRL is specific to the individual materials rather than a technology research area. Lower rated technologies (TRL 1-3) include materials using porous carbon fibers and the development of continuous, single-walled carbon nanotube (CNT) fibers. Integration of these materials into composites and structures holds promise for 20 to 30 percent weight reduction and a concomitant increase in strength and stiffness. Higher TRL technologies (TRL 4-6) include using carbon nanotubes for cabling and tapes that are intended for near-term insertion into aircraft as a replacement for conventional copper-based wires and cables.

Lightweight materials and structures enhanced by nanotechnology to reduce weight while improving material performance aligns well with NASA's expertise and capabilities. NASA has technology requirements that are unique due to long-term deep-space missions and human transport. NASA's internal development program should focus on the testing and use of nano-enhanced materials and structures and understanding their performance in the extreme environment found in space. Technology development in this area is a worthwhile investment for all launch and space systems. In addition, it will have a broad impact across non-NASA aerospace and non-aerospace applications.

Direct access to the International Space Station is likely to be beneficial to the development of lightweight materials and structures. The performance of these materials and structures can be designed, tested, and evaluated on Earth, however their evaluations will benefit substantially from multiple launch and re-entry tests, and long-term exposure to the space environment.

This technology area received the highest rating of all the level 3 technologies in the QFD matrix analysis. Lightweight materials and structures would represent a large return on investment. This technology is game-changing because reductions in the structural and payload weight of a space vehicle allow for higher efficiency

Priority	TA 10 Technologies, Listed by Priority	Top Technology Challenges: 1. Nano-Enhanced Materials: Reduce spacecraft and launch vehicle mass through the development of lightweight and/or multifunctional materials and structures enhanced by nanotechnologies.	2. Increased Power: Increase power for future space missions by developing higher-efficiency, lower-mass, and smaller energy systems using nanotechnologies.	3. Propulsion Systems: Improve launch and in-space propulsion systems by using nanotechnologies.	4. Sensors and Instrumentation: Develop sensors and instrumentation with unique capabilities and better performance using nanotechnologies.	5. Thermal Management: Improve performance of thermal management systems by using nanotechnology.
H	10.1.1. (Nano) Lightweight Materials and Structures	●	●	○		●
H	10.2.1. (Nano) Energy Generation	●	●	●	○	
H	10.3.1. Nanopropellants	○	○	●		○
H	10.4.1. (Nano) Sensors and Actuators	●	●	○	●	●
M	10.3.2. (Nano) Propulsion Systems	●		●		
M	10.3.3. (Nano) In-Space Propulsion			●		
M	10.4.3. Miniature Instrumentation				●	
M	10.1.5. (Nano) Thermal Protection and Control	●		○	○	●
M	10.1.2. (Nano) Damage Tolerant Systems	○			○	
L	10.1.4. (Nano) Adhesives	○				
L	10.1.3. (Nano) Coatings	○	○		○	○
L	10.2.2. (Nano) Energy Storage		○	○		
L	10.4.2. (Nano) Electronics					
L	10.2.3. (Nano) Energy Distribution					

●	Strong Linkage: Investments by NASA in this technology would likely have a major impact in addressing this challenge.
○	Moderate Linkage: Investments by NASA in this technology would likely have a moderate impact in addressing this challenge.
[blank]	Weak/No Linkage: Investments by NASA in this technology would likely have little or no impact in addressing the challenge.

FIGURE M.3 Level of support that the technologies provide to the top technical challenges for TA10 Nanotechnology.

launches with increased payload capacity. The reduced total launch weight allows NASA greater flexibility in mission design. The expected weight reductions are in the 20 percent-30 percent range, but could be higher especially when including the weight savings due to multifunctionality. This technology will also impact TA12 Top Technical Challenge 2—Reduced Mass, and 4—Large Aperture Systems, and supports 1—Multifunctional Structures. The technology also impacts TA14 Top Technical Challenges 1-4, Thermal Protection Systems, Zero Boil-Off Storage, Radiators, and Multifunctional Materials.

Lack of research into the fabrication methodologies related to scale will slow development of lightweight materials and structures. There is significant cost involved in scaling manufacturing processes to commercial levels if the customer set is limited or the application space for terrestrial uses of the material is limited. Insufficient research into fabrication methodologies will slow development and use of lightweight materials and structures. Additionally, strength and performance gains may not be achieved if control of the nanoparticle dispersion, ordering, and interface properties are not achieved.

Technology 10.2.1, Energy Generation

Energy generation for spaceflight can be improved by leveraging nanotechnology. Current deep space missions rely primarily on stored energy (nuclear or chemical) to power systems aboard spacecraft. Improvements in current energy generation technologies, and the development of other methods of energy generation, can enhance future missions by reducing mass, improving reliability, and increasing mission durations.

The technology readiness level for nanotechnology-enhanced energy generation is dependent on the area of implementation. Lower TRL technologies discussed in the roadmap include quantum dots connected with carbon nanotubes and CNT-enhanced flexible organic photovoltaics (TRL 2-3), while higher TRL technologies involve nano-enhanced electrode materials and optimized nano-engineered structures for anode performance increases (TRL 4-6).

There is excellent alignment between NASA needs and expertise in the area of nanotechnology enhanced power generation. However, the area of research is quite broad and there is already a substantial private and public investment in this area. Consequently, NASA can partner with industry, academia, or other government agencies (e.g., U.S. Air Force, Department of Energy) and leverage promising technologies to deliver extreme environment energy generation systems and energy scavenging methods needed for spaceflight.

The evaluation of energy generation technologies would benefit from access to the ISS. Long-duration testing of energy generating technologies in an extreme environment is critical.

Nanotechnology impacts energy generation by improving the material systems of existing energy storage and generation systems. The QFD matrix evaluation indicates a clear benefit of energy generation research to NASA and excellent alignment with NASA's needs. As with 10.1.1 Lightweight Materials and Structures, advances in energy generation through nanotechnology is game-changing because lighter, stronger materials and structures allow for more payload devoted to energy generation and power storage, and more efficient energy generation allows for lighter payloads at launch. When improvements are made in materials used for energy generation to increase reliability, performance in extreme environments, and power density, mission space is expanded or extended and costs drop. The ability to scavenge energy such that nano-sensors can be self-powered provides system self-monitoring capability at a fraction of the weight cost and with better sensor coverage. Without this development, the prospect of realizing a virtual digital fleet leader is remote. There is moderate risk involved with development of this technology area for NASA due to the breadth of the area and the numerous technologies that can be leveraged for successful space missions. However, there is substantial risk that commercial technologies developed cannot be adapted to operate with sufficient performance in extreme environments.

Technology 10.3.1, Nanopropellants

Nanopropellants include the use of nano-sized materials as a component of the propellant and as gelling agents for liquid fuels. In both cases, the nano-size provides a material with enormous reactive surface area. The use of nano-sized materials as a component of the propellant can solve several problems associated with current propellant systems. The problems of most concern are the toxicity and environmental hazards of hypergolic and solid propellants, and the handling requirements for cryogenics. Nanopropellants have the potential to have higher combustion efficiency. If the nanoparticle shape and size can be controlled, the ignition and reaction rates can be tailored. In addition, this technology has the potential to provide long-term propellant storage capabilities in space, and propulsion systems for nano- and pico-satellites. Using the nanoparticle as a gelling agent makes a propellant system easier to handle, but brings problems with pumping and injection. Issues to be resolved for the successful implementation of nanopropellants include passivation and dispersion technologies, scale up of materials and manufacturing, reduction in cost of the nano-sized particles, and system implementation. There are also the safety and health issues associated with use of any new material.

Hydrogen storage is the subject of much research. Nano-structured materials with extremely high surface areas may potentially have a sorption capacity of 8 weight percent. Much work is needed to increase the sorption capacity at operational temperatures. The Department of Energy has done significant work in hydrogen storage technology, which can be leveraged by NASA.

The TRL for nanopropellants is relatively low, with nano-sized fuel components, such as are found in "ALICE," a mixture of nano-scale aluminum particles and an ice slurry, being the highest (TRL 4). Nano-gelled propellants and nanostructures for hydrogen storage are at a lower level (TRL 2-3).

Research on nanopropellants aligns well with NASA's expertise and capabilities. NASA can partner with other agencies, such as the U.S. Air Force and the Department of Energy, that also have interests in nanopropellants. Non-aerospace applications of nanopropellants are in the future, and have little industry support at this time.

Access to the International Space Station is not required.

Propellants are critical to NASA's space mission success. The use of nanopropellants is game-changing because they can provide a 15 to 40 percent increase in efficiency, resulting in a decrease in system weight. In addition, nanopropellants can be multifunctional; that is, the propellant can act as a structural component that is consumed, as well as providing energy storage for in-space power. The development risk of nanopropellants is relatively low.

Technology 10.4.1, Sensors and Actuators

Nano-scaled sensors and actuators are an important research area for NASA's space mission needs. The nanoscale allows for improvements in sensitivity and detection capability while operating at substantially lower power levels. Nanosensors can be made of variety of materials such as biological materials, inorganic, or polymeric materials, in addition to CNTs or combinations of materials. Due to the wide variety of material choices, nanosensors can be easily integrated with sensor electronics to produce compact and low power systems with ultra-sensitive response to mechanical, thermal, radiation, and molecular perturbations.

The promise of nanosensors and sensor systems is the integration of nano-electronics and nano-power sources to deliver arrays of autonomous sensors suitable for structural health monitoring and other distributed sensing activities. Structural monitoring of the space vehicle and self-monitoring of the internal systems, in addition to astronaut health monitoring, will be required as vehicle complexity and mission durations increase. The ability of systems or structures to alert operators and spacecraft systems to changing conditions allows for a proactive approach to maintain capability. Nanosensors are smaller, more energy efficient, and more sensitive, allowing for more complete and accurate health assessments. Nanotechnology also allows for targeted sensor applications, which improves functional efficiency.

The TRL for nanosensors and nano-actuators is high for specific applications. Trace gas nanosensors have already flown on space missions and on the ISS (TRL 6). However, there is much work to be done in the area of distributive sensing, therefore this TRL is much lower (TRL 2-3). Possible limitations to progress in distributed autonomous sensing are developments in sampling, sensor cleaning, and waste removal if that is required.

Nanotechnology enhanced sensors and actuators are well aligned to NASA's needs and expertise. The reduction in scale, increase in performance, and concomitant power reductions from these sensors and actuators are required for long-term human mission success. Because NASA requires sensors that can function reliably in extreme environments, it can partner with others to perform joint research to adapt appropriate technologies to fit these specific needs. Research on nanosensors and nanoactuators is widespread in industry, academia, and government labs, and the rate of progress for this particular technology area is very rapid. Likely aerospace and non-aerospace industries having an interest in this technology include structural monitoring of aircraft and infrastructure such as buildings and bridges. In addition, nanosensors will play a large role in future health care.

Testing and evaluation of nano-enhanced sensors and actuators would benefit from access to the ISS. The knowledge gained from nanosensors embedded into a composite panel to evaluate its structural health during exposure to the extreme space environment would be of tremendous use.

The panel overrode the QFD score for this technology to designate it as a high-priority technology because the QFD scores did not capture the value of this technology in terms of overall benefit to all missions. This technology was given three high-QFD scores (all 9s) for alignment with needs for NASA, non-NASA aerospace, and non-aerospace national goals. NASA uses a variety of sensors for guidance, monitoring of structural health, crew health, engine health, debris damage, fuel and leak detection, detection of life on planets, gas sensing, atmospheric sensing, and for scientific studies, and nanotechnology is expected to significantly impact sensor technology.

MEDIUM- AND LOW-PRIORITY TECHNOLOGIES

TA10 contains 14 level 3 technologies, of which 10 were determined to be of medium or low priority. The QFD matrix indicates that for all the level 3 technologies evaluated, the alignment with NASA needs and non-NASA aerospace technology needs were high. Significant differences appeared with the technology alignment to non-aerospace needs, as well as perceived benefit to NASA.

Four technologies were determined to be medium priority. The overall alignment with non-NASA aerospace technology needs and non-aerospace national goals, while greater than the low-priority technologies, were less than that of the high-priority technologies. The medium-priority nanotechnology areas are: 10.3.2 Propulsion Systems, 10.4.3 Miniature Instrumentation, 10.3.3 In-Space Propulsion, 10.1.5 Thermal Protection and Control, and 10.1.2 Damage Tolerant Systems. It was determined that advances in the 10.1.3 Coatings, 10.1.4 Adhesives, 10.2.1 Energy Storage, 10.2.3 Energy Distribution, and 10.4.2 Electronics level 3 technology areas would not result in game-changing or major benefits to NASA, and thus they were rated low priority. There is an extensive existing infrastructure for development of these technologies in the commercial sector, and the national need is not high. The panel believes that NASA can remain effective in these areas by leveraging and testing commercial technology in extreme environments or by partnering with commercial vendors or other agencies to deliver timely solutions.

All of the technologies had some demonstrated risk and difficulty associated with them.

DEVELOPMENT AND SCHEDULE CHANGES FOR THE TECHNOLOGIES COVERED BY THE ROADMAP

Future NASA missions depend highly on advancements such as lighter and stronger materials, increased reliability, and reduced manufacturing and operating costs. All of these will be impacted by the incorporation of nanotechnology. The fold-out map included in the TA10 roadmap, Figure R, details various technologies and when they will be needed, but it is unclear from the mission descriptions why specific technologies are required and why a specific insertion date was used. This makes it challenging to suggest any specific modifications to the roadmap schedule. However, development of specific nanotechnologies may be faster than assumed, due to the large international research efforts in this area.

OTHER GENERAL COMMENTS ON THE ROADMAP

It is worth mentioning again that major challenges to the broad use and incorporation of nano-engineered materials into useful products are the limited availability of certain raw nanomaterials and their variable quality. The lack of sufficient materials will negatively impact the proposed roadmap. However, the panel is not advocating that NASA should be active in the large-scale production of nanomaterials.

Nanotechnology is a very broad area of research, and (as can be seen from the table on p. TA10-22 of NASA's TA10 roadmap), it is crosscutting with and impacts every other roadmap. As stated earlier, research in nanotechnology will also impact TA12 Top Technical Challenge 2—Reduced Mass, and 4—Large Aperture Systems, and supports 1, Multifunctional Structures. Nanotechnology also impacts TA14 Top Technical Challenges 1-4, Thermal Protection Systems, Zero Boil-Off Storage, Radiators, and Multifunctional Materials.

Furthermore, recognizing that much work on a national R&D effort in nanotechnology is underway in government labs, universities, and industry sponsored by NSF and other agencies, the NASA research for space applications should be well coordinated with this national effort.

Even within NASA, its nanotechnology research does not seem to be centrally coordinated, and thus the potential exists for substantial duplication of effort. For example, nanosensor research is being done at NASA-Glenn, NASA-Ames, and JPL, according to information presented to the panel at its January 2011 meeting. The panel suggests that there be substantial coordination between the nanotechnology researchers at the various NASA centers, the national R&D effort, and specific NASA mission end users. In addition, the panel suggests that the nanotechnology researchers collaborate as closely as possible with the researchers involved with the other roadmap level 3 technologies.

PUBLIC WORKSHOP SUMMARY

The workshop for the TA10 Nanotechnology technology area was conducted by the Materials Panel on March 9, 2011, at the Keck Center of the National Academies, Washington, D.C. The discussion was led by panel chair Mool Gupta. He started the day by giving a general overview of the roadmaps and the NRC's task to evaluate them. He also provided some direction for what topics the invited speakers should cover in their presentations. After the introduction, the day started with an overview of the NASA roadmap by the NASA authors, followed by several sessions addressing the key areas of each roadmap. For each of these sessions, experts from industry, academia, and/or government provided a 35 minute presentation/discussion of their comments on the NASA roadmap. At the end of the day, there was approximately 1 hour for open discussion by the workshop attendees, followed by a concluding discussion by the panel chair summarizing the key points observed during the day's discussion.

Roadmap Overview by NASA

The presentation by the NASA team provided an overview of the benefits of nanotechnology, including the ability to reduce vehicle mass, increase durability of materials, and improve the performance of propulsion systems, sensors, photovoltaics, and other electronics systems. It was noted that while some areas such as aeronautics—emissions and exploration can provide a technology pull, in general nanotechnology falls into the push category. After providing an overview of the roadmap TABS and the roadmap itself, there was a detailed discussion of several key capabilities. These include 30 percent lighter cryogenic propellant tanks, enabling extreme environment operations, lighter and more efficient thermal protection and management, "smart" airframe and propulsion concepts, adaptive gossamer structures, high-efficiency flexible photovoltaics, enhanced power and energy storage, miniature instruments, enabling low-mass smart satellites capable of formation flying, low-power radiation-hard reconfigurable electronics, improved astronaut health management, and ultrasensitive selective sensing.

Three different time periods were used to outline the top technical challenges facing nanotechnology:

- Next 5 years
  - Grand challenge: Controlled growth and stabilization of nanopropellants
  - Development of long-life, reliable emission sources
  - Development of characterization tools and methodologies to measure coupled properties of nanostructured materials, including non-destructive and in situ techniques
  - Optimization of bulk properties of nanomaterials

- Next 5 to 10 years
  - Grand challenge: Development of nanostructured materials 50 percent lighter than conventional composites with equivalent or better properties and durability
  - Grand challenge: Heirarchical system integration across length scales (from nano to macro)
  - Grand challenge: Development of integrated energy generation, harvesting and scavenging technologies
  - Development of manufacturing methods, including net-shape fabrication to produce nanomaterials and devices on large scales with controlled structure, morphology and quality
- Next 20 years and beyond
  - Grand challenge: Development of graphene electronics
  - Development of high-fidelity, high-reliability multiscale models to predict the properties of nanoscale materials and devices/structures
  - Development of single molecule detection methods with high specificity (2029+)

The presentation concluded with some discussion of how nanotechnology crosscuts among the majority of the NASA technology roadmaps, as well as outlining how NASA is collaborating in nanotechnology research with the Department of Defense for materials and photovoltaics, and with the Department of Homeland Security for sensors. Also discussed was the National Nanotechnology Initiative (NNI), and how there is a potential avenue for further collaborations in NNI signature initiatives.

Session 1: Nanomaterials

Brian Wardle (MIT) started with a brief introduction, followed by a quick overview of the work his group does at MIT. He commented that the TA10 roadmap provides an "accurate and compelling vision for nanomaterials development with NASA's broader mission," and that the main focus of his presentation would be on areas of potential improvement. Wardle noted that the five grand challenges identified by NASA in their roadmap are compelling, there is a mix of specific application areas and crosscutting game changers; Wardle is concerned that the crosscutting aspect is very important and emphasized that this should not get lost in the evaluation process. He sees NASA as a better fit for an integrator role rather than lead developer on nanotechnology with a few small exceptions such as nanopropellants. He commented on several specific areas, including the importance of multiscale modeling and interphase/morphology considerations, as well as questioning what role NASA should take in understanding the environmental health and safety aspects of nanomaterials. Finally, Wardle also noted that there was minimal discussion in the NASA roadmap on nanostructured metals and ceramics, considering NASA's needs and strengths in these areas.

In the question and answer session following his presentation, Wardle noted that there are reports of nanocomposites having seen 30 percent higher ultimate strength compared with current composite materials and that as a structural material, it is already being incorporated into planned missions like Juno. It was noted that controlling morphology (i.e., alignment, treating nanotubes as structural elements, etc.) currently seems to be trial and error, rather than based on predictive modeling. Wardle indicated that such modeling is beyond current capabilities and is a potential high-value area to invest in but might not be an appropriate area for NASA. Wardle commented that while production volumes are currently significant, the morphology is mixed.

Wade Adams (Rice University) followed Wardle's presentation with a discussion on observations generated by himself and a small group of colleagues at Rice University. Overall, Adams indicated that the TA10 roadmap was an excellent update and expansion of the 2000 nanotube roadmap that some of his team helped create. While Adams indicated he believed the five grand challenges to be useful, he suggested that the "Hierarchical Systems Integration" challenge merits the most attention, as it is critical to a broad implementation of nano/micro-technology into future systems, and is an area where NASA could potentially show leadership. Adams also highlighted the need for internal expertise, and expressed some concern that limited budget may keep NASA personnel from staying up-to-date on the technologies such that they can be better "smarter buyers." Adams observed that the NASA roadmap seems to have a longer-term focus, therefore identifying some short-term payoffs might be a good approach. In the end, he noted that NASA will not have enough funding to expect significant NASA leadership in many of the technical areas, but that focusing funding on a few critical needs might enable NASA leadership in these areas. Finally, Adams strongly encouraged NASA teaming with initiatives from other groups (e.g., the National Nanotechnology Initiative, Department of Defense, Department of Energy, public-private partnerships).

Session 2: Sensors and Nanoelectronics

Avik Ghosh (University of Virginia) started his discussion with a few introductory charts. He noted that for nanomaterials, near-equilibrium properties are reasonably well understood, but that the non-equilibrium regime is less so. He noted that it is becoming more difficult to avoid noise around switching areas, and recent research is attempting to identify ways to use this noise. In terms of the NASA roadmap, Ghosh indicated it was good, but that NASA needs to better spell out the needs versus capabilities; he also agreed with other presenters in that limited funding means it is important for NASA to make sure specific areas do not fall through the cracks. Relative to key challenges, Ghosh discussed that heat dissipation is a primary focus area in electronics, spintronics (and multiferroics) are promising but receiving little funding, and new materials require further research to address current limitations (e.g., graphene is good for some applications, but not for others such as switching). Ghosh commented that there are opportunities in trap dynamics and in improving the engineering of thermal transistors, but also noted that there appear to be gaps in the NASA roadmap as well in thermal conductivity (e.g., significantly less range versus electrical conductivity) and biological computing. He concluded his presentation suggesting that, in addition to partnering with others, NASA could choose a few areas not addressed by other agencies and focus on these niche applications as well.

During the discussion period after Ghosh's presentation, one of the panel members noted that companies such as Intel and IBM will likely drive technology development in these areas, and asked what NASA's focus should be. Ghosh responded that the industry appears to be focused on near-term applications, whereas the National Science Foundation is more interested in far-term exploratory research; NASA could look to bridge these differences. In responding to a question on specific technologies that NASA should focus on, Ghosh agreed that significant investment in this field is required, and that one possibility is for them to target niche areas that are not currently well funded, as well as trying to identify partnership areas to participate in. As an example of a niche area that NASA might look into, Ghosh noted that utilizing noise in sensor design is not seeing significant funding from other organizations currently. During some discussion on the current TRL of many of the technologies in this roadmap, Ghosh agreed that these typically are in the fundamental/exploratory research phase. He did note, however, that STTRAM does seem closer to feeding into products (e.g., non-volatile memory). He also stated that nanoelectronics technology will be led by industries like IBM, Intel, etc., and NASA can benefit from such developments.

Ashraf Alam (Purdue University) next gave a presentation on his perspectives on the TA10 roadmap. In terms of his overall assessment of the roadmap, Alam noted that a systematic, multi-metric evaluation of available sensor technologies would be very helpful, and that NASA could work with internal and external researchers to achieve this. Over the short term, Alam indicated that sensitivity, variability, and selectivity concerns make many new sensor technologies unsuitable for rapid deployment, and that working to improve this is a worthwhile investment. He also noted that biobarcode sensor and nanonet sensors appear to be especially promising and worthy of future work. Alam identified that one of the challenges for sensors is that the sensitivity line assumes infinitely long sensors; he noted that in reality, the sensor will fail after some period of time (e.g., 1 hour) due to the environment (e.g., salt). Another interesting application, according to Alam, is to mix the sensor with what is being measured, rather than having it at the bottom of the solution, for example. He indicates that these types of sensors can provide much higher sensitivity without much noise. Finally, Alam concluded with three key points: (1) going for high sensitivity can lead to high fluctuations, (2) under extreme environments, sensors can respond very differently, and (3) selectivity is a fundamental issue, in that it is important to cover up the gaps/spaces in sensors if possible. He stated that nanotechnology will play an important role for biosensors.

Session 3: Propulsion

Steve Son (Purdue University) noted in his presentation that the high surface areas in nanopropellants provides benefits, but also has potential negatives as well (e.g., requiring more binder in solid rockets). Regarding the technologies, he indicated that some characteristics do not change at the nanoscale, whereas other characteristics such as reaction rate do change. Son commented that these properties can be used to tailor ignition and reaction rates, control heat release location, and can affect combustion stability. As an example, he indicated that nanotechnology could be used to increase the fuel regression rate of hybrid rocket motors. According to Son, moving from the micro to nano scale can affect rheology—castings, for example, can become more brittle. Relative to green propellants, he indicated that ammonium nitrate might be better than ammonium perchlorate, and that nanoscale techniques might help ammonium nitrate improve in terms of burn rate. As for multifunctional capabilities, Son noted that solid rockets are multifunctional, and investigating liquids and slurries that could potentially be used for things like energy storage (e.g., fuel cell use) could be done. Son did identify several challenges, however, including the rheology of adding high-surface-area particles, controlling the distribution of nanoparticles (where large agglomerations can lead to loss of benefits), and passivation. He also commented that cost is always a challenge, as well as dealing with health and safety issues.

During the discussion period after Son's presentation, one of the panel members asked what reasonable efficiency increases might be expected in this area. Son responded that it is system dependent; for example, in some cases like a nano-aluminum-ice (nAl-ice) rocket, the propellant is not necessarily higher performance, but enables doing something differently. Another workshop attendee built upon this comment by noting that in the case of nAl-ice, the nano scale is what enables it work, as the propellant will not burn at the micro scale. Son did provide further comments, though, that for hybrid fuels, introducing nanoparticles to increase the regression rate could lead to potentially higher performance. When asked about the status of modeling capability for nanopropellant systems, Son responded that while there have been collaborations with modeling efforts, there is certainly room to do more, and that modeling needs to be tied to experiments to be useful. Finally, in addressing a panel member's question about TRL, Son indicated that some areas (e.g., making green propellants more viable) are near term, but other areas (e.g., using clusters of engines) is much longer term (i.e., 10 years or more).

Richard Yetter (Pennsylvania State University) next gave a presentation on his views of nanotechnology and propulsion. Regarding nanopropellants, he noted that nanopropellants will not necessarily provide higher energy densities, but can provide improved usage of the stored chemical energy. For example, he indicated that nano-ingredients could produce new gelled and solid propellants, and that nanopropellants may be usable in non-conventional applications. Yetter did note that there have been critical technology issues, where self-assembly and supramolecular chemistry of the fuel and oxidizer elements of energetic materials have lagged far behind chemistries in other disciplines (such as pharmaceuticals, microelectronics, microbiology). He indicated that this has led to limited fundamental understanding of what type of supramolecular structures provide desirable performance in combustion, mechanical, and hazard characteristics. Yetter then went in to some detailed discussion regarding self-assembly, surface passivation, graphene catalysts (e.g., to attain higher reaction rates), nano-engineered energetic materials (e.g., physical vapor deposition), and MEMS devices (e.g., micro-igniter). In terms of multifunctionality, Yetter commented that one application might be having a power system built-into the propulsion system.

During the discussion period after Yetter's presentation, he was asked about his thoughts on NASA leading in this technology area. Yetter responded that the U.S. Air Force is putting substantial efforts into this area, and that NASA's efforts are best tuned with a particular mission. He indicated that a good start might be to focus on items at the MEMS scale for self-assembly before moving to much bigger systems. Regarding the top technical challenges, Yetter noted that there is a high risk on passivation and assembly. He also suggested looking at the system implementation—how does using nano affect the system as a whole? Finally, Yetter identified several potential technology gaps, including mass/volume improvements with sensors, the use of graphene as fuel/catalyst, and the fact that safety issues are critical.

Session 4: Energy Generation and Storage

Gary Rubloff (University of Maryland) started his presentation noting that there were three areas in particular that he wanted to address: power generation, integration of nanocomponents, and hierarchical systems. Rubloff provided many detailed comments regarding "random" and "regular" heterogeneous 3D nanostructures, exposed and embedded nanostructures (e.g., applications and differences between the two approaches), engineering aspects of 3D structures (e.g., keeping impurities from impacting performance), process and device integration (e.g., integration at the nano-level to improve volume/weight), and multifunctional nanosystems (e.g., nano-integrated photovoltaic and energy storage systems). Rubloff also discussed what he refers to as the "three self's"—self-assembly, self-alignment, and self-limiting reactions—and how these can be used to keep costs down. In terms of the NASA roadmap, Rubloff indicated that for the top technical challenges, he saw mechanisms for identifying defects to be lacking the most. He also identified some technology gaps, including using modeling and simulation to guide systems design and prioritization, as well as system level strategies for managing defects and reliability. Correspondingly, Rubloff noted that the high-priority areas that NASA should focus on are defect and reliability mechanisms, integrated systems, and model-based system design. Finally, relative to the time horizon of the NASA roadmap, Rubloff commented that the manufacturing equipment for these materials/systems will need to be there in time.

Public Comment and Discussion Session

The following are views expressed during the public comment and discussion session by either presenters, members of the panel, or others in attendance.

- *Technology investment process.* One presenter suggested forming a working group across NASA centers that meets regularly and generates consensus on areas with the largest payoffs with some external review. He mentioned that, out of a $1 billion program, spending $100 million on two or three specific areas to demonstrate significant advancement/commitment might make sense.
- *Areas for NASA to lead.* It was suggested that crosscutting areas like nanotech will be a challenge, and that there is work going in other groups, so it is not clear what NASA should be doing. One response suggested that NASA should take the lead in nanotechnology research related to extreme environments and multifunctional systems. It was raised that on the sensors side, NASA could be a real user/customer but not necessarily a development leader. Another area of suggested emphasis are opportunities in propulsion (e.g., microthrusters). Energy storage was mentioned as another area to look at, but that this might be led by the Department of Energy.
- *Nanotechnology research at other agencies.* There was a discussion on the substantial research in nanotechnology going on at other groups, and it was suggested NASA should look to collaborate and benefit from the large amount invested by nine different agencies. Other agencies were said to be looking at the use of nanotechnology in radiation-hardened electronics, carbon nanotube memory, and ASICs, but no specific agency is coordinating the research. It was noted that the USAF perspective is to compromise between short/high-quality versus long/low-quality carbon nanotubes.
- *Timeframe incorporation of nanotubes into structures.* Clay nanocomposites have been incorporated into structures in the automotive industry for more than 10 years, whereas nanofiber structures have not been used much in commercial applications yet (ceramics for ballistic protection may be coming online in the next 1 to 2 years). There are examples of layers being used, but that they are not there yet for structural applications. This was seen as similar to the history with carbon fiber—initially it appeared in low-complexity sports equipment, and eventually made its way to more complex systems (e.g., aerospace). It was estimated that high-performance structural applications are likely to happen in the next 10 to 15 years.

- *Morphology*. It was stated that performance improvements are very dependent on the fiber being used, and single wall nanotube SWNT has been measured at 50 GPa (i.e., 10 × better than carbon fiber PAN). There are currently groups attempting to develop materials at the needed sizes that can be put together as composites are today. Regarding the question on weight savings in composites, it was noted that there is the ability to use nanotubes to control morphology but that alignment requires nanometer scale control, which is something that does not exist currently. Potential gains may not be that great; i.e., potentially up to 50 percent with substantial investment. Another area to look at was the composite matrix—where it was suggested that there is a significant opportunity in affecting performance via this area (e.g., gains at the 30 percent level).
- *Thermal system applications.* In terms of thermal conductivity, examples were given of nanotubes being used to help carry heat away from electronics/processors. Some groups claim to have seen a 10 percent reduction in the coefficient of thermal expansion (CTE), but that current research has not been focused on this area. It was noted that SWNT have a small but negative CTE, and that designing isotropic materials may be challenging (but also an area in which NASA might benefit).

N

TA11 Modeling, Simulation, and Information Technology and Processing

INTRODUCTION

The draft roadmap for technology area (TA) 11, Modeling, Simulation, and Information Technology and Processing, consists of four technology subareas:[1]

- 11.1 Computing
- 11.2 Modeling
- 11.3 Simulation
- 11.4 Information Processing

NASA's ability to make engineering breakthroughs and scientific discoveries is limited not only by human, robotic, and remotely sensed observation, but also by the ability to transport data and transform the data into scientific and engineering knowledge through sophisticated models. But those data management and utilization steps can tax the information technology and processing capacity of the institution. With data volumes exponentially increasing into the petabyte and exabyte ranges, modeling, simulation, and information technology and processing requirements demand advanced supercomputing capabilities.

Handling and archiving rapidly growing data sets, including analyzing and parsing the data using appropriate metadata, pose significant new demands on information systems technology. The amount of data from observations and simulations is growing much more rapidly than the speed of networks, thus requiring new paradigms: rather than bringing massive data to scientists' workstations for analysis, analysis algorithms will increasingly have to be run on remote databases.

There are also important spacecraft computer technology requirements, including intelligent data understanding, development of radiation-hard multicore chips and GPUs, fault tolerant codes and hardware,[2] and software that runs efficiently on such systems. Another important challenge is developing improved software for reliably simulating and testing complete NASA missions including human components.

[1] The draft space technology roadmaps are available at http://www.nasa.gov/offices/oct/strategic_integration/technology_roadmap.html.

[2] Intelligent adaptive systems technologies for autonomous spacecraft operations are discussed under TA 4.5.1 (vehicle systems management and fault detection and isolation and recovery).

TABLE N.1 Technology Area Breakdown Structure for TA11, Modeling, Simulation, and Information Technology and Processing

NASA Draft Roadmap (Revision 10)	Steering Committee-Recommended Changes
TA11 Modeling, Simulation, Information Technology, and Processing	One technology has been split into two parts.
11.1. Computing	
11.1.1. Flight Computing	
11.1.2. Ground Computing	
11.2. Modeling	
11.2.1. Software Modeling and Model-Checking	
11.2.2. Integrated Hardware and Software Modeling	
11.2.3. Human-System Performance Modeling	
11.2.4. Science and Engineering Modeling	Split 11.2.4 to create two separate technologies: 11.2.4a Science Modeling and Simulation 11.2.4b Aerospace Engineering Modeling and Simulation
11.2.5. Frameworks, Languages, Tools, and Standards	
11.3. Simulation	
11.3.1. Distributed Simulation	
11.3.2. Integrated System Life Cycle Simulation	
11.3.3. Simulation-Based Systems Engineering	
11.3.4. Simulation-Based Training and Decision Support Systems	
11.4. Information Processing	
11.4.1. Science, Engineering, and Mission Data Life Cycle	
11.4.2. Intelligent Data Understanding	
11.4.3. Semantic Technologies	
11.4.4. Collaborative Science and Engineering	
11.4.5. Advanced Mission Systems	

Before prioritizing the level 3 technologies included in TA11, one technology was split into two parts. The changes are explained below and illustrated in Table N.1. The complete, revised technology area breakdown structure (TABS) for all 14 TAs is shown in Appendix B.

Technology 11.2.4, Science & Engineering Modeling (which is actually titled Science and Aerospace Engineering Modeling in the text of the TA11 roadmap), was considered to be too broad. It has been split in two:

- 11.2.4a, Science Modeling and Simulation, and
- 11.2.4b, Aerospace Engineering Modeling and Simulation.

The content of these two technologies is as described in the TA11 roadmap under Section 11.2.4 Science and Aerospace Engineering Modeling, in the subsections titled Science Modeling and Aerospace Engineering, respectively.

TOP TECHNICAL CHALLENGES

The panel identified four top technical challenges for TA11, listed below in priority order.

1. Flight-capable devices and software. Develop advanced flight-capable (e.g., low-power, high-performance, radiation-hard, fault-tolerant) devices and system software for flight computing (e.g., for real-time, autonomous hazard avoidance in landing on planetary surfaces; adaptive telescope mirror technology; smart rovers; and autonomous rendezvous).

The application of increasingly powerful computational capabilities will support more ambitious undertakings, many of which rely on autonomous smart systems. However, many of the advanced devices developed for commercial terrestrial applications are not suited to the space environment. Space applications require devices that are immune, or at least tolerant, of radiation-induced effects, within tightly constrained resources of mass and power. The software design that runs on these advanced devices, with architectures different than current space-qualified devices, also requires new approaches. The criticality and complexity of the software needed for these demanding applications requires further development to manage this complexity at low risk.

2. New Software Tools. Develop new flight and ground computing software tools (and engage trained computer scientists) to take advantage of new computing technologies by keeping pace with computing hardware evolution, eliminating the multi-core "programmability gap," and permitting the porting of legacy codes.

Since about 2004 the increase in computer power has come about because of increases in the number of cores per chip ("multi-core") and use of very fast vector graphical processor units (GPUs) rather than increases in processor speed. NASA has a large budget for new computer hardware, but the challenge of developing efficient new codes for these new computer architectures has not yet been addressed. Major codes are developed over decades, but computers change every few years, so NASA's vast inventory of legacy engineering and scientific codes will need to be re-engineered to make effective use of the rapidly changing advanced computational systems. This re-engineering needs to anticipate future architectures now being developed such as Many Integrated Core (MIC) and other advanced processors with the goals of portability, reliability, scalability, and simplicity. The effort will require both the engagement of a large number of computer scientists and professional programmers and the development of new software tools to facilitate the porting of legacy codes and the creation of new more efficient codes for these new systems. Additional issues that arise as computer systems evolve to millions of cores include the need for redundancy or other defenses against hardware failures, the need to create software and operating systems that prevent load imbalance from slowing the performance of codes, and the need to make large computers more energy efficient as they consume a growing fraction of available electricity.

3. Testing. Improve reliability and effectiveness of hardware and software testing and enhance mission robustness via new generations of affordable simulation software tools.

The complexity of systems comprised of advanced hardware and software must be managed in order to ensure the systems' reliability and robustness. New software tools that allow insight into the design of complex systems will support the development of systems with well understood, predictable behavior while minimizing or eliminating undesirable responses.

4. Simulation Tools. Develop scientific simulation and modeling software tools to fully utilize the capabilities of new generations of scientific computers (e.g., for cross-scale simulations and data assimilation and visualization in Earth science, astrophysics, heliophysics, and planetary science).

Supercomputers have become increasingly powerful, often enabling realistic multi-resolution simulations of complex astrophysical, geophysical, and aerodynamic phenomena. The sort of phenomena that are now being simulated include the evolution of circumstellar disks into planetary systems, the formation of stars in giant molecular clouds in galaxies, and the evolution of entire galaxies including the feedback from supernovas and supermassive black holes. These are also multi-resolution problems, since (for example) one can't really understand galaxy evolution without understanding smaller-scale phenomena such as star formation. Other multi-scale phenomena that are being simulated on NASA's supercomputers include entry of spacecraft into planetary atmospheres and the ocean-atmosphere interactions that affect the evolution of climate on Earth. However, efficient new codes that use the full capabilities of these new computer architectures are still under development.

	Benefit	Alignment with NASA Needs	Alignment with Non-NASA Aero Tech Needs	Alignment with Non-Aerospace National Goals	Technical Risk and Reasonableness	Sequencing and Timing	Time and Effort	QFD Score (Weighted)	Panel Priority
Multiplier	**27**	**5**	**2**	**2**	**10**	**4**	**4**		
	0/1/3/9	**0/1/3/9**	**0/1/3/9**	**0/1/3/9**	**1/3/9**	**-9/-3/-1/1**	**-9/-3/-1/0**		
Technology Name	**Benefit**		**Alignment**			**Risk/Difficulty**			
11.1.1. Flight Computing	9	9	9	3	9	1	-3	394	H
11.1.2. Ground Computing	9	9	9	9	3	1	-1	354	H
11.2.1. Software Modeling and Model-Checking	3	9	9	9	3	-3	-1	176	M
11.2.2. Integrated Hardware and Software Modeling	3	9	9	9	3	1	-1	192	M
11.2.3. Human-System Performance Modeling	1	9	3	3	3	1	-1	114	L
11.2.4a. Science Modeling and Simulation	9	9	9	9	3	1	-1	354	H
11.2.4b. Aerospace Engineering Modeling and Simulation	3	9	9	1	3	-1	-3	160	M
11.2.5. Frameworks, Languages, Tools, and Standards	1	9	3	1	1	1	-1	90	L
11.3.1. Distributed Simulation	3	9	9	9	3	1	-1	192	H*
11.3.2. Integrated System Lifecycle Simulation	1	9	1	0	3	-9	-1	64	L
11.3.3. Simulation-Based Systems Engineering	1	3	9	9	1	-1	-3	72	L
11.3.4. Simulation-Based Training and Decision Support Systems	1	1	1	1	3	1	0	70	L
11.4.1. Science, Engineering, and Mission Data Lifecycle	3	9	9	0	3	1	-1	174	M
11.4.2 Intelligent Data Understanding	1	3	1	0	1	-3	-1	38	L
11.4.3 Semantic Technologies	3	9	1	1	3	1	-1	160	M
11.4.4 Collaborative Science and Engineering	0	9	3	9	3	-3	-9	51	L
11.4.5. Advanced Mission Systems	3	9	9	1	9	-9	-3	188	M

FIGURE N.1 Quality function deployment (QFD) summary matrix for TA11 Modeling, Simulation, and Information Technology and Processing. The justification for the high-priority designation of all high-priority technologies appears in the section "High-Priority Level 3 Technologies." H = High Priority; H* = High Priority, QFD score override; M = Medium Priority; L = Low Priority.

QFD MATRIX AND NUMERICAL RESULTS FOR TA11

Assessment of computing-related technologies is difficult owing to the fact that developments will in many cases be primarily motivated and utilized outside of NASA. NASA is primarily a consumer, adopter, and/or adapter of advanced information technology facilities, with the exception of spacecraft on-board processing. As a result only four technologies rank as highest priority. This does not mean that that other technologies are unimportant to NASA, only that NASA is not viewed as the primary resource for the development of these technologies.

Figures N.1 and N.2 show the relative ranking of each technology. The panel assessed four of the technologies as high priority. Three of these were selected based on their QFD scores, which significantly exceeded the scores of lower-ranked technologies. After careful consideration, the panel also designated one additional technology as a high-priority technology.[3]

Figure N.2 displays the TA11 technologies in order of priority.

The panel's assessment of linkages between the level 3 technologies and top technical challenges is summarized in Figure N.3.

[3] In recognition that the QFD process could not accurately quantify all of the attributes of a given technology, after the QFD scores were compiled, the panels in some cases designated some technologies as high priority even if their scores were not comparable to the scores of other high-priority technologies. The justification for the high-priority designation of all the high-priority technologies for TA11 appears below in the section "High-Priority Level 3 Technologies."

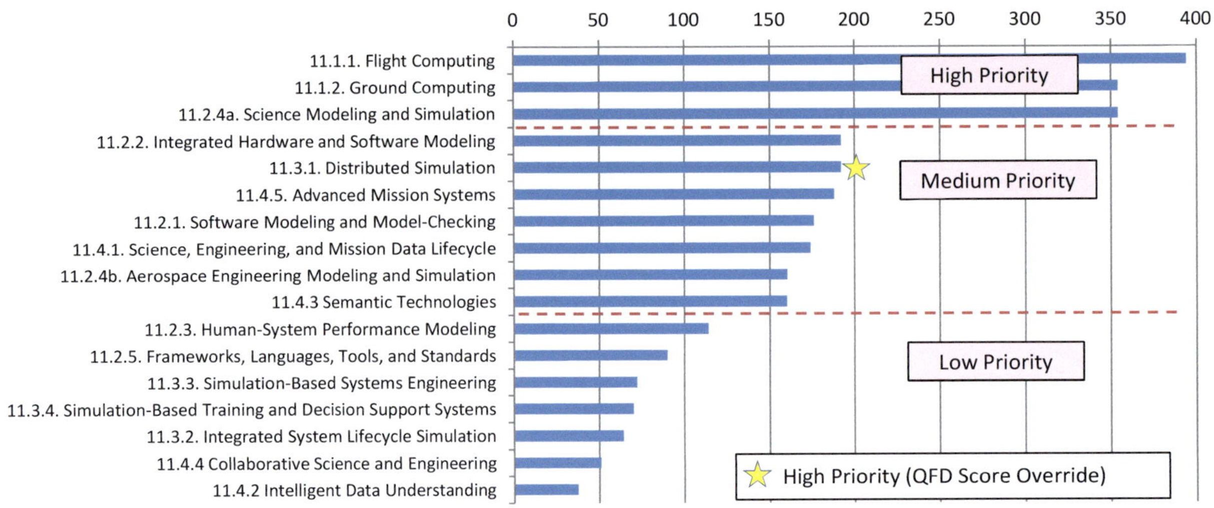

FIGURE N.2 Quality function deployment rankings for TA11 Modeling, Simulation, and Information Technology and Processing.

HIGH-PRIORITY LEVEL 3 TECHNOLOGIES

Panel 3 identified four high priority technologies in TA11. The justification for ranking each of these technologies as a high priority is discussed below.

Technology 11.1.1, Flight Computing

Flight computing technology encompasses low-power, radiation-hardened, high-performance processors. These will continue to be in demand for general application in the space community. Special operations, such as autonomous landing and hazard avoidance, are made practical by these high-performance processors. Processors with the desired performance (e.g., multi-core processors) are readily available for terrestrial applications; however, radiation-hardened versions of these are not.

A major concern is ensuring the continued availability of radiation-hardened integrated circuits for space. As the feature size of commercial integrated circuits decreases, radiation susceptibility increases. Maintaining production lines for radiation-hardened devices is not profitable. Action may be required if NASA and other government organizations wish to maintain a domestic source for these devices, or a technology development effort may be required to determine how to apply commercial devices in the space environment. For example, using multi-core processors with the ability to isolate cores that have experienced an upset may be one approach. It is not unreasonable to assume such a course may be the only option to continue to fly high-performance processors in the future.

The associated risk ranges from moderate to high, depending on the approach taken to ensure continued access to these devices. Maintaining existing production lines may be prohibitively expensive and may result in performance-constrained devices. However, such action may be necessary to maintain device availability until safe, reliable application of commercial multi-core products is achieved.

This technology is well aligned with NASA's expertise and capabilities, as evidenced by NASA's long collaboration with industry partners to develop such devices. This technology has applications throughout all aspects of the space community: civil government, national security, and commercial space. Access to the space station would not benefit this technology development.

This technology will have significant impact because advanced computer architectures, when eventually incorporated into radiation-hardened flight processors, can be expected to yield major performance improvements in on-board computing throughput, fault management, and intelligent decision making and science data acquisition, and will enable autonomous landing, hazard avoidance. Its use is anticipated across all classes of NASA missions.

Priority	TA 11 Technologies, Listed by Priority	Top Technology Challenges			
		1. Flight-Capable Devices and Software: Develop advanced flight-capable (e.g., low-power, high-performance, radiation-hard, fault-tolerant) devices and system software for flight computing (e.g., for real-time, autonomous hazard avoidance in landing on planetary surfaces; adaptive telescope mirror technology; smart rovers; and autonomous rendezvous).	2. New Software Tools: Develop new flight and ground computing software tools (and engage trained computer scientists) to take advantage of new computing technologies by keeping pace with computing hardware evolution, eliminating the multi-core "programmability gap," and permitting the porting of legacy codes.	3. Testing: Improve reliability and effectiveness of hardware and software testing and enhance mission robustness via new generations of affordable simulation software tools.	4. Simulation Tools: Develop scientific simulation and modeling software tools to fully utilize the capabilities of new generations of scientific computers (e.g. for cross-scale simulations and data assimilation and visualization in Earth science, astrophysics, heliophysics, and planetary science).
H	11.1.1. Flight Computing	●	○		
H	11.1.2. Ground Computing		●		○
H	11.2.4a. Science Modeling and Simulation		●		●
H	11.3.1. Distributed Simulation		○	○	●
M	11.2.2. Integrated Hardware and Software Modeling		○	●	
M	11.4.5. Advanced Mission Systems	○	○	○	
M	11.2.1. Software Modeling and Model-Checking		○	●	
M	11.4.1. Science, Engineering, and Mission Data Lifecycle		○		
M	11.2.4b. Aerospace Engineering Modeling and Simulation	○	○	○	
M	11.4.3. Semantic Technologies				
L	11.2.3. Human-System Performance Modeling		○	○	
L	11.2.5. Frameworks, Languages, Tools, and Standards		○	○	○
L	11.3.3. Simulation-Based Systems Engineering		○	○	
L	11.3.4. Simulation-Based Training and Decision Support Systems		○	○	
L	11.3.2. Integrated System Lifecycle Simulation		○	○	
L	11.4.4. Collaborative Science and Engineering		○	○	○
L	11.4.2. Intelligent Data Understanding		○		○

●	Strong Linkage: Investments by NASA in this technology would likely have a major impact in addressing this challenge.
○	Moderate Linkage: Investments by NASA in this technology would likely have a moderate impact in addressing this challenge.
[blank]	Weak/No Linkage: Investments by NASA in this technology would likely have little or no impact in addressing the challenge.

FIGURE N.3 Level of support that the technologies provide to the top technical challenges for TA11 Modeling, Simulation, and Information Technology and Processing.

Technology 11.1.2, Ground Computing

Ground computing technology consists of programmability for multi-core/hybrid/accelerated computer architectures, including developing tools to help port existing codes to these new architectures.

After about 2004, major improvements in computation have come from increasing numbers of compute cores per chip and improvements in accelerated processors (vector graphic processor units, GPUs), while before 2004 improvements came mainly from steadily increasing clock speed of individual compute cores. However, the vast library of legacy engineering and scientific codes does not run efficiently on the new computer architectures. Technology development is needed to create software tools to help programmers convert legacy codes and new algorithms so that they run efficiently on these new computer systems. Related challenges are developing improved compilers and run-time algorithms that improve load balancing in these new computer architectures, and developing methods to prevent computer hardware failures in systems with hundreds of thousands to millions of cores from impacting computational reliability. All users of high-performance computers face these challenges, and NASA can work on these issues with other agencies and industrial partners. Continuous technology improvements will be required as computer system architectures steadily change. Improved technology is likely to be widely applicable across NASA, the aerospace community, and beyond. Access to the space station is not needed.

This technology is game-changing because computer hardware capability has been increasing exponentially with new multi-core and accelerator hardware, but software has not been keeping pace with hardware. Solving the programmability gap has the potential to give 2 to 3 orders of magnitude improvement in computing capability, with a wide range of impact.

Technology 11.2.4a, Science Modeling and Simulation

Panel 3 split the original 11.2.4 Science and Engineering Modeling technology into two technologies, which were rated separately:

- 11.2.4a Science Modeling and Simulation: high priority.
- 11.2.4b Aerospace Engineering Modeling and Simulation: medium priority.

The 11.2.4a technology consists of multi-scale modeling, which is required to deal with complex astrophysical and geophysical systems with a wide range of length scales or other physical variables. Better methods also need to be developed to compare simulations with observations to improve physical understanding of the implications of rapidly growing NASA data sets.

Developing multi-scale models and simulations is an ongoing challenge that impacts many areas of science. Progress in this technology will require steady improvement in codes and methodology. As scientists attempt to understand increasingly complex astrophysical and geophysical systems using constantly improving data, the challenge grows to develop better methods to integrate data from diverse sources and compare these data with simulations in order to improve scientific understanding. Alignment with NASA is very good. Access to the space station is not needed.

This technology will have significant impact because it optimizes the value of observations by elucidating the physical principles involved. This capability could impact many NASA missions, and improved modeling and simulation technology is likely to have wide impacts in diverse mission areas.

Technology 11.3.1, Distributed Simulation

Distributed simulation technologies create the ability to share simulations between software developers, scientists, and data analysts, and thus, greatly enhance the value of the large investments of the simulation, which currently can require tens of millions of CPU hours. There is a need for large-scale, shared, secure, distributed environments with sufficient interconnect bandwidth and display capabilities to enable distributed simulation (processing) as well as distributed analysis and visualization of data produced by simulations. (Large simulations can typically generate terabytes of data resulting in the need for advanced data management and data mining technologies.)

The panel overrode the QFD score for this technology to designate it as a high-priority technology because the QFD scores did not capture the value this technology could provide in terms of major efficiency improvement supporting collaborations, particularly interdisciplinary studies that would benefit numerous NASA missions in multiple areas. In addition, it would have a broad impact within non-NASA aerospace as well as a broad impact on many non-aerospace communities.

MEDIUM- AND LOW-PRIORITY TECHNOLOGIES

One group of medium-priority technologies includes 11.2.2 Integrated Hardware and Software Modeling; 11.4.5 Advanced Mission Systems; and 11.2.1 Software Modeling and Model Checking. These could each provide a major (but not game changing) benefit, did not have the lowest possible scores in any category, and have a common theme that development cost can be lowered.

A second group of medium-priority technologies includes 11.4.1 Science, Engineering, and Mission Data Lifecycle; 11.2.4b Engineering Modeling and Simulation; and 11.4.3 Semantic Technologies. As system modeling and simulation capabilities advance, the modeling of system safety performance due to complex interactions within the system, with the external environment, and including anomalous behaviors are also expected to improve. The technologies in this group could each provide a major (but not game changing) benefit, did not have the lowest possible scores in any category, and have a common theme that no significant new technology is required.

One group of low-priority technologies includes 11.2.3 Human-System Performance Modeling (again including system safety performance as noted above); 11.2.5 Frameworks, Languages, Tools, and Standards; and 11.3.3 Simulation-Based System Engineering. These could each provide a minor benefit, had no low scores in any category, and have a common theme that NASA OCT does not need to invest to further development, which is already underway.

A second group of low-priority technologies includes 11.3.4 Simulation-Based Training and Decision Support Systems; 11.3.2 Integrated System Lifecycle Simulation; and 11.4.4 Intelligent Data Understanding. The latter two each had a low score in one category. These could each provide a minor benefit and have a common theme that part of all of the technology already exists.

The remaining low-priority technology was 11.4.2 Collaborative Science and Engineering. It had a low score of zero for benefit; in the view of Panel 3, it would provide no significant benefit, and much of this is being done today.

WORKSHOP SUMMARY

The Instruments and Computing Panel (Panel 3) for the NASA Technology Roadmaps study held a workshop on May 10, 2011, at the National Academies Keck Center in Washington, D.C. It focused on Modeling, Simulation, Information Technology, and Processing (NASA Technology Roadmap TA11). The workshop was attended by members of Panel 3, one or more members of the Steering Committee for the NASA Technology Roadmaps study, invited workshop participants, study staff, and members of the public who attended the open sessions. The workshop began with a short introduction by the Panel 3 chair, followed by a series of four panel discussions, then a session for public comment and general discussion, and finally a short summary and wrap-up by the Panel 3 chair. Each panel discussion was moderated by a Panel 3 member. Experts from industry, academia, and/or government were invited to present.

Panel Discussion 1: Simulation of Engineering Systems

The first session of the day focused on simulation of engineering systems and was moderated by Alan Title.

Greg Zacharias (Charles River Analytics) gave a presentation on simulation-based systems engineering. He noted that systems engineering is a very human-intensive process often with very informal and implied specifications and requirements. He believes the technical challenge of automating the processes is getting computers to think more like humans or getting human concepts into machine-readable forms. He thinks that this capability would be game changing. He also noted that another game changing potential gap in the roadmap is improved modeling of human operators in simulations. Full end-to-end systems modeling, or "digital twin," requires human operator modeling at the right level of perceptual/cognitive/motor fidelity. This also ties in with another point that he raised of the need to improve multi-resolution modeling and simulations of systems (including human operators).

Amy Pritchett (Georgia Institute of Technology) gave the second presentation focused on aviation, but with discussions of how it relates to space-based missions. Safety drivers include coupled interdependent behaviors (hardware, software, and human dynamics). She noted that the NRC decadal survey for aeronautics rated developing complex interactive adaptive systems as the most significant challenge in developing flight critical systems. She believes that the goal should be to simulate overall systems and processes (including all the people in the loop, all the vehicles, components, etc.). Interdependencies among systems need to be modeled—not just among hardware and software systems, but human operators and also organizational aspects. Modeling and simulation consists not only of physics-based modeling, but also computational systems, communication behavior, and the cognitive behaviors of humans in the system. She sees components of such an integrated model in the roadmap individually, but there doesn't appear to be anything about bringing these things together. She also agrees that there is a significant challenge of properly scaling individual models when they are linked together to model complete architectures.

Panel Discussion 2: Re-Engineering Simulation, Analysis, and Processing Codes

The next session focused on the new classes of programming languages and how to adapt to new multi-core computers and was moderated by Joel Primack.

Bill Matthaeus (Bartol Research Institute and University of Delaware) gave the first presentation, in which he related his views as both a theoretical and computational physicist. He described the challenges of dividing the responsibilities between computational scientists and end users. He noted the evolving computational paradigm shifts with the latest being the move to multicore processors working in clusters. He discussed the issue of to what degree the end users need to retool for each paradigm shift and how software compilers can ease the transitions. He noted the trend of end users becoming more detached from the details of the codes and software packages and perhaps putting too much trust in them without verifying. He described his concern that modern compilers have become less robust and have produced unstable code, code that is not transportable, or code that produces different results on different computers. As an example of the complexity of the simulations, he referred to his work on developing three dimensional magnetohydrodynamic models of the heliosphere (e.g., space weather) and compared it to terrestrial global circulation models for climate or calculating the flow around a 747 from first principles.

Bronson Messer (Oak Ridge National Laboratory) gave the second presentation and started with a review of ORNL's current capabilities. The lab has three petascale platforms in a single room and the world's second most powerful computer with over 2 petaflops (2 quadrillion floating point operations per second) of computing power. ORNL is in the midst of building a 10 to 20 petaflops computer, and by end of the decade is looking toward an exaflop computer. He reported that currently the most difficult physical challenge in large supercomputers is the power needed to move data between memory and processors and from node to node. This will change with some of the upcoming advancements. He sees the proposed exascale systems (beyond existing petascale systems) being significantly different by becoming more hierarchical and heterogeneous with increasing on-chip parallelism used to improve performance. He sees a significant challenge in dealing with the complexity of these systems and developing new programming models. He said the solution lies in a programming model that abstracts some of the architectural details from software developers. In a recent survey of potential users, there was a strong preference of evolutionary developments using current languages and tools such as MPI (Message Passing Interface).

After Messer's presentation, there was a lengthy discussion period in which Messer answered audience and panel questions. He said that NASA does not currently invest a significant amount in advancing the state of the art of supercomputing. He discussed the challenges of dealing with the massive amount of data generated and the best way of transferring the results back to the users. Finally there was some discussion that NASA will need to address the new computing paradigms of multi-core systems for their flight computers.

Panel Discussion 3: Intelligent Data Understanding, Autonomous S/C Operations, and On-Board Computing

The next session focused on on-board spacecraft processing and improving autonomy of operations and data processing. The session was moderated by David Kusnierkiewicz.

George Cancro (JHU/APL) gave the first presentation in which he provided several criticisms of the draft roadmap. His main suggestion was that the roadmap should be more holistic showing the linkages between modeling, simulation, autonomy, and operational software advancements as they are all interrelated. He also saw a lack of specific benefits and purposes for technology concepts across the entire roadmap, and believes that the fundamental question "Why is this needed for NASA?" is not addressed correctly throughout. As an example, he said the entire section for on-board computing is focused on multi-core processors, but there does not appear to be anything regarding what specific missions require multi-core processors. He identified gaps in the roadmap as: no discussions on on-board computing for large data flows; no discussion on virtual observatories, clearing houses, search engines and other tools for NASA science data necessary to perform multi-mission data analysis or anchor models; no discussions of frameworks or processes to enable modeling and simulation. He noted the significant potential for autonomous and adaptive systems and said their single biggest challenge is testing. He referenced the new Air Force roadmap "Technology Horizons" which identified "trusted autonomy" as a top issue, which he believes can only be solved through advances in testing of autonomous systems.

Noah Clemons (Intel) gave the second presentation. He felt that the roadmap is too general and instead needs to address four domains: efficient performance; essential performance (application coding with today's multi-core processors); advanced performance (more advanced, cross platform); and distributed performance (high-performance MPI clusters). He noted that a lot of existing applications were never intended to be run on parallel processors and there is a significant challenge in converting code to run on parallel processors, either by changing serial codes or by writing new codes. He emphasized the need to target these features: portability, reliability, scalability, and simplicity. He warned against putting too much focus on one particular computational architecture. He sees having to recode for a new architecture or paradigm every few years. He sees things currently heading toward one computer processor that is heterogeneous—CPU and GPU all embedded together—but in the future, that whole model will likely disappear. He believes that one game changing parallel programming technology is the idea of programming in tasks rather than managing individual threads. With this type of programming technology, programs would be structured to take advantage of highly parallel hardware by focusing on scaling (with cores) and vectorization by coding at a high level. There are several parallel programming solutions embracing "tasks rather than threads" that are built into the compiler; others are built in libraries. He thinks that structured parallel patterns that can be used as building blocks with little or no cost required (i.e., you don't have to write everything from scratch) are near a tipping point. Many of these tools already have some portability component built in. He also feels that a small investment should be made in adapting analysis tools, as some are already available to assist on how to parallelize code, optimize/improve code, and tune code.

Panel Discussion 4: Data Mining, Data Management, Distributed Processing

The final session of the day focused on using and managing data and was moderated by Robert Hanisch.

Peter Fox (Rensselaer Polytechnic Institute) gave the first presentation in which he expressed his views on how the roadmap covered data management and processing. He believes there are some gaps in the roadmap including data integration or integrate-ability and handling data fitness (quality, uncertainty, and bias). He suggests a modest investment up-front in terms of how to integrate lots of data sets from different spacecraft—as opposed to a much more difficult process after the fact. He expressed a need for collaborative development between data people (rarely in the picture up-front), algorithm developers, instrument designers, etc. There was a question from the audience regarding whether this is technology or management. Fox's response was that it is both and gave an example of a system called Giovanni which greatly improves the productivity of scientists. In terms of data fitness, from his experiences with Earth science, he seemed most concerned with bias. He noted that bias can be systematic errors resulting from distortion of measurement data; or it can be bias of people using/processing/understanding the data.

Finally, he felt that the roadmap does not correctly capture the status and future of semantic technologies as they are already in widespread production in NASA but advancements may revolutionize how science is done.

Arcot Rajasekar (University of North Carolina at Chapel Hill) gave the second presentation which focused on integrating data life cycles with mission life cycles. He discussed the challenges in end-to-end capability for exascale data orchestration. He noted that NASA has massive amounts of mission data and there is a need to share all this data over long timeframes without loss. He suggests an integrated data and metadata system, so that the data are useful for future users but currently there is no coherent technology in the roadmap to meet these needs. He believes that the current roadmap showcases the need for data-intensive capability at various levels but provides limited guidance on how to pull and push this technology. He said the information processing roadmap is very impressive but needs a corresponding "evolutionary" data orchestration roadmap. In his view, game-changing challenges to NASA include policy-oriented data life-cycle management (manage the policies, and let the policy engine manage the bytes and files); agnostic data visualization technologies; service-oriented data operations; and distributed cloud storage and computing. However, he thinks the greatest challenge for NASA is a comprehensive data management system (as opposed to doing a stove-piped approach for each mission) and noted that the technology is out there; it just needs to be done. It would be a paradigm shift to go toward an exascale data system that is data-oriented, policy-oriented, and outcome-oriented (i.e., a system that captures behavior in terms of data outcomes).

Neal Hurlburt (Lockheed Martin) gave the third and final presentation on data systems. He started with a summary of a recent study which found a need for community oversight of emerging, integrated data systems. He believes that the top challenges for NASA include current data services are not sufficiently interoperable; the cost of future data systems will be dominated by software development rather than computing and storage; uncoordinated development and an unpredictable support lifecycle for infrastructure and data analysis tools; and the need for a more coordinated approach to data systems software. However, he thinks that NASA can exploit emerging technologies for most of their needs in this area without investing in development. He believes that NASA's role should be to develop infrastructure for virtual observatories, establish reference architectures/standards, encourage semantic technologies to integrate with astronomy and geophysics communities and provide support for integrated data analysis tools. He sees the widespread use of consistent metadata/semantic annotation as near a tipping point.

Public Comment Session and General Discussion

At the end of the workshop there was some time set aside for general discussion and to hear comments from the audience. This session was moderated by Carl Wunsch (MIT).

Discussion started along the lines of NASA's role in information technology and processing technology development. It was noted that a lot of these topics are not unique to NASA and there are significant efforts initiating elsewhere, for instance in industry and commercial companies. Some expressed their view that NASA is more of a beneficiary than a key player in the technology development.

It was noted that a key difference for NASA relative to commercial endeavors is NASA's focus on minimizing risk, particularly with regard to flight systems. There was some agreement that much of what is commercially available is not compact enough, reliable enough, low power enough, etc., to fly in space. An example was given of radiation-hardened computing power or CCDs; industry is way ahead technologically, but it can no longer be used in space. It was noted that space technology is so far behind in those areas that the old technology cannot be purchased in the marketplace. It was suggested that NASA needs to team with DOD, which has deeper pockets and similar objectives. Someone also warned that if NASA does not develop something because it assume commercial interests will do it—but commercial will only do it if there's economic payoff—there is a risk that NASA/science will be at the mercy of the market.

The example of radiation hardened electronics led to some further detailed discussion. It was noted that a lot of computing now is being done with radiation-hardened FPGAs and ASICs, which are readily available. It was mentioned that FPGAs are harder to validate and every ASIC manufacturer has its own set of simulators, compilers, etc. It was suggested that fault tolerance could be approached in a different way. A proposed technology challenge was made of developing radiation-hardened design using current technology for integrated circuits that does not need specialized facilities to produce.

Some discussion in the workshop focused on data systems (especially regarding science data), and the presenters handled engineering data and science data interchangeably, but NASA handles these two domains very differently. Every mission has its own Context-Driven Content Management system to do configuration management, which is part of product data lifecycle management. It was suggested that there should be an effort to advance the state-of-the-art and share that technology across missions for systems engineering and intelligent data understanding. It was said that the solution does not have to be homogenous, as missions really are different. Finally, it was noted that technology for smaller missions has not been addressed, i.e., common buses. There was agreement that there will not be a large mission in the coming years, and smaller missions are becoming more and more expensive.

O

TA12 Materials, Structures, Mechanical Systems, and Manufacturing

INTRODUCTION

The draft roadmap for technology area (TA) 12 Materials, Structures, Mechanical Systems, and Manufacturing, is organized into five level 2 technology areas:[1]

- 12.1 Materials
- 12.2 Structures
- 12.3 Mechanical Systems
- 12.4 Manufacturing
- 12.5 Cross-Cutting

The TA12 portfolio is extremely broad and differs from most other TAs in that it consists of enabling core disciplines and encompasses fundamental new capabilities that directly impact the increasingly stringent demands of NASA science and exploration missions. These missions depend highly on advancements such as lighter and stronger materials and structures, with increased reliability and with reduced manufacturing and operating costs. Identified technologies are truly interdisciplinary and support virtually all of the other TAs.

In TA12, NASA identified two critical areas: human radiation protection and reliability technologies. Long-term human exploration will require new radiation protection technology, i.e., lightweight radiation-shielding materials, multifunctional structural design and innovative manufacturing. Crosscutting technologies will be required to ensure extremely reliable vehicles and systems for safe travel to destinations millions of miles from Earth.

Before prioritizing the level 3 technologies included in TA12, the panel considered whether to rename, delete, or move technologies in the technology area breakdown structure (TABS). No changes were recommended for TA12. The TABS for TA12 is shown in Table O.1, and the complete, revised TABS for all 14 TAs is shown in Appendix B.

[1] The draft space technology roadmaps are available at http://www.nasa.gov/offices/oct/strategic_integration/technology_roadmap.html.

TABLE O.1 Technology Area Breakdown Structure for TA12, Materials, Structures, Mechanical Systems, and Manufacturing

NASA Draft Roadmap (Revision 10)	Steering Committee-Recommended Changes
TA12 Materials, Structures, Mechanical Systems, and Manufacturing	The structure of this roadmap remains unchanged.
12.1. Materials	
12.1.1. Lightweight Structure	
12.1.2. Computational Design	
12.1.3. Flexible Material Systems	
12.1.4. Environment	
12.1.5. Special Materials	
12.2. Structures	
12.2.1. Lightweight Concepts	
12.2.2. Design and Certification Methods	
12.2.3. Reliability and Sustainment	
12.2.4. Test Tools and Methods	
12.2.5. Innovative, Multifunctional Concepts	
12.3. Mechanical Systems	
12.3.1. Deployables, Docking, and Interfaces	
12.3.2. Mechanism Life Extension Systems	
12.3.3. Electro-mechanical, Mechanical, and Micromechanisms	
12.3.4. Design and Analysis Tools and Methods	
12.3.5. Reliability/Life Assessment/Health Monitoring	
12.3.6. Certification Methods	
12.4. Manufacturing	
12.4.1. Manufacturing Processes	
12.4.2. Intelligent Integrated Manufacturing and Cyber Physical Systems	
12.4.3. Electronics and Optics Manufacturing Process	
12.4.4. Sustainable Manufacturing	
12.5. Crosscutting	
12.5.1. Nondestructive Evaluation and Sensors	
12.5.2. Model-Based Certification and Sustainment Methods	
12.5.3. Loads and Environments	

TOP TECHNICAL CHALLENGES

The panel identified six top technical challenges for TA12. These are described briefly below in priority order. While not inconsistent with those identified in the NASA roadmap document itself, they differ in that there was no attempt to explicitly include challenges in each of the level 2 areas represented; that is: materials, structures, mechanical systems, manufacturing, and crosscutting.

1. Multifunctional Structures. Conceive and develop multifunctional structures, including shielding, to enable new mission capabilities such as long-duration human spaceflight, and to reduce mass.

Structures carry load and maintain shape. To the extent that a structure can simultaneously perform additional functions, especially those that would normally require add-on systems, mission capability can be increased with decreased mass. Integral shielding to reduce radiation exposure and micrometeoroid and orbital debris (MMOD) risk for human spaceflight missions would be game-changing, and the ISS would be useful to verify such concepts. Other advanced multifunctional structures concepts would enable structures, including joints, to provide thermal protection and control, electrical signal and power transmission, electrical energy and fuel storage, self-sensing and healing, and active shape control. Improved cryogenic boil-off protection, for instance, would considerably reduce the mass required for a Mars mission. Such multifunctional materials and structures will require new design analysis tools and might exhibit new failure modes; these should be understood for use in systems design and space systems operations.

2. Reduced Mass. Reduce mass of launch vehicle, spacecraft, and propulsion structures to increase payload mass fraction, improve mission performance, and reduce cost.

Lightweight materials and structures are required to enhance mission performance and enable new mission opportunities. Advanced composites and revolutionary structural concepts would substantially reduce structural weight in launch vehicles, cryo-tanks, propulsion systems, and spacecraft, increasing the payload mass fraction. More energetic propellants would reduce fuel mass in solid motors, and higher-temperature and lower-erosion materials would reduce the weight of engine nozzles. Reduced mass of inflatable habitats and space structures, deployable space systems, and large-scale structures would enable new exploration and science missions.

3. Computational Modeling. Advance new validated computational design, analysis, and simulation methods for materials and structural design, certification, and reliability.

First-principles physics models offer the game-changing potential to guide tailored computational materials design. Multi-scale models are needed to encompass composite materials, interfaces, failure, multi-component and deployable structures, and integrated control systems; multi-physics models are needed to address manufacturing processes, operation in extreme environments, and active materials. Conservatism is embedded in established design methodology in several ways, including statistics-based material allowables and traditional factors of safety. Uncertainty management and quantification, if supported by an experimental foundation, offers the potential to reduce weight as well as certification and life-cycle costs by rationalizing sometimes excessive conservatism. Physics-based and computation-based errors can be quantified and compared to required accuracy and confidence levels. A validated computational modeling methodology could provide the basis for certification by analysis, with experimental evidence, as available, used to verify and improve confidence in the suitability of a design. Computational models will be needed to design-in improved reliability, as well as to interpret measurements made by health-monitoring systems. Structures may need to be designed differently to accommodate health monitoring, including unobtrusive sensors and sensor integration, and to enable materials and structures health assessment and sustainability for long-duration missions.

4. Large-Aperture Systems. Develop reliable mechanisms and structures for large-aperture systems. These must be stowed compactly for launch, yet achieve high-precision final shapes.

Numerous NASA missions employ mechanical systems and structures that must deploy reliably in extreme environments, often to achieve a desired shape with high precision. Such systems include instrument arms, antennas, optical surfaces, solar sails, and some re-entry thermal protection systems. These can be deployed, assembled, or manufactured in space, and may involve flexible materials. Modularity and scalability are desirable features of such concepts, and may require development of autonomous adaptive control systems and technology to address critical functional elements and materials. Concerns include sliding joints and bearings, friction and tribology, coatings and lubrication, as well as their performance and durability over extended periods in storage and extreme operational environments. Performance of large precise space systems cannot be directly verified in the 1-*g* ground environment, so the ISS would be useful for verification of such concepts.

5. Structural Health Monitoring. Enable structural health monitoring and sustainability for long-duration missions, including integration of unobtrusive sensors, and responsive on-board systems.

Mission assurance would be enhanced by an integrated structural health monitoring system that could detect and assess the criticality of in-service damage or fault, then define an amelioration process or trigger a repair in self-healing structures. Such a system requires light, reliable, rugged, unobtrusive, and multifunctional sensors that can be integrated into the structure along with power and data transmission capability. Software to combine disparate data, to diagnose and predict structural health, and to enable the necessary repairs is also a significant challenge. An autonomous integrated on-board systems capability would be game-changing for long-duration, remote missions.

6. Manufacturing. Enable cost-effective manufacturing for reliable high-performance structures and mechanisms made in low-unit production, including in-space manufacturing.

Advanced NASA space missions need affordable structures, electronics systems, and optical payloads. Affordable high-performance structures require advances in manufacturing technology. Such advances include automation using reusable flexible tooling; database- and model-based simulation to ensure selection of the lowest-cost yet reliable and scalable approach; non-autoclave processes for polymer matrix composites to minimize infrastructure investment; in-space manufacture and assembly of large structures such as fuel depots; and means for cost-effective manufacture of lightweight precision optical systems for large structures. In-space manufacturing offers the potential for game-changing weight savings and new mission opportunities; as an example, NASA and DARPA recently pursued the possibility of manufacturing large optical systems in space. The ISS could be used to demonstrate lightweight in-space structures manufacturing capability.

QFD MATRIX AND NUMERICAL RESULTS FOR TA12

Figure O.1 shows the panel's consensus ratings of the 23 level 3 technologies for the TA12 roadmap.

Clearly, benefit is the major discriminator among these technologies, while technical risk and reasonableness is the second most important discriminator. Alignment is a less significant discriminator. Most TA12 technologies have the potential to impact multiple NASA missions in multiple areas because *every* mission would benefit from reduced structural mass, and most would benefit from improved structural reliability and reduced cost.

Figure O.2 shows the consensus rankings of the level 3 technologies. As shown in the figure, 12.2.5 Innovative Multifunctional Concepts received the highest QFD score. A couple of break points in the consensus scores facilitate sorting into relative high, medium, and low-priority categories. The two top medium-ranked technologies were promoted into the high-priority category because of their close relationship to other high-priority technologies. 12.3.5 Reliability/Life Assessment/Health Monitoring is important in its own right, and it closely supports 12.3.1 Deployables, Docking and Interfaces. 12.4.2 Intelligent Integrated Manufacturing and Cyber Physical Systems[2] supports a number of other high-priority technologies and NASA missions.

CHALLENGES VERSUS TECHNOLOGIES

Figure O.3 shows the relationship between the 23 individual level 3 TA12 technologies and the top technical challenges.

Note that the lowest-priority technologies as determined by the QFD rankings tend not to be strongly connected to the top technical challenges. (These are identified by an "L" in the left-most column, and are linked to the top challenges mainly by open circles.) All of the high-priority technologies and many of the medium-priority ones have a strong connection to at least one of the top technical challenges. This shows a good level of consistency in the evaluations by the panel.

[2] A cyber-physical system (CPS) features tight coordination between computational and physical elements. A CPS typically involves a network of interacting elements, and is closely tied to concepts of robotics and sensor networks.

Technology Name	Benefit	Alignment with NASA Needs	Alignment with Non-NASA Aero Tech Needs	Alignment with Non-Aerospace National Goals	Technical Risk and Reasonableness	Sequencing and Timing	Time and Effort	QFD Score (Weighted)	Panel Priority
Multiplier									
	0/1/3/9	0/1/3/9	0/1/3/9	0/1/3/9	1/3/9	-9/-3/-1/1	-9/-3/-1/0		
	Benefit	Alignment			Risk/Difficulty				
12.1.1. (Materials) Lightweight Structure	3	9	9	9	9	-1	-3	236	H
12.1.2. (Materials) Computational Design	3	9	9	9	1	1	-3	164	M
12.1.3. Flexible Material Systems	3	9	3	3	3	1	-1	168	M
12.1.4. (Materials) Environment	3	9	3	1	3	1	-3	156	M
12.1.5. Special Materials	1	9	1	1	3	-1	-1	98	L
12.2.1. (Structures) Lightweight Concepts	3	9	9	9	9	1	-3	244	H
12.2.2. (Structures) Design and Certification Methods	3	9	9	9	9	-1	-3	236	H
12.2.3. (Structures) Reliability and Sustainment	1	9	3	3	3	-1	-1	106	L
12.2.4. (Structures) Test Tools and Methods	1	9	3	1	3	-3	-1	94	L
12.2.5. (Structures) Innovative, Multifunctional Concepts	9	9	9	9	3	1	-3	346	H
12.3.1. Deployables, Docking, and Interfaces	3	9	3	1	9	-1	-1	216	H
12.3.2. Mechanism Life Extension Systems	1	9	1	1	3	-3	-1	90	L
12.3.3. Electro-mechanical, Mechanical and Micromechanisms	1	9	1	1	3	-3	-1	90	L
12.3.4. (Mechanisms) Design and Analysis Tools and Methods	3	9	9	9	9	-3	-3	228	H
12.3.5. (Mechanisms) Reliability / Life Assessment / Health Monitoring	3	9	9	9	3	-1	-1	184	H*
12.3.6. (Mechanisms) Certification Methods	3	9	9	9	3	-1	-3	176	M
12.4.1. Manufacturing Processes	3	9	9	9	3	-3	-1	176	M
12.4.2. Intelligent Integrated Manufacturing and Cyber Physical Systems	3	9	9	9	3	-1	-1	184	H*
12.4.3. Electronics and Optics Manufacturing Process	1	9	3	3	3	-3	-1	98	L
12.4.4. Sustainable Manufacturing	1	9	3	3	1	-3	-1	78	L
12.5.1. Nondestructive Evaluation and Sensors	3	9	9	9	9	-1	-3	236	H
12.5.2. Model-Based Certification and Sustainment Methods	3	9	9	3	3	-1	-3	164	M
12.5.3. Loads and Environments	1	9	3	3	3	-1	-3	98	L

FIGURE O.1 Quality function deployment (QFD) summary matrix for TA12 Materials, Structures, Mechanical Systems, and Manufacturing. The justification for the high-priority designation of all high-priority technologies appears in the section "High-Priority Level 3 Technologies." H = High Priority; H* = High Priority, QFD score override; M = Medium Priority; L = Low Priority.

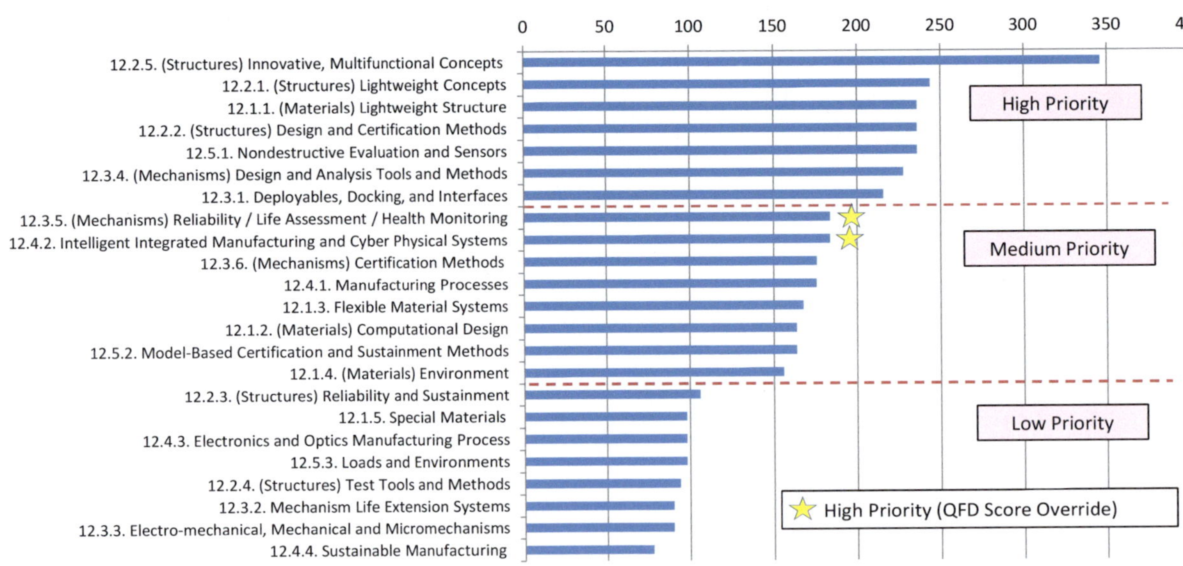

FIGURE O.2 Quality function deployment rankings for TA12 Materials, Structures, Mechanical Systems, and Manufacturing.

Priority	TA 12 Technologies, Listed by Priority	Top Technology Challenges: 1. Multifunctional Structures: Conceive and develop multifunctional structures, including shielding, to enable new mission capabilities such as long-duration human space flight, and to reduce mass.	2. Reduced Mass: Reduce mass of launch vehicle, spacecraft, and propulsion structures to increase payload mass fraction, improve mission performance, and reduce cost.	3. Computational Modeling: Advance new validated computational design, analysis and simulation methods for materials and structural design, certification, and reliability.	4. Large-Aperture Systems: Develop reliable mechanisms and structures for large-aperture systems. These must be stowed compactly for launch, yet achieve high-precision final shapes.	5. Structural Health Monitoring: Enable structural health monitoring and sustainability for long-duration missions, including integration of unobtrusive sensors, and responsive on-board systems.	6. Manufacturing: Enable cost-effective manufacturing for reliable high-performance structures and mechanisms made in low-unit production, including in-space manufacturing.
H	12.2.5. (Structures) Innovative, Multifunctional Concepts	●	○				
H	12.2.1. (Structures) Lightweight Concepts	○	●		●		
H	12.1.1. (Materials) Lightweight Structure	○	●				
H	12.2.2. (Structures) Design and Certification Methods	○	○	●	○	○	
H	12.5.1. Nondestructive Evaluation and Sensors					●	○
H	12.3.4. (Mechanisms) Design and Analysis Tools and Methods			●	●		
H	12.3.1. Deployables, Docking, and Interfaces				●		
H	12.3.5. (Mechanisms) Reliability / Life Assessment / Health Monitoring			○	●	●	
H	12.4.2. Intelligent Integrated Manufacturing and Cyber Physical Systems	○				○	●
M	12.3.6. (Mechanisms) Certification Methods			○	●		
M	12.4.1. Manufacturing Processes		○				●
M	12.1.3. Flexible Material Systems	●	○		○		
M	12.1.2. (Materials) Computational Design	○	○	●			
M	12.5.2. Model-Based Certification and Sustainment Methods			●		●	
M	12.1.4. (Materials) Environment	●	○			○	
L	12.2.3. (Structures) Reliability and Sustainment			○	○	●	
L	12.1.5. Special Materials		○			○	
L	12.4.3. Electronics and Optics Manufacturing Process						○
L	12.5.3. Loads and Environments			○		●	
L	12.2.4. (Structures) Test Tools and Methods			○	○		
L	12.3.2. Mechanism Life Extension Systems				○	○	
L	12.3.3. Electro-mechanical, Mechanical and Micromechanisms				○		
L	12.4.4. Sustainable Manufacturing						○

●	Strong Linkage: Investments by NASA in this technology would likely have a major impact in addressing this challenge.
○	Moderate Linkage: Investments by NASA in this technology would likely have a moderate impact in addressing this challenge.
[blank]	Weak/No Linkage: Investments by NASA in this technology would likely have little or no impact in addressing the challenge.

FIGURE O.3 Level of support that the technologies provide to the top technical challenges for TA12 Materials, Structures, Mechanical Systems, and Manufacturing.

Furthermore, many of the TA12 roadmap technologies are connected to each other in support of a common top technical challenge or a crosscutting roadmap technology. For instance, many of the roadmap technologies support challenges related to reliability, health monitoring, and sustainability.

HIGH-PRIORITY LEVEL 3 TECHNOLOGIES

Panel 5 identified nine high-priority technologies in TA12. The justification for ranking each of these technologies as a high priority is discussed below.

Technology 12.2.5, Innovative, Multifunctional Concepts (Structures)

Structures that perform functions in addition to carrying load and maintaining shape can increase mission capability while decreasing mass and volume, potentially benefitting all future space missions. Multifunctional structural concepts involve increasing levels of system integration and provide a foundation for increased autonomy. Habitat structures with integral shielding would reduce radiation exposure and MMOD risk for long-duration human spaceflight missions; these might involve flexible materials and inflatable structures. Other innovative multifunctional concepts would enable load-bearing structures to provide thermal isolation, control and protection in cryo-tanks, habitats, sensor supports and TPS, and address joints as well as primary structure. These concepts would enable on-orbit fuel storage depots and benefit human exploration and science missions by reducing the mass and complexity of thermal control systems. (For instance, improved cryogenic boil-off protection would reduce mass for a Mars mission by 50 percent; Braun, 2011). Sensory and controlled structures would benefit from the ability to conduct electrical signals and power, enabling health monitoring and adaptation. Other multifunctional structures might store energy and autonomously repair damage.

Multifunctional structures technology is considered to be at TRL 2 for many level 4 technology items, TRL 3 in several cases, such as integrated MMOD protection, and up to TRL 5 for integrated windows and active control of structural response. The highest-priority technologies are at TRL 2-3.

The human spaceflight applications of multifunctional structures technology are unique to NASA and dictate that NASA lead associated technology development. Some multifunctional structures concepts, such as those involving thermal-structural and electrical-structural functionality, are likely to find broader applications in multiple areas and multiple missions, and beyond the aerospace field, including electronics and aircraft. NASA would benefit from partnerships in the development of these technology concepts. One example of a potential cooperative effort might be the commercial development and demonstration of thermally conductive electronics support structures.

Some elements of multifunctional structures concepts would benefit from access to the ISS. Specifically, demonstration of habitat structures with integral radiation and MMOD shielding, including long-term exposure to the space environment would increase the TRL of such concepts. While beneficial for other multifunctional technology concepts, access to the ISS would not be required.

This technology is game-changing because multifunctional habitat structures with integral shielding could reduce radiation exposure and MMOD risk for human spaceflight, bringing risk levels into acceptable ranges with reduced structural mass and launch vehicle volume. Other multifunctional structures technologies are likely to impact multiple areas and multiple missions and find uses beyond the aerospace field. The development risk is moderate-to-high, perhaps exceeding that of past efforts to develop comparable technology.

Technology 12.2.1, Lightweight Concepts (Structures)

Lightweight structural concepts could significantly enhance future exploration and science missions and enable new missions. Improved performance of reduced mass launch vehicle systems with increased payload mass fraction could provide benefits for all future space missions. Lightweight cryo-tank concepts could improve launch vehicle performance and potentially enable on-orbit fuel storage depots (crosscutting with TA14). Small-scale inflatable space systems concepts have been demonstrated and commercial scale-up for inflatable crewed systems are planned for later this decade. These concepts and lightweight inflatable ground habitats could enable future exploration missions. Lightweight concepts for deployable solar sails, precision space structures, and inflatable, deployable heat shields could provide opportunities for new missions or significantly benefit planned science missions. Advanced composite materials play an important role in developing lightweight structural concepts. Integration of advanced materials and structures technology provide the maximum benefit in development and optimization of lightweight concepts.

Lightweight (structural) concepts are considered to be at TRL 2-3 to 5. Some inflatable space systems have been demonstrated at TRL 6. Examples of lightweight structures concepts at TRL 2-3 include cold hibernating elastic memory self-deployable structures and partially flexible composites with shape memory wire.

Lightweight structural concepts developed by NASA and the aerospace industry have found extensive applications in transportation, commercial aircraft and military systems. Some space applications of lightweight concepts, such as aluminum-lithium cryo-tank structures, solid rocket motor cases, and payload structures, have been demonstrated; however, there are significant new opportunities for adoption of lightweight concepts for future space missions. NASA can partner with other government agencies and/or industry where possible to develop and demonstrate lightweight concepts that will support future NASA missions. An example of a potential cooperative effort is the commercial development and demonstration of an inflatable space habitat that would further NASA's exploration goals.

Some elements of lightweight concepts would benefit from access to the ISS. Specifically, demonstration of in-space manufacturing of lightweight structures, deployment of an inflatable module, and long-term exposure of materials used in these concepts would increase the TRL of lightweight concept technologies. While beneficial, access to the ISS is not required.

Lightweight concepts technology could significantly benefit all exploration and science missions and is aligned with NASA's goals and objectives. The level of risk for lightweight concepts technology ranges from moderate to high depending on the specific technology and application. Many of the lightweight concepts beyond TRL 2-3 are mission dependent, and the timing and effort required to advance from lower TRLs to TRL 6 will depend on the specific application.

Weight reductions from lightweight concepts technology could significantly enhance planned exploration and science missions and have the potential to enable new missions. Lightweight structural concepts for habitats, safe havens, and ground-based infrastructure, particularly those technologies that satisfy multifunctional requirements, could enable new human exploration missions to the Moon or Mars. Lightweight deployable structures can enable future science missions with requirements for large-scale structures, precision deployment, and shielding.

Technology 12.1.1, Lightweight Structure (Materials)

Advanced composite, metallic, and ceramic materials, as well as cost-effective processing and manufacturing methods, are required to develop lightweight structures for future space systems. Further advances are needed if increased benefits from lightweight structures are to be attained. The application of non-autoclave-cured large composite structures to launch vehicles would likely reduce structural weight by more than 30 percent compared to metallic structures. Advanced material systems could enable multifunctional structural designs to reduce radiation levels, improve MMOD protection, and enhance thermal management. Incorporation of nanotechnology-engineered materials in lightweight structures offers the potential for game-changing weight saving and performance improvements (crosscutting with TA10). Materials technology for lightweight structures is relevant to all of NASA's planned and future missions.

Lightweight materials are considered to be generally at TRL 2-3, and higher in select areas. Moderate effort is required to reach TRL 6, comparable to that of previous efforts.

Lightweight structural materials developed by NASA and other government agencies, academia, and the aerospace industry have found extensive applications in transportation, commercial aircraft, and military systems. Continued NASA leadership in materials development for space applications could result in new materials systems with significant benefits in weight reduction and cost savings. NASA will likely have opportunities to pursue these materials in partnership with other federal agencies and industry.

Access to the ISS is not required for development of lightweight materials; however, the ISS could serve as a test bed for evaluation of the exposure of such materials to the space environment.

This technology has the potential to significantly reduce the mass of virtually all launch vehicles and payloads—creating opportunities for new missions, improved performance, and reduced cost. The level of risk for materials development and lightweight structures ranges from moderate to high, with non-autoclave-cured composites as a moderate risk, and development and incorporation of nanotechnology materials in high-performance lightweight structures a modestly higher risk.

Technology 12.2.2, Design and Certification Methods (Structures)

Current structural certification approaches rely on a conservative combination of statistics-based material qualification and experience-based load factors and factors of safety, followed by design development and qualification testing. Verification testing and mission history indicates that structures tend to be over-designed and thus heavier than necessary. A model-based "virtual digital certification" methodology could be developed to design and certify space structures more cost-effectively. Advanced physics-based models that predict structural response, failure modes, and reliability using deterministic and probabilistic approaches are a key requirement for such a methodology. This methodology and associated models should be verified and validated with test data at all necessary levels of scale and complexity to ensure confidence in their application. A design and certification methodology based on validated high-fidelity analytical models promises payoffs in weight savings by reducing excess conservatism in the current methodology and in cost reduction by eliminating the large-scale structural tests that are currently required.

Methods for advanced design and certification are considered to be generally at TRL 3. This is determined by the availability of validated models for virtual digital certification.

NASA has been a leader in developing this technology. Investments from the Air Force Research Labs in similar technologies have contributed significantly and are expected to continue. Several national labs have significant programs in uncertainty management and quantification. The technology to be developed is not only critical in terms of weight reduction and affordability improvements to NASA's space missions but also to DOD space structures. NASA can partner with other federal agencies that also have interest in this technology, such as DOD and DOE, to leverage existing expertise.

Access to the ISS is not required for this technology development. However, ISS design development and qualification test data may be useful in validating the new models and methodologies resulting from this technology development.

This technology provides another path to lighter and more affordable space structures while assuring adequate reliability. A verified and validated model-based design and certification methodology offers payoffs in lightweight structural designs and affordable certification without extensive testing, while ensuring long-term reliability of space structures. Physics-based models will be required to simulate structural response in a virtual digital fleet leader (VDFL) that would include a digital representation of a vehicle and a real time system to assess vehicle health and identify action necessary to address vehicle performance. Overall, the benefits of this technology rank below those of multifunctional and lightweight structures and materials. Since multiple NASA missions would benefit from improved structural design and analysis capability, the technology alignment was among the highest in this technology area. This high ranking carried over to non-NASA structures as well, since improvements in lightweight structures design, probabilistic design methods, and simulation will also benefit DOD, DOE, and other advanced structural applications. The risk and level of difficulty associated with this technology is high, since significant effort from NASA, industry, academia, and other government agencies will be required to advance the current state of the art. Further, although the objectives have been identified, several challenges need to be overcome, particularly in model development and virtual testing, to reach TRL 6.

This technology is applicable to all NASA space vehicles including uncrewed, robotic and human-rated vehicles for use in science missions, and human exploration over extended periods of time.

Technology 12.5.1, Nondestructive Evaluation and Sensors (Crosscutting)

Non-destructive evaluation (NDE) has evolved from its early uses for quality control, product acceptance, and periodic inspection to include continuous health monitoring and autonomous inspection. New NDE and sensor technology, including in situ embedded sensor arrays to assess vehicle and space systems health, integrated analysis to predict vehicle and on-board systems operational capability, and autonomous NDE and sensor operations, will be required for long-duration space missions. Early detection, localization, and mitigation of critical conditions will enhance mission safety and reliability. NASA has proposed an integrated NDE and sensor technology capability in a VDFL that would include a digital representation of a vehicle with real time assessment of vehicle structural health to predict performance and identify operational actions necessary to address vehicle performance. VDFL is an initial step in an overall systems approach to monitor, identify, assess, and respond to on-orbit conditions that impact mission success.

Non-destructive evaluation and sensor technology is considered to be at TRL 2-3 for many level 4 technology items. However, some sensor technology is at a higher level and will require integration into vehicle systems to achieve an overall TRL 6.

Nondestructive evaluation and sensor development by NASA and other government agencies, industry, and academia has led to improved product quality and reduced failures of space structures. Partnership opportunities exist with academia, industry, and other organizations in the development of new NDE and sensor technology

Access to the ISS is not generally required for continued development of nondestructive evaluation and sensor technology.

NDE and sensor technology can result in a major increase in reliability of missions. NDE and sensor technology has numerous crosscutting applications and the potential for significant enhancement of safety and mission assurance of future long-duration space missions. NASA missions would benefit from an integrated NDE approach to monitor, identify, assess, and respond to on-orbit conditions that impact human exploration and science missions. NASA has proposed a VDFL as an eventual technology development. This concept could be expanded to address not only structural integrity of space vehicles, but to include overall vehicle system performance and operation. The VDFL concept has the potential to be game-changing, though not in a 20-year horizon.

NDE and sensor technologies are likely to impact multiple areas and multiple missions, especially as mission durations continue to increase. Assessing and maintaining vehicle integrity with minimal human intervention will be essential for long-duration missions involving complex vehicles and for finding uses beyond the aerospace field. The development risk is moderate-to-high, and consistent with that of past efforts to develop comparable technology. Judgment suggests a clear utility for this technology but no specifically identified users, though the opportunity may exist for partnerships with other agencies.

Technology 12.3.4, Design and Analysis Tools and Methods (Mechanical Systems)

High-fidelity kinematics and dynamics design and analysis tools and methods are essential for modeling, designing, and certifying advanced space structures and mechanical systems including turbomachinery, landing systems and deployment mechanisms. This technology includes the tools and interfaces required to increase data flow rates between various systems to enable real time use of mechanical system data. A mechanism interrelation/correlation analysis methodology would enable creation of a single model of spacecraft mechanical systems and would reduce the stack-up of margins across disciplines, e.g., aero-loads, vehicle dynamics, and structural response. Such models could be integrated into a health-management system for diagnosis, prognosis, and performance assessment and in a VDFL system.

This technology includes control design techniques for achieving deployment, stiffness control, shape control, and disturbance rejection. This involves perhaps iterative technology development, since the models that yield the best control results are not the same models used for other purposes (stress analysis, for instance). The most appropriate model should be used for control design, and such models may not be totally physics-based.

Methods for advanced design and analysis are considered to be generally at TRL 2. This is determined by the availability of interrelation/correlation analysis systems.

NASA has been actively developing design and analysis tools and methods for kinematics and rotor dynamics analyses and precursor flight high-data-rate technologies for space vehicle mechanisms. The Air Force Research Laboratory has also invested in deployable mechanisms modeling and testing. The technology is required for both NASA and Air Force space vehicles. NASA could lead or partner with other federal agencies that also have interest in this technology.

Access to the ISS is not required for this technology development. However, deployable systems tested on-orbit can provide valuable data for development of analysis tools.

This technology can enable a dramatic increase in the reliability of mechanical systems, such as those required for separation, release, and deployment. Improved predictive modeling of spacecraft mechanical systems will reduce overall stack-up of margins across disciplines leading to reduced weight and better performance of concepts with minimal ground testing. The overall benefit of this technology is in the same class as 12.2.2 Design and Certification Methods for structures. Since multiple NASA missions would benefit from improved mechanisms design and analysis capability, the technology alignment was among the highest in this technology area. The risk and level of difficulty associated with this technology were rated as high since significant effort from NASA, industry, academia, and other government agencies would be required to advance the state of the art.

This technology is applicable to all NASA space vehicles including uncrewed, robotic and human-rated vehicles for use in science missions, and human exploration over extended periods of time.

Technology 12.3.1, Deployables, Docking, and Interfaces (Mechanical Systems)

Many future science missions involving imaging and scientific data collection will benefit from the combination of a large aperture and precision geometry. Achieving such structures within the constraints of anticipated launch vehicles will most likely involve deployment, possibly including flexible materials, although other approaches including assembly or in-space manufacturing can be considered. Docking and the associated interfaces provide another approach to building up larger platforms from smaller ones, and these are encountered in human spaceflight missions, along with habitats deployed from flexible materials. These mechanical systems and structures must deploy reliably in extreme environments and achieve a desired shape with high precision; some systems may require the use of a control system to maintain a precise shape under operational disturbances. Such systems include antennas, optical elements, and solar sails. Modularity and scalability are desirable features of such concepts.

Deployables, docking, and interfaces technology beyond the current applications for antennas, solar panels, sun shields, and landing systems for science missions and docking systems on the ISS is considered to be at TRL 2-6 for many level 4 technology items, and nominally at TRL 4. Advanced deployables and docking systems have been developed to TRL 6, but typically for smaller systems. The highest-priority technologies are at TRL 3-4.

Large precise aperture systems are critical to some NASA science missions as well as to some DOD surveillance missions, enabling advanced mission performance. This suggests that NASA lead associated technology development, finding partners when feasible.

Some aspects of deployable structures and docking concepts would benefit from access to the ISS. If the systems are relatively large and flexible, their performance cannot be directly verified in the 1-*g* and 1-atmosphere ground environment. In these cases, the ISS could be used to verify such concepts or to validate design and certification models.

This technology will assure the reliable deployment and expected high performance of large precision structures. Without demonstrations associated with this technology, there would continue to be considerable uncertainty and risk involved in fielding such systems. These systems will provide major increases in performance for NASA science missions. Many aspects of precision deployable structures and mechanisms technology are likely to find broader applications in multiple areas and multiple missions, and to a large subset of the aerospace field that requires precise structural geometry. The development risk is moderate-to-high, similar to that of past efforts to develop comparable technology. Judgment suggests a clear utility and clear users, with some possibility of partnerships with other agencies. Space missions have not infrequently failed as the result of failure of a separation, release or deployment system. The pursuit of improvements in the reliability of such systems is a critical technology development area.

Technology 12.3.5, Reliability/Life Assessment/Health Monitoring (Mechanical Systems)

In recent experience, the reliability of mechanical systems, including deployment, separation and release, and motorized systems, has been a more significant contributor to the failure of space missions than the reliability of structures designed to meet current certification standards. Important technical concerns include sliding joints and bearings, friction and tribology, coatings, and lubrication, as well as their performance and durability over extended periods in storage and extreme operational environments. An integrated sensor system would provide a basis for determining the current state of a mechanical system, as well as prediction of future behavior. To be most effective in assuring mission reliability, the ability to take corrective action must also be designed into the system.

Reliability, life assessment, and health monitoring technology is considered to be at TRL 2-3 for many level 4 technology items, TRL 4 for environmental durability testing, and TRL 1 for general life extension prediction and the VDFL concept. Reliability can be advanced significantly for specific classes of mechanical systems at a time.

Mission success requires highly reliable spacecraft mechanical systems, especially for long-duration missions. Some elements of mechanical systems reliability would benefit from access to the ISS. For instance, long-term exposure of materials and operation of devices would increase the TRL. While beneficial, access to the ISS is not required.

This technology could enable a dramatic increase in the reliability of mechanical systems and structures, especially for long-duration space missions. The intrinsic risk associated with such missions could be reduced through the development of health monitoring systems. This technology area is closely linked with the area of deployables, docking, and interfaces, which itself was ranked high in the QFD evaluation. Significant improvement in the reliability of mechanical systems would have a major benefit on assurance of space mission success. Many aspects of reliable mechanisms technology are likely to find applications in multiple areas and multiple missions, to the broad aerospace field, and in some non-aerospace fields. The development risk is moderate-to-high, perhaps exceeding that of past efforts to develop comparable technology. There is a good possibility for outside partnerships.

Technology 12.4.2, Intelligent Integrated Manufacturing and Cyber Physical Systems (Manufacturing)

As a rule, the fielding of high-performance materials, structures, and mechanisms for space applications requires specialized manufacturing capability. Through advances in technology, largely IT-based, more general but flexible manufacturing methods can be adapted to produce specialized components and systems. A database and data-mining capability would be useful to support a terrestrial and interplanetary design, manufacturing, and operations supply chain. High-fidelity manufacturing process models could be used to simulate various manufacturing scenarios to enable rapid evaluation of process alternatives. An intelligent product definition model could be used to simulate the full behavior of components through all stages of their life cycle. Hardware and software technologies will need to be coordinated to develop the next generation of robotics and automation for space structures. This will require the development of cyber-physical systems that enable adaptable and autonomous manufacturing for long-duration crewed spaceflights, including direct digital manufacturing (DDM). In-space manufacturing has the potential to be game-changing by reducing the structural mass that must be delivered to orbit or to the surface of other worlds.

Intelligent integrated manufacturing technology is considered to be at TRL 4, as determined by the availability of validated product definition, and manufacturing process models. In-space manufacturing is at a considerably lower TRL, perhaps 1-2.

There are existing industrial capabilities in production process modeling, factory automation and simulation, and product life-cycle modeling. Investments from the Air Force Research Labs in similar technologies have contributed significantly and are expected to continue because of the potential impacts on affordability. Manufacturing is an area in which NASA can benefit from monitoring developments in hardware, software, and supply chain management. There is potential to form government, university, and industry consortia to pursue these ends.

Access to the ISS is not required for this technology development. However, the ISS could be a useful platform to test in-space manufacturing processes.

This technology would enable physical components to be manufactured in space, on long-duration human missions if necessary. For some exploration missions, this could reduce the mass that must be carried into space. Furthermore, this technology promises improved affordability of one-off structures made from high-performance materials. Multiple NASA missions, especially science missions with constrained budgets, would benefit from cradle-to-grave product life-cycle and manufacturing simulation to select affordable designs. For instance, non-autoclave processes would substantially reduce the infrastructure investment needed to manufacture small runs of large polymer matrix composite structures. Additionally, NASA and DARPA recently conducted a study focused on developing larger (>100 m), lighter space-based optical systems using in-space manufacturing. Small-scale (mm) manufacturing concepts were demonstrated, but significant effort would be required to scale up to meaningful optical systems. This technology is perhaps at TRL 2, and has potential for ISS demonstration. Therefore, the technology alignment for NASA applications was among the highest in this technology area. This high ranking, however, did not carry over to non-NASA applications, where amortization over multiple units changes the manufacturing approach required to ensure affordability.

The risk and level of difficulty associated with this technology were rated as high since significant effort from NASA, industry, academia, and other government agencies will be required to advance the current state of the art.

This technology is applicable to all NASA space vehicles including uncrewed, robotic and human-rated vehicles for use in science missions, and human exploration over extended periods of time.

MEDIUM- AND LOW-PRIORITY TECHNOLOGIES

TA12 contains 23 level 3 technologies, of which 14 were determined to be of medium or low priority. Six technologies were rated medium priority, not including the two that were originally rated medium priority but were promoted to the high-priority category, and eight were rated low priority.

The ranked QFD results, shown in Figure O.1, provide some insight into the reasons that these technologies did not receive high-priority ratings.

For these six medium-priority technologies, the technical risk was considered to be either too low or the required effort was considered unreasonable. A second factor for the lower half of these technologies was reduced alignment with non-NASA aerospace technology and national goals. In the medium-priority technologies there are significant efforts underway in the aerospace industry and other agencies related to manufacturing processes, flexible structures, certification methods for mechanical systems, model-based certification and sustainment methods, materials computational design, and environmental materials characterization.

Most of the level 4 technology items associated with 12.3.6 Certification Methods are at a low TRL. NASA should be able to partner with others in the development of many of the level 4 items in 12.4.1 Manufacturing Processes, and the panel has included 12.4.1(d) In-Space Assembly, Fabrication, and Repair with the high-priority technology 12.4.2 Intelligent Integrated Manufacturing and Cyber Physical Systems. 12.1.3 Flexible Materials Systems supports 12.2.5 Innovative Multifunctional Concepts, as well as 12.3.1 Deployables, Docking, and Interfaces. The benefits of research in 12.1.2 Computational Design of Materials are unlikely to be realized within the timeframe addressed by this study. 12.5.2 Model-Based Certification and Sustainment was also regarded as a valuable goal, but with benefits unlikely to be realized within the study timeframe.

The likely benefit of pursuing the eight low-priority technologies, broadly defined, was considered to be smaller than that of pursuing the high- and medium-priority technologies. Furthermore, the technical risk associated with these technologies was considered to be either too low, or the required effort was considered unreasonable. Finally, these technologies generally exhibited reduced alignment with non-NASA aerospace technology and national goals. While important in selected aerospace applications, technologies such as special materials, electronics and optics manufacturing processes, electromechanical and micromechanical systems and sustainable manufacturing were identified as areas where industry, other agencies, and academia could partner with NASA in selected technology development related to future NASA missions. Technology areas including loads and environment, test tools and methods, and mechanical life extension methods were rated similarly as technology efforts best conducted through industry and academia partnerships.

The monitoring aspects of 12.5.3 Loads and Environments might be considered to be included with 12.5.1 Nondestructive Evaluation and Sensors. 12.5.1 Special Materials is a kind of "grab-bag" of unrelated technologies that did not generally fit well with this roadmap. The panel suggests that the associated level 4 technology items be supported as needed by likely users. NASA could partner with others in the development of some of the level 4 items of 12.4.3 Electronics and Optics Manufacturing Process, while the technology related to large ultra-light precision optical structures fits well with other high-priority technology areas. 12.3.3 Electro-mechanical, Mechanical, and Micromechanisms includes a variety of perhaps unrelated level 4 technology items.

DEVELOPMENT AND SCHEDULE CHANGES FOR THE TECHNOLOGIES COVERED BY THE ROADMAP

Perhaps as a result of the need to address such a broad range of technologies in a summary document, the TA12 roadmap devotes little space to discussion of the assumed mission model, or to the inter-dependence of technology development. To some degree, it can be read as a catalog of technology items as much as it can be read as a plan. While such information is included in the Figure 2 foldout in the draft roadmap, detailed interpretation is left to the reader. This makes it challenging to suggest specific modifications to the schedule.

OTHER GENERAL COMMENTS ON THE ROADMAP

The TA12 roadmap addresses neither improved understanding of the intense vibroacoustic environment of launch nor novel approaches that could reduce structural dynamic response. These extreme loads frequently drive the structural design of spacecraft. This is most closely associated with the following level 3 technologies: Loads and Environments 12.5.3. (Crosscutting); and Design and Certification Methods 12.2.2. (Structures). There is also a crosscutting aspect with active control of vibroacoustic environments and response (TA04).

PUBLIC WORKSHOP SUMMARY

The workshop for the TA12 Materials, Structures, Mechanical Systems, and Manufacturing technology area was conducted by the Materials Panel on March 10, 2011, at the Keck Center of the National Academies, Washington, D.C. The discussion was led by panel chair Mool Gupta. He started the day by giving a general overview of the roadmaps and the NRC's task to evaluate them. He also provided some direction for what topics the invited speakers should cover in their presentations. After the introduction, the day started with an overview of the NASA roadmap by the NASA authors, followed by several sessions addressing the key areas of each roadmap. For each of these sessions, experts from industry, academia, and/or government provided a 35 minute presentation/discussion of their comments on the NASA roadmap. At the end of the day, there was approximately 1 hour for open discussion by the workshop attendees, followed by a concluding discussion by the panel chair summarizing the key points observed during the day's discussion.

Roadmap Overview by NASA

The NASA team presented an overview of the TA12 roadmap. They noted that in developing the roadmap, they focused on innovating and game changing areas instead of incremental improvements. The team also indicated that they looked at both push areas (e.g., physics-based methods, materials, intelligent manufacturing, sustainment, reliability) as well as pull areas (e.g., affordability, multifunctionality, lightweight, environmentally friendly). Overall, the team identified 23 different technologies in the roadmap, and noted that many of these are cut across different disciplines outside the TA12 roadmap. The team also noted that during the roadmap development, they had substantial interaction with the other NASA individuals developing the other TA roadmaps. Finally, the team highlighted that they believed the TA12 roadmap was aligned with the NASA strategic objectives.

One topic that the NASA team indicated was a key focus of their roadmap was the VDFL. This technology includes high-fidelity modeling and simulation, design and certification methods, situational awareness, and life prediction and sustainment. According to the team, the VDFL is needed for future NASA endeavors such as deep space travel, where it is difficult to do resupply or provide safe havens in case of emergencies. Essentially, they indicated that the VDFL is a long-term technology aimed at lowering costs and improving reliability for future NASA missions.

The NASA team also spent some time discussing the top technical challenges that they developed in the TA12 roadmap. In terms of the overall top challenges, the team noted that radiation protection for humans and reliability rose to the top of the list. They also identified top challenges for specific areas, including: materials (e.g., new tailored materials, computational materials technologies), structures (e.g., robust lightweight/multifunctional structures, VDFL), mechanical systems (e.g., higher reliability and predictable performance, precision deployable mechanisms for large space structures), and manufacturing (e.g., advanced manufacturing processes, sustainable manufacturing).

After the NASA presentation, a discussion period followed in which several panel members asked the team questions. In responding to a question how the NASA team views the role of nanomaterials in structures, the team responded that they identified products out of the TA10 roadmap that they could use. They noted that materials, manufacturing, and structures work at a larger scale, and there is a need to figure out how to use/implement nanomaterials at this larger level. The team noted that areas such as using nanoclays as a toughener and permeability barrier are likely nearer term. Other areas, such as negative CTE materials and platelet materials (to decrease permeability and increase life) have already seen some use in consumer products or flight systems. One workshop attendee also concurred with this viewpoint that applications are becoming nearer term, and that in the next 5 to 10 years much more exploitation should be possible.

The concept of the VDFL also generated some discussion. One of the panel members indicated that he viewed the VDFL as a kind of systems engineering technology, rather than a materials/structures technology. He also commented that he felt the VDFL did not go far enough, as it should also include propulsion, guidance/navigation and control (GN&C), on-board sensors, and other features to both monitor and transmit health on all subsystems. Another panel member noted that the NASA team had mentioned certifying models would be part of VDFL, and asked whether this is the overall approach to validation and verification (V&V). The NASA team responded that models for qualification and certification already exist, and that the VDFL is really about making a transition to certifying the models, rather than certifying the program/mission. The team also noted that each vehicle using VDFL will be a test case for improving the models over time.

Session 1: Materials

Tia Benson Tolle (Air Force Research Laboratory) presented her comments on the NASA roadmap. She indicated she was encouraged to see the acknowledgement in the roadmap of the need for a long-term investment strategy, as well as the push/pull tension built into the roadmap portfolio. She noted that multiple studies have concluded that building in such tension is a proven approach for maximizing innovation and improving product development. Benson Tolle also emphasized that computational design/methods are key to accelerated maturation of complex engineered materials, and they will be relied on more in the future. She noted that while there have been good individual efforts focused on improving computational methods, there is still a need to have a broader and more integrated approach. Benson Tolle also discussed several other materials areas, including hybrid materials, morphing materials, emerging energy harvesting technologies, leveraging TPS investments, and digital manufacturing processes. In terms of the top technical challenges, she commented that for the exploitation of nanotailoring, the role of the interface (and its effect on matrix material) is only generally understood, and could use more focus to advance this. Some high-priority technology areas that Benson Tolle emphasized include multifunctional materials, and integrated computational materials science and engineering (ICMSE). Finally, she noted that nanotailored composites and three-dimensional fiber architectures are near the tipping point where additional investments can help mature these areas.

After Benson Tolle's presentation, workshop participants asked her whether AFRL had formal programs for interacting with NASA research programs. Benson Tolle responded that for her area (i.e., materials), she was not aware of any particular forums for interaction, but that other areas (e.g., engine development) have formal processes. She did note, however, that there certainly appears to be room for further collaboration with NASA (i.e., in taking advanced materials to the point where they can be exploited for air and space applications). Later in the discussion, another participant asked Benson Tolle about AFRL's experience in the tradeoffs of multifunctional materials. Benson Tolle answered that first, optimization at the system level must be done early on. Second, she noted that as researchers engineer materials further, the opportunity to change the property tradeoff space may open up.

Byron Pipes (Purdue University) followed with a presentation of his views of the NASA roadmap. He noted that while the industry has used composites for more than 30 years, there is still an inability to accurately predict failure modes, and that this results in overdesigning composite structures. Relative to computational materials, Pipes indicated that there are both the "design aided by experiments" and "certification aided by experiments" aspects to consider. He commented that multiscale modeling provides a way to certify materials in ways that do not require experiments. Pipes then suggested that NASA's goal should be to think about simulation driven materials and structures certification, as well as about simulation driven materials and structures design. He also noted that while NASA serves both aero and space with very different goals (i.e., pervasiveness in aero versus unique solutions in space), human safety is a central issue to both. Pipes indicated that some areas to emphasize might include materials (e.g., computational design materials), structures (e.g., design and certification), cross-cutting model-based certification, and manufacturing (e.g., manufacturing processes). He also noted that micro design models are an area to emphasize with high priority. Pipes then asked the question that, for virtual digital certification, how do you get the FAA and other groups to think more about this? He highlighted this as an area that impacts all missions. Finally, Pipes concluded noting smart materials and devices is another area with a low TRL that might provide benefits to NASA (e.g., health monitoring).

After Pipes' presentation, workshop participants noted that it is important to incorporate early in the process the mechanism for integrating sensors into the structure, and commented that NASA has projects looking into this (e.g., MEMS). Pipes also suggested that there is more to be done in terms of data acquisition analysis and that it is more than just building the sensor into the structure. On another topic, a participant commented that the Boeing 787 symbolizes the state of the art in certification, and asked Pipes what he felt the next steps were. Pipes responded that there are many significant capabilities coming out of the labs that can be taken advantage of. He also indicated that it would be desirable to take some of the uncertainty and empiricism out of the models, as it is becoming increasingly unaffordable to test every piece of structure in every vehicle in the future. Pipes concluded that the science is there, but only recently have has computational power advanced to allow full use of this he indicated that he believes there will be many more improvements in the next 10 years.

Session 2: Structures

Les Lee (Air Force Office of Scientific Research; AFOSR) started with a brief overview of AFOSR and its research portfolio. He noted that one of the key areas of focus is in multifunctional design and materials. He notes that in some cases the performance for these may be less than a unifunctional part, but that this is acceptable as long as the overall system improves—system metrics are needed to quantify this. In terms of the NASA roadmap, Lee indicated that the roadmap appeared to be well laid out and contained a good balance between push and pull technologies. He did suggest that the roadmap could use more emphasis on the integration between materials and structures for multifunctional design, as well as providing more coverage of "weakest links" (e.g., joints, discontinuities). On this latter point in particular, a workshop attendee concurred that 90 percent of the issues she deals with are in the interface. Lee also commented that the roadmap coverage on predictive capabilities and VDFL integration appeared to be optimistic; while he indicated he thought this would be useful, he did not think it should be used as an excuse to skip verification testing. Also, Lee noted that although the roadmap coverage of reliability analysis was good, there also needed to be some focus on game changing areas such as autonomic systems. Finally, Lee indicated that self-healing technology is important (e.g., repeated healing), and is critical for deep space missions.

Lisa Hill (Northrop Grumman) noted up front in her presentation that cost and affordability are key considerations for technology investments, yet this only shows up in the NASA roadmap at a high level. While she indicated that the roadmap does a good job in laying out where we could go, it would also benefit from more quantification of why. Hill also commented that the push/pull discussion was done well, as was the concept of linking technologies to a long-term goal (e.g., VDFL). On the other hand, she identified potential gaps (e.g., digital direct manufacturing), as well as areas that could use additional clarification (e.g., the various structures and materials technologies in the roadmap with similar names, the significant connectivity with the TA10 roadmap). Hill also questioned why solar sails were listed as a mission in the roadmap in 2020, as they are already flying at small scales. In terms of VDFL, she noted that currently minimal data is obtained from structures and mechanisms, and that unless forced into the system design, contractors typically will not include these. Hill also commented that the VDFL tools sound useful, but if they are available and used in the design, then the VDFL would not be needed later. Some of the top technical challenges that Hill mentioned include mass producible (e.g., hundreds to tens of thousands annually), modularity, scalability, and obtaining useful performance data on structures (e.g., joints). She did identify some technology gaps in the roadmap, however, including minimal discussion of the production aspects of modular structures, materials and deployments for large optical systems, and concepts for dealing with launch loads in different ways (e.g., friction in joints for damping).

Session 3: Mechanical Systems

Rakesh Kapania (Virginia Polytechnic Institute and State University) started his presentation by commenting that materials are very important, but so is how they are placed (i.e., direction, geometry). According to Kapania, there are several technology areas that he sees as key to the roadmap: deployables, dockings, and interfaces (e.g., extensibility, correlation between scaled and full-size models), mechanism life extension systems (e.g., understanding the response of structures to non-stationary random excitations), electro-mechanical systems, design and analysis tools and methods (e.g., connecting analysis with health monitoring, modeling for multifunctional structures, importance of numerical ill-conditioning), reliability/life assessment/health monitoring (e.g., miniaturization, reliability-based structural optimization), and certification methods (e.g., computational-based certification, need for reduced-order modeling, reliability of computers and software). Kapania identified several top technical challenges, including: miniaturization to reduce weight, lack of information on loading environment, effects of radiation on material properties, and analysis and design of multidisciplinary systems (e.g., information management, reliable software). In terms of gaps in the roadmap, Kapania noted that energy requirements, energy harvesting, fiber optics based sensors, and distributed sensing could all use more attention. He also commented that there are several high-priority areas for NASA specifically, including: miniaturization and optimization, reliable software for multi-system analysis, and understanding failure modes of multifunctional materials. Kapania also suggested that some aspects of fabrication (e.g., from computer file to three-dimensional object, with sensing, actuation, computing, damage detection, and self-repairing all built in) as well as reliable software able to perform multi-system analysis accurately without conditioning problems are game changing areas. When asked after his presentation on what areas NASA should lead, Kapania responded that one area is in characterizing the space environment—i.e., what it is, what the loads are. He noted that once the requirements are understood, progressing to a solution is straightforward. Kapania also commented that structural health monitoring (which he considers to be near a "tipping point" of significant benefit with additional investment) is something desired by most industries, including those outside aerospace (e.g., the automotive industry).

Session 4: Manufacturing and Crosscutting Areas

Ming Leu (Missouri University of Science and Technology) started his presentation by identifying technologies in the manufacturing and crosscutting areas contained in the NASA roadmap. For in-space assembly, fabrication, and repair, he suggested that NASA could potentially take the lead in this. Leu also added two new areas to the Manufacturing Process area: multi-scale modeling and simulation (as large increases in computer power make this important), and nanomanufacturing (where NASA should look to leverage existing National Science Foundation investments). Leu provided detailed commentary on several technology areas, including: Laser Assisted Material Processing (which is an advanced type of three-dimensional printing applicable to in-space manufacturing and repair), intelligent integrated manufacturing and cyber physical systems, sustainable manufacturing (including consideration of environment, economy, and energy—E^3—aspects), Nondestructive Evaluation (NDE), and loads and environments. On NDE in particular, Leu noted that the roadmap appears to focus primarily on ultrasound techniques; he suggested that other methods (e.g., eddy current, microwave, millimeter wave) should also be looked at, and that sensor fusion is another aspect deserving attention. Relative to the top technical challenges in the NASA roadmap, Leu commented that making accurate predictions based on multi-scale modeling will take a long time, and that trying to make complex three-dimensional parts with high precision is difficult. He also commented that there appeared to be some gaps in the roadmap, including: multi-scale modeling and simulation, nanomanufacturing, and lifecycle product and process design (or E^3 technologies). Leu indicated several areas that he views as high-priority for NASA, including autonomous fabrication, repair, and assembly at point of use, advanced robotics, functionally gradients composites capable of surviving very-high-temperature environments. In terms of technologies close to a tipping point, Leu noted that composites manufacturing (and polymer matrix composites in particular) could benefit substantially from additional investment. Finally, Leu commented after his presentation that in his view, it is important for NASA to get involved in these areas as the industry is typically not willing to invest.

Glenn Light (Southwest Research Institute) followed Leu with a presentation on his perspectives on the NASA roadmap. Light noted that the roadmap stated its goals well in terms of how NDE technologies can feed into the safety/reliability of long-duration space missions and the assessment/maintenance of vehicle integrity with minimal human intervention. He also commented that the roadmap provided a good discussion on prognostics (i.e., the ability to detect defects, assess the situation, and provide a prognosis of remaining life or usage), and that this is an area deserving of attention. On the other hand, Light also highlighted some areas that he felt the NASA roadmap did not cover as well, including: types of defects and damage that might be anticipated, practical aspects and effective integration of sensors and sensor life, sensing and monitoring the fields/environment around the structure, technology to route repairing materials through the structure, and wireless power transfer to sensors. In terms of top technical challenges in NDE and sensor systems, Light indicated that these include sensor integration with minimal detrimental effects, sensitivity to early damage, increasing sensor life, sensors and micro-circuitry embedded in the structure, environmental protection for structures (e.g., coatings), and the ability to monitor how coatings are wearing. Light also discussed several areas he felt are high-priority for NASA, including: development of sensors in practical form factors, wireless power transfer to embedded sensors (e.g., local energy harvesting), on-call repair technologies and self-healing metals and composites, and integration of embedded sensors as part of the structural design. As for alignment with NASA, he indicated that many of these areas align well with NASA's expertise, capabilities, and facilities. Light did suggest that there is a need to set clearly defined national goals for the space program, and minimize the requirement for cost sharing and need to have dual-use technologies (which many federal contracts do). Light also highlighted several technologies that he considers game-changing, including capillary repair materials, practically embedded sensor arrays that positivity impact structural strength, and sending power to all sensors wirelessly. He additionally noted that remote sensing of the environment around the structure is near a tipping point and may benefit from further investment. Regarding embedded sensors, Light indicated that this requires a new concept for teaching structural design. Finally, in responding to a question from a workshop participant on the use of fiber optics embedded for sensing purposes, Light noted that this is good for some things (e.g., stress analysis), but the main issue to date has been the detrimental impact to the structure.

Public Comment and Discussion Session

The following are views expressed during the public comment and discussion session by either presenters, members of the panel, or others in attendance.

- *Individual observations on technology development.* One workshop attendee provided a relatively detailed set of observations based on his being involved in the technology development efforts at a U.S. government agency. He noted that materials are very different for space systems, as issues with radiation resistance and aggressive environments need to be accounted for. He also commented that space systems are typically low-rate production programs. This attendee then suggested that in computational modeling, most methods do not account for manufacturing variability, and that there have been some spectacular failures when this was used as the basis for validating the system. He emphasized that there is no substitute for testing; while there are many physics-based models out there, there is an issue with not knowing everything that might impact a system. An example he provided of this is for impurities in lithium ion batteries, and the fact that these components still need to be tested for 10 years to understand their 10 year lifetime. Finally this attendee noted the need to think politically to be successful in receiving technology funding; he suggested that looking for dual-use technologies at low manufacturing readiness, and looking at productizing these, might be a way to do this.
- *Collaboration with other agencies.* During the discussions in the day, several presenters and attendees highlighted the need for NASA to look at how to take advantage of other agencies' investments in technology development. One workshop attendee indicated that there is a lack of a formal interchange process for technology development among these different groups, and suggested that the NASA Office of the Chief Technologist look to stand up a process like this. This attendee also commented that NASA should look at the National Research Laboratories, AFRL, and other groups, and perform a gap analysis to find out which areas might be most applicable to and worthy of NASA investment.
- *Radiation protection.* One of the panel members noted that NASA had identified radiation protection as one of its top technical challenges in the roadmap. This spurred many comments, including some from the NASA team suggesting that they are looking at materials like metal organic foam in tanks to help as shielding for habitat modules. Another workshop attendee noted that for protecting electronics, there are two approaches: radiation-hardened electronics, or radiation protection for non-radiation-hardened electronics. He commented that if NASA develops better ways to shield spacecraft from radiation, it could have a large benefit for uncrewed spacecraft systems as well. Building on this comment, another attendee noted that taking more of an active materials approach might be beneficial.
- *Certification of materials.* Participants commented that it appears as if there are substantial costs associated with certification, and that this frequently is a barrier to using new materials in actual systems. One participant additionally commented that in the past there were mission requirements to use technologies that were TRL 6 or higher, whereas the organizations developing push technologies frequently stop at about TRL 4—this leads to a "valley of death" that is hard to overcome. Finally, there was a comment from another attendee that improving physics-based modeling is one way to try and streamline certification, but there is a need to find a good balance between modeling and testing for certification.
- *Reliability.* Toward the end of the public discussion session, one workshop participant asked what others thought about having reliability as a grand challenge for the roadmap. In response, one of the NASA team members noted that having precise knowledge of the structural reliability is important relative to integration and saving mass/volume. A workshop attendee also suggested that it is important to consider whether these technologies are being used because they can, or because they should (e.g., embedded sensors). He commented that the benefit from embedding sensors in laminates versus metals needs to be addressed, for example, and that it is important to take a pragmatic approach in applying these technologies. Another example he provided was self-healing: if a structure has many tubes and holes, there will be a reduction in strength. He noted that there needs to be a filter applied so that these technologies make their way onto a vehicle to make it more robust.

REFERENCE

Braun, R., National Aeronautics and Space Administration. 2011. "Investments in Our Future: Exploring Space Through Innovation and Technology," presentation at the Johns Hopkins University Applied Physics Laboratory, Laurel, Md., May 25, 2011. Available at http://www.nasa.gov/pdf/553607main_APL_Bobby_5_27_11_DW2.pdf.

P

TA13 Ground and Launch Systems Processing

INTRODUCTION

The draft roadmap for Technology Area (TA) 13, Ground and Launch Systems Processing, consists of four level 2 technology subareas:[1]

- 13.1 Technologies to Optimize the Operational Life-Cycle
- 13.2 Environmental and Green Technologies
- 13.3 Technologies to Increase Reliability and Mission Availability
- 13.4 Technologies to Improve Mission Safety/Mission Risk

The goal of TA13 is to provide a flexible and sustainable U.S. capability for ground processing as well as launch, mission, and recovery operations to significantly increase safe access to space, including:

- Transportation, assembly, integration, and processing of the launch vehicle, spacecraft, and payload hardware at the launch site including launch pad operations;
- Launch processing infrastructure and its ability to support future operations;
- Range, personnel, and facility safety capabilities;
- Launch control and landing operations including weather and recovery for flight crews, flight hardware, and returned samples;
- Mission integration and control center operations and infrastructure; and
- Environmental impact mitigation for ground and launch operations.

The primary benefit derived from ground and launch processing advances is reduced cost, freeing funds for other investments. Currently, launch vehicle and payload processing and ground operations are significant contributors to mission life cycle costs.

Prior to prioritizing the level 3 technologies included in TA13, the panel considered whether to rename, delete, or move technologies in the technology area breakdown structure (TABS). No changes were recommended for

[1] The draft space technology roadmaps are available at http://www.nasa.gov/offices/oct/strategic_integration/technology_roadmap.html.

TA13. The TABS for TA13 is shown in Table P.1, and the complete, revised TABS for all 14 TAs is shown in Appendix B.

TOP TECHNICAL CHALLENGES

Advances in ground and launch systems processing implies overcoming several challenges, such as reducing the cost of maintaining and operating ground control and launch infrastructure, improving safety, and improving the timeliness, relevance, and accuracy of information provided to ground control and launch personnel (e.g., in part through improvements in inspection and anomaly detection capabilities). Although advanced technology can contribute to solving these challenges, they are most effectively addressed through improvements in management practices, engineering, and design. The study did not identify any technology challenges related to TA13 that rise to the level of the top technical challenges associated with the other roadmaps.

QFD MATRIX AND NUMERICAL RESULTS FOR TA13

Figures P.1 and P.2 show the panel's consensus ratings of the 19 level 3 technologies for the TA13 roadmap. None of them are ranked as high priority, primarily because the benefit of each technology would be a minor improvement in life cycle cost, at best.

TABLE P.1 Technology Area Breakdown Structure for TA13, Ground and Launch Systems Processing

NASA Draft Roadmap (Revision 10)	Steering Committee-Recommended Changes
TA13 Ground and Launch Systems Processing	The structure of this roadmap remains unchanged.
13.1. Technologies to Optimize the Operational Life-Cycle	
13.1.1. Storage, Distribution & Conservation of Fluids	
13.1.2. Automated Alignment, Coupling, & Assembly Systems	
13.1.3. Autonomous Command & Control for Ground and Integrated Vehicle/Ground Systems	
13.2. Environmental and Green Technologies	
13.2.1. Corrosion Prevention, Detection, & Mitigation	
13.2.2. Environmental Remediation & Site Restoration	
13.2.3. Preservation of Natural Ecosystems	
13.2.4. Alternate Energy Prototypes	
13.3. Technologies to Increase Reliability and Mission Availability	
13.3.1. Advanced Launch Technologies	
13.3.2. Environment-Hardened Materials and Structures	
13.3.3. Inspection, Anomaly Detection & Identification	
13.3.4. Fault Isolation and Diagnostics	
13.3.5. Prognostics Technologies	
13.3.6. Repair, Mitigation, and Recovery Technologies	
13.3.7. Communications, Networking, Timing & Telemetry	
13.4. Technologies to Improve Mission Safety/Mission Risk	
13.4.1. Range Tracking, Surveillance & Flight Safety Technologies	
13.4.2. Landing & Recovery Systems & Components	
13.4.3. Weather Prediction and Mitigation	
13.4.4. Robotics/Tele-robotics	
13.4.5. Safety Systems	

	Benefit	Alignment with NASA Needs	Alignment with Non-NASA Aero Tech Needs	Alignment with Non-Aerospace National Goals	Technical Risk and Reasonableness	Sequencing and Timing	Time and Effort	QFD Score (Weighted)	Panel Priority
Multiplier	27	5	2	2	10	4	4		
	0/1/3/9	0/1/3/9	0/1/3/9	0/1/3/9	1/3/9	-9/-3/-1/1	-9/-3/-1/0		
Technology Name	Benefit	Alignment			Risk/Difficulty				
13.1.1. Storage, Distribution, Conservation of Fluids	1	9	9	1	3	-3	-1	106	M
13.1.2. Automated Alignment, Coupling, and Assembly Systems	1	3	0	0	1	-1	-1	44	L
13.1.3. Autonomous Command and Control for Ground and Integrated Vehicle/Ground Systems	1	3	1	1	3	-3	-1	60	L
13.2.1. Corrosion Prevention, Detection, and Mitigation	1	3	1	9	3	1	-1	92	M
13.2.2. Environmental Remediation and Site Restoration	1	0	0	9	1	1	-1	55	L
13.2.3. Preservation of Natural Ecosystems	0	1	1	9	3	-3	-3	31	L
13.2.4. Alternate Energy Prototypes	0	1	1	3	3	-3	-3	19	L
13.3.1. Advanced Launch Technologies	1	3	3	0	3	-3	-3	54	L
13.3.2. Environment-Hardened Materials and Structures	1	3	3	3	3	-3	-1	68	L
13.3.3. Inspection, Anomaly Detection, and Identification	1	9	3	1	3	-3	-1	94	M
13.3.4. Fault Isolation and Diagnostics	1	9	3	1	3	-3	-1	94	M
13.3.5. Prognostics Technologies	1	9	3	1	3	-3	-1	94	M
13.3.6. Repair, Mitigation, and Recovery Technologies	1	9	3	1	3	-3	-1	94	M
13.3.7. Communications, Networking, Timing, and Telemetry	0	9	9	0	3	-3	-1	77	M
13.4.1. Range Tracking, Surveillance, and Flight Safety Technologies	1	9	9	0	3	1	-1	120	M
13.4.2. Landing and Recovery Systems and Components	1	3	1	0	3	-3	-1	58	L
13.4.3. Weather Prediction and Mitigation	0	9	9	1	3	-3	-1	79	M
13.4.4. Robotics / Telerobotics	0	9	3	1	1	-3	-1	47	L
13.4.5. Safety Systems	1	9	9	1	1	-1	-1	94	M

FIGURE P.1 Quality function deployment (QFD) summary matrix for TA13 Ground and Launch Systems Processing. M = Medium Priority; L = Low Priority.

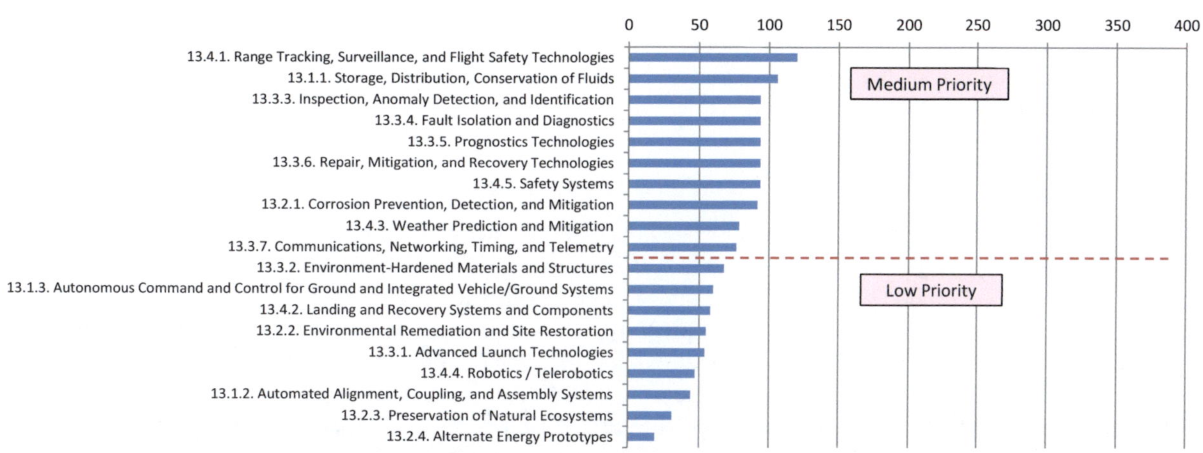

FIGURE P.2 Quality function deployment rankings for TA13 Ground and Launch Systems Processing.

HIGH-PRIORITY LEVEL 3 TECHNOLOGIES

As noted above, the panel did not identify any high-priority Level 3 technologies for TA13.

MEDIUM- AND LOW-PRIORITY TECHNOLOGIES

The TA13 roadmap is, for the most part, composed of various engineering projects that, while probably useful in reducing the cost of launching vehicles, are not primarily technology development undertakings within the TRL 1-6 threshold of this study.

The best approach for reducing the cost of future NASA ground and launch systems processing is to design new launch vehicles from the outset for low-cost operation. Much has already been learned in the commercial launch vehicle sector about how to reduce the cost of ground operations.

In considering the merit of diverting NASA resources to drive down launch operations costs through technology development, it should be noted that for the foreseeable future, the rate of launch of NASA payloads will be low and will occur largely through procurement of launch services from private companies who will conduct their own ground and launch systems processing using their own facilities. As a result, it is unlikely that NASA will be developing multiple launch systems of its own in the foreseeable future. In addition, the largest cost content in NASA ELV missions tends to reside in the payload, launch vehicle, and on-orbit operations, with launch operations costs being only a very small part of the total mission investment. Therefore, it is not advisable for NASA to make significant technology investments in advancing ground and launch systems processing.

DEVELOPMENT AND SCHEDULE CHANGES COVERED BY THE ROADMAP

The panel does not have any recommendations with regard to development and schedule changes.

PUBLIC WORKSHOP SUMMARY

The workshop for the Ground and Launch Systems Processing technology area was conducted by the Propulsion and Power Panel on March 24, 2011, on the campus of the California Institute of Technology in Pasadena, California. The discussion was led by panel member Joyce McDevitt. McDevitt started the day by giving a general overview of the roadmap and the NRC statement of task for this study. She also provided some direction for what topics the invited speakers should cover in their presentations. Experts from industry, academia, and government were invited to lead a 25 minute presentation and discussion of their perspective on the draft NASA roadmap for TA13. At the end of the session, there was a short open discussion by the workshop attendees that focused on the recent session. At the end of the day, there was a concluding discussion led by McDevitt summarizing the key points observed during the day's discussion.

Session 1: Traditional Launch Service Providers

Bernard Kutter (United Launch Alliance) began the session with traditional launch services providers by reviewing the recent development of the ground systems for the Evolved Expendable Launch Vehicle (EELV) program. He noted that ground systems technology is architecture specific and needs be incorporated in the early stages of launch system design. The cost of the EELV system was reduced by taking advantage of state-of-the-art technology to streamline operations. Kutter noted that development of some new technologies might offer modest benefits. These technologies are related to sub-cooling of cryogenic propellants, enhancing the performance of water suppression systems, enabling development of space-based ranges for launch, and facilitating automation of expert systems.

Bill Findiesen (Boeing) focused on a few areas within the draft roadmap for TA13 related to anomaly detection, isolation, and prognostics with greater autonomy. He presented the results of a previous study that concluded there was a potential for a next generation reusable launch vehicle to have ground support costs 30 percent to 40 percent lower than the ground support costs for the space shuttle. This would require incorporating automation and integration of disparate systems early in the design of the new vehicle and its ground systems. This would likely involve providing more real-time information to ground support personnel. He envisioned that advances in ground support technology would also improve mission assurance as well as launch reliability and safety, particularly if a large data set is collected during the vehicle development and early testing.

John Steinmeyer (Orbital) said that ground operations are critical to launch success, and they represent a significant portion of launch costs. He also noted that vehicle on-pad time drives operations costs; launch sites are austere environments that are tough on infrastructure. He emphasized that launch site costs are a function of requirements that driven by vehicle, payload, and mission complexities. He provided some lessons learned, such as the importance of detailed and repeatable procedures; small, experienced launch teams; highly automated fueling, and streamlined vehicle-payload integration. The specific technologies that he believes will support NASA's next heavy lift system include vehicle health monitoring, cryogenic fluid management, helium optimization, automated fueling systems, corrosion protection, and distributed mission control.

Stan Graves (ATK) emphasized the advantage provided by automated systems in reducing the human error, thereby increasing reliability and reducing costs. He said that designing a vehicle for support is better than designing support for a vehicle. He also suggested moving manufacturing operations off site and should be thought of as a commercial operation. Finally he stated that designing a more robust vehicle will lead to reduced launch delays and overall cost savings. He suggested that these improvements are mostly engineering solutions and do not require new technologies.

In the group discussion that followed several workshop participants said that much of the improvements in ground processing will involve the insertion of proven technologies, including state-of-the-art information processing and data technologies, rather than the development of new technologies. It was noted that NASA will desire to balance (1) near-term cost savings associated with reuse of existing infrastructure with (2) long-term cost savings associated with the development and deployment of new systems. Several participants also suggested that the launch site footprint and on-pad time should both be reduced as much as possible.

Session 2: New and Emerging Companies

Jeff Greason (XCOR) began the session with new and emerging companies in the suborbital and launch industries by reporting that XCOR's ground and flight operations are a significant departure from traditional launch systems and have much more in common is aircraft. XCOR's launch vehicle does not use any toxic materials or pyrotechnics, enabling all of their operations to easily meet OSHA and other government standards. XCOR does not operate on a federal space launch range, nor do they have any range safety systems. Instead, XCOR relies on the pilot to handle anomalies and aborts. Preflight operations take about 90 minutes and four people, and XCOR has demonstrated a vehicle turnaround time of less than 9 minutes.

William Pomerantz (Virgin Galactic) said that their system is also significantly different than conventional launch systems, and it also has the ability to flying multiple times per day. Their top design drivers are increasing flight rate, reducing costs, and improving safety. Pomerantz said that most of the technologies in the NASA ground and launch processing roadmap have little value to Virgin Galactic, but a few could be useful. Those include upper atmosphere weather monitoring; temperature hardened materials; and fault detection, isolation, and reconfiguration. He also suggested that some worthwhile technologies might be missing, such as lightning protection, information technologies for ground support, and integrated strain sensing systems.

Brett Alexander (Commercial Spaceflight Federation) was unable to attend the workshop, but William Pomerantz presented his material. He began by reviewing the Commercial Spaceflight Federation's mission, which is to make commercial human spaceflight a reality through advocacy and sharing best industry practices to improve safety. He encouraged NASA to balance its traditional mission with the ability to adapt to a much more diverse launch industry with new concepts and quicker launch tempos. He believes the most challenging future systems will be those that are new to NASA, such as a reusable rockets with powered vertical descent.

Gordon Woodcock (formerly of Boeing) said that each technology's importance should be considered in the context of its ability to support potential exploration architectures. He presented an architecture proposal that takes advantage of several in-space technology developments to reduce launch requirements and eliminate the need to develop a large heavy-lift launch vehicle. This system would rely on advanced electric propulsion systems, reusable flight elements, and in-space storage and transfer of propellants. To make such an architecture feasible and affordable, the flight rate of large boosters would need to be increased through technology that reduces on-pad time and automates the checkout process. He also noted the importance of significantly reducing the amount of helium used for launch.

In the group session there was some discussion on what technologies could be of use to the emerging launch companies. There was little agreement. For example, on the topic of non-destructive evaluation, one speaker said that an emerging company would rather build a stronger structure than incorporate internal health monitoring. When asked how NASA can streamline operations, several participants said NASA should simplify systems and procedures as much as possible and make the flight system much more self contained and autonomous, all of which might require starting from stretch. There was also some discussion on the architecture proposed by Woodcock. He said the single most important technology advance would be advanced electric propulsion systems. Another participant noted that architectures that take into account the full cost of operations do not look like the Apollo architecture, but NASA keeps relying on Apollo-like architectures.

Session 3: Other Interested Parties

Emmett Peter (Walt Disney Company) began the session by presenting an overview of the engineering that the Walt Disney Company uses to ensure the safety and operability of its amusement park rides. These rides carry millions of passengers each year, which means that one-in-a-million incidents are more likely and system life and operational standards must be at a high level to prevent them. The systems are rarely able to incorporate off-the-shelf equipment. They are designed to be mistake proof, and they rely heavily on autonomy. For example, some systems use automated coupling and transfer of high-pressure gasses and fluids, with hundreds of coupling cycles each day. Peter also reviewed Disney's automated maintenance verification system, which features clear responsibilities and timelines and incorporates handheld wireless units for technicians. After reviewing the NASA roadmaps, he said that consistency and commonality are useful but difficult to achieve. Also, classifying and qualifying parts in order of criticality is a good practice to assure that inspection is focused on the correct parts; the highest payoff for ground systems might be found in structural health monitoring and corrosion technologies.

Brian Wilcox (Jet Propulsion Laboratory) presented an architecture that offered a radical departure from traditional systems. His architecture would use a high-altitude, equatorially tethered balloon to winch rocket stages above to an altitude above 95 percent of the atmosphere. Launching rockets from a position above the atmosphere improves the performance of the small rocket systems that are featured in Wilcox's architecture. Wilcox believes that mass producing small rockets at a rate of more than 5,000 per year would lead to significant cost savings over producing a few traditional rockets. The payloads from 222 of these small rockets would be assembled on orbit into large propulsion stages.

Edward Bowles (General Atomics) reviewed the development of the electromagnetic aircraft launch system (EMALS) and discussed the potential for using the technology for space launch. The next generation of U.S. aircraft carriers will use EMALS instead of traditional steam-powered catapults. EMALS will use linear electric motors to provide up to 300,000 pounds of force to launch aircraft and recover naval aircraft. Adapting this technology to develop a rocket launch assist system would require much longer linear motors to accelerate much larger vehicles to much higher velocities. The most expensive part of the space-launch EMALS system would be the power generation systems. The goal would be to accelerate a vehicle weighing 500 tons to a speed of Mach 3 using a track 8 miles long. Bowles suggested such a system could break even after about 40 launches.

In the group discussion, Peter said that the Disney Corporation designs their systems for constancy and interoperability from the beginning. He also said that they design with margins based on the part class with critical parts having significant margins. He also said that Disney uses stainless steel and a lot of brushing and coating to fight corrosion. Wilcox said that his system does not require any significant technology investment, and the concept could be demonstrated with a subscale balloon. Bowles said that the terminal velocity of his system is limited by the frequency of the electronics. He also said that his system could be developed without significant technology advances, although advances in power system technology would be beneficial.

Session 4: Safety Experts and Non-NASA Government Personnel

The final session of the day was aimed capturing the views of safety experts and non-NASA government agencies.

John Schmidt (Naval Research Laboratory) began the session with safety experts and non-NASA government personnel by noting that many safety systems can be generalized across multiple aviation, maritime, and space applications. He spent the majority of his talk reviewing efforts by the U.S. Navy to mitigate corrosion more effectively. Recent emphasis has been on improving and optimizing the human efforts to combat corrosion. That can be accomplished by improving tools, work environments, and training. The Navy is also working on improving incident data capture.

Michael Kelly (Federal Aviation Administration) endorsed NASA efforts to improve automated on-board checkout and vehicle health management. As a result, he said, U.S. systems will be operated in a manner similar to the operation of Russian systems, albeit with more advanced technology. Kelly also supported improvements in inspection, anomaly detection and identification, and telemetry and tracking. He agreed with an emphasis on commonality of communication, although he suggested that commonality will not be beneficial for most other systems because of the challenge of achieving "one-size-fits-all" solutions. This is particularly true with the diverse new vehicle designs being developed by emerging, nontraditional space launch companies. Kelly also foresees that new affordable vehicles will be tightly integrated systems with streamlined operations. He also noted that all of his suggestions could be achieved without much in the way of technology advances. Instead, the major challenges are associated with engineering and implementation.

Q

TA14 Thermal Management Systems

INTRODUCTION

The draft roadmap for Technology Area (TA) 14, Thermal Management Systems, consists of three technology subareas:[1]

- 14.1 Cryogenic Systems
- 14.2 Thermal Control Systems
- 14.3 Thermal Protection Systems

Thermal Management Systems are systems and technologies that that are capable of handling high thermal loads with excellent temperature control, with a goal of decreasing the mass of existing systems. TA14 is concerned with three broad areas of application of thermal management: Cryogenic Systems, which are systems operating below −150°C; Thermal Control Systems, operating near room temperature; and Thermal Protection Systems, which operate above about 500°C.

Before prioritizing the level 3 technologies included in TA14, the panel considered whether to rename, delete, or move technologies in the technology area breakdown structure (TABS). No changes were recommended for TA14, although corrections were made to the names of two technologies. The TABS for TA14 is shown in Table Q.1, and the complete, revised TABS for all 14 TAs is shown in Appendix B.

TOP TECHNICAL CHALLENGES

The panel identified seven top technical challenges for TA14, presented here in priority order.

1. Thermal Protection Systems: Develop a range of rigid ablative and inflatable/flexible/deployable Thermal Protection Systems (TPS) for both human and robotic advanced high-velocity return missions, either novel or reconstituted legacy systems.

[1] The draft space technology roadmaps are available at http://www.nasa.gov/offices/oct/strategic_integration/technology_roadmap.html.

TABLE Q.1 Technology Area Breakdown Structure for TA14, Thermal Management Systems

NASA Draft Roadmap (Revision 10)	Steering Committee-Recommended Changes
TA14 Thermal Management Systems	The steering committee made no changes to the structure of this roadmap, although NASA's draft roadmap had a different name for two technologies.
14.1. Cryogenic Systems	
14.1.1. Passive Thermal Control	
14.1.2. Active Thermal Control	
14.1.3. Integration and Modeling	Rename: 14.1.3. Systems Integration
14.2. Thermal Control Systems	
14.2.1. Heat Acquisition	
14.2.2. Heat Transfer	
14.2.3. Heat Rejection & Energy Storage	
14.3. Thermal Protection Systems	
14.3.1. Entry/Ascent TPS	Rename: 14.3.1 Ascent/Entry TPS
14.3.2. Plume Shielding (Convective & Radiative)	
14.3.3. Sensor Systems & Measurement Technologies	

TPS is mission critical for all future human and robotic missions that require planetary entry or reentry. The current availability of high-TRL rigid ablative TPS is adequate for LEO re-entry, but is inadequate for high-energy re-entries to Earth or planetary missions. Ablative materials are enabling for all NASA, military, and commercial missions that require high-mach number re-entry, such as near-Earth asteroid visits and Mars missions, whether human or robotic (Venkatapathy, 2009a, 2009b). System studies have shown that large entry heat shields provide a potentially enabling means to increase landed mass on a planetary (Mars) surface (Jamshid et al., 2011; McGuire et al., 2011). In many cases, updating existing obsolete TPS materials and processes that were developed in the past may be faster and cheaper than the development of new materials or methods. Some are not now available either because of lost technology, new restrictions on material, or other factors. For example, carbon-phenolic recertification is needed before it can be used for future missions. Other new materials show considerable promise.

2. Zero Boil-Off Storage: Accelerate research on advanced active and passive systems to approach near-zero boil-off in long-term cryogenic storage.

Long-term missions that require cryogenic life-support supplies (e.g., LOX), cryogenic propellants (LH2), or very low temperatures for scientific instrument support will require near-zero boil-off rates. Multiple technologies are proposed in the TA14 roadmap, some of which provide incremental but desirable improvements in cryogenic technology. Emphasis should be on reliable, repairable, supportable active and passive systems that can be integrated into many missions. Many of the technologies are parallel in their impact. Some will emerge as top candidates.

3. Radiators: Develop improved space radiators with reduced mass.

Radiators are used for energy removal from spacecraft and planetary base systems, and are mission-critical for many proposed missions. To reduce radiator mass, area, and pumping power, research is needed on variable emissivity, very low absorptivity-to-emissivity ratio, self-cleaning, and high-temperature coatings, as well as research on lightweight radiators or compact storage systems for extending EVA capability.

4. Multifunctional Materials: Develop high-temperature multifunctional materials that combine structural strength good insulating ability, and possibly other functions.

Multifunctional systems can provide significant mass savings due to combining thermal and structural functions, allowing increased payload weight. Presently, these functions are separately incorporated in spacecraft design. Multi-functional TPS and multi-layer insulation (MLI) systems that combine thermal, structural, micrometeoroid and orbital debris (MMOD), and crew radiation protection could provide significant weight savings and enable long-duration missions, and can also be used for planetary habitat thermal and multifunctional protection. This challenge is also ranked third by TA12.

5. Verification and Validation: Develop, verify, validate, and quantify uncertainty analysis requirements for new or improved comprehensive computer codes for thermal analysis.

Upgrades to predictive codes for ablation during re-entry heating are needed to include closely coupled multiphase ablation and radiative heating into the flow simulations, with careful attention given to verification, validation, and uncertainty quantification. All thermal analysis codes should include (1) verification that the codes have no internal errors, and accurately code the equations used for modeling and analysis; (2) predictions validating against all available experimental data, accounting for experimental error bands; and (3) quantifying the confidence in code predictions, accounting for uncertainties in the data used as model input, uncertainties in the mathematical models used in the analysis, and uncertainties caused by the numerical technique that is implemented (e.g., discretization errors in time and space). Without these attributes, the results generated by the codes are unreliable for design. This challenge is also addressed by TA10 and TA12.

6. Repair Capability: Develop in-space Thermal Protection System (TPS) repair capability.

Repair capability is especially important for long-duration missions, where no safe-haven repair facilities will be available. TPS repair developed for Space Shuttle Orbiter TPS (reinforced carbon-carbon/tiles) should be continued and expanded to provide a repair method for future spacecraft, both NASA and commercial.

7. Thermal Sensors. Enhance thermal sensor systems and measurement technologies.

Operational instrumentation is necessary to understand anomalies, material or performance degradation and performance enhancements, as well as for advanced science mission measurements. Ultra-lightweight sensor systems may provide data needed to identify on-orbit damage, measurement of liquid levels in a microgravity environment, in situ or self-repairing mechanisms, or adaptive control algorithms that can compensate for damage without repairing. Accurate instrumentation to monitor reentry TPS performance is necessary to validate emerging predictive codes for heat shield design. Each of these would improve flight safety and the probability of mission success.

QFD MATRIX AND NUMERICAL RESULTS FOR TA14

The averaged quality function deployment (QFD) matrix for the nine level 3 technologies in TA14 is given in Figure Q.1.

The weighted scores for all level 3 technologies evaluated with the QFD approach are listed in Figure Q.2. 14.3.1 Ascent/Entry TPS received a much higher score than all other level 3 technologies, creating an obvious break point in assigning the high rating. 14.1.2 Active Thermal Control is a needed technology to support zero boil off of cryogenic fluids. Though 14.1.2 Active Thermal Control did not achieve the high rating of 14.3.1 Ascent/Entry TPS, it is considered an enabling technology for a wide variety of long-duration missions, and was thus also assigned high priority. These two technologies are therefore discussed at length. The other seven technologies were rated as "Medium" or "Low."

CHALLENGES VERSUS TECHNOLOGIES

In Figure Q.3, the technologies are listed in descending priority on the vertical columns, and the challenges are shown in the horizontal top row. The correlation between the two is indicated by high correlation (solid symbols), weak correlation (open symbols) or little or no correlation (no symbols). It is seen that the challenges correlate to

Technology Name	Benefit	Alignment with NASA Needs	Alignment with Non-NASA Aero Tech Needs	Alignment with Non-Aerospace National Goals	Technical Risk and Reasonableness	Sequencing and Timing	Time and Effort	QFD Score (Weighted)	Panel Priority
Multiplier	27	5	2	2	10	4	4		
	0/1/3/9	0/1/3/9	0/1/3/9	0/1/3/9	1/3/9	-9/-3/-1/1	-9/-3/-1/0		
	Benefit	Alignment			Risk/Difficulty				
14.1.1. Passive Thermal Control	3	3	1	1	3	-3	-3	106	M
14.1.2. Active Thermal Control	3	9	3	3	3	-3	-1	152	H*
14.1.3. Systems Integration (Thermal Management)	3	9	1	1	3	-3	-3	136	M
14.2.1. Heat Acquisition	1	3	3	1	3	-3	-1	64	L
14.2.2. Heat Transfer	1	3	3	3	3	-3	-1	68	L
14.2.3. Heat Rejection and Energy Storage	3	9	1	1	3	-3	-1	144	M
14.3.1. Ascent/Entry TPS	9	9	1	1	9	-1	-3	366	H
14.3.2. Plume Shielding (Convective and Radiative)	1	9	3	1	3	-3	-1	94	M
14.3.3. Sensor Systems and Measurement Technologies (Thermal Management)	1	9	3	3	3	-1	-1	106	M

FIGURE Q.1 Quality function deployment (QFD) summary matrix for TA14 Thermal Management Systems. The justification for the high-priority designation of all high-priority technologies appears in the section "High-Priority Level 3 Technologies." H = High Priority; H* = High Priority, QFD Score override; M = Medium Priority; L = Low Priority.

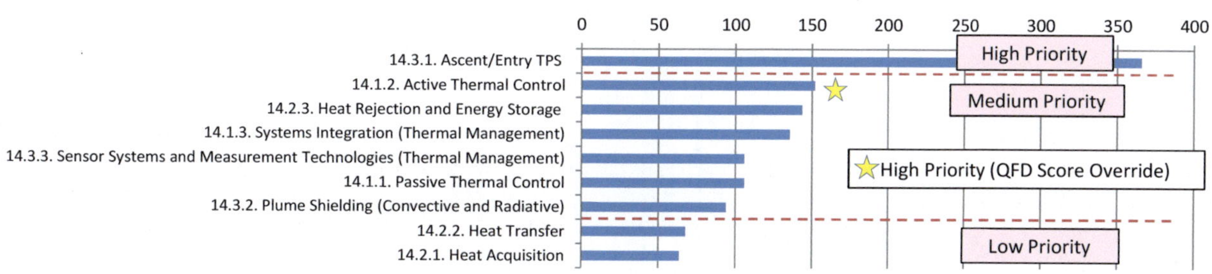

FIGURE Q.2 Quality function deployment rankings for TA14 Thermal Management Systems.

some degree with the priorities of the technologies, as seen by the roughly diagonal pattern of high and moderate blocks.

HIGH-PRIORITY LEVEL 3 TECHNOLOGIES

Panel 5 identified two high-priority technologies in TA14. The justification for ranking each of these technologies as a high priority is discussed below.

Technology 14.3.1, Ascent/Entry TPS

Effective heat shields and thermal insulation during ascent and atmospheric entry are mission-critical for all robotic and human missions that require entry into a planetary atmosphere. This is an area that has suffered the loss of previous technology (from the Apollo era) due to the ageing and retirement of personnel. Newer safety and environmental regulations have also required changes to earlier TPS fabrication and formulation processes.

Priority	TA 14 Technologies, Listed by Priority	Top Technology Challenges: 1. Thermal Protection Systems: Develop a range of rigid ablative and inflatable/ flexible/deployable thermal protection systems for both human and robotic advanced high-velocity return missions, either novel or reconstituted legacy systems.	2. Zero Boil-Off Storage: Accelerate research on advanced active and passive systems to approach near-zero boil-off in long-term cryogenic storage.	3. Radiators: Develop improved space radiators with reduced mass.	4. Multifunctional Materials: Develop high-temperature multifunctional materials that combine structural strength, good insulating ability, and possibly other functions.	5. Verification and Validation: Develop, verify, validate, and quantify uncertainty analysis requirements for new or improved comprehensive computer codes for thermal analysis.	6. Repair Capability: Develop in-space thermal protection system repair capability.	7. Thermal Sensors: Enhance thermal sensor systems and measurement technologies.
H	14.3.1. Ascent/Entry TPS	●			○	○	○	○
H	14.1.2. Active Thermal Control		●	●				○
M	14.2.3. Heat Rejection and Energy Storage	○	○	●	○			
M	14.1.3. Systems Integration (Thermal Management)				●	●		
M	14.1.1. Passive Thermal Control		●	○				
M	14.3.3. Sensor Systems and Measurement Technologies (Thermal Management)	○				○	●	●
M	14.3.2. Plume Shielding (Convective and Radiative)				○		○	
L	14.2.2. Heat Transfer	○	○	○	○	○		
L	14.2.1. Heat Acquisition			○	○	○		○

Symbol	Meaning
●	Strong Linkage: Investments by NASA in this technology would likely have a major impact in addressing this challenge.
○	Moderate Linkage: Investments by NASA in this technology would likely have a moderate impact in addressing this challenge.
[blank]	Weak/No Linkage: Investments by NASA in this technology would likely have little or no impact in addressing the challenge.

FIGURE Q.3 Level of support that the technologies provide to the top technical challenges for TA14 Thermal Management Systems.

Further, higher velocity re-entry for more advanced missions requires development of more TPS with higher limits on temperature, and radiative and total heat flux.

Current TRL is approximately 3 for all except LEO missions. New approaches such as using multiple layers with thermal protection gradients, inclusion of various additives (e.g., various nanotube or nanoparticle materials) to promote anisotropic conduction are below TRL 3, but show promise for improved performance.

NASA maintains the test facilities necessary for qualifying new systems, but the test facilities must be modified to incorporate high radiative heat fluxes to simulate conditions expected for high-velocity re-entry.

Other potential users are USAF and possibly commercial space developers.

There is little or no need for the Space Station in developing this technical area, although one could envision using the station as a base for preparing a high-Mach test re-entry mission using an accelerated return trajectory.

Ascent/Entry TPS is game-changing because it is necessary for every planetary atmospheric ascent and/or entry mission, including every mission for return to Earth. Because the necessary level of effort for developing appropriate TPS is high, a joint NASA-industry development and testing program with careful coordination would maximize efficiency for NASA.

Particularly critical level 4 technology items are Rigid Ablative TPS, Obsolescence-Driven TPS Materials and Process Development, Multi-Functional TPS, and Flexible TPS (crosscutting with TA09-EDL and TA12). Supportive are In-Space TPS Repair and Self-Diagnosing/Self Repairing TPS.

This assessment of 14.3.1 Ascent/Entry TPS as a high priority agrees with the TA09 report (Appendix L), which also lists Rigid TPS as a high priority.

Technology 14.1.2, Active Thermal Control of Cryogenic Systems

Low to zero boil-off of cryogenic fluids will be mission-critical for long-duration missions, and cannot be achieved with present technology. Active thermal control will enable long-term storage of consumables such as LOX for human missions, for cryogenic propellants for both human and robotic missions, for supporting lunar or planetary surface stations, and for supporting scientific instruments that require cryogenic conditions. Active control (recondensation) of cryogenic systems will be necessary to counter remaining heat leaks after effective passive thermal control technologies are applied. A goal of this technology is to develop an overall cryogenic system design that integrates active and passive technologies into an optimal system, as well as instrumentation and sensors to monitor fluid mass. Minimization of active system capacity through effective use of passive control should help increase overall system reliability.

Many level 4 technology items are proposed in the roadmap for active thermal control, but most provide incremental improvement over existing technology. Taken as a whole, they may provide significant reduction in boil-off rates. Most are at TRL 3.

NASA will be the de facto lead in guiding improvements in this technology because of the need for the technology for long-term missions although there are many potential users in non-NASA aerospace who can benefit. However, their needs are less critical to mission fulfillment; generally they can accept some loss rate, unlike NASA.

The Space Station can provide a platform for testing in actual conditions, although less costly Earth-based testing in cryogenic vacuum test chambers can be used for most initial testing.

The panel overrode the QFD score for this technology to designate it as a high-priority technology because the QFD scores did not capture the value of this technology in terms of its ability to enable a wide variety of long-duration missions. This technology generally received high scores because it is mission critical, but lower scores in some areas because the projected gains are incremental for many of the level 4 technology items.

Many of the proposed technologies are inter-related, and careful monitoring and systems integration possibilities should be developed to allow continuing support of those that appear most promising.

MEDIUM- AND LOW-PRIORITY TECHNOLOGIES

Five of the nine level 3 technologies in TA14 ranked medium priority (14.2.3 Heat Rejection and Energy Storage, 14.1.3 Systems Integration, 14.1.1 Passive Thermal Control, 14.3.3 Sensor Systems and Measurement

Technologies, and 14.3.2 Plume Shielding). The two remaining level 3 technologies scored low priority (14.2.2 Heat Transfer and 14.2.1 Heat Acquisition).

The Thermal Management Systems technologies that were ranked as medium and low priority are useful in supporting future NASA spacecraft and missions. These technologies apply to all or nearly all NASA and non-NASA space missions in all or most mission classes, but in a supporting role. These technologies can provide incremental improvements in overall thermal management system performance, but they do not appear to be mission critical or game-changing. 14.1.3 Systems Integration, 14.3.3 Sensor Systems and Measurement Technologies, and 14.1.1 Passive Thermal Control received medium ratings because the technology items are incremental improvements without breakthrough ideas. Passive Thermal Control is a necessary adjunct to 14.1.2 Active Thermal Control, which is listed as a high priority, and passive control improvements can reduce the capacity needed in the active systems, but by themselves cannot achieve the zero-boil-off goal.

If breakthrough ideas come forth in some of these medium- and low-ranked technologies, then they can be pursued more vigorously.

OTHER GENERAL COMMENTS ON THE ROADMAP

Software validation and the use of ground test facilities are two overarching, crosscutting issues pertinent to TA14 that are addressed in detail in Chapter 4.

Budgetary and staffing restraints make it impossible for NASA to carry out all of the tasks proposed in the roadmaps. It will be necessary to coordinate and cooperate with other organizations that can fund and carry out major parts of the research for their own purposes. NASA can then piggy-back on this work. However, it will be necessary to proactively interact with these organizations. Determining which organizations can best help NASA carry out its missions is a daunting task, and will require significant management effort.

The choice of tasks for direct NASA research support will depend to some extent on which tasks can be expected to be performed by others, making NASA research moot. However, there will always be areas where NASA needs will not correlate with external research agencies, and NASA would maximize its return on investment by focusing its funding support in these areas. Such areas as re-entry thermal protection systems for high-velocity re-entry, radiation shielding, reduced mass structures, low-temperature cryogenic radiators, etc. will probably require either internal research teams or funding for contracted work.

The Office of the Chief Technologist should continuously monitor progress on the technology items outlined in the roadmap for it to remain relevant. Some items will prove to be unfeasible; others will progress faster than expected, so priorities for support and funding will shift in out years. NASA does conduct such reviews, and is encouraged to continue and expand this oversight.

Many of the tasks could (and perhaps should) be combined. The draft roadmap breaks Technology 14.1.1 Passive Thermal Control into eight items: large-scale MLI, advanced MLI systems, multifunctional MLI/MMOD, ground-to-flight insulation, low conductivity supports, low conductivity tanks, in situ insulation, and low-temperature radiators. All of these items deal with minimizing heat leaks, and the research should be attacked as an overall systems problem rather than on a technology item-by-technology item basis.

TA14 is very interdependent with other research roadmaps, and coordination across these lines will require careful management to assure cooperation and avoid duplication. In particular, many of the TA14 technologies are dependent on or synergistic with TA01 (Launch Propulsion Systems), TA02 (In-Space Propulsion Systems), TA09 (Entry, Descent, and Landing Systems), TA10 (Nanomaterials), and TA12 (Materials, Structural and Mechanical Systems, and Manufacturing), and have significant interactions with the others.

PUBLIC WORKSHOP SUMMARY

The workshop for the TA14 Thermal Management Systems technology area was conducted by the Materials Panel on March 11, 2011, at the Keck Center of the National Academies, Washington, D.C. The discussion was led by panel chair Mool Gupta. Gupta started the day by giving a general overview of the roadmaps and the NRC's task to evaluate them. He also provided some direction for what topics the invited speakers should cover in their presentations. After the introduction, the day started with an overview of the NASA roadmap by the NASA authors, followed by several sessions addressing the key areas of each roadmap. For each of these sessions, experts from industry, academia, and/or government provided a 35 minute presentation/discussion of their comments on the NASA roadmap. At the end of the day, there was approximately 1 hour for open discussion by the workshop attendees, followed by a concluding discussion by the panel chair summarizing the key points observed during the day's discussion.

Roadmap Overview by NASA

During the overview of the NASA TA14 roadmap, the NASA team noted that they had a large trade space to cover, ranging from milliwatt cryogenic systems to zero boil-off (ZBO) to crew/vehicle thermal management and more. For these reason, they split the roadmap into three primary categories based on temperature: Cryogenic Systems for temperatures less than −150°C, Thermal Control Systems for temperatures between −150°C and a few hundred degrees C above zero, and Thermal Protection Systems for temperatures above several hundred degrees C. The NASA team also noted how they tied their roadmap to the NASA Strategic Goals and Agency Mission Planning Manifest, as well as utilizing the design reference missions from the NASA Human Exploration Framework Team (HEFT) efforts.

In terms of the top technical challenges, the NASA team categorized these into different categories based on timing:

- Near-term
  - Mid-density ablator materials and systems for exo-LEO missions (>11 km/s entries)
  - Innovative thermal components and loop architecture
  - 20 K cryocoolers and propellant tank integration
  - Low conductivity structures/supports
  - Two-phase heat transfer loops
  - Obsolescence driven TPS materials and processes
  - Supplemental Heat Rejection Devices (SHReDs)
- Mid-term
  - Hot structures
  - Low-temperature/power cryocoolers for science applications
- Far-term
  - Inflatable/flexible/deployable heat shields

The NASA team also indicated that this roadmap is crosscutting with several others, and that they had discussions with the teams for TA6, TA9, TA11, and TA12. The NASA team also described how many of these technologies can provide a benefit to NASA:

- Reduced mass
  - 20 K cryocoolers with 20 W capacity
  - Single-loop thermal control systems (elimination of interface heat exchanger)
  - Supplemental Heat Rejection Devices (both vehicle and EVA)
  - Large-scale multi-layer insulation (MLI) and low conductivity supports and tanks for cryogenic systems
- Increased reliability
  - Single-loop thermal control systems
  - Reliable heaters/controllers reduce multiple strings
- Improved performance
  - 150 W/cm^2 flexible TPS
  - Liquid metal heat pipes
  - Two-phase flow systems for tight temperature control

Additionally, the NASA team highlighted how these technologies could benefit the energy, construction, environmental, and automotive sectors.

During the discussion following the NASA team's presentation, there was some discussion of the types of collaboration going on between different groups regarding the technical properties of mid-density ablators, flexible heat shields, and multifunctional systems. On the latter in particular, some follow-on comments from the NASA team were that for multifunctionality incorporating radiation protection, it is important to make sure material properties are not degraded by the multifunctionality. One workshop participant asked the NASA team for their views on the near- and long-term impacts of nanotechnology in thermal management. The response was that thermal straps and phase change materials are areas where mass savings may be realized with carbon nanotubes; in general, though the NASA team indicated that many of their thermal technologies could benefit from nanotechnology. Additionally, there was some discussion on radiators, in particular related to MMOD impacts, redundancy, and reliability (e.g., it was commented that the reliability for the ISS was 0.9999 over 10 years of life).

Regarding technologies near a "tipping point," the NASA team indicated that many of their technologies are in the TRL 4-5 range and have already experienced small on-ground demos, and that the next step is to flight test some of these. When one of the workshop participants asked about the kinds of flights required, the NASA team responded that identifying push missions has been difficult, and that some technologies might benefit from using the ISS or suborbital vehicles, while others (e.g., cryocoolers) can be advanced with additional ground testing for integration and other aspects.

Finally, there was substantial discussion on facilities. One of the workshop participants commented that while he understood that the NASA team was asked not to address facility issues in their roadmap, advancing some technologies to higher TRLs (e.g., mid-density ablators) requires facilities that do not exist. In responding to his question on what NASA was doing about this, the NASA team responded that there have been two teams looking at arcjet facilities initially, but that this has now morphed into an oversight/implementation group. In general, though, one NASA team member indicated that if demand/throughput is not there, than sustaining the business case for these facilities is difficult, and each NASA center is struggling with this. Another team member concurred, and supplemented that you need to have assured capability, facilities to test in relevant environments, and throughput. Based on the agency risk posture, this team member noted that having multiple facilities spreads out the risk, while also allowing different physics to be investigated at different locations.

Session 1: Cryogenic Systems

Ray Radebaugh (NIST, retired) provided a presentation on his experiences and views on cryogenic systems. He started with an overview of the benefits/applications, which in particular for NASA he highlighted as densification (e.g., liquefaction and separation), quantum effects (e.g., fluids and superconductivity), and low thermal noise (e.g., for sensors). Regarding insulation materials, Radebaugh highlighted the need to investigate ways of reducing the mass of multi-layer insulation (MLI), as well as noting that while foams and aerogels may be lower cost, they might not be as effective as MLI. For radiators, he showed a graph indicating how there is a lower temperature limit to radiating in space, and that development is needed to get to lower temperatures. Radebaugh then talked about active thermal control, and how it is important to look for ways to reduce the specific power, mass, and vibration for 20 K cryocoolers. He also provided examples of how Hubble uses Turbo-Brayton cryocoolers and Plank uses Joule-Thomson cryocoolers, and that investments could allow cryocoolers for scientific instruments to work with higher input temperatures. He also noted that Turbo-Brayton designs need to move away from using Neon to other fluids, and that pulse-tube cryocoolers require improvements in performance and efficiency. Radebaugh later observed how the NASA roadmap did not appear to address using O_2 and CH_4 (useful for ISRU) for high-power liquefiers, and that there some trade space exploration is required between low-temperature radiators and active cooling. Another area he felt the roadmap did not address was the need for thermal expansion matching over wide temperature ranges (e.g., matching materials). Other gaps in the roadmap identified by Radebaugh included cryogenics for zero boil-off and liquefaction applications, as well as a technology path for cold compressors. The top technical challenges that Radebaugh highlighted included reducing the mass of cryocoolers, increasing cryocooler efficiency, lightweight insulations in a wide range of atmospheres, flexible radiators, and heat transport over long distances. Radebaugh concluded his presentation noting that NASA is good in performing overall system studies, and has expertise in specific areas as well; he also suggested utilizing expertise at other groups (e.g., USAF, NIST, private industry).

After his presentation, Radebaugh was asked about his thoughts on the importance of thermal interfaces. Radebaugh responded that many of the applications of interest (e.g., cryocoolers) do not generate much heat, but that for some applications heat spreaders or similar items may be required to transfer heat into the system. Radebaugh was also asked about cryocooler vibration, and whether this is an aspect of compressor design. In this case, Radebaugh indicated that vibration can be a significant issue for space observatories. He noted that the Hubble Turbo-Brayton cryocooler is a rotary system with very low vibrations, but that pulse-tube and Stirling cryocoolers may have reciprocating parts that may generate vibrations, and that typically space applications will use multiple pistons to damp these vibrations.

Session 2: Thermal Control Systems and Modeling/Simulations

The session on Thermal Control Systems started with a teleconference discussion with David Gilmore (The Aerospace Corporation). A workshop participant asked Gilmore about his observations on the roadmap. Gilmore indicated that the roadmap was largely consistent what he had seen in NASA centers and DOD, and that many of the technologies outlined seemed to be applicable to DOD as well. On the other hand, Gilmore noted that there appeared to be gaps, including (1) the need for ultra-reliable thermal management, which is required for deep space missions and will drive thermal design; structural-thermal-optical analysis codes, because faster, integrated codes could reduce analysis cycle times from months down to much lower durations and (2) science applications, because many scientific missions have unique thermal requirements (e.g., thermal stability requirements in order to maintain the sensitivity on future decadal survey space observatories, and techniques for thermal balance testing of large passive cryogenic observatories). Gilmore responded to a question about which decadal survey missions might be enabled by these technologies, and indicated that a Venus lander might require some of the insulation and phase change technologies, as do probes to Jupiter; he also noted that applications such as terrestrial planet finding and imaging require significant cryogenic technologies. Gilmore then suggested that a focus on how widespread the utility is might help in prioritizing technologies. For example, he noted that radiators see wide usage. Other high-payoff technologies he discussed included two-phase pump loops (e.g., enabling for high-power space systems), and advanced pumps (both low- and high-power applications). Another participant asked about reverse cooling for habitats, to which Gilmore responded that there does not appear to be much research on this, and that in general the design philosophy is to keep things as simple as possible for reliability. When asked to comment on the state of the art and future directions for MLI and insulations, Gilmore noted that this area is important for science missions. He also commented that today these materials are custom made for each application, and that simplifying the process of building and installing insulations could lead to cost savings on missions. Finally, there was some discussion of moving absorbance down to 0.01. Gilmore indicated that while this is desirable, it is uncertain how far down this can go. He noted that coatings with low absorbance can be developed, but then methods to keep the materials/coatings clean are also required (e.g., lotus coatings to minimize dust issues). There was some discussion that these coatings can make radiators smaller, and have the potential to reduce mass and cost for missions.

Next, Robert Moser (University of Texas at Austin) provided a presentation on "Modeling and Simulation: Verification, Validation, and Uncertainty Quantification." Moser started out by taking several quotes from the NASA roadmap text, noting that many statements deal with modeling and simulation. He indicated that modeling and simulation is important, as it is used to develop science-based predictions to support decision making. He also suggested it is also valuable when experiments in specific regimes cannot be performed, but then this leads to the need to understand the uncertainty in modeling. Moser mentioned that the roadmap appears to pay minimal attention to quantifying or improving the reliability of computational predictions, and that this may be a gap. Next, Moser talked about uncertainty quantification, and the need for verification and validation. He said that code verification is critically important, but generally not given enough attention; some methods he suggested include good software practices, doing end-to-end testing of models, and potentially using manufactured solutions. On the validation side,

Moser indicated that there are several needs, including: math models for uncertainty, algorithms and software tools for computing with uncertainty, and characterization of experimental uncertainties. He presented an example of a heat flux gage where he was able to get an idea of the uncertainty in the measurements of that system. Another example he provided was on the NASA Orion vehicle, where uncertainty quantified in results lets NASA make more rational decisions regarding margins in the system design. Moser concluded with several recommendations: (1) ensure that rigorous code validation is applied to computational simulations; (2) develop modeling software with modern a posteriori analysis and adaptivity; (3) develop/adopt formulations and software tools for uncertainty quantification; and (4) tightly integrate physical modeling, uncertainty analysis and experimental programs to ensure reliable uncertainty assessments. During the discussion following the presentation, Moser was asked by a workshop participant how one deals with the absence of some physics. Moser responded that this is a challenge, and that in general you calibrate with the data available to the extent that you can do so. When asked about his comments on the NASA roadmap, Moser indicated that what he thought was missing was defining what is needed to simulate, and how it should be done. He commented that obtaining data to quantify uncertainty and reliability calculations can be difficult. Finally, responding to another question on the role of modeling and simulation as part of the design process, Moser suggested that in some situations modeling and simulation might be used to provide confidence in the system to be fielded.

Session 3: Thermal Protection Systems

The session on Thermal Protection Systems started with a presentation by one of the panel members, Don Curry (Boeing). Curry started with a table showing historical thermal protection system (TPS) mass fractions for several human-rated vehicles. In general, this was on the order of around 10 percent. Curry provided some discussion on different ablative materials, including the Apollo AVCO and Ames' PICA materials. For carbon phenolic TPS, Curry noted that in many cases this is the only feasible TPS material for specific missions, yet the difficulty obtaining aerospace-grade Rayon is a significant issue for future missions. In terms of TPS testing, Curry mentioned how reusable TPS materials have had mission lifespans quantified via arcjets and other testing. He also provided some data on AVCO used for Apollo; in this case thousands of hours of testing were performed, along with multiple facilities. Curry noted that this was all necessary in order for the material to be available in time. Curry discussed the importance of testing, noting that the final design values for many properties (e.g., thermal conductivity of char) come from arcjet testing; likewise understanding material properties such as compression, shear, etc., is required. Related to TPS design, Curry highlighted many important considerations, including: aerothermal environment, strength/stiffness, thermal gradients, venting characteristics, outgassing, space environment, damage tolerance, repairability, and refurbishment. Finally, Curry concluded by emphasizing that test facilities are important to TPS development, and that eliminating facilities will lead to significantly increased risk. After his presentation, Curry was asked why Orion did not use PICA, as it has a low density but high heat of ablation. Curry's response was that PICA is a tile system, and can potentially crack due to tension in the structure (there are also typically gaps between tiles to account for this). Curry also noted that in the past, PICA has had problems with shockloads during separation pyros.

Next, Bill Willcockson (Lockheed Martin) gave a presentation on TPS materials. Starting out his discussion talking about past experience in robotic missions, Willcockson noted that Viking had hundreds of tests (potentially up to a 1,000), and that tests might be a good metric relative to human missions. For Jupiter/Venus entry (e.g., 10,000 W/cm^2 class), he noted that carbon phenolic can no longer be made, and that these types of materials cannot be tested without arcjet facilities. Relative to affordability, he commented that PICA is three times the cost (process-wise) versus SLA-561V, and that Lockheed Martin has been developing new materials such as MonA to address this. He noted that while SLA-561V was developed in the 1970s, it is important to keep older technologies like this up to date to avoid obsolescence issues. Regarding flexible materials, Willcockson suggested that these are at relatively low TRLs and maturing slowly. During his wrap-up, Willcockson highlighted the importance of investment in TPS; in particular: (1) the need for a resurrection of carbon phenolic, which may require rebuilding facilities as well; (2) the need for a mechanism to take advantage of experienced folks at large companies (e.g., similar to the SBIR program); (3) continued funding to maintain existing TPS materials; and (4) NASA program support needed to offset arcjet costs. Willcockson also noted that the NASA roadmap does not mention in-flight instrumentation use, but that this was done on Pathfinder and is being done on MSL because funding was available. After Willcockson's presentation, a workshop participant provided an additional comment that shock loads can force design changes (e.g., the use of RCC instead of tiles for attach points on the space shuttle), and that testing and modeling need to go together.

Chris Mangun (CU Aerospace) next provided a presentation with a materials perspective on the NASA roadmap. For rigid ablative TPS, he noted that PICA is the current state of the art, but posed the question on whether it will work for the next generation. Mangun noted that for reentry with high heating rates, the thermosetting resin must char, and outgassing of resin is advantageous, as it thickens the boundary layer and reduces heat flux. He listed several desired TPS properties, including low thermal conductivity, high heat of ablation, mechanically tough—not brittle (i.e., resin must adhere well to reinforcement)—and monolithic construction (i.e., avoiding tiles). He provided some discussion on aromatic thermosetting copolyesters, and noted several benefits and potential future applications of this material. Another topic that Mangun commented on was the use of AlB_2 as a planar reinforcement for metal matrix composites (MMCs). Regarding self-healing materials for applications such as micrometeoroid and orbital debris protection, structural recovery, and self-sealing cryotanks, Mangun noted that dual-microcapsule systems in composites are one option; he also mentioned that new microvascular approaches can continuously deliver healing agents. (Note that a microvascular network in a structural composite can also introduce dynamic, reconfigurable functionality, such as damage sensing, thermal management, and radiation protection.) Mangun concluded his presentation noting that it may be possible to accelerate some technologies (e.g., multifunctional TPS, structurally integrated TPS, and self-repairing composites).

Public Comment and Discussion Session

The following are views expressed during the public comment and discussion session by either presenters, members of the panel, or others in attendance. (Note that due to an early end time for the last day of the workshop, there was limited time available for the public discussion period.)

- *Roadmap funding assumptions.* A participant asked the NASA team what the funding assumptions were, as the roadmap lists timeframes to specific TRL numbers for some of the technologies. The NASA team responded that there was no guidance on this, but in general they asked their staff to develop the details of each technology development assuming a "reasonable" funding profile.
- *Importance of dual-use technologies.* One workshop attendee posed the question on how much importance NASA puts on dual-use of the technologies, i.e., the applicability for a technology to benefit others outside NASA. The NASA team responded that while they are always looking for potential spinoffs, that will not drive the development of a specific technology.

REFERENCES

Jamshid, A., Samareh, J.A., and Komar, D.A. 2011. Parametric mass modeling for Mars entry, descent and landing system analysis study. AIAA Paper 2011-1038. 49th AIAA Aerospace Sciences Meeting, Orlando, January 4-7, 2011. American Institute of Aeronautics and Astronautics, Reston, Va.

McGuire, M.K., Arnold, J.O., Covington, M.A., and Dupzyk, I.C. 2011. Flexible ablative thermal protection sizing on inflatable aerodynamic decelerator for human Mars entry descent and landing. AIAA Paper 2011-344. 49th AIAA Aerospace Sciences Meeting, Orlando, January 4-7, 2011. American Institute of Aeronautics and Astronautics, Reston, Va.

Venkatapathy, E. 2009a. Thermal Protection System Technologies for Enabling Future Sample Return Missions. White paper submitted to the Planetary Science Decadal Survey, National Research Council, Washington, D.C.

Venkatapathy, E. 2009b. Thermal Protection System Technologies for Enabling Future Outer Planet Missions. White paper submitted to the Planetary Science Decadal Survey, National Research Council, Washington, D.C.

R

Steering Committee, Panel, and Staff Biographical Information

STEERING COMMITTEE

RAYMOND S. COLLADAY, *Chair,* is the president of RC Space Enterprises, Inc., an aerospace consulting company. He is a retired corporate officer of the Lockheed Martin Corporation and the former president of the Lockheed Martin Astronautics Company in Denver. He has taught leadership and ethics for the Colorado School of Mines; and has served on a number of steering committees, boards, and commissions. Before entering the private sector, he was director of the Defense Advanced Research Projects Agency (DARPA) of the U.S. Department of Defense (DOD) and was associate administrator of NASA, where he had senior executive responsibility for the agency's aeronautics and space research and technology development including operations oversight of Ames, Langley, Dryden, and Glenn Research Centers. Dr. Colladay started his aerospace career at NASA Glenn Research Center in propulsion research and development (R&D) before moving to NASA Headquarters where he moved up through a number of leadership positions before being appointed associate administrator. He has been a member of the Air Force Scientific Advisory Board and various Defense Science Board summer studies. He is a fellow of the American Institute of Aeronautics and Astronautics (AIAA) and of the American Astronautical Society. He earned his Ph.D. in mechanical engineering from Michigan State University. Dr. Colladay has served on many National Research Council (NRC) committees, including the Committee on the Rationale and Goals of the U.S. Civil Space Program and the Planning Committee for Workshop on U.S. Civil Space Policy Committee for the Review of NASA's Capability Roadmaps. Dr. Colladay also serves as chair of the Aeronautics and Space Engineering Board.

JOHN D. ANDERSON, JR., is the curator of aerodynamics at the National Air and Space Museum and is professor emeritus of aerospace engineering at the University of Maryland (UM). At UM he was department chair from 1973 to 1980, and a distinguished scholar/teacher until his retirement in 1999. Prior to that, from 1959 to 1962 he served as lieutenant and task scientist at the Aerospace Research Laboratory at Wright-Patterson Air Force Base (AFB) working in hypersonic aerodynamics. From 1966 to 1972 he was chief of the Hypersonics Group at the U.S. Naval Ordnance Laboratory. At UM, he conducted research in hypersonic and high temperature gas dynamics, atmospheric entry vehicles, and hypersonic scramjet engines. At the National Air and Space Museum he conducts research on the history of aeronautical engineering. He has published ten book titles with McGraw-Hill, Cambridge University Press, Academic Press, and Johns Hopkins University Press, as well as over 120 papers on radiative gas dynamics, entry aerothermodynamics, gas dynamic and chemical lasers, computational fluid dynamics, applied aerodynamics, hypersonic flow, and the history of aerodynamics. He is a member for the National Academy of Engineering (NAE), a fellow of the Royal Aeronautical Society, and an honorary fellow of the AIAA. He earned his Ph.D. in aeronautical and astronautical engineering from Ohio State University. He has served on three NRC Panels to Review Air Force Office of Scientific Research (AFOSR) Proposals in Fluids (1996, 2002, and 2004).

JAMES B. ARMOR, JR., is vice president, strategy and business development, at ATK Spacecraft Systems & Services, where he is responsible for small satellite, satellite component, and engineering services business areas. Major General Armor is also on the Board of Directors of NAVSYS Corporation, Colorado Springs, Colorado, a firm providing advanced research and development products and services in Global Positioning System (GPS) and other timing and navigation systems. He is additionally on the Board of Advisors of the Secure World Foundation, a not-for-profit advocacy and think tank for sustainable space. Major General Armor retired from the USAF in 2008, where his last position was as director of the National Security Space Office (NSSO) in the Office of the Under Secretary of the Air Force, Washington, D.C. While there he was responsible for coordinating all defense and intelligence space activities. Prior to the NSSO, he was director, signals intelligence systems acquisition and operations at the National Reconnaissance Office, vice commander of the Warner Robins Air Logistics Center at Robins AFB, and program director of the NAVSTAR GPS at Los Angeles AFB. He earlier served as a combat crew missile launch officer, a laser signal intelligence analyst, and a satellite launch system integrator. In addition, he was selected and qualified as a DOD space shuttle payload specialist, and was first to study information warfare while a research fellow at the National War College. He is an associate fellow of AIAA. He recently was a member of two NRC committees, Rationale and Goals for U.S. Civil Space Program, and AF Scientific, Technical, Engineering and Math (STEM) Workforce Needs; and a reader in a third, Defending Planet Earth: Near-Earth Object Surveys and Hazard Mitigation Strategies.

EDWARD F. CRAWLEY is the Ford Professor of Engineering at the Massachusetts Institute of Technology (MIT), and is a professor of aeronautics and astronautics and of engineering systems. He received an S.B., S.M., and Sc.D. in aerospace engineering from MIT. He was a founder of the Systems Design and Management Program at MIT, and has served as the department head of aeronautics and astronautics at MIT, the executive director of the Cambridge – MIT Institute, and currently serves as the director of the Bernard M. Gordon – MIT Engineering Leadership Program. His research focuses on the domain of architecture, design, and decision support in complex technical systems that involve economic and stakeholder issues. His current domains of architectural research include energy systems, Earth observation and human spaceflight. Dr. Crawley is a fellow of the AIAA and the Royal Aeronautical Society and is a member of three national academies of engineering, in Sweden, the United Kingdom, and the NAE in the United States. He has served as chair of the NASA Technology and Commercialization Advisory Committee, and was a member of the 1993 Presidential Advisory Committee on the Space Station Redesign, and the 2009 U.S. Human Spaceflight Plans (Augustine) Committee. He recently co-chaired the NRC committee reviewing the NASA Exploration Technology Development Program. He was a visiting lecturer at the Moscow Aviation Institute, and is a guest professor at Tsinghua University in Beijing. He was a finalist in the NASA astronaut selection in 1980. In 2004 he received the Distinguished Eagle Scout Award of the Boy Scouts of America. He has founded three entrepreneurial companies and currently sits on several corporate boards.

RAVI B. DEO is founder and principal of EMBR, an aerospace engineering and technology services company specializing in strategic planning, business development, program management and structural engineering. Dr. Deo formerly served as the director, technology, space systems market segment at Northrop Grumman Corporation's Integrated Systems Sector. He has worked as a program and functional manager for government-sponsored projects on cryotanks, integrated airplane and space vehicle systems health management, and structures and materials, thermal protection systems, and software development. He has extensive experience in road mapping technologies, program planning, technical program execution, scheduling, budgeting, proposal preparation, and business management of technology development contracts. Among his significant accomplishments are the NASA-funded Space Launch Initiative, Next Generation Launch Technology, Orbital Space Plane, and High Speed Research programs, where he was responsible for the development of multidisciplinary technologies. Dr. Deo is the author of more than 50 technical publications and is the editor of one book. He served on the NRC Panel C: Structures and Materials of the Steering Committee on Decadal Survey of Civil Aeronautics and Panel J: High-Energy Power and Propulsion and In-space Transportation of the Committee for the Review of NASA's Capability Roadmaps. He has also served on the Scientific Advisory Board to the Air Force Research Laboratory.

WALT FAULCONER is president of Strategic Space Solutions, LLC, an aerospace consulting company that he started in 2010 to advise NASA, the National Oceanic and Atmospheric Administration, DOD, and commercial companies on strategic planning, business development, systems engineering, and management. Previously, he was the business area executive for civilian space at the Johns Hopkins University Applied Physics Laboratory (APL), responsible for all of the NASA missions at APL, including MESSENGER, the first orbiter of Mercury; New Horizons, the first mission to Pluto; and STEREO, the twin spacecraft investigating coronal mass ejections from the Sun. Prior to joining APL, Mr. Faulconer was with Lockheed Martin for 26 years and served in a variety positions, including director for strategic planning for the Space Systems Company, business development director for human spaceflight and space transportation, and project manager for advanced technology space transportation programs including the X-33 Military Spaceplane and the Crew the Transfer Vehicle program. He served as a systems engineer and mission operations lead on the Space Shuttle program and a variety of classified space programs. He has a master's degree in systems management from the University of Southern California and a bachelor's degree in space science from the Florida Institute of Technology.

PHILIP D. HATTIS holds the position of laboratory technical staff at the Draper Laboratory (the laboratory's highest technical position), with 36 years of aerospace system design, development, integration and test experience. His responsibilities have included technical leadership roles for small and large projects requiring challenging Guidance, Navigation, and Control (GN&C) and avionics system development, including responsibility for assuring robust, integrated GN&C/avionics fault management capabilities. His GN&C/avionics technical leadership has been applied to the space shuttle, the International Space Station, the Orion spacecraft, advanced Earth observation systems, autonomous air and spaceflight systems, uncrewed aerial vehicles, reusable launch vehicles, hypersonic vehicles, precision Mars landing systems, ballistic missile defense systems, precision delivery airdrop systems, ground warrior systems, and helicopter fire control systems. He has served on and led major program red team reviews for NASA, other government agencies, and for aerospace contractors. He is a lifetime fellow and past board member of the AIAA, as well as a past AIAA vice president for public policy. He is a recipient of the Draper Distinguished Performance Award, the AIAA Distinguished Service Award, and NASA recognition for his contributions to the STS-1 and STS-8 missions. He received his Ph.D. from MIT and has served as thesis advisor to numerous MIT graduate students and is an occasional technical and technology policy lecturer at MIT.

TAMARA E. JERNIGAN currently serves as the deputy principal associate director for the Weapons and Complex Integration (WCI) and principal associate directorate at Lawrence Livermore National Laboratory (LLNL). WCI is responsible for ensuring the safety, reliability, and security of the U.S. nuclear stockpile in the absence of testing through a comprehensive science-based program. Dr. Jernigan initially joined LLNL as the principal deputy associate director for the Physics and Advanced Technologies Directorate and later became the associate director for strategic human capital management. Prior to joining LLNL, Dr. Jernigan was selected as a NASA astronaut in 1985. She is a veteran of five Space Shuttle missions where she supervised the pre-flight planning and in-flight execution of critical activities aboard STS-40, 52, 67, 80, and 96. On STS-67, Dr. Jernigan served as payload commander where the crew conducted continuous ultraviolet observations of a variety of stars, planets, and distant galaxies. During Dr. Jernigan's last flight, STS-96, the crew performed the first docking to the International Space Station and Dr. Jernigan executed a spacewalk of nearly eight hours to attach equipment to the exterior of the station. Dr. Jernigan is the recipient of numerous awards including Outstanding Woman of the Year in Science for Alameda County (2004), the NASA Distinguished Service Medal (2000), the Lowell Thomas Award, Explorer's Club (2000), five NASA Space Flight Medals (2000, 1996, 1995, 1992, 1991), the NASA Distinguished Service Medal (1997), the NASA Group Achievement Award—EVA Developmental Test Team (1997), the Federation Aeronautique Internationale Vladimir Komorov Diploma (1997 and 1996), the NASA Outstanding Leadership

Medal (1996), the NASA Outstanding Performance Award (1993), the NASA Exceptional Service Medal (1993), and the Laurels Award, Aviation Week (1991). Dr. Jernigan earned her Ph.D. in space physics at Rice University. She has served as a member of the Space Studies Board, and on the Committee on the Scientific Context for Space Exploration.

JOHN C. KARAS is vice president and general manager for human spaceflight for Lockheed Martin Space Systems Company where he is responsible for coordinating the corporation's capabilities and assets for human space exploration. This includes space shuttle operations on the company's external tank program, and he serves on the advisory board of United Space Alliance. Likewise, exploration programs such as the Orion multi-purpose crew vehicle are under his direction. Previously, he served as vice president of business development and was responsible for strategic planning, advanced technology concepts. Mr. Karas was also director of the Advanced Space Systems and Technology Department, where he was responsible for management of operations research, system predesign, and technology development. Under his direction, the department focused on structures and propulsion technologies, including single stage to orbit and national aerospace plane cryogenic systems and contracted R&D. Mr. Karas also served as manager of Advanced Avionics Systems, the group responsible for new technology demonstration. These new technologies included developments such as adaptive GN&C, multiple fault-tolerant controls, a totally electric vehicle using electromechanical actuators and artificial intelligence. Mr. Karas earned his B.S. in electrical engineering from the Georgia Institute of Technology and has taken advanced course work toward a master's degree in engineering and a master's degree in business administration. Mr. Karas has not served on any previous National Academies studies.

JOHN M. KLINEBERG is the former CEO of Swales Aerospace and retired president of Space Systems/Loral (SS/L). Before assuming the presidency of SS/L, Dr. Klineberg served as executive vice president for Loral's Globalstar program where he successfully led the development, production, and deployment of the Globalstar satellite constellation for cellular telephone services. Prior to joining Loral in 1995, Dr. Klineberg spent 25 years at NASA where he served in a variety of management and technical positions. He was the director of the Goddard Space Flight Center, director of the Lewis (now Glenn) Research Center, deputy associate administrator for aeronautics and space technology at NASA Headquarters, and a research scientist at the Ames Research Center. Before beginning his career at NASA, he conducted fundamental studies in fluid dynamics at the California Institute of Technology (Caltech) and worked at the Douglas Aircraft Company and the Grumman Aircraft Company. Dr. Klineberg has a B.S. in engineering from Princeton University and M.S. and Ph.D. degrees from Caltech. He is the vice chair of the NRC Space Studies Board, the former chair of two NRC study committees, including the Committee to Review the NASA Astrobiology Institute, a former member of two other NRC committees, and a former member of the Aeronautics and Space Engineering Board.

IVETT A. LEYVA is a senior aerospace engineer in the Aerophysics Branch of the Space and Missile Propulsion Division of the Air Force Research Laboratory, where she focuses on the design of liquid rocket engines. Dr. Leyva is an experimentalist and currently studies the effects of acoustic fields on liquid rocket injectors and also works in the area of hypersonic boundary layer transition. Previously, she was a senior aerodynamicist at Microcosm, Inc., where she was responsible for the development of ablative chambers and also performed numerical/analytical studies of Microcosm's launch vehicles' subcomponents. Prior to Microcosm she was employed at General Electric's Global Research Center where she led the design, development, and testing of several pulse detonation concepts and coordinated joint projects with scientists from the former Soviet Union. Dr. Leyva holds several patents in the United States and Europe in the area of propulsion. She received her Ph. D. in aeronautics from Caltech. She served on the Committee to Review NASA's Exploration Technology Development Programs, the Committee on Air Force/Department of Defense Aerospace Propulsion, and the Steering Committee on Decadal Survey of Civil Aeronautics. She currently serves on the Aeronautics and Engineering Board.

LESTER L. LYLES is a general in the U.S. Air Force (retired). While on active duty, his assignments included program director of the medium-launch vehicles program and space-launch systems offices during the recovery from the Challenger Space Shuttle accident, vice commander and commander of Ogden Air Logistics Center, commander of the Space and Missile Systems Center, director of the Ballistic Missile Defense Organization, vice chief of staff at Headquarters USAF, and commander, Air Force Materiel Command. General Lyles currently serves on the board of directors for several corporations, including General Dynamics Corp., Dayton Power & Light, KBR Corp., Precision CastParts Corp., Battelle Laboratories, and United Service Automobile Association (USAA). He has received many awards and decorations including the Defense Distinguished Service Medal, the Distinguished Service Medal, the Defense Superior Service Medal, and the Legion of Merit with oak leaf cluster. He was named Astronautics Engineer of the Year by the National Space Club in 1990 and received the Roy Wilkins Renown Service Award for outstanding contributions to military equal opportunity policies and programs from the NAACP in 1994. General Lyles is a member of the National Academy of Engineering. In 2003, was named Black Engineer of the Year/Lifetime Achievement and he received an Honorary Doctorate from New Mexico State University. General Lyles also has a B.S. in mechanical engineering from Howard University and an M.S. in mechanical/nuclear engineering from New Mexico State University. He is a graduate of the Defense Systems Management College, the Armed Forces Staff College, the National War College, and the National and International Security Management Course at Harvard University. General Lyles served on the Augustine Space Committee in 2009, developing the agenda for the Human Space Flight missions of NASA. He also chaired the NRC's "Roles and Rationale Study of the U.S. Civil Space Programs." General Lyles serves on the Defense Science Board and the President's Intelligence Advisory Board.

H. JAY MELOSH is a Distinguished Professor of Earth and Atmospheric Sciences, Physics, and Aerospace Engineering at Purdue University. Dr. Melosh's previous positions include professor of planetary sciences at the Lunar and Planetary Laboratory, University of Arizona, associate professor of planetary science at Caltech, and associate professor of geophysics at State University of New York. He has made many important contributions to Earth and planetary sciences, including definitive studies of the collisional origin of the Moon and the process of impact cratering. His other major contributions include acoustic fluidization, dynamic topography, and planetary tectonics. He is active in astrobiological studies relating chiefly to microorganism exchange between the terrestrial planets. Dr. Melosh is a member of the National Academy of Sciences. He received his A.B. in physics from Princeton University and his Ph.D. in physics and geology from Caltech. Dr. Melosh has served on the Committee on Planetary and Lunar Exploration and on both the Steering Committee and the Mitigation Panel for the Review of Near-Earth Object Surveys and Hazard Mitigation Strategies.

DANIEL R. MULVILLE is an independent consultant in aerospace systems, engineering, and management. He has led a number of technical reviews for NASA including the recent Near Earth Object Study and Lunar Robotics Architecture Study. He also led technical and management studies of the National Polar-Orbiting Operational Environmental Satellite System (NPOESS), N-Prime and DAWN programs, and served on the team assessing the Japan Aerospace Exploration Agency's space program. At NASA he was the associate deputy administrator responsible for directing and managing NASA's daily operations. He also served as NASA's chief engineer responsible for the overall review of technical readiness and execution of NASA programs and was deputy director of the Materials and Structures Division in the Office of Aeronautics and Space Technology. He directed the composites technology program and structures elements of the advanced launch systems program. Prior to his employment with NASA he served as the structures technology manager for the Naval Air Systems Command and directed the structures research, technology and exploratory development for Navy aircraft and air-launched missile systems. He led the composites fuselage and empennage development for the AV-8B and propulsion structures for the F/A-18. He was a program manager for structures research at the Office of Naval Research and a research engineer at the Naval Research Laboratory developing design and failure analysis methods. Dan Mulville has been awarded NASA's Distinguished Service Medal and Outstanding Leadership Medal and research publication awards by NRL. He received his Ph.D. in structural mechanics from Catholic University. Dr. Mulville has not served on any previous National Academies studies.

DAVA J. NEWMAN is a professor in the Department of Aeronautics and Astronautics and Engineering Systems at MIT and affiliate faculty in the Harvard-MIT Health Sciences and Technology Program. Dr. Newman is also a MacVicar Faculty Fellow and director of the Technology and Policy Program at MIT. She specializes in investigating human performance across the spectrum of gravity. She is an expert in the areas of extravehicular activity (EVA), human movement, physics-based modeling, biomechanics, energetics, and human-robotic cooperation. Dr. Newman's finite element modeling work provided NASA the first three-dimensional representation of bone loss and loading applicable for long-duration missions. She has an active research program in advanced EVA including advanced space suit design, human-robotic cooperation, and biomedical devices. Dr. Newman also focuses on engineering education involving active learning, hands-on design, and information technology implementation to enhance student learning. She was named one of the Best Inventors of 2007 for her BioSuit™ system by *Time* magazine. Dr. Newman received a B.S. in aerospace engineering from the University of Notre Dame, master's degrees in aeronautics and astronautics and technology and policy from MIT, and a Ph.D. in aerospace biomedical engineering from MIT. She is a former member of the NRC Aeronautics and Space Engineering Board and has served on numerous NRC study committees.

RICHARD R. PAUL is an independent consultant with 40 years of R&D-related management experience. Major General Paul retired from the USAF in 2000 after 33 years of service and retired from the Boeing Company in 2007 after 7 years of service. He served as a vice president in Boeing's centralized research and technology organization that develops advanced technologies for Boeing's family of commercial aircraft and defense-related aerospace products. In that assignment, he led the organization's strategic development and modeling and simulation activities, and was the executive manager of Boeing's 2,000-person technical fellowship. During his Air Force career, Major General Paul served in three Air Force laboratories, a product center, two major command headquarters, USAF Headquarters in the Pentagon, and a joint staff assignment. His latter three jobs were aligned with the Air Force science and technology enterprise, where he served in his final assignment as the commander of the Air Force Research Laboratory. Major General Paul is currently a member of the Air Force's Air University Board of Visitors, the National Research Council (NRC) Board on Army Science and Technology, the National Science Foundation Small Business Innovation Research Advisory Committee, and the Wright State University Research Institute Advisory Board. He is a former chair of the Industrial Research Institute (a consortium of 200 companies conducting R&D) and a former advisor to the Sandia National Laboratories Board of Directors Missions Committee. Major General Paul has served on several NRC committees, including the Committee on the Role and Scope of Mission-Enabling Activities in NASA's Space and Earth Science Missions, the Steering Committee of the NASA Technology Roadmaps, and the Committee on Making the Soldier Decisive on Future Battlefields.

LISELOTTE J. SCHIOLER is responsible for research program development for new non-NASA Langley Research Center clients at the National Institute of Aerospace (NIA). She has almost 30 years of experience in fundamental research, as well as program and proposal development, proposal consulting, and program management. Prior to her employment at NIA, she worked for the federal government as a researcher in high-temperature structural ceramics (U.S. Army) and as a program manager for ceramics/high-temperature materials (USAF Office of Scientific Research and the National Science Foundation [NSF]), as well as at a large aerospace company, a small hi-tech business, and herself, running a consulting company. She has participated on many advisory committees, including for DOE and NASA, as well as running review panels for proposals submitted to the NASA Microgravity Materials Program. Dr. Schioler is a fellow of the American Ceramic Society and has held several editorial positions for their publications. She holds a Sc.D. in ceramic science from MIT.

GERALD SCHUBERT is a professor in the Department of Earth and Space at the University of California, Los Angeles. Dr. Schubert's research interests include theoretical studies of the internal structures of the giant planets and their major satellites. He also studies the interiors of the terrestrial planets and the atmospheres of Venus and the outer planets. He has been associated with many spacecraft missions: interdisciplinary scientist and co-investigator for the Atmospheric Structure Experiment on Galileo; member of the Magellan Radar Investigation Group; interdisciplinary scientist for Pioneer Venus; co-investigator for Apollo 16's Lunar Surface Magnetometer;

and co-investigator for Apollo 15 and 16's subsatellite magnetometers. Dr. Schubert has served as a member of the NASA Planetary Geology and Geophysics Management Operations Working Group; the Lunar and Planetary Geoscience Review Panel and Geophysics Group Chief; and the Planetary Atmospheres Review Panel and Dynamics Group Chief (1995). He received his B.E.P and M.A.E. in engineering physics and aeronautical engineering from Cornell University and his Ph.D. in engineering and aeronautical sciences from the University of California, Berkeley. Dr. Schubert is a member of the NAS. He previously served on the NRC Committee on Planetary and Lunar Exploration, the 2002 solar system decadal survey New Frontiers in Solar System Exploration, the Committee on New Opportunities in Solar System Exploration, and the Planetary Science Decadal Survey (Satellites Panel).

PROPULSION AND POWER PANEL

JOHN R. ROGACKI, *Chair,* is associate director of the Florida Institute for Human and Machine Cognition (IHMC). Prior to joining IHMC, Dr. Rogacki served as director of the University of Florida's Research and Engineering Education Facility (REEF), a unique educational facility in Northwest Florida supporting U.S. Air Force research and education needs through graduate degree programs in mechanical, aerospace, electrical, computer, industrial, and systems engineering. Under Dr. Rogacki's leadership, the REEF grew into a highly capable and internationally respected research and education facility. Among his past experiences, Dr. Rogacki has served as: NASA's deputy associate administrator for space transportation technology (in charge of the Space Launch Initiative); program director for the Orbital Space Plane and Next Generation Launch Technology Programs; co-chair of the NASA/DOD Integrated High Payoff Rocket Propulsion Technology Program; director of NASA's Marshall Space Flight Center's Space Transportation Directorate; director of the Propulsion and Power Directorate for the USAF Research Laboratory; director of the USAF Phillips Laboratory Space and Rocket Propulsion Directorate; and deputy director of the Flight Dynamics Directorate of the USAF Wright Laboratory. He has served as primary NASA liaison for the National Aerospace Initiative; co-chair, DOD Future Propulsion Technology Advisory Group; co-chair, DOD Ground and Sea Vehicles Technology Area Readiness Assessment Panel; member of the National High Cycle Fatigue Coordinating Committee; and senior NASA representative to the Joint Aeronautical Commanders Group. He was associate professor of engineering mechanics (and chief of the Materials Division) at the USAF Academy. In 2005, he graduated from the Senior Executives Program in National and International Security at Harvard's JFK School of Government. An accomplished pilot, Rogacki has logged over 3300 flying hours as pilot, instructor pilot, and flight examiner in aircraft ranging from motorized gliders to heavy bombers. Dr. Rogacki earned his Ph.D. and M.S. in mechanical engineering from the University of Washington, and his B.S. in engineering mechanics from the U.S. Air Force Academy.

DOUGLAS M. ALLEN is an independent consultant. Mr. Allen has 30 years of experience in advanced aerospace technology research, development, and testing. He is an expert in space power technology, achievements include leading the successful first flight of multi-junction solar cells, leading the successful first flight of modular concentrator solar arrays, teaching AIAA's Space Power Systems Design short course, leading development of high specific energy batteries, managing development of nuclear space power systems, and leading development of solar power systems designed to survive hostile threats. He was awarded AIAA's "Aerospace Power Systems Award" for outstanding career achievements. Previously, Mr. Allen worked for the Schafer Corporation from 1992 through 2010. He led multiple modeling and simulation efforts including development of SPECTTRA for AFRL's Space Vehicles Directorate to model satellite power systems and to perform technology trades to show payoffs of advanced technologies at the system level when applied to specific satellite systems. Mr. Allen was Schafer's chief engineer for a NASA contract that included developing a concept for Moon and Mars exploration and conceptual design of a Crew Exploration Vehicle. Prior to that, Mr. Allen managed launch vehicle and power technology programs for the Strategic Defense Initiative Organization in the Pentagon. Mr. Allen received his M.S. in mechanical engineering/energy conversion in 1982 and his B.S. in mechanical engineering in 1980 from the University of Dayton. His previous NRC membership service includes the Committee on Radioisotope Power Systems Project and the Committee on Thermionic Research and Technology.

HENRY W. BRANDHORST, JR., is president and chief technology officer of Carbon-Free Energy, LLC and is a visiting professor at Auburn University, charged with developing a Nuclear Power Engineering minor course of study for the Samuel Ginn College of Engineering. Dr. Brandhorst retired as director of the space research institute at Auburn University in June 2010 and as chief of the power technology division at the NASA Lewis Research Center in 1996. He has developed lightweight, high efficiency solar cells for space missions, advanced lightweight solar array technologies as well as Stirling and Brayton Dynamic space (as well as for terrestrial) power systems. He participated in the first flight of a concentrating photovoltaic power system on Deep Space 1. He has demonstrated "direct drive" solar electric propulsion with a terrestrial concentrator solar array. Dr. Brandhorst has been involved with the ASRG Stirling radioisotope power generating system and the development of a 5 kW free-piston Stirling convertor for a fission surface power system. He has been awarded several NASA medals for Exceptional Engineering Achievement and Outstanding Leadership. He received his Ph.D. in nuclear chemistry in 1961 from Purdue University. Dr. Brandhorst was a member on the NRC Committee on Review of the NASA Institute for Advanced Concepts.

DAVID E. CROW is retired senior vice president of engineering at Pratt and Whitney Aircraft Engine Company and professor emeritus of mechanical engineering at the University of Connecticut. At Pratt and Whitney he was influential in the design, development, test, and manufacturing in support of a full line of engines for aerospace and industrial applications. He was involved with products that include high-thrust turbofans for large commercial and military aircraft; turboprops and small turbofans for regional and corporate aircraft and helicopters; booster engines and upper stage propulsion systems for advanced launch vehicles; turbopumps for the Space Shuttle; and industrial engines for land-based power generation. His involvement included sophisticated computer modeling and standard work to bring constant improvements in the performance and reliability of the company's products, while at the same time reducing noise and emissions. Dr. Crow is a member of the NAE. He received his Ph.D. in mechanical engineering in 1972 from the University of Missouri-Rolla, his M.S. in mechanical engineering in 1970 from Rensselaer Polytechnic Institute, and his B.S. in mechanical engineering in 1966 from University of Missouri-Rolla. Dr. Crow's current NRC service includes chair of the Panel on Air and Ground Vehicle Technology–2011 (member in 2009), as a member on the Committee on Examination of the U.S. Air Force's Aircraft Sustainment Needs in the Future and its Strategy to Meet Those Needs, the Army Research Laboratory Technical Assessment Board, and the Board on Manufacturing and Engineering Design. His previous service includes vice chair of the Aerospace Engineering Peer Committee and membership on the Committee for the Review of Proposals to the 2009 Engineering and Physical Science Research and Commercialization Program (ERCP) of the Ohio Third Frontier Program, the Panel on Air and Ground Vehicle Technology–2007, the Committee for the Evaluation of NASA's Fundamental Aeronautics Research Program, the Committee on Analysis of Air Force Engine Efficiency Improvement Options for Large Non-Fighter Aircraft, the Committee on Air Force/Department of Defense Aerospace Propulsion, the Panel B: Propulsion and Power, the 2005 NAS Award in Aeronautical Engineering Selection Committee, the NAS Award in Aeronautical Engineering Selection Committee, and the Aerospace Engineering Peer Committee.

ALEC D. GALLIMORE is an Arthur F. Thurnau Professor in the Department of Aerospace Engineering at the University of Michigan where he directs the Plasmadynamics and Electric Propulsion Laboratory. Dr. Gallimore is also the associate dean for research and graduate education in Michigan's College of Engineering. His primary research interests include electric propulsion and plasma diagnostics. He has experience with a wide array of electric propulsion technologies including Hall thrusters, ion thrusters, arcjets, RF plasma sources, 100-kW-class steady MPD thrusters, and MW-level quasi-steady MPD thrusters. Dr. Gallimore has implemented a variety of probe, microwave, and optical/laser plasma diagnostics and is the author of over 280 journal and conference papers on electric propulsion and plasma physics. He has graduated 30 Ph.D. students and 12 M.S. students. He serves on the AIAA Electric Propulsion Technical Committee and is a fellow of AIAA. Dr. Gallimore is an associate editor for the *Journal of Propulsion and Power* and for the *JANNAF Journal* (propulsion). He received his B.S. in aeronautical engineering from Rensselaer, and his M.A. and Ph.D. degrees in aerospace engineering from Princeton. Dr. Gallimore's current NRC service includes membership on the Committee on an Assessment of Concepts and Systems for U.S. Boost-Phase Missile Defense in Comparison to Other Alternatives and previous membership on the Committee on Conventional Prompt Global Strike Capability, the Committee on Directed Energy Technology for Countering Indirect Weapons, the Committee on Future Air Force Needs for Survivability, the Panel on Engineering, the Panel J: High-Energy Power and Propulsion and In-space Transportation, the Committee on Technology for Human/Robotic Exploration and Development of Space, and the Committee for Undersea Weapons Science and Technology.

MARK W. HENLEY is a senior engineer and program manager at Boeing Research and Technology, in advanced technology concepts for future space transportation and energy systems. His most recent energy-related work at Boeing has focused on new technology for use on Earth, But his background includes applications of solar, thermal, chemical, and nuclear energy sources in orbit and on the moon (and Mars). Mr. Henley managed Boeing's Space Solar Power studies for NASA in 1998 through 2005, coordinating a dozen interrelated research contracts. He also served as principal investigator to demonstrate laser-photovoltaic power transmission technology that could enable operations in the permanently shadowed craters near the moon's South Pole, where ice resources have recently been discovered. Mr. Henley previously managed advanced programs at Rockwell, evaluating commercialization of space launch systems from the former Soviet Union, studying an "Inspector" sub-satellite for the International Space Station and leading cryogenic upper stage design activities. Prior to Boeing/Rockwell, he spent over 10 years at General Dynamics, planning and developing Atlas commercial launch systems, and performing advanced space research and technology studies. He began his career at the California Space Institute, part of the University of California. Mr. Henley received his B.A. in physics in 1982 and his M.S. in aerospace engineering in 1988, both from the University of California, San Diego.

ANTHONY K. HYDER is a professor of physics at the University of Notre Dame. Dr. Hyder's research is in the interaction of spacecraft with the space environment. He is also a member of the Joint Institute for Nuclear Astrophysics. His recent work has focused on the design of spacecraft systems, especially the electrical power and thermal management subsystems, and on the operation of high sensitivity IR sensors aboard spacecraft. He has continued work also in the physics of high-brightness particle accelerators. He has served on a number of national and international panels and advisory boards including the NATO Sensors panel, the Defense Intelligence Agency Scientific Advisory Board, the Advisory Board for the Missile Defense Agency, and the Army Science Board. Dr. Hyder is nominated for his background in military weapons systems development, electronics, sensors, non-lethal weapons, WMD, space systems, and data fusion. Dr. Hyder received his B.S. in physics for the University of Notre Dame, his M.S. in space physics and Ph.D. in nuclear physics from the Air Force Institute of Technology. He received the Air Force Institute of Technology (AFIT) Distinguished Alumnus title in 2005. He has extensive NRC membership service including the Committee on Forecasting Future Disruptive Technologies, the Committee on Research, Development, and Acquisition Options for U.S. Special Operations Command, the Panel on Engineering, the Panel on Enabling Concepts and Technologies, the Committee for Materials, Structures, and Aeronautics for Advanced Uninhabited Air Vehicles, and the Committee on the TOPAZ International Program.

IVETT A. LEYVA. *See the steering committee listing above.*

PAULO C. LOZANO is the H.N. Slater Associate Professor of Aeronautics and Astronautics at the Massachusetts Institute of Technology (MIT). His main interests are plasma physics, space propulsion, ion beam physics, small satellites, and nanotechnology. Part of Dr. Lozano's research portfolio includes the development of highly efficient and compact ion propulsion systems for pico/nano-satellites. In 2007, he received MIT's Karl Chang Innovation award for his work in electrochemical microfabrication on porous metals. In 2008 he received the Young Investigator Program Award from the U.S. Air Force for his work on micro-propulsion and in 2011 received the Future Mind award from the *Quo* science magazine and the Discovery Channel. He has received the Outstanding Faculty UROP Mentor Award for his contributions to the research experience of undergraduate students at MIT. Dr. Lozano has published three patents and over 60 conference and journal publications. He teaches subjects in space and rocket propulsion, fluid mechanics, and plasma physics. Dr. Lozano is a senior member of the AIAA and the American

Physical Society. He earned his M.S. and Ph.D. in space propulsion from MIT. Dr. Lozano's previous NRC service includes membership on the Mitigation Panel.

JOYCE A. McDEVITT is an independent consultant of systems safety. Currently she is a member of NASA's Aerospace Safety Advisory Panel and recently served as project safety engineer for JHU/APL's project to develop and launch the Pluto-New Horizons spacecraft. Mrs. McDevitt previously served as a program manager with Futron Corporation and Computer Sciences Corporation, where she furnished range safety and system safety support to government and commercial clients and held project safety responsibilities for JHU/APL's Midcourse Space Experiment spacecraft. She led a team to provide support to the Commercial Space Transportation Licensing and Safety Division of the Federal Aviation Administration. During her nearly 30 years of civil service at NASA Headquarters, the Air Force Systems Command, and the Naval Ordnance Station, Mrs. McDevitt acquired and applied safety expertise in space, aeronautical, facility, and weapons systems and in propellant, explosive, and chemical processes. She is a registered professional engineer in safety engineering and a senior member of the International System Safety Society. She earned a B.S. in chemical engineering from the University of New Hampshire and an M.S. in engineering from Catholic University. Her previous NRC membership service includes the Committee on Space Launch Range Safety and the Committee on Assessing Passenger Submersible Safety.

ROGER M. MYERS is the deputy lead for the space and launch systems business unit and executive director, electric propulsion and integrated systems for Aerojet General Corporation, providing strategy, program management, and business management oversight for Aerojet's space systems. Prior to this appointment, Dr. Myers was the general manager of Aerojet, Redmond Operations, a 430-person organization focused on in-space propulsion. As the executive director of electric propulsion and integrated systems department at the Redmond facility, Dr. Myers leads programs and strategic planning for advanced spacecraft systems development. Prior to this appointment he served as the executive director, systems and technology development, focused on the development, qualification, and first-article flight production of leading-edge chemical and electric in-space propulsion systems for Aerojet. Prior to joining Aerojet Redmond Operations (then Olin Aerospace) in 1996 as the director, electric propulsion. Dr. Myers held various supervisory and research positions at the NASA Glenn Research Center (then the NASA Lewis Research Center) and Princeton University. Dr. Myers earned his Ph.D. in mechanical and aerospace engineering at Princeton University and a B.S. in aerospace engineering, summa cum laude, from the University of Michigan. His previous NRC membership service includes the Panel to Review Air Force Office of Scientific Research (AFOSR) Proposals in Propulsion–2005 and Panel J: High-Energy Power and Propulsion and In-space Transportation.

LAWRENCE J. ROSS is the chief executive officer for Aerospace Engineering Associates, LLC. In this role he has performed technical studies of aerospace related issues for the purpose of formulating strategic plans and investment strategies. Mr. Ross examines and makes recommendations with respect to specific organizational and management problems being encountered by a client organization and provides assistance in proposal formulation and conducting due diligence reviews. He has assessed the launch readiness of launch vehicles, assessing the state and viability of specific aerospace projects. In January 2007, he co-founded the Aerospace Engineering Associates, LLC. Mr. Ross was the director of the NASA Wind Tunnel Program Office from 1994 to 1995. This was an ad hoc assignment reporting to the administrator of NASA to set up, organize, and direct a task force responsible for planning a proposed $2.5B National Wind Tunnel Complex. From 1963 through 1994, he held the roles of director, deputy director, and space director at NASA's Lewis Research Center, a $1 billion operation responsible for delivering a diverse product line of research, technology, and systems development for the nation's space and aeronautical undertakings. He was the chairman of the Delta Rocket #178 Flight Accident Review Board, an ad hoc assignment to organize and manage an in-depth investigation into the cause of the flight failure and to formulate a set of corrective actions to preclude future flight failures. In addition, he held various positions at the Lewis Research Center associated with the development and launch of Titan and Atlas based launch vehicles including assignments as design engineer, project engineer, chief engineer, and project manager. He earned his B.E.E. from Manhattan College in 1963 and completed the Harvard Senior Managers Program in 1991.

RAYMOND J. SEDWICK is an associate professor of aerospace engineering and director of the Space Power and Propulsion Laboratory in the A. James Clark College of Engineering at the University of Maryland (UM) where he has been since 2007. Prior to this appointment, Dr. Sedwick was a principal research scientist and associate director of the MIT Space Systems Laboratory in the MIT Department of Aeronautics and Astronautics for a period of 10 years. At UM, Dr. Sedwick's current research includes RF plasma sources for space propulsion, plasma-assisted combustion, long-range resonant inductive power transfer, and novel fusion confinement for space and terrestrial power applications. His research interests include most forms of in-space power generation and propulsion with particular interest in nuclear systems and the applications of plasmas. Dr. Sedwick was the inaugural recipient of the Bepi Colombo Prize, as well as the recipient of an NSF CAREER award on the development of compact helicon sources. He is an associate fellow of the AIAA and serves on the Nuclear and Future Flight Technical Committee. Dr. Sedwick earned his B.S. in aerospace engineering at Pennsylvania State University, his S.M. and Ph.D. in 1994 and 1997, respectively, in aeronautics and astronautics from MIT.

GEORGE F. SOWERS is vice president of business development and advanced programs for United Launch Alliance (ULA) headquartered in Denver, Colorado. He is responsible for strategic planning, advanced technology development, advanced concept development and new business acquisition efforts. Before joining ULA, Dr. Sowers was director of business development & advanced programs for Lockheed Martin Space Systems Company, Space Transportation line of business located in Denver, Colorado. Dr. Sowers previously served as director of Mission Integration for the Atlas launch vehicle program. In this role, he was responsible for all activities to integrate and fly satellites on Atlas launch vehicles. This included interface requirements development, mission design, dynamics and systems analysis and flight software development. Prior to this assignment, Dr. Sowers was the chief systems engineer and director of the Systems Engineering and Integration Team (SEIT) for the Atlas V development program. This group was responsible for systems requirements development and verification, systems test, systems integration, and systems analysis. Dr. Sowers served on the Atlas V development program from near its inception through the first flight in 2002. Dr. Sowers began his career in the aerospace industry with Martin Marietta in 1981 on the Titan program as a flight design engineer. He left the company in 1983 to obtain his Ph.D. Upon his return to Martin Marietta in 1988, Dr. Sowers assumed a number of increasingly responsible positions on the Titan program culminating in the role of deputy chief engineer. Dr. Sowers received his B.S. in physics from Georgia Tech in 1980 and obtained his Ph.D. in physics from the University of Colorado in 1988.

ROBOTICS, COMMUNICATIONS, AND NAVIGATION PANEL

STEPHEN P. GOREVAN, *Chair,* is the chairman and co-founder of Honeybee Robotics Spacecraft Mechanisms Corporation of New York. Honeybee is a NASA supplier of advanced robotics research and development engineering as well as a supplier of spacecraft subsystems that range from robotic devices such as the Mars Exploration Rover Rock Abrasion Tool or RAT all the way to deployment devices for spacecraft solar arrays. Mr. Gorevan has guided Honeybee to act as a close industry companion to the planetary science community. This technological support to the planetary science community has led to the development of robotics devices to be found on the Mars Exploration Rovers, the Phoenix Lander, the Mars Science Laboratory and for future missions to Venus (Honeybee has developed a high temperature motor), a small planetary body (sampling systems), Titan (anchoring systems), and the moon. Mr. Gorevan has also guided Honeybee to support NASA and the DARPA in the use of robotics for on-orbit servicing operations, an ongoing interest. Mr. Gorevan has a B.A. in music from New York University and a B.S. in mechanical engineering from the City College of New York. He has previously served as a member of the NRC Steering Committee for Workshops on Issues of Technology Development for Human and Robotic Exploration and Development of Space.

JULIE A. ADAMS is an associate professor of computer science and computer engineering in the Electrical Engineering and Computer Science Department at Vanderbilt University. Dr. Adams directs the Human-Machine Teaming Laboratory. Her research focuses on distributed artificially intelligent algorithms for autonomous multiple robot coalition formation and the development of complex human-machine systems for large human and robotic teams. Dr. Adams is a recipient of a NSF CAREER award. She earned her Ph.D. in computer science from the University of Pennsylvania. Dr. Adams served as a member of the NRC's Army Research Laboratory Technical Assessment Board's Soldier Systems Panel.

EDWARD J. GROTH III is a professor of physics at Princeton University. His research has included IR astronomy, high-speed optical photometry including timing of the Crab Pulsar, and studies of large-scale structure and cosmology. In 1977 he was selected as the data and operations team leader for what became the Hubble Space Telescope. After launch in 1990, he was appointed the deputy principal investigator for the Wide Field and Planetary Camera Instrument. He also served on the ad-hoc committee to characterize the error in the primary mirror; a prerequisite for the fixes put in place at the time of the first servicing mission in late 1993. His research included carrying out the first HST survey, now known as the "Groth Strip," and the first weak lensing analysis of HST data. He also participated in Keck observations to obtain spectroscopy for the objects in the survey. He has participated in an Optical SETI project and has served (2004-2009) on the External Independent Readiness Board for NASA's Navigator program which seeks to discover and characterize Earth-like planets orbiting in the habitable zones of nearby stars. He was the associate chair of the Princeton Physics Department from 2001-2008. He has served as Princeton's representative to USRA for a number of years; as vice chair of the USRA Council of Institutions, 2006-2008, and chair of the Council and member of the Board of Trustees, 2008-2010. Dr. Groth received a B.S. from Caltech and a Ph.D. from Princeton University, both in physics. Dr. Groth served on the NRC Task Group on the Scope of the Space Telescope Science Institute (STScI).

PHILIP D. HATTIS. *See the steering committee listing above.*

JONATHAN P. HOW is the Richard Cockburn Maclaurin Professor of Aeronautics and Astronautics at MIT. At both Stanford University and MIT, Dr. How led the development of navigation, control, and autonomy algorithms for systems comprised of multiple vehicles. His research interests include (1) design and implementation of distributed robust planning algorithms to coordinate multiple autonomous vehicles in dynamic uncertain environments; (2) developing distributed navigation (including estimation using differential GPS and RF ranging sensors), planning, and control algorithms for formation-flying spacecraft; and (3) adaptive flight control to enable autonomous agile flight and aerobatics. Dr. How was the planning and control lead for the MIT DARPA Urban Challenge team that placed fourth in the 2007 race. He was the recipient of the 2002 Institute of Navigation Burka Award, a recipient of a Boeing Special Invention award in 2008, is an associate fellow of AIAA, and a senior member of IEEE. In the past Dr. How has served on the SAB Readiness Review Board for the Air Force Research Lab/ Vehicles Directorate and he was a member of the Gravitational Wave Visiting Committee (GSFC Lab. of High Energy Astrophysics). He earned his Ph.D. in aeronautics and astronautics from MIT.

JAMES W. LOWRIE is the former director of autonomous systems (retired, January 31, 2010) at Lockheed Martin Missiles and Fire Control Systems. Mr. Lowrie has an extensive technical background in research, development, transition, and activation of major autonomous systems for military, civil space, and commercial applications. He has served as chief engineer on multiple programs including DARPA advanced robotics projects, space station robotics elements, Mars exploration spacecraft, and military unmanned systems. Mr. Lowrie also has broad management experience with both small and large business structures. He founded, grew, and sold a small business to a Fortune 500 company and served in an executive capacity with Lockheed Martin for over 20 years. Mr. Lowrie has an in-depth knowledge of both government and commercial contracting and business operations and has experience in a broad range of customer environments including NASA, DOD, Department of Homeland Security, and numerous commercial enterprises. Mr. Lowrie has a B.S. in electrical engineering from Johns Hopkins University.

DAVID P. MILLER is a professor of space science and robotics in the School of Aerospace & Mechanical Engineering at the University of Oklahoma (OU) with additional appointments in the School of Computer Science and the bioengineering programs at OU and the College of Teachers at the International Space University. While at JPL, Dr. Miller led the design and prototyping of the lab's small rover program which eventually led to the Sojourner rover on the Mars Pathfinder Mission. Miller was one of the founders of iRobot (then known as ISRobotics) and was a co-founder of KIPR, a robotics outreach non-profit. Dr. Miller's research interests include planetary robot mobility and the interplay between mechanics and intelligence, development of assistive technologies related to human mobility and technology education. Dr. Miller's space robotics work has been recognized with numerous NASA Certificates of Recognition, NASA Group Achievement Awards, a NASA Space Act Board Award, the JPL Lew Allen Award and the NASA Exceptional Service Medal. His outreach work resulted in receiving the Ames Research Center Dave Lavery Technology Award. He earned his Ph.D. in computer science from Yale University.

JONATHAN R. SALTON is a distinguished member of technical staff at Sandia National Laboratories in the intelligent systems, robotics and cybernetics group. He has a varied background including systems engineering in the power industry, mechanical design engineering in the aerospace industry, and robotics research and development duties at Sandia since 2000. Presently, he is the PI on and is managing five large R&D projects at Sandia (DARPA Urban Hopper—a shoebox-sized hopping robot, a Hopper transition program, a DARPA antisubmarine warfare project, the development of a highly mobile robot for emergency miner rescue operations, and a small scale hydrocarbon power generation project). Duties at Sandia have also included the thermodynamic modeling and analysis of a developmental air processing system, the automation of the processes involved with harvesting and processing chile peppers, the development of an analytical prediction tool for magnetic vehicle mobility, the development of a mobility analysis tool to predict off-road mobility, and various other development projects involving one-of-a-kind small-scale mechanisms. Prior to his position at Sandia, Mr. Salton was at NASA where he was the design lead for multiple specialty EVA tools used for NASA's shuttle and ISS programs for which he received NASA's prestigious Silver Snoopy award. All of the tools that he helped to design and develop are currently still being used in the assembly and maintenance of the International Space Station and several have been used on Hubble Space Telescope repair/upgrade missions. He received B.S. and M.S. degrees in mechanical engineering from the University of Wisconsin-Milwaukee, where his research centered on electromagnetic effects on dynamically agitated viscous fluids. Mr. Salton is a registered professional engineer in the state of New Mexico. He has previously served as a member of the NRC Panel on Human Exploration Systems and Mobility and Autonomous Systems and Robotics.

DONNA L. SHIRLEY is president of Managing Creativity, a consulting and training firm. Prior to that she served as assistant dean of engineering for advanced program development and as an instructor in aerospace mechanical engineering at the University of Oklahoma. These positions followed a 33-year career with the Jet Propulsion Laboratory in Pasadena, California, which culminated with her management of the NASA Mars Exploration Program. This included the Pathfinder and Mars Global Surveyor missions and the Sojourner Rover program. Ms. Shirley has experience in aerospace engineering, space science, government technical program management, and systems engineering. Ms. Shirley received a BA in journalism and a B.S. in aerospace engineering from the University of Oklahoma and an M.S. in aerospace engineering from the University of Southern California. She served as a member of the NRC Committee on New Opportunities in Solar System Exploration, the Committee on the National Aerospace Initiative, and the Task Group on the Availability and Usefulness of NASA's Space Mission Data.

GEORGE W. SWENSON, JR., is professor emeritus of electrical engineering and astronomy at the University of Illinois, Urbana-Champaign, having joined the Illinois faculty in 1956. Prior to that he served short terms and earned tenure at Washington University (St. Louis) and Michigan State University and spent a visiting year at the University of Alaska. Dr. Swenson has taught a wide variety of courses in electrical engineering and applied mathematics, has supervised approximately 10 Ph.D. and 40 M.S. students in electrical engineering and astronomy, and has served in turn as head of the Astronomy Department and the Electrical and Computer Engineering Department at Illinois. With the advent of artificial Earth satellites in 1957 he organized a program of ionosphere research at Illinois, designed several satellite-borne instrument packages for NASA and the U.S. Air Force, and co-authored some of the earliest publications on satellite-enabled ionosphere research. Dr. Swenson designed and led the construction of two pioneering radio telescopes at Illinois, each the largest of its type at the time, and established and directed the Vermilion River Observatory from 1957 to 1982. From 1964 through 1968 he was on leave from

Illinois as visiting scientist at the National Radio Astronomy Observatory where he was manager of the team that produced the conceptual design and the proposal for the Very Large Array radio telescope. Dr. Swenson has served on numerous national and international scientific boards and commissions. Since retiring from the University of Illinois in 1988 he has continued to pursue research in radio engineering and physical acoustics and has supervised two Ph.D. and 23 M.S. theses in those topics. He is a member of the NAE, a fellow of the IEEE and the AAAS, and a Guggenheim Fellow. Dr. Swenson received his Ph.D. from the University of Wisconsin, Madison. He has served on a number of NRC studies, most recently as a member of the Committee on Commercial Aviation Security and as chair of the Panel on Airport Passenger Screening.

INSTRUMENTS AND COMPUTING PANEL

JAMES L. BURCH, *Chair,* is vice president of the division of space science and engineering at the Southwest Research Institute in San Antonio, Texas. He is an expert in the design and use of space plasma physics instruments. He has served as principal investigator on the IMAGE, Rosetta, Dynamics Explorer 1, and ATLAS-1 space science missions, and he is principal investigator of the instrument suite science team for the NASA Magnetospheric Multiscale mission. He received his B.S. in physics from St. Mary's University, his Ph.D. in space science from Rice University, and an M.S.A. in R&D management from George Washington University. He has an extensive history with the NRC having served as a chair on the Committee on Distributed Arrays of Small Instruments for Research and Monitoring in Solar-Terrestrial Physics: A Workshop, the Committee on Exploration of Outer Heliosphere: A Workshop, the Committee on Solar and Space Physics, and as a member on the Committee on the Scientific Context for the Exploration of the Moon, the Committee for the Review of NASA Science Mission Directorate Science Plan, the Committee on the Assessment of the Role of Solar and Space Physics in NASA's Space Exploration Initiative, and the Space Studies Board, the Committee on Solar and Space Physics: A Community Assessment and Strategy for the Future, the Panel on Solar-Wind-Magnetosphere Interactions, the Committee on Solar and Space Physics, and the AFOSR Atmospheric Sciences Review Panel.

PHILIP E. ARDANUY, a principal engineering fellow at Raytheon Intelligence and Information Systems, serves as chief technologist and chief scientist on multiple active Raytheon contracts with NASA, NOAA, and EPA. He specializes in developing integrated mission concepts through government-industry-academic partnerships. His research and development career extends across net-centric and system-of-systems concepts; remote sensing applications and systems engineering; the research-to-operational transition; telepresence-telescience-telerobotics, tropical meteorology and modeling; the Earth's radiation budget (ERB) and climate as member of the Nimbus-7 ERB science team; satellite instrument calibration, characterization, and validation; STEM education; and public outreach. Dr. Ardanuy joined Hughes Aircraft Company in 1995 as manager for Earth sciences. Raytheon acquired Hughes in 1999, and he took on broad-ranging engineering, scientific, and business development responsibilities. Dr. Ardanuy's prior NRC service includes being a member of the NRC Committee on a Strategy to Mitigate the Impact of Sensor De-scopes and De-manifests on the NPOESS and GOES-R Spacecraft. He also served on the NRC Committee on Environmental Satellite Data Utilization and on the Panel on Earth Science Applications and Societal Needs for the 2007 Decadal Survey of Earth Science and Applications from Space, and on the Panel on Options to Ensure the Climate Record from the NPOESS and GOES-R Spacecraft. Dr. Ardanuy received his doctorate in meteorology from Florida State University (FSU). His professional affiliations include the NAS/NRC Committee on Earth Studies, the American Meteorology Society (AMS), NOAA Science Advisory Board's Environmental Information Services Working Group, Maryland Space Business Roundtable Board of Directors and President Emeritus, UCAR Weather Coalition, SPIE Remote Sensing System Engineering Conference, NOAA CREST Institute External Advisory Board, and chair of the AMS Satellite Meteorology, Oceanography, and Climatology Committee. Dr. Ardanuy's honors and awards include his 2011 designation as Fellow of the AMS, recipient of multiple NASA group achievement awards, recipient of the Raytheon Excellence in Business Development Award, and the Raytheon Peer Award. He has over 100 publications to his name, including articles in peer-reviewed journals, book chapters, and conference presentations.

WEBSTER CASH is professor of astrophysics and planetary sciences and of aerospace engineering sciences at the University of Colorado in Boulder. His research interests focus on the design, fabrication, and use of space instrumentation for astronomical purposes. His current concentration is in development of new techniques for imaging and spectroscopy in the x-ray, direct observation of exoplanets in visible wavelengths, and adaptation of space experiments to the new generation of suborbital vehicles. Dr. Cash has served on the Panel on Astronomy and Astrophysics for the Committee on Priorities for Space Science Enabled by Nuclear Power and Propulsion: A Vision for Beyond 2015, and on the Infrastructure Panel of the New Worlds, New Horizons Decadal Survey.

JOHN A. HACKWELL is principal director of the sensor systems subdivision at The Aerospace Corporation (CA) where he specializes in earth remote sensing. At Aerospace he led the development of the Spatially Enhanced Broadband Array Spectrometer (SEBASS), the first high-sensitivity airborne imaging spectrometer to operate in the 3–13.5 micron spectral region, which first flew in 1995. Since then Dr. Hackwell has led the development of a series of increasingly capable infrared imaging spectrometers. Before moving to Aerospace, he was a faculty member at the University of Wyoming where he co-developed the 2.3-m Wyoming Infrared Telescope and served as its director from 1977 to 1985. He also developed the Broadband Array Spectrometer System, an astronomical instrument that was used at multiple observatories and flew on the NASA Kuiper Astronomical Observatory and that is still in use at the NASA 3-m Infrared Telescope Facility on Mauna Kea. Dr. Hackwell earned his Ph.D. in physics from University College London.

ROBERT J. HANISCH is a senior scientist at the Space Telescope Science Institute (STScI) and director of the U.S. Virtual Astronomical Observatory. He has led many efforts in the astronomy community in the area of information systems and services, focusing particularly on efforts to improve the accessibility and interoperability of data archives and catalogs. From 2000 to 2002 he served as chief information officer (CIO) at STScI, overseeing all computing, networking, and information services for the Institute and participating as a member of the Director's Office staff. Prior to becoming CIO at STScI he had oversight responsibility for the HST Data Archive and was the leader of the effort to establish the Multimission Archive at Space Telescope as the active optical/UV archive center for NASA astrophysics missions. Dr. Hanisch received his Ph.D. in astronomy in 1981 from the University of Maryland, College Park.

DAVID Y. KUSNIERKIEWICZ is chief engineer of the space department at the Johns Hopkins University APL. He has an extensive background in designing, integrating, and testing power system electronics for spacecraft. He held the position of mission system engineer for the NASA New Horizons Pluto-Kuiper-Belt Mission and is still the mission and spacecraft system engineer for the NASA Thermosphere, Ionosphere, Mesosphere, Energetics and Dynamics program. He has served on numerous review boards for NASA missions, including Lunar Reconnaissance Obiter; Lunar Robotic Explorer; Dawn; Juno; and ST-8 (part of the New Millennium Program technology development program). Mr. Kusnierkiewicz has received a B.S. and M.S. in electrical engineering from the University of Michigan. He was a member of the Mitigation Panel for the NRC Committee to Review Near-Earth Object Surveys and Hazard Mitigation Strategies.

JOEL R. PRIMACK is Distinguished Professor of Physics at the University of California, Santa Cruz, and director of the University of California systemwide High Performance Astro-Computing Center. He specializes in the formation and evolution of galaxies and the nature of the dark matter that makes up most of the matter in the universe. He is one of the principal originators and developers of the theory of Cold Dark Matter, which has become the basis for the standard modern picture of structure formation in the universe. He is currently using supercomputers to simulate and visualize the evolution of the universe and the formation of galaxies under various assumptions, and comparing the predictions of these theories to the latest observational data. He received a Ph.D. from Stanford University. Dr. Primack is a fellow of both the American Association for the Advancement of Science and the American Physical Society. He was a member of the NRC Committee on NASA's Beyond Einstein Program: An Architecture for Implementation.

GERALD SCHUBERT. *See the steering committee listing above.*

DANIEL A. SCHWARTZ is a senior physicist at the Harvard-Smithsonian Center for Astrophysics. His research specialties are X-ray astronomy, studies of active galactic nuclei and extragalactic jets, observational cosmology, and X-ray mirror and detector instrumentation. He served as the Smithsonian Astrophysical Observatory project scientist during development of the Chandra X-ray Observatory, and he continues to lead the science operations team and the instrument support team for the Chandra mission as it operates in its 2nd decade on-orbit. He has served on and chaired the Rossi X-ray Timing Explorer science working group, and served on the Space Infra-Red Telescope Facility independent external review panel, participated in the Vision Mission Study for the Generation-X X-ray observatory, and he was coordinator for the technology section of the Generation-X Astrophysics Strategic Mission Concept Study.

ALAN M. TITLE (NAS/NAE) is a senior fellow at the Lockheed Martin Advanced Technology Center in Palo Alto, California. He is a leading expert in the development of advanced solar astronomy instruments and sensors. He has been either the principal investigator or responsible scientist for the development of seven space science missions, which have flown on Skylab, the space shuttle, JAXA, ESA, and NASA missions. Dr. Title received his Ph.D. in physics from the California Institute of Technology in 1967. He has had the experience of serving on numerous NASA, NSF, National Laboratory, and university advisory committees. Dr. Title is a member of the NAS, NAE, and IAA. He has an served on an extensive list of NRC committees, he chaired the 2010 Arctowski Medal Selection Committee, served as vice-chair on the Panel on the Sun and Heliospheric Physics, and the Panel on Solar Astronomy, the executive committees of the decadal surveys of Astrophysics and Heliophysics, and as a member on the Committee on PI-led Missions in the Space Sciences: Lessons Learned, the Committee on Solar and Space Physics: A Community Assessment and Strategy for the Future, the Astronomy and Astrophysics Survey Committee, the Space Studies Board, the Task Group on Ground-Based Solar Research, and the Panel for Review of the Explorer Program

DANIEL WINTERHALTER is a principal scientist at the Jet Propulsion Laboratory (JPL) in Pasadena, California. He is also chief scientist of JPL's program scientist for human/robotic mission systems. Further, he is the NASA Engineering and Safety Center's (NASA Langley) chief scientist. His research interests include the detection of low frequency radio emissions from extra solar planets, the spatial evolution of the solar wind, and the solar wind interaction with planetary environments (particularly with Mars and the moon). He has been involved in planning and/or implementation of numerous space missions including Voyager, Ulysses, Mars Orbiter, Mars Global Surveyor, Cassini, Mercury Orbiter, and Mars Science Orbiter. Dr. Winterhalter received his Ph.D. in geophysics and space physics from the University of California, Los Angeles. He is a committee member on the NRC Panel on Solar and Heliospheric Physics.

CARL WUNSCH is the Cecil and Ida Green Professor of Physical Oceanography at MIT. His research has focused on estimating the time varying ocean circulation and its implications for climate and pale climate by combining global general circulation models and the recently available global data sets. His work has included using the mathematical tools such as "inverse methods" and the general mathematical methods of estimation and control theory with large-scale general circulation models of the ocean. Dr. Winch received his Ph.D. in geophysics from MIT. He is a member of the NAS. His extensive NRC service includes membership on the Committee on National Security Implications of Climate Change on U.S. Naval Forces, previously as chair on the 1998 Alexander Aggassiz Medal Selection Committee, the Ocean Studies Board, the Panel on the World Ocean Circulation Experiment, and the Steering Committee for Workshop on Global Observations and Understanding of the General Circulation of the Oceans, a member-at-large on the 2004 NAS Class I Membership Committee, as ex officio member on the Committee on Radio Frequencies, the U.S. National Committee for the International Union of Geodesy and Geophysics, and previous membership on the Committee on Science Opportunities Enabled by NASA's Constellation System, the Committee for the Review of NASA Science Mission Directorate Mission Plan, the Report Review Committee, the Committee on Metrics for Global Change Research, the 2005 Arthur L. Day Prize and

Lectureship Selection Committee, the Committee for Review of the Science Implementation Plan of the NASA Office of Earth Science, the Committee of Geophysical and Environmental Data, the 1995 Alexander Aggassiz Medal Selection Committee, the 1989 Alexander Aggassiz Medal Selection Committee, the Ocean Studies Board, the Briefing Panel on Earth Viewing Remote Sensing, the Ocean Climate Research Committee, the Ad Hoc Group on Ocean Flux Experiments, and the Space Studies Board.

HUMAN HEALTH AND SURFACE EXPLORATION PANEL

BONNIE J. DUNBAR, *Chair,* is an independent aerospace consultant and president of Dunbar International LLC. Prior to establishing her own business, she served 5 years as president and CEO of the Museum of Flight and subsequently was under contract to them for expansion of their space collection, gallery, and STEM education. Dr. Dunbar was a practicing engineer with the Boeing Company (Boeing Computer Services) and the Rockwell International Space Division (Space Shuttle) before she began her extensive career with NASA. She accepted a position as a payload officer/flight controller at the Johnson Space Center for 2 years before she was selected as a NASA astronaut, flying five Space Shuttle flights. She also served two tours in Washington, D.C., as support for the Challenger accident Rogers Commission and then as deputy associate administrator for the Office of Life and Microgravity Sciences at NASA Headquarters. In 1994-1995, Dr. Dunbar trained in Star City, Russia, as a backup crew member for a 3-month flight on the Russian space station, MIR, and was certified by the Russian Gagarin Cosmonaut Training Center to fly on long duration MIR space station flights. In 1995 and 1996, she was detailed to the NASA JSC Mission Operations Directorate as assistant director, where she was responsible for chairing the International Space Station Training Readiness Reviews and facilitating Russian-U.S. operations and training strategies. Dr. Dunbar has also served as assistant director of the Johnson Space Center, responsible for university research oversight, and as deputy director for the Biological Sciences and Applications Division, as associate director to the Space and Life Sciences Directorate, responsible for technology integration and risk management. Dr. Dunbar retired from NASA in 2005. She is a member of the National Academy of Engineering; a fellow of AIAA, the American Ceramic Society, and the Royal Aeronautical Society; and is an elected corresponding member of the Royal Society of Edinburgh. Dr. Dunbar has B.S. and M.S. degrees in ceramic engineering from the University of Washington and a Ph.D. in mechanical/biomedical engineering from the University of Houston. Dr. Dunbar recently served on the Committee on Human Spaceflight Crew Operations, served as co-chair for the NRC Committee to Review NASA's Exploration Technology Development Programs in 2008, and has served as a member of the Committee on Engineering Education, the Aerospace Engineering Peer Committee, the Bernard M. Gordon Prize, and the Awards Committee.

DAVID L. AKIN is an associate professor in the Department of Aerospace Engineering at the University of Maryland, where he is also the director of the Space Systems Laboratory and of the Institute for Dexterous Space Robotics. Dr. Akin's current research includes space systems design and space human factors, focusing on advanced technologies for human/robot collaborations, as well as integrated robotic systems for space, undersea, and medical rehabilitation. He was also the principal investigator for the Experimental Assembly of Structures in EVA (EASE) flight experiment on STS 61-B, and the ParaShield flight test vehicle with the American Rocket Company. He was a member of the NASA Space Science Advisory Committee, the NASA Independent Review Team for the Mars 2003 Rover mission, and currently serves on the AIAA Space Automation and Robotics Technical Committee. He has served on the NASA Telerobotics and EVA Working Groups and the NASA Advisory Council on the Role of Humans in Geostationary Orbit. He has written over 100 papers on aerospace systems design, EVA, teleoperation, robotics, and space simulation. He received SB, SM, and ScD from the Department of Aeronautics and Astronautics at MIT.

DALLAS G. BIENHOFF is manager of in-space and surface systems for The Boeing Company. At Boeing, Mr. Bienhoff has led contract and IRAD studies on space transportation and space exploration architecture studies, lunar habitats, propellant depots, cislunar transportation systems, and technology demonstration mission concepts. He was also a member of the Space Shuttle main engine development team at Rocketdyne. Programs of note on which he has been involved include Boeing Vision for Space Exploration Concept Exploration & Refinement study, Minimum Functionality Habitation Element; Rockwell X-33 Reusable Launch Vehicle, X-38 Crew Return Vehicle, Advanced Launch System, and Shuttle-C studies. Mr. Bienhoff was also the Boeing co-lead with NASA for the Russian FGB module on NASA's International Space Station Russian Integration Team; and participated in several Access to Space studies as contract manager and a member of multiple NASA-industry teams. Mr. Bienhoff has a M.S. in engineering from California State University, Northridge, and a BSME from Florida Institute of Technology, Melbourne, Florida.

ROBERT L. CURBEAM, JR., is the president of the Aerospace and Defense Group of ARES Corporation. This division performs high-end systems engineering, safety and mission assurance, risk management, and program/project management for numerous NASA centers and several DOD clients. Captain Curbeam, USN (retired), served in various capacities during his active duty service including operational flying as an F-14 radar intercept officer, project manager for the F-14 air-to-ground weapons separation program, and 13 years with NASA. His time with NASA included numerous technical assignments including deputy associate administrator for Safety and Mission Assurance (S&MA), director of the Constellation Program S&MA, and service in the Astronaut Corps during which he completed three spaceflights and performed seven spacewalks. He is also a graduate of TOPGUN and the U.S. Navy Test Pilot School. Captain Curbeam received an M.S. in aeronautical engineering from the Naval Postgraduate School.

GREGORY J. HARBAUGH is the president and CEO of the Sigma Chi Foundation. In this position he is responsible for all business, operations, and strategic elements of the organization. Following work as a space shuttle flight operations engineer and technical manager in Mission Control he was selected as a NASA astronaut in June 1987. He served as a crewmember on four shuttle missions, logging a total of 818 hours in space, including 18 hours 29 minutes performing spacewalks. He then served as manager of the Extravehicular Activity (space walk) Project Office, where he managed advanced spacesuit technology research and development for future planetary (Moon and Mars) missions. Mr. Harbaugh retired from NASA in March 2001. He has an M.S. in physical science from the University of Houston, Clearlake. He served on the NRC Committee on the Assessment of Options for Extending the Life of the Hubble Space Telescope and the Panel on Human Exploration Systems and Mobility and Autonomous Systems and Robotics.

TAMARA E. JERNIGAN. *See the steering committee listing above.*

DANIEL R. MASYS is affiliate professor of Biomedical and Health Informatics at the University of Washington, Seattle. Previously he served as professor and chair of the Department of Biomedical Informatics, and professor of Medicine at the University of California San Diego. He served as chief of the International Cancer Research Data Bank of the National Cancer Institute, National Institutes of Health, and from 1986 through 1994 was director of the Lister Hill National Center for Biomedical Communications, which is the computer research and development division of the National Library of Medicine. Dr. Masys is a member of the Institute of Medicine. He received his M.D. from The Ohio State University College of Medicine. Dr. Masys is a former member of the Committee to Review NASA's Exploration Technology Development Programs, the Committee on Aerospace Medicine and Medicine of Extreme Environments, and chaired the Committee on NASA's Research on Human Health Risks.

ERIC E. RICE is the CEO and chairman of ORBITEC (Orbital Technologies Corporation). He has over 44 years of aerospace business experience. He has been leading the development of ORBITEC as an important contributor to the nation's space program. He led the development of an AIAA position paper related to In-Situ Resource Utilization (ISRU). He is in the process of developing uses for in-situ materials for lunar base infrastructure fabrication and missions, including propellants, gases, materials, life support fluids, composites, ceramics, and concrete. He is also interested in lunar mining, excavation, and lunar dust mitigation. He was the founding chairman of the AIAA Space Colonization Technical Committee (SCTC) in 2002, where he has advocated work on space resources and initiated formation of a new Space Resources Utilization Technical Committee (SRUTC) through the SCTC. The SCTC also focuses on space tourism, bases, exploration, colonization/settlements, and terraforming. He currently serves as the PI on the Phase 3 Universal Space Launch Vehicle (USLV) project for the USAF/AFRL. He has served as PI for a NASA project dealing with the in situ acquisition and processing of lunar volatile gases. He served as a NIAC Fellow, conducting advanced concept studies including (1) development of lunar water ice/hydrogen recovery system architecture, and (2) advanced system concept for total ISRU-based propulsion and power systems for unmanned and manned Mars exploration. He was the PI for the NASA/MSFC program involving a revolutionary approach to the carbothermal reduction of lunar oxides to produce lunar oxygen. Dr. Rice was also the PI for two ISRU projects funded by NASA/JSC; namely developing reactors to recover water from planetary dust particles and development of a ground-based lunar ice simulator. He also has been involved in lunar and Mars base concept development. In addition, he has completed a NASA/GRC program that demonstrated carbon monoxide/oxygen-based propulsion systems. Dr. Rice has also served as an adjunct professor at the University of Wisconsin, teaching space transportation and propulsion. Dr. Rice is leading industrial activities of the Wisconsin Space Grant Consortium (WSGC) as the associate director of industry programs and serves on the Space Grant Advisory Board. Dr. Rice earned a Ph.D. in aeronautical and astronautical engineering from The Ohio State University. He also holds a B.S. in chemistry from the University of Wisconsin, Madison. Dr. Rice served on the NRC Steering Committee for Workshops on Issues of Technology Development for Human and Robotic Exploration and Development of Space.

RONALD E. TURNER is a fellow with Analytic Services Inc. He is an internationally recognized expert in radiation risk management for astronauts, particularly in response to solar storms. He is the senior science advisor to the NASA Innovative Advanced Concepts (NIAC) program. For nine years he was the ANSER point of contact to the NASA Institute for Advanced Concepts (NIAC), an independent institute charged with creating a vision of future space opportunities to lead NASA into the twenty-first century. He was a participating scientist on the Mars Odyssey program. He serves on the Advisory Council to the National Space Biomedical Research Institute Center for Acute Radiation Research. Dr. Turner earned his Ph.D. in physics from the Ohio State University. He has served on several NRC committees; currently he is serving on the NRC Committee on Solar and Space Physics, and he recently served on the Committee for the Evaluation of Radiation Shielding for Space Exploration.

MATERIALS PANEL

MOOL C. GUPTA, *Chair,* is a Langley Distinguished Professor and director of NSF I/UCRC Center for Lasers and Plasmas at the University of Virginia. Previously, he was director of the Applied Research Center, program director for materials science and engineering and a research professor in the Department of Electrical and Computer Engineering at Old Dominion University. He worked at the Research Laboratories of Eastman Kodak Company for 17 years as a senior scientist and group leader. Before joining Kodak he was senior scientist at the Jet Propulsion Laboratory, Caltech. Dr. Gupta's research interests include nanomaterials, solar energy, sensors, and photon processing of materials. Other professional activities include Materials Research Society short course instructor for optoelectronic materials, processes, and devices course for over six years; adjunct professor, Department of Materials Science and Engineering, Cornell University for over eight years; conference chair for 1996 SPIE Conference on Nonlinear Frequency Conversion. He is editor-in-chief for the CRC Handbook of Photonics 1st and 2nd edition. He has over 140 research publications and 26 patents and was inducted in Kodak's Inventors Gallery. He has been reviewer and principal investigator for many government agencies. He has a Ph.D. in physics from Washington State University. Dr. Gupta was a member of the NRC Committee on the Ohio Third Frontier Program: Proposal and Progress Reviews, Ohio Research Scholars Program.

GREGORY R. BOGART is currently a principal member of the technical staff in the Integrated Microdevice Systems Organization of Sandia National Laboratories. Dr. Bogart led the surface design, development, and manufacturing efforts for BioStar, Inc. and delivered the first low-cost, disposable, physician office based, silicon biosensor for infectious diseases. At Lucent Technologies-Bell Laboratories, Dr. Bogart was responsible for deep reactive ion etching and scaling MEMS processes from 6-inch to 8-inch wafers and was the first to deliver large area, thin membranes using dry etch techniques. He was also responsible for fabrication of a MEMS-based nano-*g* accelerometer with displacement sensitivity of 12 fm/ Hertz$^{1/2}$. Until late 2009, Dr. Bogart was vice president of engineering for Symphony Acoustics, Inc., which designed and manufactured optical based sensors for seismic and audio applications. His research interests include process integration of new materials and techniques to deliver unique micro and nanometer-sized structures along with large-scale manufacturing techniques for producing those structures. Current research involves energy harvesting materials, stamped metamaterials, large-scale photonic and phononic crystal fabrication with unique materials. Dr. Bogart earned his Ph.D. in analytical chemistry from Colorado State University in Fort Collins, Colorado, and has 14 patents.

DONALD M. CURRY is a thermal analyst at the Boeing Company, supporting thermal protection system activities. Prior to that Dr. Curry was with the NASA Johnson Space Center (JSC) from 1963 until his retirement in January 2007. Dr. Curry has 44 years of experience in the areas of entry heating and thermal protection systems starting with the Gemini spacecraft through the space shuttle. He was the subsystem manager of the Space Shuttle Orbiter Leading Edge Structural System (LESS), which consists of the reinforced carbon-carbon (RCC) wing leading edge, nose cap, chin panel, and forward external plate attachments, responsible for the direction, coordination, technical design, development, testing, analysis, and flight operational support. Dr. Curry was the JSC aeroassist flight experiment (AFE) project area manager (PAM) with responsibility for the AFE Aerobrake structure and thermal protection system (TPS). He participated in the Orbiter return to flight (RTF) program as the Orbiter LESS/RCC NASA Systems Engineer (NSE), responsible for insight and oversight of contractor activities pertaining to the operation and maintenance of the Orbiter LESS. He also served as the JSC technical lead for evaluation of hot structure and ablator TPS for advanced NASA programs. Dr. Curry received his Ph.D. in mechanical engineering from the University of Houston.

JOHN R. HOWELL is the Ernest Cockrell, Jr., Memorial Chair Emeritus, Department of Mechanical Engineering at the University of Texas at Austin where he was a faculty member since 1978. He previously taught at the University of Houston and prior to his teaching career was an aerospace technologist at the NASA Lewis (now Glenn) Research Center. His research career has centered on radiative heat transfer. Dr. Howell received his B.S. and M.S. in chemical engineering and his Ph.D. in engineering from the Case Institute of Technology. He is a member of the NAE and a foreign member of the Russian Academy of Science. Dr. Howell served on the NRC Panel on Benchmarking the Research Competitiveness of the U.S. in Mechanical Engineering; the Panel on a Constrained Space Exploration Technology Program; two panels on chemical demilitarization technology and closure planning; and currently serves on the NRC Standing Committee on Chemical Demilitarization.

GEORGE A. LESIEUTRE is professor and head of the Department of Aerospace Engineering, and director of the Center for Acoustics and Vibration at Pennsylvania State University. Prior to joining Penn State in 1989, he held positions at SPARTA (Space Technology) and Rockwell International (Satellite Systems). At SPARTA, he developed and performed research programs involving composite materials and structures for precision space applications. At Rockwell, he analyzed and characterized spacecraft structures, including truss sizing optimization, composite stress analysis, fracture control, design of damping treatments, and control-structure interaction. His present research interests include structural dynamics, passive damping, active structures, and energy harvesting; his publications are highly-cited and his research has had an impact on the practice. Dr. Lesieutre served as principal investigator of a number of major DARPA programs in adaptive structures. He is the recipient of the Zarem Educator Award from the AIAA, an AIAA Sustained Service Award, and has received five society best paper awards, as well as an Outstanding Research award from Penn State. He is a fellow of AIAA, and presently serves on the AIAA Board of Directors. He earned a B.S. in aeronautics and astronautics from MIT, and a Ph.D. in aerospace engineering from UCLA. Dr. Lesieutre has served on two NRC Panels for the Review of Air Force Office of Scientific Research (AFOSR) Mechanics Research Proposals.

LISELOTTE J. SCHIOLER. *See the steering committee listing above.*

ROBERT E. SKELTON is professor emeritus at University of California, San Diego in the Department of Mechanical and Aerospace Engineering. As an engineer in Huntsville, he designed control systems for SKYLAB and other spacecraft from 1963-1975. At the School of Aeronautics and Astronautics at Purdue he developed new theories of model reduction and control design for flexible spacecraft (1975-1977). As professor and endowed chair of UCSD, he founded the current dynamic systems and control program. His technical contributions are described in 5 books and 400 papers and include algorithms for integrating structure and control design, control of flexible structures, optimization of sensor and actuation resources in large-scale systems, and new structural designs that allow integration of control functions (most recent book *Tensegrity Systems,* 2009). For his interdisciplinary work he received honors from three engineering societies: AIAA (Fellow), IEEE (Fellow) ASCE (co-recipient of the Norman Medal). He received the Senior Scientist award from the JSPS (Japan Society for the Promotion of Science), a Research Award from the Alexander von Humboldt Foundation, the Russell Severance Springer Chair from UCB in 1991, and a letter of appreciation from the NASA director for contributions to Hubble service and repair missions (serving on EIRR panel for 3 Hubble servicing missions). He earned his Ph.D. from UCLA. He served as a member of NRC's Aeronautics and Space Engineering Board from 1985-1988.

GEORGE W. SUTTON is currently retired. He was previously a part-time senior research scientist at Cobham Analytic Solutions/SPARTA where he has been instrumental in providing guidance for and reviews of new concepts for ballistic missile defense and the initiation of advanced systems for advanced sensors and weapons for ballistic missile defense. Prior to joining SPARTA, Inc., Dr. Sutton was a principal engineer with ANSER (a not-for-profit corporation), where he was a member of the SETA team for BMDO for interceptor technology and high-energy lasers. He performed and published original analyses of aero-optical performance of externally cooled windows, uncooled optical dome and window thermal radiance, stresses, and optical aberrations; discrimination capability of 1-, 2-, and 3-color passive optical and laser measurements; interceptor test bed flight test planning; testing techniques image motion compensation for strap down seekers; performance of various FPAs for target acquisition; and supported the Space Based Laser project. From 1992-1996, he was director of the Washington office and chief scientist for Aero Thermo Technology, Inc. Dr. Sutton was a member of the SETA team for BMDO theater ballistic missile interceptor technology, concentrating on aerothermochemistry, aero-optics, and structures for BMDO hit-to-kill ballistic missile interceptors. He wrote the original interceptor flyout computer program that included window heating, window emission noise, and target signal-to-noise ratio. He also wrote the original computer program for end-game guidance and control to determine seeker resolution and accuracy effect on miss distance. He has also completed postdoctoral courses in supersonic aerodynamics, boundary layers, turbulence, plasma physics, and program management. Prior to that, he worked at Helionetics on excimer and blue-green lasers for communications. Before that he worked at the Avco-Everett Research Laboratory on gasdynamic lasers (a name he coined) electric carbon dioxide lasers, and excimer lasers. Dr. Sutton modeled laser beam propagation through atmospheric turbulence with molecular absorption and fog/clouds. He also modeled the distortion of laser mirrors due to heat absorbed by the reflective coating, and modeled and performed experiments on material damage. He performed laser systems studies and wavelength optimization including propagation and threat lethality using the statistics of atmospheric turbulence, absorption, and fog. He invented the hypersonic reentry heat protection (ablation) material that was used successfully on ICBM reentry vehicles, the Corona Satellite Recovery Vehicle, and the Mercury manned reentry capsule. Dr. Sutton is a member of the NAE. He was editor-in-chief of the *AIAA Journal* for almost 30 years, and has received numerous medals and awards. He received a Ph.D. in mechanical engineering and physics from Caltech. He previously served as a member of six NRC committees, most recently the Committee on Directed Energy Technology for Countering Indirect Weapons.

ENTRY, DESCENT, AND LANDING PANEL

TODD J. MOSHER, *Chair,* is currently the director of design and development for the Dream Chaser, Sierra Nevada Corporation's (SNC) commercial crew vehicle, which has been awarded two phases of funding in the NASA Commercial Crew Development Program. As such, he is responsible for the design and development of all of the subsystems that constitute the Dream Chaser lifting body space vehicle. Before that he was the director of spacecraft business development at SNC, where he helped win the Orbcomm Second Generation program with a satellite order to build 18 satellites with an option for 30 more. He also was the program manager for the Operationally Responsive Space Multi-Mission Space Vehicle. Before working at SNC, Dr. Mosher worked at Lockheed Martin on NASA's plans to return to the Moon, served as an assistant professor at Utah State University in the Mechanical and Aerospace Engineering Department, worked at the Aerospace Corporation, taught at the University of California, Los Angeles, and also worked for General Dynamics. He has authored 50 professional publications (journal and conference papers). Dr. Mosher has taught students from around the world and advised several winning student competition teams. As an associate fellow he held many leadership positions in the AIAA. He was a finalist in the 2009 NASA astronaut selection. Dr. Mosher has a Ph.D. and an M.S. in aerospace engineering from the University of Colorado; an M.S. in systems engineering from the University of Alabama, Huntsville; and a B.S. in aerospace engineering from San Diego State University. He has an extensive NRC membership record, including the Committee on Assessment of NASA Laboratory Capabilities, the Committee to Review NASA's Exploration Technology Development Programs, the Committee to Review NASA's Space Communications Program, and the Committee for the Review of NASA's Pioneering Revolutionary Technology (PRT) Program.

JOHN D. ANDERSON, JR. *See the steering committee listing above.*

TYE M. BRADY is a principal member of the technical staff and space systems engineering group leader at Draper Laboratory in Cambridge, Massachusetts. He has worked over the past 22 years on spacecraft instrumentation, design, and integration on a wide variety of space programs including HETE, HETE-II, ASCA, CHANDRA, ASTRO-E, and TACSAT-2. At Draper, he has led the development of a novel, fully successful, on-orbit attitude sensor that marked the first successful operation of a MEMS gyro and Active Pixel Sensor star camera in space. He currently is technical director for lunar landing at Draper developing a next generation landing system capable of safe and highly precise global landing. His research interests include advanced landing systems, GNC instrumentation, systems engineering process, autonomous systems, and star camera design. In 2009, Mr. Brady was awarded NASA's Exceptional Public Service Medal for outstanding technical leadership, a prestigious award given to nongovernment employees for exceptional contributions to NASA's mission. He earned his B.S. in aerospace engineering from Boston University and his S.M. of aeronautics and astronautics from MIT.

BASIL HASSAN is manager of the Aerospace Systems Analysis Department in the Integrated Military Systems Center at Sandia National Laboratories. He has been employed at Sandia since 1993 as a senior and principal member of technical staff and as a manager. Previously, he served as manager of the Aerosciences Department and the Computational Thermal and Fluid Mechanics Department, as well as acting senior manager for the Thermal, Fluid, and Aerosciences Group and acting chief of staff to the laboratory president. He has primarily worked in research and development in nonequilibrium computational fluid dynamics with application to aerodynamics and aerothermodynamics of high-speed flight vehicles. Dr. Hassan has also worked in ablation for hypersonic reentry vehicles, drag reduction for low-speed ground transportation vehicles, and high-velocity oxygen fuel thermal sprays. He has managed aerosciences research, code development, and analysis, both in the computational and experimental areas, including having responsibility over Sandia's transonic and hypersonic wind tunnels and its associated diagnostics development. Dr. Hassan is also the lead for the National Nuclear Security Agency Tri-Lab Support Team for the PECOS Center for Hypersonic Re-entry at University of Texas, Austin. He has been an active member in AIAA for more than 26 years and is currently an associate fellow. He has held a variety of leadership positions at AIAA and is currently vice president for technical activities for the AIAA board of directors. Previously he held the position of director-technical for engineering and technology management. Dr. Hassan has extensive knowledge of NASA's capabilities and facilities and has served on a variety of external review boards for NASA. He has also served on a variety of university educational advisory boards, including the Aerospace Engineering Department at Texas A&M University and the Mechanical and Aerospace Engineering Department at New Mexico State University. Dr. Hassan received his B.S., M.S., and Ph.D. degrees in aerospace engineering from North Carolina State University. He served on the NRC Committee on Assessment of NASA Laboratory Capabilities and the Panel to Review Air Force Office of Scientific Research (AFOSR) Proposal in Fluids.

STEPHEN RUFFIN is an associate professor in the School of Aerospace Engineering at Georgia Institute of Technology, director of NASA's Georgia Space Grant Consortium, head of the Aerothermodynamics Research and Technology Laboratory, and chair of the Aerodynamics and Fluid Mechanics Group. He is a specialist in high temperature gas dynamics, compressible flow aerodynamics, and airframe propulsion integration. He is leading development of a 3-D Cartesian-grid-based Navier-Stokes solver for design applications and development of Cartesian-grid approaches for chemically reacting flows. Dr. Ruffin is also conducting research on high speed, high temperature flows in which vibrational energy modes are substantially excited and in which chemical non-equilibrium exists. He has developed a novel thermochemical model which provides improved predictions in these types of flows. Dr. Ruffin led several computational and experimental studies in a NASA ballistic range of blunted reentry vehicles and noses employing this concept. As principal investigator of the NASA Ames 3-D NASP Nozzle Simulation Project he developed a 3-D Navier-Stokes computer program for accurately predicting the propulsive exhaust flow and its interaction with a generic afterbody region. He received his Ph.D. in aeronautics and astronautics from Stanford University in 1993, a M.S. in aeronautics and astronautics from MIT in 1987, and a B.S. in mechanical and aerospace engineering from Princeton University in 1985. Dr. Ruffin's previous NRC committee membership includes the Panel on Air and Ground Vehicle Technology–2007 and Panel A: Aerodynamics and Aeroacoustics.

ROBERT J. SINCLAIR is the chief engineer at Airborne Systems North America, Space Systems, in Santa Ana, California. Mr. Sinclair has been involved in the design and development of decelerator systems for over 3 decades. He was the lead engineer for the Huygens Descent Control Sub-System and the Beagle 2 EDLS. Mr. Sinclair is currently leading the design team for the NASA Orion Earth Landing System as well as leading the design teams for a number of the NASA Commercial Crew Development systems. Mr. Sinclair has spent his entire career working on deceleration systems and is passionate about the subject. Many of his designs are in service with the U.K. Ministry of Defence, the U.S. DOD, and many agencies throughout the world. Mr. Sinclair received his Higher National Diploma in mechanical engineering from Stevenage College (U.K.) in 1987.

BYRON D. TAPLEY (NAE) holds the Clare Cockrell Williams Centennial Chair in Engineering and is director of the Center for Space Research at the University of Texas in Austin. His research interests include orbit mechanics, precision orbit determination, guidance and navigation, nonlinear parameter estimation, satellite data analysis and the uses of methods from these areas to study the Earth and planetary system. Currently, he is the mission principal investigator for the Gravity Research and Climate Experiment (GRACE) Mission, which is the first NASA Earth System Pathfinder Mission. A recent focus of his research has been directed to applying the GRACE measurements to determine accurate models for the Earth's gravity field and using these measurements for studies of climate driven mass exchange between the Earth's dynamic system components. He is a member of the NAE and a fellow member of AIAA, the American Geophysical Union (AGU), and the American Association for the Advancement of Science. The NASA Medal for Exceptional Scientific Achievement, the NASA Public Service Medal, the AAS Brouwer Award, the AIAA Mechanics and Control of Flight Award, the NASA Exceptional Public Service Medal, and the AGU Charles A. Whitten Medal are among the awards he has received. He has been a principal investigator for seven NASA and international missions. He is a registered professional engineer in the state of Texas. He has served as a member of the NASA Advisory Council (NAC), vice chair of the NAC Science Committee, and chair of the NAC Earth Science Subcommittee. He earned his Ph.D. in engineering mechanics, his M.S. in engineering mechanics, and his B.S. in mechanical engineering from the University of Texas in Austin. His previous membership service includes the Panel on Climate Variability and Change, the Space Studies Board, the Panel to Review NASA's Earth Observing System in the Context of the USGCRP, the Committee on NASA's Space Station Engineering and Technology Development, the NASA Technical Roadmaps Study, and the Aeronautics and Space Engineering Board, the Geophysics Research Forum, and the Steering Committee for the Study and Workshop on NASA's Space Research and Technology Program.

BETH E. WAHL is an independent consultant in Littleton, Colorado, with over 30 years of experience in aerospace systems development and space mission systems engineering. While employed at the Jet Propulsion Laboratory,

Ms. Wahl was the cognizant engineer responsible for the entry aeroshell for the Mars Pathfinder Lander and was engineering lead for instrument integration and landed operations development for the Mars Polar Lander. Additionally at JPL, she led the investigation and development for several enabling technologies for NASA's Pluto mission, which served as pathfinders for the eventual New Horizons mission to Pluto, as well as subsequent Mars missions. More recently, Ms. Wahl supported Lockheed Martin, conducting feasibility studies, risk assessments, and providing systems engineering analysis for the Constellation Program's Orion Crew Capsule. Since 2001, she has supported NASA, conducting technical studies and providing technical assessment of both proposed and development projects for space science missions. Ms. Wahl earned a B.S.M.E from California State University, Long Beach in 1986 and her M.B.A. from the University of La Verne in 1991. She served as a member on the NRC Committee on Technology for Human/Robotic Exploration and Development of Space.

GERALD D. WALBERG is professor emeritus of the Department of Mechanical and Aerospace Engineering, North Carolina State University. From 2000 to 2009, he operated Walberg Aerospace, a research company specializing in entry aerothermodynamics, trajectory optimization, and planetary mission analysis. After establishing Walberg Aerospace, Dr. Walberg worked for NASA Langley on the Revolutionary Aerospace Concepts Program and carried out reentry safety analyses on the Stirling Radioisotope Power System for Teledyne Energy Systems and the Multimission Radioisotope Thermoelectric Generator System for Boeing/Rocketdyne. While at NASA Langley, Dr. Walberg played a lead role in the analysis and testing of the Apollo Heat Shield, led a team that developed some of the first rigorous analyses of radiatively-coupled flow fields, applied these analyses in supporting the Viking, Pioneer Venus, and Galileo Probe missions and chaired the Flight Readiness Revenue for the Galileo heat shield. Following retirement from NASA, he taught at the NASA/George Washington University Joint Institute for Advancement of Flight Sciences. In 1999, he retired from teaching where he was the director of the North Carolina State University Mars Mission Research Center in the Department of Mechanical and Aerospace Engineering from 1991 through 1999. From 1957 to 1989, Dr. Walberg was employed at the NASA Langley Research Center where he held positions ranging from research engineer to deputy director for space. Dr. Walberg received his Ph.D. in aerospace engineering from North Carolina State University in 1974, an M.S. in aerospace engineering from Virginia Polytechnic Institute and State University in 1961, and his B.S. in aeronautical engineering from Virginia Polytechnic Institute and State University in 1956. His NRC service includes previous membership on the Committee to Review NASA's Exploration Technology Development Programs and the Committee on Space Facilities.

STAFF

ALAN C. ANGLEMAN, *Study Director,* has been a senior program officer for the Aeronautics and Space Engineering Board (ASEB) since 1993, directing studies on the modernization of the U.S. air transportation system, system engineering and design systems, aviation weather systems, aircraft certification standards and procedures, commercial supersonic aircraft, the safety of space launch systems, radioisotope power systems, cost growth of NASA Earth and space science missions, and other aspects of aeronautics and space research and technology. Previously, Mr. Angleman worked for consulting firms in the Washington area providing engineering support services to the Department of Defense and NASA Headquarters. His professional career began with the U.S. Navy, where he served for nine years as a nuclear-trained submarine officer. He has a B.S. in engineering physics from the U.S. Naval Academy and an M.S. in applied physics from the Johns Hopkins University.

JOSEPH K. ALEXANDER is a senior program officer for the Space Studies Board (SSB). He served as SSB director from 1998 to 2005. He was previously Deputy Assistant Administrator for Science in EPA's Office of Research and Development (1994-98), Associate Director of Space Sciences at the NASA Goddard Space Flight Center (1993-94), and Assistant Associate Administrator for Space Sciences and Applications in the NASA Office of Space Science and Applications (1987-93). Other positions have included Deputy NASA Chief Scientist and Senior Policy Analyst at the White House Office of Science and Technology Policy. Mr. Alexander's own research work has been in radio astronomy and space physics. He received B.S. and M.A. degrees in physics from the College of William and Mary and completed the Advanced Management Program at the Harvard Business School.

IAN W. PRYKE is a senior program officer with the SSB. Mr. Pryke, who retired from the European Space Agency (ESA) in 2003, is also a senior fellow/assistant professor at the Center for Aerospace Policy Research in the School of Public Policy at George Mason University. While at ESA, he first worked in the areas of data processing and satellite communications and then the Earth Observation Programme Office, where he was involved in the formulation of ESA's Remote Sensing program. In 1979, he moved to the ESA Washington, D.C., office, where he served as a liaison to both government and industry in the United States and Canada. He became head of the office in 1983. Mr. Pryke holds a B.S. in physics from the University of London and an M.A. in space electronics and communications from the University of Kent. He is a fellow of the American Astronautical Society, the AIAA, and the British Interplanetary Society. He is also a member of the International Academy of Astronautics and an associate founder and trustee of the International Space University.

ROBERT L. RIEMER joined the NRC in 1985. He is a staff member for the Board on Physics and Astronomy (BPA) who is shared with SSB and ASEB. He served as senior program officer for the two most recent decadal surveys of astronomy and astrophysics and has worked on studies in many areas of physics and astronomy for the BPA (where he served as associate director from 1988-2000). Prior to joining the NRC, Dr. Riemer was a senior project geophysicist with Chevron Corporation. He received his Ph.D. in experimental high-energy physics from the University of Kansas-Lawrence and his B.S. in physics and astrophysics from the University of Wisconsin-Madison.

JOHN WENDT joined the ASEB as a part-time, off-site senior program officer in 2002. He has served as study director on proposal evaluations for the Air Force Office of Scientific Research and the State of Ohio and has participated in NASA-sponsored studies associated with the Exploration Technology Development Program, NIAC, and NASA's laboratory facilities. From 1990 to 1999, Dr. Wendt served as director of the von Karman Institute (VKI) for Fluid Dynamics, a NATO-affiliated international postgraduate and research establishment located in Brussels, Belgium. He joined the VKI in 1964 and served as head of the Aeronautics/Aerospace Department and dean of the faculty prior to becoming director. His research interests were rarefied gas dynamics, transonics, high angle of attack aerodynamics, and hypersonic reentry, including major inputs to the European Hermes space shuttle program. Dr. Wendt has consulted for the U.S. Air Force, NATO, and the European Space Agency. He is a fellow of the American Institute of Aeronautics and Astronautics (AIAA). Dr. Wendt holds a B.S. in chemical engineering from the University of Wisconsin and a Ph.D. in mechanical engineering and astronautical sciences from Northwestern University.

MAUREEN MELLODY has been a program officer with the ASEB since 2002, where she has worked on studies related to NASA's aeronautics research and development program, servicing options for the Hubble Space Telescope, and other projects in space and aeronautics. Previously, she served as the 2001-2002 AIP Congressional Science Fellow in the office of Congressman Howard L. Berman (D-CA), focusing on intellectual property and technology transfer. Dr. Mellody also worked as a postdoctoral research scientist at the University of Michigan in 2001. She received a Ph.D. in applied physics from the University of Michigan in 2000, an M.S. in applied physics from the University of Michigan in 1997, and a B.S. in physics in 1995 from Virginia Tech. Her research specialties include acoustics and audio signal processing.

CATHERINE A. GRUBER is an editor with the SSB. She joined the SSB as a senior program assistant in 1995. Ms. Gruber first came to the NRC in 1988 as a senior secretary for the Computer Science and Telecommunications Board and has also worked as an outreach assistant for the National Academy of Sciences-Smithsonian Institution's National Science Resources Center. She was a research assistant (chemist) in the National Institute of Mental Health's Laboratory of Cell Biology for 2 years. She has a B.A. in natural science from St. Mary's College of Maryland.

AMANDA R. THIBAULT, a research associate, joined the ASEB in 2011. Ms. Thibault is a graduate of Creighton University, where she earned her B.S. in atmospheric science in 2008. From there she went on to Texas Tech University where she studied lightning trends in tornadic and non-tornadic supercell thunderstorms and worked as a teaching and research assistant. She participated in the VORTEX 2 field project from 2009 to 2010 and graduated with an M.S. in atmospheric science from Texas Tech University in 2010. She is a member of the American Meteorological Society.

DIONNA WILLIAMS is a program associate with the SSB, having previously worked for the National Academies' Division of Behavioral and Social Sciences and Education for 5 years. Ms. Williams has a long career in office administration, having worked as a supervisor in a number of capacities and fields. Ms. Williams attended the University of Colorado, Colorado Springs, and majored in psychology.

TERRI BAKER joined the SSB in June 2009 as a senior program assistant, having worked previously at the National Academies' Center for Education. She has held numerous managerial, administrative, and coordinative positions, in which she focused on improving productivity and organization. Mrs. Baker is working on her B.A. in business management.

RODNEY HOWARD joined the SSB as a senior project assistant in 2002. Before joining SSB, most of his vocational life was spent in the health profession—as a pharmacy technologist at Doctor's Hospital in Lanham, Maryland, and as an interim center administrator at the Concentra Medical Center in Jessup, Maryland. During that time, he participated in a number of Quality Circle Initiatives which were designed to improve relations between management and staff. Mr. Howard obtained his B.A. in communications from the University of Baltimore County in 1983.

LINDA WALKER has been with the National Academies since 2007. Before her assignment with the SSB, she was on assignment with the National Academies Press. Prior to her working at the National Academies, she was with the Association for Healthcare Philanthropy in Falls Church, Virginia. Ms. Walker has 28 years of administrative experience.

ANNA B. WILLIAMS, a Fall 2011 Christine Mirzayan Science and Technology Policy Fellow, received her Ph.D. in organic chemistry from Northeastern University. Her doctoral work focused on the inhibition of protein-protein interactions between nuclear receptors and coactivator proteins, which are known to play a key role in a number of pathologies including hormone- responsive cancers. Another aspect of her work was in the development of synthetic methodology toward the efficient radiolabeling of compounds of known biological activity for use as radioactive tracers. Dr. Williams received her bachelor's degree in chemistry with a minor in philosophy from Dickinson College, where she became particularly interested in the interface between science and the public sphere.

MICHAEL H. MOLONEY is the director of the SSB and the ASEB at the NRC. Since joining the NRC in 2001, Dr. Moloney has served as a study director at the National Materials Advisory Board, the BPA, the Board on Manufacturing and Engineering Design, and the Center for Economic, Governance, and International Studies. Before joining the SSB and ASEB in April 2010, he was associate director of the BPA and study director for the Astro2010 decadal survey for astronomy and astrophysics. In addition to his professional experience at the NRC, Dr. Moloney has more than 7 years experience as a foreign-service officer for the Irish government and served in that capacity at the Embassy of Ireland in Washington, D.C., the Mission of Ireland to the United Nations in New York, and the Department of Foreign Affairs in Dublin, Ireland. A physicist, Dr. Moloney did his graduate Ph.D. work at Trinity College Dublin in Ireland. He received his undergraduate degree in experimental physics at University College Dublin, where he was awarded the Nevin Medal for Physics.